AF327641

Pharmacological Sciences: Perspectives for Research and Therapy in the Late 1990s

Edited by

A. C. Cuello

B. Collier

Birkhäuser Verlag
Basel · Boston · Berlin

Editors:

Professor A. Claudio Cuello
Department of Pharmacology and Therapeutics
McIntyre Medical Sciences Building
McGill University
3655 Drummond Street
Montreal, Quebec
Canada H3G 1Y6

Professor Brian Collier
Department of Pharmacology and Therapeutics
McIntyre Medical Sciences Building
McGill University
3655 Drummond Street
Montreal, Quebec
Canada H3G 1Y6

Library of Congress Cataloging-in-Publication Data

Pharmacological sciences : perspectives for research and therapy in
  the late 1990s / edited by A.C. Cuello, B. Collier.
    "This volume arose from the scientific program of the XIIth
International Congress of Pharmacology, held in Montreal, Canada,
July 24–29, 1994" – Pref..
    Includes bibliographical references and index.
    ISBN 3-7643-5072-5 (hardcover : acid-free paper). ISBN 0-8176-5072-5
(hardcover : acid-free paper)
    1. Pharmacology – Congresses. I. Cuello, A. C. (A. Claudio)
II. Collier, B. (Brian), 1940– . III. International Congress of
Pharmacology (12th : 1994 : Montréal, Quebec)
    [DNLM: 1. Pharmacology-trends-congresses. 2. Drug Therapy-
trends-congresses. 3. Research–trends-congresses. QV 4 P5355
1995]
RM300.P515 1995
615'.1 – dc20
DNLM/DLC
for Library of Congress

Die Deutsche Bibliothek - CIP - Einheitsaufnahme

**Pharmacological sciences: perspectives for research and
therapy in the late 1990s** / ed. by A. C. Cuello; B. Collier. -
Basel ; Boston ; Berlin : Birkhäuser, 1995
    ISBN 3-7643-5072-5 (Basel...)
    ISBN 0-8176-5072-5 (Boston)
NE: Cuello, A. Claudio [Hrsg.]

© 1995 Birkhäuser Verlag
P.O. Box 133
CH-4010 Basel/Switzerland
Printed on acid-free paper produced from chlorine-free pulp ∞ .
Printed in Germany

ISBN 3-7643-5072-5
ISBN 0-8176-5072-5

9 8 7 6 5 4 3 2 1

Contents

Pharmacology of Ion Channels

Drug Metabolism

Neuropharmacology

Purines

Cardiovascular Pharmacology

Nitric Oxide

Endocrine Pharmacology and Immunopharmacology

The Pharmacology of Gene Expression

Pulmonary Pharmacology

Chemotherapy

Toxicology

Regulatory Requirements for Drug Registration

Pharmacological Methods

Instructional Methods

Preface

This volume arose from the scientific program of the XIIth International Congress of Pharmacology, held in Montreal, Canada, July 24–29, 1994. The scientific program included plenary lectures and symposia, in addition to poster presentations and colloquia. The abstracts of the Congress presentations were published as supplement 1 of volume 72 by the *Canadian Journal of Physiology & Pharmacology*. The Congress organizers sought a more expansive treatment of the Congress proceedings and appointed Dr. A. Claudio Cuello to coordinate preparation of the present volume; Dr. Brian Collier was chair of the scientific program committee and, thus, also collaborated on this work.

The objective that we pursued was to produce a volume of reasonable size which would feature all of the plenary lectures and symposia from those authors who agreed to participate. To this end, we solicited mini-reviews from plenary lecturers and asked symposia organizers to coordinate a single short-review covering the individual topics within their event. Those who accepted this challenge are evident in this volume. We express our gratitude to these authors for doing so, and for exercising considerable ingenuity in completing their task within a reasonable time.

As is obvious from the above, the volume reflects the work of the authors of each chapter as well as that of the whole scientific program committee who toiled to develop a somewhat balanced set of lectures and symposia reflective of modern pharmacology. For this, we thank our colleagues of the program committee. Their work was complemented by that of the IUPHAR advisory board, to whom we also express our gratitude. Of course, there would be no volume and no program without the IUPHAR executives who entrusted the organization of their Congress to our Canadian colleagues; thank you all. We trust that this volume will serve as a reference publication to those who want to familiarize themselves with recent developments in a wide variety of fields within the pharmacological sciences.

Finally, our gratitude must be extended to Sid Parkinson, who did all the hard work associated with manuscript collection, editing and organizing, as well as to Dr. J.C.F. Habicht of Birkhäuser Verlag who supported this enterprise with considerable enthusiasm.

A. Claudio Cuello
Brian Collier

Contributors

G. Alvan, Department of Pharmacology, University of Toronto, Canada

M. W. Anders, Department of Pharmacology, University of Rochester, Rochester, New York, USA

Raymond Ardaillou, Service d'Explorations Fonctionnelles, INSERM 64, Hôpital Tenon, Paris, France

Y. Asaoka, Biosignal Research Center, Kobe University, Kobe 659, Japan

Jeffrey Atkinson, Laboratoire de Pharmacologie, Faculté de Pharmacie de l'Université Henri Poincaré, Nancy 1, 54000 Nancy, France

K.L. Audus, Department of Pharmaceutical Chemistry, The University of Kansas, Lawrence, KS, USA

K.D. Bagshawe, Charing Cross Hospital, London, UK

David G. Bailey, Department of Medicine, Victoria Hospital, London, Ontario, Canada N6A 4G5

J.C. Barnes, Pharmacology Division, Glaxo Research and Development Ltd., Ware, Hertfordshire SG12 0DP, UK

Peter J. Barnes, Department of Thoracic Medicine, National Heart and Lung Institute, London SW3 6LY, UK

Philip M. Beart, Department of Pharmacology, Monash University, Clayton, Victoria 3168, Australia

Nicolau Beckmann, Preclinical Research, Sandoz Pharma Ltd, CH-4002 Basel, Switzerland

Giorgio Bellomo, Clinica Medica I, Policlinico S. Matteo, Università di Pavia, I-27100 Pavia, Italy

Ruth E. Billings, Department of Environmental Health, Colorado State University, Fort Collins, Colorado 80523, USA

J. Edwin Blalock, Department of Physiology and Biophysics, University of Alabama, Birmingham, Alabama 35294, USA

Joel G. Bockaert, CNRS UPR 9023, 34094 Montpellier, France

Ronald T. Borchardt, Department of Pharmaceutical Chemistry, The University of Kansas, Lawrence, KS, USA

N.G. Bowery, Department of Pharmacology, The School of Pharmacy, University of London, London WC1N 1AX, UK

John Caldwell, Department of Pharmacology and Toxicology, St. Mary's Hospital Medical School, London W2 1PG, UK

Paul A. Cann, Middlesbrough Hospital, Middlesbrough, Cleveland TS5 5AZ, UK

Ernesto Carafoli, Laboratorium für Biochemie, ETH-Zentrum, Ch-8092 Zürich, Switzerland

E. Carmeliet, Laboratory of Physiology, University of Leuven, Gasthuisberg, Herestraat 49, 3000 Leuven, Belgium

Marc G. Caron, Department of Physiology and Howard Hughes Medical Institute, Duke University Medical School, Durham, North Carolina, USA

J.R. Carpenter, Gardiner-Caldwell Communications Ltd., Macclesfield, England

Carol E. Cass, Department of Biochemistry, University of Alberta, Edmonton, Alberta, Canada

Icilio Cavero, Rhône-Poulenc Rorer, Centre de Recherche de Vitry-Alfortville, 13 Quai Jules Guesde, 94400 Vitry Sur Seine, France

Marie Odile Christen, Solvay Pharma, Laboratoires de thérapeutique moderne L.T.M., 92151 Suresnes Cedex, France

George P. Chrousos, NICHD, Pediatric Endocrinology, National Institutes of Health, Bethesda, Maryland 20820-892, USA

D.E. Clarke, Institute of Pharmacology, Syntex Research, Palo Alto, CA 94303, USA

Michael G. Collis, Cardiovascular Biology, Pfizer Central Research, Sandwich, Kent, UK

D.J. Crankshaw, Department of Obstetrics and Gynaecology, McMaster University, Hamilton, Ontario, Canada

Stanley T. Crooke, Isis Pharmaceuticals Inc., Carlsbad, California, USA

A. Claudio Cuello, Department of Pharmacology and Therapeutics, McGill University, Montreal, Quebec, Canada H3G 1Y6

J. de Champlain, Faculté de Médecine, Département de Physiologie, Université de Montréal, Montréal, Canada H3C 3J7

S.M.F. de Morais, NIEHS, North Carolina, USA

Annette C. Dolphin, Department of Pharmacology, Royal Free Hospital Medical School, London, UK

Andy Dray, Sandoz Institute for Medical Research, London WC1E 6BN, UK

Michael J. Dunn, Department of Physiology and Biophysics, Case Western Reserve University, Cleveland, Ohio, USA

Erik Dybing, Department of Environmental Medicine, National Institute of Public Health, N-0462 Oslo, Norway

A.B. Ebeigbe, Departments of Physiology and Pharmacology, University of Benin, Benin City, Nigeria

Frances A. Edwards, Department of Pharmacology, University of Sydney, Sydney, NSW 2006, Australia

Gillian Edwards, School of Biological Sciences, University of Manchester, Stopford Building, Oxford Road, Manchester M13 9PT, UK

John M. Essigmann, Whitaker College of Health Sciences and Technology, Massachusetts Institute of Technology, Cambridge, Massachusetts, USA

Ronald W. Estabrook, Department of Biochemistry, University of Texas Southwestern Medical Center at Dallas, 5323 Harry Hines Blvd., Dallas, Texas 75235-9038, USA

Marc Feldmann, Kennedy Institute of Rheumatology, Hammersmith, London, UK

Stephen K. Fisher, Neuroscience Laboratory, Mental Health Research Institute and Department of Pharmacology, University of Michigan, Ann Arbor, Michigan 48104-1687, USA

A.P.D.W. Ford, Institute of Pharmacology, Syntex Research, Palo Alto, CA 94303, USA

Bertil B. Fredholm, Department of Pharmacology, Karolinska Institute, S-104 01 Stockholm, Sweden

A.G. García, Department of Pharmacology, Universidad Autónoma de Madrid, Madrid 28029, Spain

Vincent Geenen, Centre Hospitalier Université de Liege, Service d'Endocrinologie, Domaine Universitaire du Sart Tilman 35, B-4000 Liege 1, Belgium

Hansruedi Glatt, Institute of Toxicology, University of Mainz, Mainz, Germany

Martin J. Glennie, The Lymphoma Unit, Tenovus Laboratory, General Hospital, Southampton, UK

Théophile Godfraind, Laboratoire de Pharmacologie, Université Catholique de Louvain, FARL 5410, 1200 Brussels, Belgium

J.A. Goldstein, NIEHS, North Carolina, USA

Frank J. Gonzalez, National Cancer Institute, National Institutes of Health, Bethesda, Maryland 20892, USA

Jay I. Goodman, Department of Pharmacology and Toxicology, Michigan State University, East Lansing, Michigan, USA

F. Peter Guengerich, Department of Biochemistry and Center of Molecular Toxicology, Vanderbilt University, School of Medicine, Nashville, Tennessee 37232-0146, USA

Ji-Sheng Han, Neuroscience Research Center, Beijing Medical University, 38 Xue Yuan Road, Beijing 100083, China

Stephen F. Heinemann, Molecular Neurobiology, Salk Institute, San Diego, California 92186-5800, USA

Claus W. Heizmann, Department of Pediatrics, University of Zürich, Steinweisstrasse 75, CH-8032 Zürich, Switzerland

Guy A. Higgins, Glaxo Research and Development Ltd., Ware, Hertfordshire SG12 0DP, UK

E.A. Higgs, The Wellcome Foundation Ltd., Beckenham, Kent BR3 3BS, UK

D. Hillaire-Buys, Laboratoire de Pharmacologie, Institute de Biologie, F-34060 Montpellier, France

Barry H. Hirst, Department of Physiological Sciences, University of Newcastle, Newcastle upon Tyne, UK

Morley D. Hollenberg, Endocrine Research Group, Departments of Pharmacology and Therapeutics and Medicine, The University of Calgary, Faculty of Medicine, Calgary, Canada T2N 4N1

Alan Horton, Department of Biochemistry, University of Birmingham, Birmingham B15 2TT, UK

D. Hoyer, Sandoz Pharma Ltd., CH-4002 Basel, Switzerland

Jan D. Huizinga, Department of Biomedical Sciences, McMaster University, Hamilton, Ontario, Canada L8N 3Z5

F. Hullin, Biosignal Research Center, Kobe University, Kobe 659, Japan

Eigill F. Hvidberg, Department of Clinical Pharmacology, University Hospital, Copenhagen, Denmark

T. Inaba, Department of Pharmacology, University of Toronto, Toronto M5S 1A8 Canada

Tadishi Inagami, Department of Biochemistry, Vanderbilt University School of Medicine, Nashville, TN 37232, USA

Ryuji Inoue, Department of Pharmacology, Kyushu University, Fukuoka 812, Japan

Graham A.R. Johnston, Adrien Albert Laboratory of Medicinal Chemistry, Department of Pharmacology, The University of Sydney D06, NSW 2006, Australia

Gregory J. Kaczorowski, Merck Sharp and Dohme Research Laboratories, Rahway, New Jersey 07065-0900, USA

Akira Karasawa, Pharmaceutical Research Laboratories, Kyowa Hakka Kogyo Ltd., Shimotagari, Japan

R. Kato, Department of Pharmacology, Keio University, Japan

Paul A. Kelly, INSERM Unité 344, Faculté de Médecine Necker, 75730 Paris, France

Chris J. Kemp, CRC Beatson Laboratories, Beatson Institute for Cancer Research, Bearsden, Glasgow, Scotland, UK

Sadao Kimura, Centre for Biomedical Science, Chiba University School of Medicine, Chiba, Japan

Kathleen M. Knights, Department of Clinical Pharmacology, School of Medicine, Flinders University of South Australia, Bedford Park, Australia 5042

Brian K. Kobilka, Division of Cardiovascular Medicine and Howard Hughes Medical Institute, Stanford University Medical School, Stanford, California, USA

George F. Koob, Department of Neuropharmacology, The Scripps Research Institute, La Jolla, California 92037, USA

K. Kuriyama, Kyoto Prefectural University of Medicine, Kamikyo-ku, Kyoto 602, Japan

Toshio Kuroki, Department of Cancer Cell Research, Institute of Medical Science, University of Tokyo, Tokyo, Japan

J. Lambert, Department of Pharmacology and Clinical Pharmacology, Ninewells Hospital and Medical School, Dundee DD1 9SY, Scotland, UK

Harold E. Lane, Corning Costar Corporation, Cambridge, Massachusetts, USA

Bernhard H. Lauterburg, Department of Clinical Pharmacology, University of Berne, 3010 Berne, Switzerland

Gérard Le Fur, SANOFI Recherche, 75008 Paris, France

Chun Guang Li, Pharmacology Research Laboratory, Department of Medical Laboratory Science, Royal Melbourne Institute of Technology, Melbourne 3001, Victoria, Australia

Charles S. Lieber, The Alcohol Research and Treatment Center and the Section of Liver Diseases and Nutrition, Bronx Veterans Administration Medical Center and Mount Sinai School of Medicine, New York NY 10468, USA

David Lodge, Lilly Research Center, Erl Wood Manor, Windlesham, Surrey GU20 6PH, UK

Peter I. Mackenzie, Department of Clinical Pharmacology, School of Medicine, Flinders University of South Australia, Bedford Park, Australia 5042

Alberto Mantovani, Istituto di Richerche Farmacologiche, Mario Negri, Milano, Italy

Tomoh Masaki, Department of Pharmacology, Faculty of Medicine, Kyoto University, Kyoto 606, Japan

Michael E. McManus, Department of Physiology and Pharmacology, University of Queensland, Queensland 4072, Australia

Frederick A.O. Mendelsohn, Department of Medicine, University of Melbourne, Austin Hospital, Heidelberg 3084, Victoria, Australia

Roland Mertelsman, Department of Hematology/Oncology, Medical Center, University of Freiburg, Freiburg, Germany

Urs A. Meyer, Department of Pharmacology, Biozentrum of the University of Basel, CH-4056 Basel, Switzerland

Enrico Mihich, Roswell Park Cancer Institute, Grace Cancer Drug Center, Buffalo, NY 14263, USA

Leonard P. Miller, Gensia Inc., 9360 Towne Centre Drive, San Diego, California, USA

Anis Mir, Preclinical Research, Sandoz Pharma Ltd, CH-4002 Basel, Switzerland

S. Moncada, The Wellcome Foundation Ltd., Beckenham, Kent BR3 3BS, UK

Flavio Moroni, Dipartimento di Farmacologia Preclinica e Clinica Mario Aiazzi Mancini, Università degli Studi di Firenze, 50134 Firenze, Italy

E.J. Mylecharane, Department of Pharmacology, University of Sydney, New South Wales, Australia

S. Nakamura, Department of Biochemistry, Kobe University School of Medicine, Kobe 650, Japan

Shigetada Nakanishi, Institute for Immunology, Medical Faculty, Kyoto University, Yoshida, Sakyo-ku, Kyoto 606, Japan

Claudio A. Naranjo, Psychopharmacology Research Program, Sunnybrook Health Science Centre, Toronto, Ontario, Canada M4N 3M5

Pierluigi Nicotera, Institute of Environmental Medicine, Karolinska Institute, S-171 77 Stockholm, Sweden

Y. Nishizuka, Department of Biochemistry, Kobe University School of Medicine, Kobe 650, Japan

Franz Oesch, Institute of Toxicology, University of Mainz, Mainz, Germany

Barbara Oesch-Bartlomowicz, Institute of Toxicology, University of Mainz, Mainz, Germany

Kyo-Ichiro Okuda, Department of Surgery I, Miyazaki Medical College, Kiyotake, Miyazaki 889-16, Japan

R.W. Olsen, Department of Pharmacology, University of California School of Medicine, Los Angeles, California 90024, USA

E.K.I. Omogbai, Departments of Physiology and Pharmacology, University of Benin, Benin City, Nigeria

Sten Orrenius, Institute of Environmental Medicine, Karolinska Institute, S-171 77 Stockholm, Sweden

Masanori Otsuka, Department of Pharmacology, Faculty of Medicine, Tokyo Medical and Dental University, Tokyo 113, Japan

J.M. Palacios, Almirall Laboratories, Research Center, Cardener 68–74, 08024 Barcelona, Spain

Rodolfo Paoletti, Institute of Pharmacological Sciences, University of Milan, Via Balzaretti 9, 20122 Milan, Italy

Domenico Pellegrini-Giampietro, Dipartimento di Farmacologia Preclinica e Clinica Mario Aiazzi Mancini, Università degli Studi di Firenze, 50134 Firenze, Italy

John W. Phillis, Department of Physiology, Wayne State University, Detroit, Michigan 48201, USA

Olaf Pongs, Zentrum für Molekuläre Neurobiologie, Institut für Neurale Signalverarbeitung, D-20246 Hamburg, Germany

Oliver W. Press, University of Washington Cancer Center, Seattle, Washington 98195, USA

Uli Quast, Pharmakologisches Institut, Eberhard-Karls-Universität, D-72074 Tübingen, Germany

Michael J. Rand, Pharmacology Research Laboratory, Department of Medical Laboratory Science, Royal Melbourne Institute of Technology, Melbourne 3001, Victoria, Australia

P.K. Rangachari, Department of Medicine, McMaster University, Hamilton, Ontario, Canada L8N 3Z5

Mark M. Rasenick, Department of Physiology and Biophysics, University of Illinois College of Medicine, Chicago, Illinois 60612-7342, USA

J.A. Ribeiro, Laboratory of Pharmacology, Gulbenkian Institute of Science, 2781 Oeiras, Portugal

Charles R. Rosenfeld, Department of Pediatrics, University of Texas, Southwestern Medical School, Dallas, Texas 75235, USA

Elliott M. Ross, Department of Pharmacology, University of Texas, Dallas, Texas 75235-9041, USA

William Rostene, INSERM U.339, Hôpital St. Antoine, 75571 Paris Cedex 12, France

Markus Rudin, Preclinical Research, Sandoz Pharma Ltd, CH-4002 Basel, Switzerland

Juan M. Saavedra, National Institutes of Health, Department of Health and Human Services, Bethesda, Maryland 20892, USA

Kenton M. Sanders, University of Nevada School of Medicine, Reno, Nevada 89557, USA

B.I. Sasyniuk, Department of Pharmacology, McGill University, 3655 Drummond Street, Montreal, Quebec, Canada H4A 3N7

André Sauter, Preclinical Research, Sandoz Pharma Ltd, CH-4002 Basel, Switzerland

P.R. Saxena, Department of Pharmacology, Faculty of Medicine and Health Sciences, Erasmus University Rotterdam, 3000 DR Rotterdam, The Netherlands

Gunter Schultz, Department of Pharmacology, Free University of Berlin, Berlin, Germany

Paul B. Sigler, Howard Hughes Medical Institute, Yale University, New Haven, CT 06510, USA

T.G. Smart, Department of Pharmacology, The School of Pharmacy, University of London, London WC1N 1AX, UK

Philip L. Smith, Drug Delivery, SmithKline Beecham Pharmaceuticals, King of Prussia, Pennsylvania, USA

Maurizio R. Soma, Institute of Pharmacological Sciences, University of Milan, Via Balzaretti 9, 20122 Milan, Italy

Duncan J. Stewart, McGill Vascular Biology Group, McGill University, Montreal, Canada

Trevor W. Stone, School of Pharmacology, Institute of Biomedical and Life Sciences, University of Glasgow, Scotland, UK

Peter D. Suzdak, Novo Nordisk, 2760 Malov, Denmark

Brian D. Sykes, Department of Biochemistry, University of Alberta, Edmonton, Alberta, Canada T6G 2H7

Norio Taira, Tohoku University School of Medicine, Aoba-ku 980, Sendai, Japan

Palmer Taylor, Department of Pharmacology, University of California, San Diego, La Jolla, California 92093-0636, USA

Hans Thoenen, Department of Neurochemistry, Max Planck Institute for Psychiatry, D-82152 Martinsried, Germany

Ronald G. Thurman, Department of Pharmacology, University of North Carolina, Chapel Hill, North Carolina 27599, USA

Pieter B.M.W.M. Timmermans, Discovery Research, DuPont Merck Pharmaceutical Company, Experimental Station, Wilmington, DE 19880-0400, USA

J.-M. Trifaró, Department of Pharmacology, University of Ottawa, Ottawa, Canada K1H 8M5

Akira Tsuji, Faculty of Pharmaceutical Sciences, Kanazawa University, Kanazawa, Japan

G.T. Tucker, Department of Medicine and Therapeutics, The Royal Hallamshire Hospital, University of Sheffield, Sheffield, UK

M.D. Tyers, Pharmacology Division, Glaxo Research and Development Ltd., Ware, Hertfordshire SG12 0DP, UK

Jack Uetrecht, Faculties of Pharmacy and Medicine, University of Toronto and Sunnybrook Health Science Centre, Toronto, Canada M5S 1A1

Casey Van Bremen, Department of Pharmacology, University of British Columbia, Vancouver, British Columbia V6T 1Z3

Jan M. van Ree, Department of Medical Pharmacology, Rudolf Magnus Institute for Neurosciences, Utrecht University, 3584 CG Utrecht, The Netherlands

P.A. van Zwieten, Departments of Pharmacotherapy and Cardiology, University of Amsterdam, Academic Medical Centre, 1105 AZ Amsterdam, The Netherlands

John R. Vane, The William Harvey Research Institute, The Medical College of St. Bartholomew's Hospital, London EC1M 6BQ, UK

Timothy D. Warner, The William Harvey Research Institute, The Medical College of St. Bartholomew's Hospital, London EC1M 6BQ, UK

Richard M. Weinshilboum, Department of Pharmacology, Mayo Medical School/Mayo Clinic, Rochester MN 55905, USA

Peter G. Welling, Warner-Lambert/Parke-Davis, Ann Arbor, Michigan 48105-2430, USA

Albrecht Wendel, Department of Biochemical Pharmacology, University of Konstanz, D-7750 Konstanz, Germany

Arthur H. Weston, School of Biological Sciences, University of Manchester, Stopford Building, Manchester M13 9PT, UK

Thomas H. White, Department of Pharmacology, Dalhousie University, Halifax, Nova Scotia, Canada B3H 4H7

John Wijdenes, Immunotherapy Laboratories, Besançon, France

Grant R. Wilkinson, Department of Pharmacology, Vanderbilt University School of Medicine, Nashville, TN 37232-6000, USA

D.W. Williams, Department of Pharmacology, University of Melbourne, Parkville, Australia

Michael Williams, Neuroscience Discovery, Abbott Laboratories, Abbot Park, IL 60064, USA

P.B. Williams, Department of Pharmacology, Eastern Virginia Medical School, Norfolk, Virginia, USA

Yasushi Yamazoe, Department of Pharmacology, School of Medicine, Keio University, 35 Shinanomachi, Shinjuku-ku, Tokyo 160, Japan

Masashi Yanagisawa, Department of Molecular Genetics, Howard Hughes Medical Institute, The University of Texas, Southwestern Medical Center at Dallas, Dallas, Texas, USA

Chung S. Yang, Laboratory for Cancer Research, College of Pharmacy, Rutgers University, Piscataway, New Jersey 08855-0789, USA

F.D. Yocca, CNS Drug Discovery, Bristol-Myers Squibb Pharmaceutical Research Institute, Wallingford, CT 06492, USA

Li Yu-Lee, Department of Medicine, Rheumatology Section, Baylor College of Medicine, Houston, Texas 77030, USA

Edwin E. Zvartau, Pavlov Medical Institute, St. Petersburg 197089, Russia

Drug Receptors

Pharmacological Sciences: Perspectives for
Research and Therapy in the Late 1990s
ed. by A.C. Cuello and B. Collier
© 1995 Birkhäuser Verlag Basel/Switzerland

Molecular Studies of Glutamate Receptors

Shigetada Nakanishi

Institute for Immunology, Kyoto University Faculty of Medicine, Yoshida, Sakyo-ku, Kyoto 606, Japan

Summary. With the aid of our novel cloning strategy that combines electrophysiology and a *Xenopus* oocyte expression system, we have cloned and characterized the NMDA receptors and metabotropic receptors (mGluRs) and have demonstrated that both receptors consist of diverse members of receptor subunits or subtypes. Our studies have indicated that mGluR6 is responsible for synaptic transmission from photoreceptors to ON-bipolar cells and is essential in segregating visual signals into ON- and OFF-pathways. We have also disclosed that in the accessory olfactory bulb, mGluR2 at the presynaptic site of granule cells modulates inhibitory GABA transmission from granule cells to mitral cells and plays an important role in discriminating olfactory stimuli. This modulation by the activation of mGluR2 also induces an olfactory memory formation. On the basis of these findings, the functional importance of second-order neurons (bipolar cells and mitral cells) in sensory information processings is discussed.

Introduction

Glutamate receptors are essential for many integrative brain functions such as learning and memory and also for neural development [1, 2]. Excitatory neurotoxicity via glutamate receptors is also critical in some pathophysiological processes such as ischemic neuronal cell death and epilepsy [1, 2]. Glutamate receptors are classified into two distinct groups of receptors termed ionotropic and metabotropic receptors. The ionotropic receptors contain glutamate-gated, cation-specific ion channels and are subdivided into NMDA receptors and AMPA/kainate receptors [1, 2]. The metabotropic receptors (mGluRs) are coupled to intracellular signal transduction via G proteins [1, 2]. We have cloned and characterized the NMDA receptors and mGluRs with the aid of our functional cloning strategy that combines electrophysiology and a *Xenopus* oocyte expression system [3–5]. This chapter deals with our molecular studies indicating the existence of diverse members of the NMDA receptors and mGluRs and discusses our recent findings of some crucial roles 'of different subtypes of mGluRs in visual and olfactory sensory transmissions.

Molecular Diversity of Glutamate Receptors

The NMDA receptors consist of two distinct types of subunits, one termed NMDAR1 and the other four termed NMDAR2A–2D [4, 6].

NMDAR1 possesses all properties characteristic of the NMDA receptor-channel complex: agonist and antagonist selectivity, glycine activation, voltage-dependent Mg^{2+} blockade, Ca^{2+} permeability and Zn^{2+} inhibition. NMDAR2A–2D show no channel activity in a homomeric structure, but potentiate the NMDA receptor activity in a heteromeric formation with NMDAR1. These subunits also confer the variability in the properties of the NMDA receptors by different heteromeric subunit configurations. The NMDAR1 mRNA is ubiquitously expressed throughout the brain regions, whereas individual NMDAR2 mRNAs are distinctly distributed in different brain regions. The functional heterogeneity of the NMDA receptors in different neuronal cells is thus generated by the functional and anatomical differences of the NMDAR2 subunits. The NMDA receptor subunits, unlike other ligand-gated ion channels, most likely possess a membrane-spanning structure with an extracellular domain on the N-terminal side and a cytoplasmic domain on the C-terminal side. The asparagine at the second transmembrane segment of NMDAR1 is within a channel pore and is responsible for governing a high Ca^{2+} permeability and the blockade of Mg^{2+} and other channel blockers [7].

The mGluRs form a family of at least seven different subtypes termed mGluR1–mGluR7 [2, 5, 8]. These receptor subtypes have seven transmembrane segments common to other members of G protein-coupled receptors, but show no sequence similarity to the conventional G protein-coupled receptors. The mGluR family also possesses an unusually large extracellular domain responsible for glutamate binding at their N-terminal regions [9] and thus represents a novel family of G protein-coupled receptors. The seven mGluR subtypes differ in their agonist selectivities and signal transduction mechanisms [2]. mGluR1 and mGluR5 are coupled to IP_3/Ca^{2+} signal transduction and efficiently respond to quisqualate. The other five are all linked to the inhibitory cAMP cascade, but mGluR2 and mGluR3 effectively interact with L-(2S, 1′S, 2′S)-(carboxycyclopropyl)glycine (L-CCG-1) and trans-1-aminocyclopentane-1,3-dicarboxylate (tACPD), whereas mGluR4, mGluR6 and mGluR7 potently react with L-2-amino-4-phosphonobutyrate (L-AP4) [10–12]. The seven subtypes of the mGluR family are thus classified into three subgroups according to their sequence similarities, signal transduction mechanisms and agonist selectivities. All but mGluR6 mRNA are widely but distinctly distributed in various brain regions [2]. These findings strongly indicate that individual mGluR subtypes have their own functions by specializing the signal transduction and expression patterns in different nerve cells.

mGluR6 in Visual Transmission

On the basis of our knowledge of the diverse members of the glutamate receptors, it is essential to explore the physiological roles of individual receptor subtypes or subunits in integrative brain functions and during neural development. We have addressed this interesting question by several different approaches. A key processing of visual signal transmission is detecting contrasts in visual signals. This processing for visual contrasts occurs at the level of bipolar cells by segregating the visual signals into ON-center and OFF-center pathways. It has been reported that synaptic transmission from photoreceptors to bipolar cells in the ON-center pathway is mediated by an L-AP4-sensitive mGluR through the coupling to the cGMP cascade [13, 14]. We have identified mGluR6 from a retinal cDNA library and indicated that this mGluR subtype is exclusively expressed in the bipolar cell layer and selectively responds to L-AP4 after DNA transfection in CHO cells [15]. Immunohistochemical and immunoelectronmicroscopic analysis has provided compelling evidence that mGluR6 is restrictedly located at the postsynaptic site of rod (ON-type) bipolar cells in the adult rat retina [16]. Furthermore, our recent study using targeted disruption of mGluR6 gene has indicated that mGluR6 is essential in evoking ON responses to light exposure. Thus, taken together with the electrophysiological characterization of mGluR-mediated signal transduction in ON-bipolar cells, our findings have introduced the following model for synaptic transmission from photoreceptors to ON-bipolar cells (Fig. 1). When light activates photoreceptors, intracellular concentrations of cGMP decrease through the stimulation of phosphodiesterase via transducin. The decrease in cGMP concentrations hyperpolarizes photoreceptors by closing the cGMP-gated ion channel and reduces glutamate release. Under this situation, the mGluR6-G protein-phosphodiesterase system in ON-bipolar cells remains inactive, and high concentrations of cGMP are maintained in ON-bipolar cells

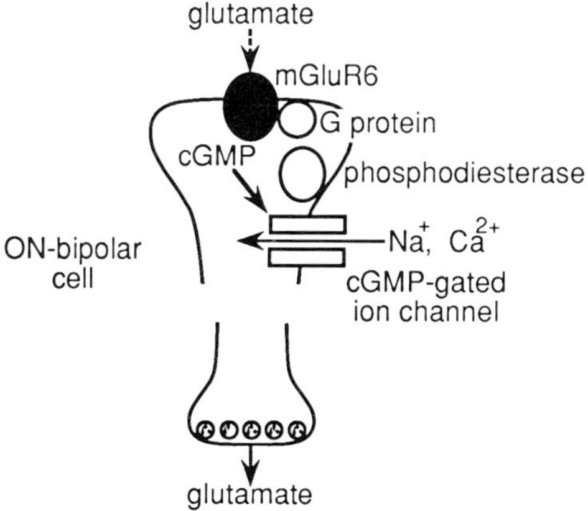

Fig. 1. mGluR6 function in the ON-pathway of the visual system.

and stimulate the cGMP-gated ion channel, thus resulting in depolarization of ON-bipolar cells. Glutamate release from ON-bipolar cells is augmented and in turn excites AII amacrine cells or ganglion cells, and this excitation is transmitted to the brain. Our investigation has provided the first indication that a specific mGluR subtype plays a critical role in synaptic transmission. Interestingly, mGluR6 is initially distributed in both the cell bodies and dendrites of the bipolar cells and is gradually concentrated at the postsynaptic site during retinal development [16]. mGluR6 in the bipolar cell thus provides an interesting example indicating the developmental specialization of a receptor localization in the central nervous system.

mGluR2 in Olfactory Transmission

We have also defined the physiological role of mGluR2 in olfactory sensory transmission of the accessory olfactory bulb (AOB). In the AOB, mitral cells receive afferent inputs from the vomeronasal nerve and transmit excitatory outputs to various brain regions (see Fig. 3). Granule cells are inhibitory interneurons that form typical dendrodendritic synaptic contacts with mitral cells. These synapses undergo reciprocal regulation; the granule cell is excited by glutamate from the mitral cell and in turn inhibits the mitral cell by GABA. Our analysis with *in situ* hybridization and immunohistochemistry has indicated that mGluR2 is highly expressed and localized at the dendrites of granule cells of the AOB [17]. Because we also identified 2-(2, 3-dicarboxycyclopropyl)glycine (DCG-IV) as a selective and potent agonist for mGluR2 [17], we have examined the effect of the DCG-IV-mediated activation of mGluR2 on GABA transmission from the granule cell to the mitral cell with the use of slice-patch clamp recording techniques. This analysis of the role of mGluR2 in olfactory sensory transmission has indicated that

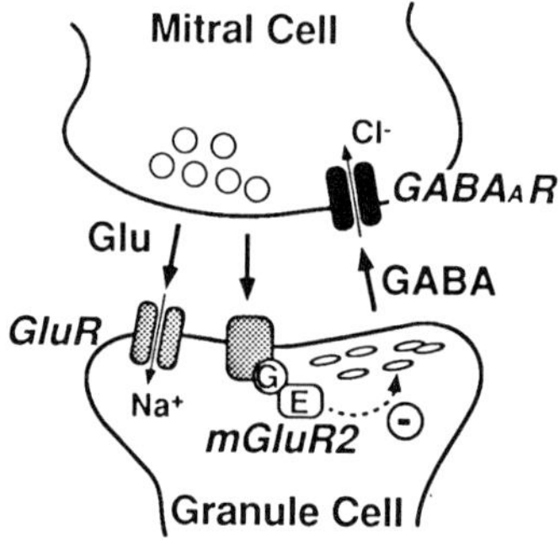

Fig. 2. Modulatory function of mGluR2 in the accessory olfactory transmission. Reprinted with permission from [17].

glutamate released from the mitral cell activates mGluR2 at the presynaptic site of the granule cell and relieves the mitral cell from GABA-mediated inhibition [17] (Fig. 2). The granule cell forms divergent synaptic contacts with not only the original mitral cell, but also with a large number of neighboring mitral cells, and thus causes both recurrent and lateral inhibition via inhibitory GABA transmission. Under this synaptic network, glutamate released from excited mitral cells relieves GABA-mediated self-inhibition via the activation of mGluR2, but would maintain the lateral inhibition of unexcited neighboring mitral cells. This mechanism should evidently enhance the signal-to-noise ratio between the excited mitral cells and their neighboring mitral cells and would contribute to the discrimination and resolution of different olfactory stimuli (Fig. 3).

To further substantiate the role of mGluR2 in olfactory sensory transmission, we have investigated the effects of the activation of mGluR2 on olfactory memory functions by animal behavioral analysis. Female mice form an olfactory memory at mating, evoking pregnancy block after exposure to unfamiliar male pheromones of a different strain, but offsetting this effect to familiar phermones of the stud male. This olfactory block of pregnancy, known as the Bruce effect, is caused by sustained enhancement of norepinephrine in the AOB after mating [18]. Enhanced norepinephrine persistently excites mitral cells by reducing GABA transmission from granule cells and results in olfactory

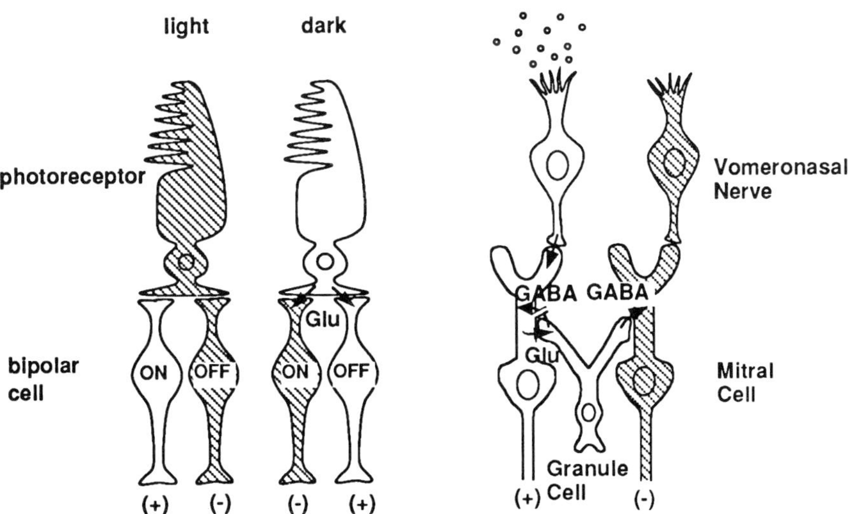

Fig. 3. Signal processing at the second-order neurons of the visual and olfactory systems.

memory formation specific to phermones exposed after mating. Because the activation of both the norepinephrine receptor and mGluR2 seemed to commonly reduce GABA transmission, we have examined whether the activation of mGluR2 by its specific agonists could create olfactory memory formation without mating. The results of this study have explicitly indicated that the activation of mGluR2 induces a specific olfactory memory formation [19]. Furthermore, this memory induced by the mGluR2 agonists faithfully reflects the memory formed at mating. Thus, our study demonstrates that a specific mGluR is involved in evoking neuronal plasticity responsible for the formation of recognition memory.

Our studies of both the visual and olfactory transmission systems indicate that the second-order neurons (bipolar cell and mitral cell) are crucial in discriminating or resolving external sensory stimuli (Fig. 3). This process seems to be essential in transmitting external signals to central parts of the brain. The peripheral sensory organs receive graded external signals. If these graded external signals were directly transmitted to higher centers of the brain, frequent errors in transmission would occur during several information steps to the brain. One way of minimizing the effect of transmission errors is that the sensory system uses a strategy to measure the difference in sensory stimuli rather than their absolute intensities and to transmit a signal proportional to this difference. Thus, the essential function of the second-order neurons is to enhance the signal-to-noise ratio of external stimuli and to decompose these signals into elementary components. Furthermore, at the level of the mitral cell, the olfactory transmission is modulated in combination with the regulatory function of the granule cell interneuron, and this modulation is effectively utilized in evoking neuronal plasticity for an olfactory recognition memory. Thus, our studies demonstrate not only that the mGluRs are important in synaptic transmission and modulation, but also that the second-order neurons are already essential in information processing to transmit external signals to the central parts of the brain.

Acknowledgements

This work was supported by research grants from the Ministry of Education, Science and Culture of Japan, The Ministry of Health and Welfare of Japan, the Uehara Memorial Foundation, the Senri Life Science Foundation, and the Sankyo Foundation of Life Science.

References

1. Nakanishi S. Molecular diversity of glutamate receptors and implications for brain function. Science 1992; 258: 597–603.

2. Nakanishi S, Masu M. Molecular diversity and functions of glutamate receptors. Annu. Rev. Biophys. Biomol. Struct. 1994; 23: 319–348.
3. Masu Y, Nakayama K, Tamaki H, Harada Y, Kuno M, Nakanishi S. cDNA cloning of bovine substance-K receptor through oocyte expression system. Nature 1987; 329: 836–838.
4. Moriyoshi K, Masu M, Ishii T, Shigemoto R, Mizuno N, Nakanishi S. Molecular cloning and characterization of the rat NMDA receptor. Nature 1991; 354: 31–37.
5. Masu M, Tanabe Y, Tsuchida K, Shigemoto R, Nakanishi S. Sequence and expression of a metabotropic glutamate receptor. Nature 1991; 349: 760–765.
6. Ishii T, Moriyoshi K, Sugihara H, Sakurada K, Kadotani H, Yokoi M, et al. Molecular characterization of the family of the N-methyl-D-aspartate receptor subunits. J. Biol. Chem. 1993; 268: 2836–2843.
7. Sakurada K, Masu M, Nakanishi S. Alteration of Ca^{2+} permeability and sensitivity to Mg^{2+} and channel blockers by a single amino acid substitution in the N-methyl-D-aspartate receptor. J. Biol. Chem. 1993; 268: 410–415.
8. Tanabe Y, Masu M, Ishii T, Shigemoto R, Nakanishi S. A family of metabotropic glutamate receptors. Neuron 1992; 8: 169–179.
9. Takahashi K, Tsuchida K, Tanabe Y, Masu M, Nakanishi S. Role of the large extracellular domain of metabotropic glutamate receptors in agonist selectivity determination. J. Biol. Chem. 1993; 268: 19341–19345.
10. Hayashi Y, Tanabe Y, Aramori I, Masu M, Shimamoto K, Ohfune Y, et al. Agonist analysis of 2-(carboxycyclopropyl)glycine isomers for cloned metabotropic glutamate receptor subtypes expressed in Chinese hamster ovary cells. Br. J. Pharmacol. 1992; 107: 539–543.
11. Tanabe Y, Nomura A, Masu M, Shigemoto R, Mizuno N, Nakanishi S. Signal transduction, pharmacological properties, and expression patterns of two rat metabotropic glutamate receptors, mGluR3 and mGluR4. J. Neurosci. 1993; 13: 1372–1378.
12. Okamoto N, Hori S, Akazawa C, Hayashi Y, Shigemoto R, Mizuno N, et al. Molecular characterization of a new metabotropic glutamate receptor mGluR7 coupled to inhibitory cyclic AMP signal transduction. J. Biol. Chem. 1994; 269: 1231–1236.
13. Nawy S, Jahr CE. Suppression by glutamate of cGMP-activated conductance in retinal bipolar cells. Nature 1990; 346: 269–271.
14. Shiells RA, Falk G. Glutamate receptors of rod bipolar cells are linked to a cyclic GMP cascade via a G-protein. Proc. R. Soc. Lond. (B) 1990; 242: 91–94.
15. Nakajima Y, Iwakabe H, Akazawa C, Nawa H, Shigemoto R, Mizuno N, et al. Molecular characterization of a novel retinal metabotropic glutamate receptor mGluR6 with a high agonist selectivity for L-2-amino-4-phosphonobutyrate. J. Biol. Chem. 1993; 268: 11868–11873.
16. Nomura A, Shigemoto R, Nakamura Y, Okamoto N, Mizuno N, Nakanishi S. Developmentally regulated postsynaptic localization of a metabotropic glutamate receptor in rat rod bipolar cells. Cell 1994; 77: 361–369.
17. Hayashi Y, Momiyama A, Takahashi T, Ohishi H, Ogawa-Meguro R, Shigemoto R, et al. Role of a metabotropic glutamate receptor in synaptic modulation in the accessory olfactory bulb. Nature 1993; 366: 687–690.
18. Brennan P, Kaba H, Keverne EB. Olfactory recognition: A simple memory system. Science 1990; 250: 1223–1226.
19. Kaba H, Hayashi Y, Higuchi T, Nakanishi S. Induction of an olfactory memory by the activation of a metabotropic glutamate receptor. Science 1994; 265: 262–264.

GABA Receptor Pharmacology

Graham A.R. Johnston

*Adrien Albert Laboratory of Medicinal Chemistry, Department of Pharmacology,
The University of Sydney D06, NSW 2006, Australia*

Summary. There is increasing evidence for the existence of "novel" receptors for the neuro-transmitter GABA that pharmacologically fall outside the currently accepted classification of two major subtypes. The $GABA_{A/B}$ classification, introduced in 1981, defines $GABA_A$ receptors as being sensitive to antagonism by bicuculline and insensitive to baclofen, while $GABA_B$ receptors are insensitive to antagonism by bicuculline and are activated by baclofen.

GABA receptors insensitive to both bicuculline and baclofen have been described extensively in both vertebrates and invertebrates. These "novel" GABA receptors have been given a variety of designations – $GABA_C$, $GABA_{NANB}$ (non-A, non-B), and ρ-receptors (cloned from retina). With the increasing evidence for similarities between these receptors, their subtype is most conveniently designated $GABA_C$ in an extension of the current $GABA_{A/B}$ nomenclature, though a more definitive classification of GABA receptors based on pharmacological, physiological and molecular biological considerations will emerge in due course.

$GABA_C$ receptors may constitute a third major sub-class of GABA receptors distinct in structure and function from the now classical $GABA_A$ and $GABA_B$ receptors. They may represent a relatively simple form of ligand gated ion channel which are made up of homo-oligomeric subunits, in contrast to the hetero-oligomeric $GABA_A$ receptors. The more complex $GABA_A$ receptors may have evolved from the simpler $GABA_C$ receptors.

The discovery in 1970 that the convulsant alkaloid bicuculline could specifically antagonise the inhibitory action of GABA in the cat spinal cord, provided a useful new pharmacological agent to use to support biochemical and physiological evidence that GABA was the major inhibitory neurotransmitter in CNS of vertebrates [1]. Bicuculline-sensitive synaptic inhibition was found throughout the brain and spinal cord. It became clear, however, that not all actions of GABA on nerve cells could be antagonised by bicuculline. The discovery of a specific class of bicuculline-insensitive GABA receptors that could be selectively activated by the GABA analogue baclofen led to the classification of GABA receptors into two subtypes: $GABA_A$ receptors were defined as being sensitive to antagonism by bicuculline and insensitive to baclofen, while $GABA_B$ receptors were defined as insensitive to antagonism by bicuculline and activated by baclofen [2].

$GABA_A$ and $GABA_B$ receptors differed not only in their pharmacology but also in their functionality. $GABA_A$ receptors gated chloride ion channels and were part of a superfamily of ligand gated ion channels that include nicotinic and glycine receptors, together with certain glutamate and 5HT receptor subtypes. $GABA_B$ receptors, which have thus far resisted cloning, were linked to second messenger systems and thus resembled muscarinic receptors and metabotropic glutamate receptors. $GABA_A$ and $GABA_B$ receptors have both pre- and postsynaptic locations on neurones. An extensive pharmacology for $GABA_A$ and $GABA_B$ receptors has emerged [3–5]. Both subtypes are heterogeneous. The heterogeneity of $GABA_A$ receptors may result from different subunit compositions, as these receptors consist of hetero-oligomeric membrane spanning glycoproteins in which five subunits combine to form individual ion channel complexes [5]. More than 15 different, but structurally related, gene products have been described coding for $GABA_A$ receptor protein subunits. It is likely that differing combinations of these gene products give rise to $GABA_A$ receptor complexes with differing pharmacology and physiology, e.g. differing sensitivity to modulation by benzodiazepines and ethanol and different channel properties [5]. Studies on $GABA_B$ receptors influencing transmitter release in different CNS regions have provided evidence for a multiplicity of $GABA_B$ receptors [6].

CACA as a Selective $GABA_C$ Ligand

The $GABA_C$ hypothesis arose in 1984 from studies on a range of structurally related GABA analogues that inhibited neuronal firing in a bicuculline-insensitive manner, but were unable to influence $[^3H]$-(-)-baclofen binding [7]. These studies suggested the possibility of a third subtype of GABA receptors insensitive to the classic $GABA_A$ antagonist bicuculline and the classic $GABA_B$ agonist baclofen. The lead compound was *cis*-4-aminocrotonic acid (CACA, Z-4-aminobut-2-enoic acid), an analogue of GABA held in a folded conformation. CACA was approximately one-quarter as potent as GABA or the *trans*-isomer TACA in inhibiting the firing of cat spinal neurones, and its inhibitory action could not be antagonised by bicuculline [8]. Unlike baclofen, which also had a bicuculline-insensitive neuronal action, CACA was equipotent as an inhibitor of Renshaw cells and spinal interneurones.

The effects of CACA have now been studied in a variety of preparations. It appears to be a weak but selective agonist for $GABA_C$ receptors, while GABA and TACA appear to be some 5 to 20 times more potent than CACA as agonists at $GABA_C$ receptors but non-selective [9, 10]. The binding of $[^3H]$-GABA to rat brain membranes contains a CACA-sensitive component that is insensitive to bicuculline and baclofen [11, 12]. $[^3H]$-CACA has been prepared, but the toxic nature of the preparation has limited the extent of the binding studies that have been carried out thus far [13]. It shows high specific binding to rat cerebellar membranes that can be displaced potently by GABA, TACA (IC_{50} 0.025 μM) and less potently by unlabelled CACA (IC_{50} 0.5 μM) [9]. Autoradiographic studies with $[^3H]$-CACA and the selective

GABA$_A$ ligand [^{3}H]-muscimol showed that both ligands bound to similar sites in the cerebellum in the molecular layer and to a lesser extent in the granule cell layer, but were completely different in other brain areas [10].

GABA$_C$ Receptors in Retina

Electrophysiological, pharmacological and molecular biological studies in the retina have provided extensive evidence for the existence of GABA$_C$ receptors. Patch clamp studies of cultured rod bipolar cells from rat retina showed GABA activated bicuculline- and baclofen-insensitive chloride channels that could be selectively activated by CACA [14]. Similar results were found in studies on rod horizontal cells isolated from white perch retina [15].

Gene expression studies in *Xenopus* oocytes of mRNA derived from bovine retina or from a human retinal cDNA library yielded a homo-oligomeric GABA receptor, activation of which by GABA and CACA produced a chloride current that is insensitive to bicuculline and closely resembles the GABA$_C$ responses observed in retinal neurones [16, 17]. These receptors were originally termed ρ-receptors [see 17] but more recent studies indicate the designation, GABA$_C$ receptors, is preferable [18]. In all of these studies on GABA$_C$ receptors derived from retina, CACA is a selective agonist weaker in potency than GABA and TACA, consistent with the earlier studies on cat spinal cord [8] and the recent studies on [^{3}H]-CACA binding [9]. Several GABA analogues acted as antagonists, the most potent being imidazole-4-acetic acid [18]. Retinal GABA$_C$ responses were not modulated by steroids, barbiturates or benzodiazepines known to modulate GABA$_A$ responses, and were not influenced by GABA$_A$ antagonists such as bicuculline, SR-95531 (gabazine) and (+)-hydrastine, but they were sensitive to the chloride channel blocker picrotoxin. In addition to activating chloride channels, CACA has been shown to inhibit voltage-dependent calcium channels mediating presynaptic inhibition in goldfish retina that are insensitive to both baclofen and bicuculline, suggesting that CACA interacts with G-protein linked receptors as well as ligand gated ion channels [19].

GABA$_C$-like Receptors in Invertebrates

Bicuculline-insensitive GABA receptors have been described in dissociated lobster thoracic neurones [20], the nematode *Ascaris* [21], in receptors derived from cockroach CNS mRNA expressed in *Xenopus* oocytes [22] and in receptors cloned from the fruitfly *Drosophila* [23]. The responses observed in the dissociated lobster thoracic neurones closely resemble the $GABA_C$ responses described in vertebrate retina [20], whereas those observed in receptors cloned from *Drosophila* appear to differ in their modulation by barbiturates and neurosteroids [23]. This suggests a heterogeneity of invertebrate bicuculline-insensitive GABA receptors, not all of which closely resemble vertebrate $GABA_C$ receptors.

As Complex as ABC?

Are there three major subtypes of GABA receptors, or are $GABA_C$ receptors really a subclass of $GABA_A$ receptors that are less sensitive to the classic $GABA_A$ antagonist bicuculline? It is known that a single point mutation in the $GABA_A$ α_1-subunit, substituting α_1-Phe-64 by Leu, can lead to a 200-fold reduction in the antagonist potency of bicuculline on recombinant $GABA_A$ hetero-oligomeric receptors [24], but it is not known if such a mutant receptor becomes sensitive to CACA as an agonist. This key Phe residue in $GABA_A$ α_1-subunits is absent in the ρ-subunits that could have been cloned from retina and in the *Drosophila* GABA receptor subunits that afford bicuculline-insensitive receptors.

There is molecular biological evidence that $GABA_C$ receptors in retina are distinct from $GABA_A$ receptors [18]. In oocytes expressing total retina mRNA, distinct $GABA_A$ and $GABA_C$ responses can be observed. $GABA_A$ and $GABA_C$ subunits aggregate independently of one another – $GABA_A$ subunits forming hetero-oligomeric receptors while $GABA_C$ subunits form homo-oligomeric complexes. The close similarity between the $GABA_C$ responses in retinal horizontal cells and those in oocytes expressing only the ρ_1-subunit indicates that the functional receptors normally expressed in retina could be homo-oligomeric [18]. Receptors expressed from single $GABA_A$ receptor subunits do not yield functional GABA receptors that resemble the physiology and pharmacology of $GABA_A$ receptors found in intact neuronal tissue [5]. Interestingly the bicuculline-insensitive GABA receptors cloned from *Drosophila* are fully functional as homo-oligomers [23].

$GABA_C$ receptors may represent a relatively simple form of ligand gated ion channel receptor which is made up of homo-oligomeric subunits, in contrast to the hetero-oligomeric $GABA_A$ receptors. Their overall pharmacology appears simpler than that of the classic $GABA_A$ receptors, especially with respect to modulation by allosteric modulators such as neurosteroids and benzodiazepines, and there are differences in agonist and antagonist specificity. The more complex $GABA_A$ receptors may have evolved from the simpler $GABA_C$ receptors.

Acknowledgements

The author is grateful to Robin Allan, Muallâ Akinci, Colleen Drew, Rujee Duke, Frances Edwards, Ann McGregor, Ken Mewett, and Tony Weiss for their collaboration, Lindy Holden-Dye, Jeremy Lambert and George Uhl for helpful discussions, and the Australian National Health and Medical Research Council for financial support.

References

1. Curtis DR, Duggan AW, Felix D, Johnston GAR. GABA, bicuculline and central inhibition. Nature 1970; 226: 1222–1224.
2. Hill DR, Bowery NG. [^{3}H]-Baclofen and [^{3}H]-GABA bind to bicuculline-insensitive GABA$_B$ sites in rat brain. Nature 1981; 290: 149–152.
3. Kerr DI, Ong J. GABA agonists and antagonists. Med. Res. Rev. 1992; 12: 593–636.
4. Krogsgaard-Larsen P, Frølund B, Jørgensen FS, Schousboe A. GABA$_A$ receptor agonists, partial agonists, and antagonists. Design and therapeutic prospects. J. Med. Chem. 1991; 37: 2489–2505.
5. Macdonald RL, Olsen RW. GABA$_A$ receptor channels. Annu. Rev. Neurosci. 1994; 17: 569–602.
6. Bonnanno G, Raiteri M. Multiple GABA$_B$ receptors. Trends Pharmac. Sci. 1993; 14: 259–261.
7. Drew CA, Johnston GAR, Weatherby RP. Bicuculline-insensitive GABA receptors: studies on the binding of (-)-baclofen to rat cerebellar membranes. Neurosci. Lett. 1984; 52: 317–321.
8. Johnston GAR, Curtis DR, Beart PM, Game CJA, McCulloch RM, Twitchin B. *cis*- and *trans*-4-aminocrotonic acid as GABA analogues of restricted conformation. J. Neurochem. 1975; 24: 157–160.
9. Johnston GAR. GABA$_C$ receptors. Prog. Brain Res. 1994; 100: 61–65.
10. Johnston GAR. GABA receptors: as complex as ABC? Clin. Exp. Pharmacol. Physiol. 1994; 21: 521–526.
11. Balcar VJ, Joó F, Kása P, Dammasch IE, Wolff JR. GABA receptor binding in rat cerebral cortex and superior cervical ganglion in the absence of GABAergic synapses. Neurosci. Lett. 1986; 66: 269–274.
12. Drew CA, Johnston GAR. Bicuculline- and baclofen-insensitive γ-aminobutyric acid binding to rat cerebellar membranes. J. Neurochem. 1992; 58: 1087–1092.
13. Duke RK, Allan RD, Drew CA, Johnston GAR, Mewett KN, Long MA, Than C. The preparation of tritiated *E*- and *Z*-4-aminobut-2-enoic acids, conformationally restricted analogues of the inhibitory neurotransmitter 4-aminobutanoic acid (GABA). J. Label. Compd. Radiopharmac. 1993; 33: 527–540.
14. Feigenspan A, Wässle H, Bormann J. Pharmacology of GABA receptor Cl$^-$ channels in rat retinal bipolar cells. Nature 1993; 361: 159–162.
15. Qian H, Dowling JE. Novel GABA responses from rod driven retinal horizontal cells. Nature 1993; 361: 162–164.
16. Shimada S, Cutting G, Uhl GR. γ-Aminobutyric acid A or C receptor? γ-Aminobutyric acid ρ_1 receptor RNA induces bicuculline-, barbiturate- and benzodiazepine-insensitive γ-aminobutyric reponses in *Xenopus* oocytes. J. Pharmacol. Exp. Therap. 1992; 41: 683–687.
17. Woodward RM, Polenzani L, Miledi R. Characterization of bicuculline/baclofen-insensitive (ρ-like) γ-aminobutyric acid receptors expressed in *Xenopus* oocytes. 2. Pharmacology of γ-aminobutyric acid$_A$ and γ-aminobutyric acid$_B$ receptor agonists and antagonists. Mol. Pharmacol. 1993; 43: 609–625.
18. Qian H, Dowling JE. Pharmacology of novel GABA receptors found on rod horizontal cells of the white perch retina. J. Neurosci. 1994; 14: 4299–4307.
19. Matthews G, Ayoub GS, Heidelberger R. Presynaptic inhibition by GABA is mediated via two distinct GABA receptors with novel pharmacology. J. Neurosci. 1994; 14: 1079–1090.

20. Jackel C, Krenz W-D, Nagy F. Bicuculline/baclofen-insensitive GABA response in crustacean neurones in culture. J. Exp. Biol. 1994; 191: 167–193.
21. Holden-Dye L, Willis RJ, Walker RJ. Azole compounds antagonise the bicuculline-insensitive GABA receptor on the cells of the parasitic nematode *Ascaris suum*. Br. J. Pharmacol. 1994; 111: 188P.
22. Lummis SCR. Insect GABA receptors: characterization and expression in *Xenopus* oocytes following injection of cockroach CNS mRNA. Mol. Neuropharmacol. 1992; 2: 167–172.
23. Chen R, Belelli D, Lambert JL, Peters JA, Reyes A, Lan NC. Cloning and functional expression of a *Drosophila* γ-aminobutyric acid receptor. Proc. Natl. Acad. Sci. USA 1994; 91: 6069–6073.
24. Sigel E, Baur R, Kellenberger S, Malherbe P. Point mutations affecting antagonist affinity and agonist dependent gating of $GABA_A$ receptor channels. EMBO J. 1992; 11: 2017–2023.

Pharmacological Sciences: Perspectives for
Research and Therapy in the Late 1990s
ed. by A.C. Cuello and B. Collier
© 1995 Birkhäuser Verlag Basel/Switzerland

Serotonin Receptor Subtypes: Exploiting their Therapeutic Potential

J.M. Palacios

Almirall Laboratories, Research Center, Cardener, 68-74, 08024 Barcelona, Spain

Introduction

Soon it will be the 50th anniversary of the isolation, identification and chemical synthesis of serotonin (5-hydroxy tryptamine, 5-HT). Since the discovery of this mediator, 5-HT has been the object of a large amount of work and speculation. An important drug of the serotoninergic family, LSD, was also discovered more than 50 years ago. In recent years, interest in serotonin and serotoninergic mechanisms has increased significantly as a consequence of the development in this area of many compounds with therapeutic utility. Drugs such as the gastrokinetic cinitapride, clebopride or cisapride, new anxiolytics such as buspirone, or the treatments for chemotherapy-induced vomiting such as ondansetron, ganisetron and tropisetron, the recently developed antimigraine compounds such as sumatriptan, or new antidepressants such as prozac, all share as a common denominator their interaction with serotoninergic mechanisms, either at the level of the receptors or the transporter. The purpose of this article is to assess recent advances in the understanding of serotonin receptors on the development of new compounds with therapeutic utility.

Serotonin receptors were first subdivided in the 1950s by Rocha and Silva, and by Gaddum and Piccarelli. After a long period of relative neglect, serotonin receptors came to the fore in the 1970s when Peroutka and Snyder [1], once again, subdivided serotonin receptors into two families, the 5-HT1 and the 5-HT2. This was possible thanks to the availability of radiolabeled molecules such as [³H]-5HT, [³H]-LSD and [³H]-spiperone, the latter being a newly developed neuroleptic which clearly presented a high affinity not only for dopamine D2 receptors, but also for 5-HT2 receptors [2]. Since then, the race to define serotonin receptor subtypes was undertaken at such high speed that from the beginning of the 1980s until now more than 14 different receptor subclasses have been defined. This has been possible due to the combination of what we can call "traditional" pharmacological tools and modern molecular biological pharmacology. Among the traditional pharmacological tools, the development of selective agonists and antagonists came first. These drugs were used as ligands in combination with selective or non selective compounds, making possible the study of receptor ligand interaction and rapid definition of affinity for a large number of drugs for these particular receptors, the study of receptor distribution, and the analysis of the *in vivo* effects of selective and non selective compounds. Among the new technologies, the use of molecular biological tools has allowed to clone many 5-HT receptor subtypes and, based on their sequence similarities, allowed their classification into different 5-HT receptor families. The *in vitro* mutations of these receptors allow for the examination of domains in the protein receptor molecule, which are important for ligand recognition or coupling with second messenger systems. The availability of an artificial system where the receptors are independently expressed (cell lines expressing high levels of receptors) simplifies the analysis of drug profiles and the definition of interactions with second messengers. Using *in situ* hybridization techniques, the cellular distribution of the mRNAs coding for a given receptor molecule can also be studied. That, together with receptor binding analysis, allows for the establishment of possible cellular distribution patterns. Furthermore, two new different procedures can be used to functionally analyze the lack of receptors in a living organism. One is the reduction of receptor levels by antisense oligonucleotides against a given receptor in either an entire animal or cerebral slices. The second is by disrupting the receptor expression using transgenic "knockout" animals. Both these techniques have now been used in different degrees for the study of serotonin receptors. It should be emphasized that there is no radical separation between these techniques; they are rather complementary, and without classical pharmacological tools it would still be difficult, not to say impossible, to exploit the therapeutic potential of a given receptor population.

Serotonin Receptors: Molecular Biology and the Outburst in the Number of Different Subtypes

The $\beta 2$ adrenergic receptor was the first member of the superfamily of neurotransmitter receptors (with seven transmembrane domains) to be characterized. In an attempt to identify genes homologous to the $\beta 2$ receptor, a gene coding for a protein with identical topology to that of the adrenergic receptor gene was isolated, although it was not possible to characterize the ligand corresponding to this gene; this gene was named G-21, and for some time the search went on in an attempt to find out its endogenous ligand. Finally, it was demonstrated that G-21 coded for the serotonin receptor 5-HT1A [3]. The isolation of this gene prompted a series of studies pursuing the identification of other related receptors. Separately, and based on the application of classical molecular biology methodologies (the so-called expression-cloning method), the gene of the receptor which is enriched in the choroid plexus of the rat brain, i.e., the 5-HT1C [4], was isolated and characterized. Both genes, 5-HT1A and 5-HT1C, allowed for the rapid cloning of several other serotoninergic receptors.

So far, at least 14 different genes coding for serotonin receptors have been characterized, not taking into account possible species differences, which would increase the number considerably. In Table 1 a summary of the current classification of serotonin receptors is shown. It describes the various serotoninergic receptor families which can be classified according to similarity in their aminoacid sequence, and their signal transduction mechanisms. This would include mediation of their action by either stimulations or inhibitions of adenylate cyclase, stimulation of phosphatidyl inositol hydrolysis, or interaction with ionic channels, or the receptor itself which forms an ionic channel [5]. These subgroups belong to the 5-HT1 receptor family [6] that involves members A, B, D, E, F, in which 1C is not included since it has been re-classified in a more rational manner as 2C. The 5-HT2 family is now composed of 5-HT2A (old 5-HT2), a newly characterized receptor of the rat stomach fundus, the 2F (now 5-HT2B), and 5-HT2C (old 5-HT1C). 5-HT3 receptor family, of which only one subtype has been so far characterized [7], belonging to the group of receptors forming ionic channels, similar to the nicotinic receptors of acetylcholine, or the GABA receptors for which a large variability in their units is claimed, all of which lead us to the assumption that, although it has not yet been proven, the 5-HT3 serotonin receptor family should also present a large variety of members.

The latest addition to the collection of serotonin receptor clones is the 5-HT4 receptor. The structure of the 5-HT4 receptor has been reported by Brancheck and collaborators [8]. The 5-HT4 receptor is a classical G-protein coupled receptor with seven transmembrane domains, and does not present a high degree of homology with any of the cloned 5-HT receptors linked to G-proteins.

From the Genes to the Clinic

A consequence of the proliferation of 5-HT receptor subtypes has been the identification of agents acting selectively in these receptors. In the last few years, the medicinal chemists have been able to synthesize numerous molecules which have later shown clinical utility, or are about to be introduced into human therapy. The structures of some of these molecules are illustrated in Fig. 1.

In the field of 5-HT1 agents, the first introduction of a relatively selective agonist, buspirone, for the treatment of anxiety has been

Table 1. Classification of 5-HT receptors

Nomenclature	Selective agonists	antagonists	Radioglands	Molecular structure	Distribution	Therapeutic utility
5-HT$_{1A}$	8-OH-DPAT Flesinoxane Ipsapirone Buspirone Tandospirone Gepirone	WAY100135 Spiperone	[^{3}H]8-OH-DPAT	cAMP↓ K$^+$ Channel	Limbic system (Hippocampus, raphe nuclei) neocortex, enteric nerves.	Anxiety, depression, hypertension, sexual and feeding behaviour.
5-HT$_{1B}$ (5HT1Dβ)	CP93129 5-CT RU 24969	Pindolol	[^{125}I]GTI [^{125}I]CYP	cAMP↓	Basal ganglia, substantia nigra, hippocampus, visual pathways.	Motility, depression, aggressiveness, migraine.
5-HT$_{1D}$ (5-HT-1Dα)	5-CT Metergoline Metisergide Sumatriptan	Metiotepine Mianserine Quipacine	[^{125}I]GTI	cAMP↓	Substantia nigra basal ganglia, limbic system.	Depression, obsessive/ compulsive behavior, feeding conduct, psychosis.
5-HT$_{1E}$	—	—	[^{3}H]5-HT	cAMP↓	Cerebral cortex	Unknown
5-HT$_{1F}$	sumatriptan		[^{3}I]5-HT [^{125}I]ILSD	cAMP↓	Limbic structures	Unknown
5-HT$_{2A}$	α-methyl-5-HT	Ketanserine Ritanserine Espiperone Mianserine Cinanserine Mesulergine Amitriptiline Pipamperone Setoperone Risperidone	[^{3}H]ketaserine [^{125}I]DOI	IP$_3$/DG	Neocortex, striatrum, hippocampus, basilar artery, aorta, ileum, uterus, blood platelet	Psychosis, sleep.

5-HT$_{2B}$	α-methyl-5-HT	LY53857	[^{3}H]5-HT	IP$_3$/DG	Stomach fundus in rat	Unknown
5-HT$_{2C}$	α-methyl-5-HT	Mesulergine Ritanserine Mianserine Pizotifen Ciproheptadine LY 53857	[^{3}H]-mesulergine	IP$_3$/DG	Choroid plexus, limbic system, components of the motor system.	Depression, obsessive/ compulsive behavior, feeding conduct, psychosis.
5-HT$_3$	2-methyl 5-HT m-chlorine- phenyl biguanide	Tropisetron Ondansetron Granisetron Zacopride Renzapride Metoclopramide Morphine	[^{3}H]zacopride	Cation channel	Ultimate area, vagus nerve, spinal cord, limbic and striated systems, sensitive neurons of enteric nervous system.	Emesis, memory, drug dependence.
5-HT$_4$	Cinitapride Cisapride Renzapride Zacopride BIMU8	GR113808 RS23597190	[^{3}H]GR113808	cAMP↑	Caudate putamen, substantia nigra, hippocampus, ileum.	Gastrointestinal motility.
5-ht$_{5A}$	Ergotamine 5-CT	?	[^{125}I]ILSD [^{3}H]5-CT	—	Habenule, Hippocampus	?
5-ht$_{5B}$	Ergotamine 5-CT	Metisergide ?	[^{125}I]ILSD [^{3}H]5-CT	—	Cortex, hippocampus, cerebellum,	?
5-ht$_6$	—	—	[^{125}I]ILSD [^{3}H]5-HT	cAMP↑	Striated nucleus Hippocampus	?
5-ht$_7$	5-CT Lisuride	Metiotepine ? Clozapine ?	[^{125}I]ILSD [^{3}H]5-HT	cAMP↑	Cortex	Psychosis ?

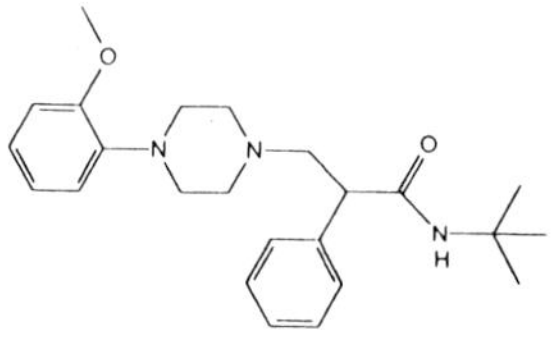

J.M. Palacios

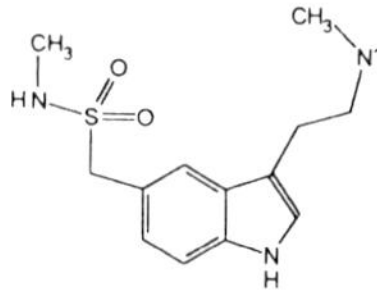

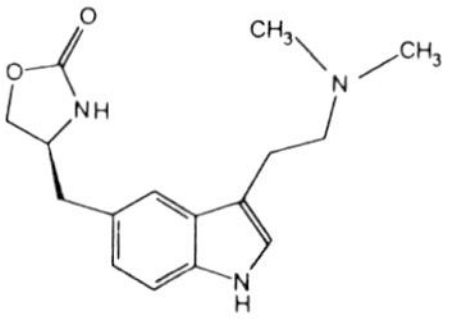

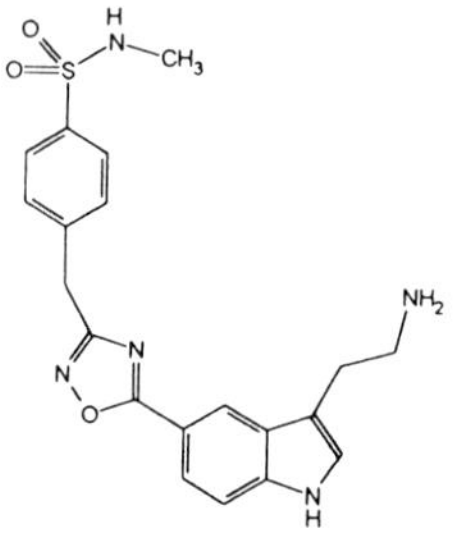

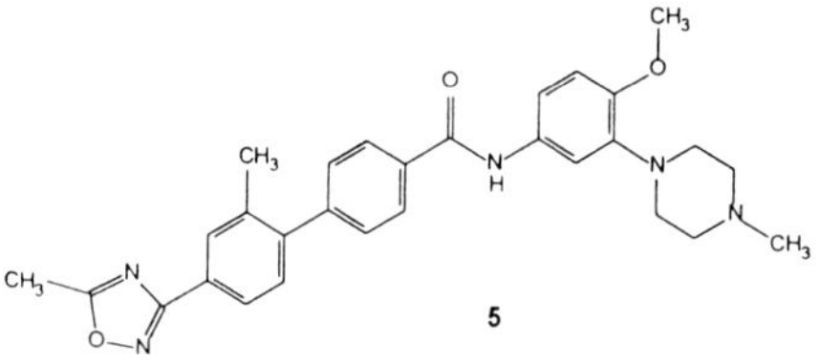

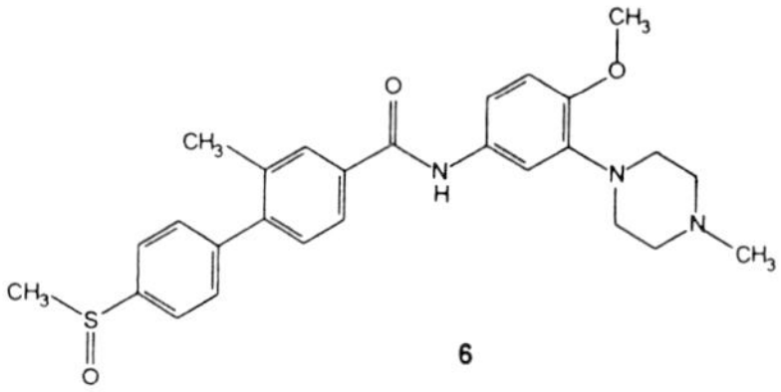

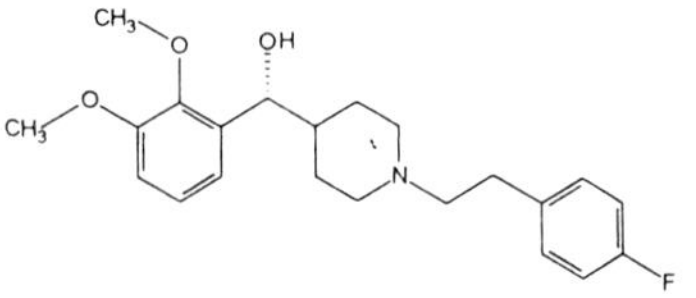

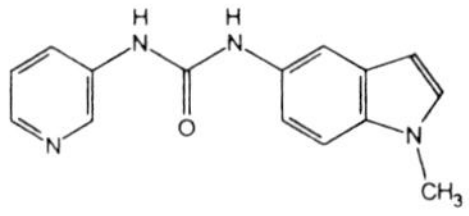

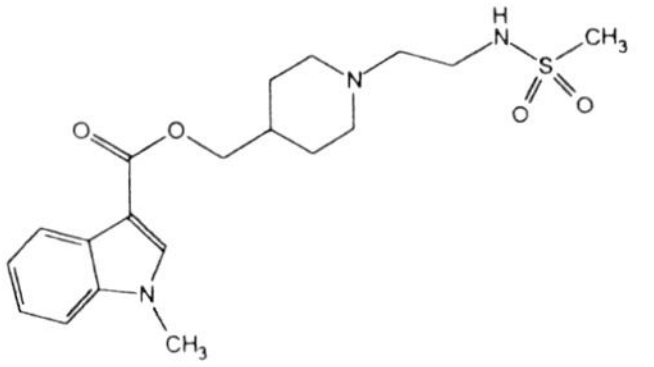

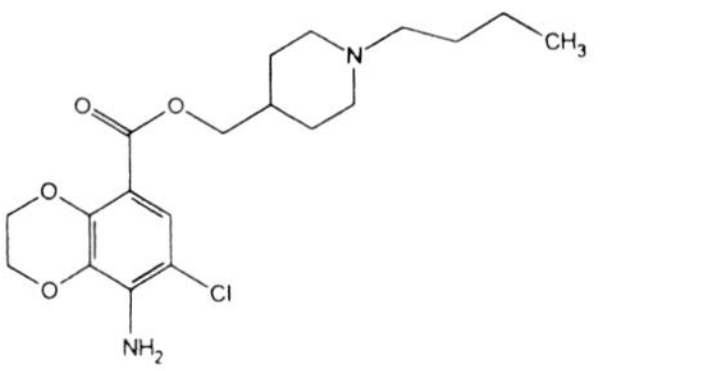

Fig. 1. Structure of some recently described serotoninergic compounds. (1) WAY 100135 a "silent" 5-HT-1A antagonist. (2, 3 and 4) are 5-HT-1Dα/β agonists; (2) is Sumatriptan; (3) is the Wellcome's compound 311C90, and (4) is from Merck and Co. (5) and (6) are 5-HT-1D antagonists from Glaxo; (5) is GR127935; (7) and (8) are 5-HT2 antagonists; (7) is the Marion Merrell Dow compound MDL-100907, a selective 5-HT-2A antagonist, and (8) is SB-200646A from Smith Kline Beecham, a 5-HT-2B/2C antagonist; (9) and (10) are 5-HT-4 antagonists; (9) is GR-113808 from Glaxo, and (10) is SB 204070 from Smith Kline Beecham.

followed by numerous other attempts to use these agents in therapy. In this sense, molecules like gepirone, ipsapirone and tandospirone are being developed for the treatment of both anxiety and depression. The fact that 5-HT1A agonists could have a potential in these diseases could be related to the enrichment of these receptors in limbic areas of the mammalian brain, and more importantly, in the raphe, where they occupy a very strategic location as somatodendritic receptors of the serotoninergic neurons controlling the release of serotonin [9, 10]. The development of "silent" serotoninergic antagonists has also been a long-standing goal for several groups. Recently, Wyeth has reported on a compound, WAY 100135 (Fig. 1), which appears to be one of the first "silent" 5-HT1A antagonists, although the claim is still controversial.

The introduction of sumatriptan for the treatment of migraine has been one of the major advances of recent years. Sumatriptan is a 5-HT1Dα/1Dβ agonist and due to its clear therapeutic utility in the treatment of migraine, several other compounds belonging to the same pharmacological class are now being investigated. No selective alpha or beta receptor agent has been described yet; this is crucial to define the site of action of these compounds. It appears that the predominant form of the receptors in terms of abundance of the mRNA or binding sites seems to be the beta form. A selective antagonist for these receptors has been recently described, GR 127935 (Fig. 1) [11]. The antagonist is providing further insights into the central functions of 5-HT1Dα/1Dβ receptors. Because of the predominant presynaptic location of these receptors in the mammalian brain, it may modulate 5-HT release, and thus it has been postulated that they could be a target for the development of new antidepressants. 5-HT1Dα/1Dβ receptors are present in high concentrations in the basal ganglia and in the limbic areas of the rat brain [9, 10]. The studies of Hen and collaborators [12] of the 1B receptor in transgenic "knock-out" mice have shown an involvement of these receptors in the modulation of aggressive behavior, making this system even more interesting as a target for new drug development.

The 5-HT2 family has recently been reorganized and re-categorized (see above and Table 1). All three members of the family are now the subject of intensive investigation. The appeal of these receptors is based on both their presence in many cortical areas [9, 10], as well as the fact that their interaction with 5-HT2 receptor could be important in the mechanism of action of some new antipsychotics (i.e., the ratio of affinities of certain antipsychotics for the 5-HT2 receptors vs the dopamine D2 receptors could be of importance in the therapeutic profile of these compounds). The new serotonin-dopamine hypothesis is being investigated. Ritanserine, a compound with 5-HT2A and dopamine D2 receptor antagonist properties is an example of this type of molecule. The 5-HT2C (old 1C) receptors are currently being proposed as a target for new drug development thanks to the development of selective antagonists for these receptors. Compounds selectively acting on these receptors have been described recently.

Clearly, more medicinal chemistry and biology work is necessary to understand the role of 5-HT2C receptors in the brain. The 5-HT2C mRNA presents a widespread distribution in the brain, and in terms of abundancy, is one of the predominant serotonin receptors in the mammalian brain [9, 10]. The implication of 5-HT2C mechanisms in many different brain functions is increasing with time, and the availability of selective compounds for these receptors (see Fig. 1) will certainly improve our understanding of their role in brain function.

The 5-HT3 system is pharmacologically mature [13], even if one of the major problems still to be resolved is the existence or not of subtypes. Until now, only pharmacological evidence has been provided to support differences in the activity of 5-HT3 agents, although species differences could be based on these pharmacological differences. Several 5-HT3 antagonists have been introduced in therapy, particularly for the treatment of chemotherapy-induced emesis. The first one was ondansetron, followed by granisetron and tropisetron.

The antiemetic activity of these compounds is supported by the abundance of 5-HT3 receptors in the pathways mediating emesis in the brain [9, 10]. On the other hand, 5-HT3 receptors are also present in forebrain areas in the mammalian brain, particularly in the limbic areas of the human brain, i.e., hippocampus, amygdala and also the basal ganglia including caudate putamen and substantia nigra pars reticulata. Investigations on the possible utility of 5-HT3 antagonists in neurological and psychiatric diseases are still ongoing, although effects have been found related to the treatment of memory deficit and also in anxiety with these compounds. Nevertheless, no 5-HT3 antagonist has yet been licensed for use in treatment of any of these indications.

As mentioned before, the most recently cloned serotonin receptor corresponds to the 5-HT4 receptor. The 5-HT4 receptor was first proposed on the basis of biochemical and pharmacological studies by Bockaert et al. [14], and it was discovered by the stimulating activity of some benzamides on the cyclase in neuronal cultures, and the inhibitory properties of tropisetron, the Sandoz 5-HT3 antagonist, in these cultures. Both agonists and antagonists [15] (Fig. 1) have been described for these particular receptors with selectivity and high affinity and, as mentioned above, the structure of the gene has been elucidated. There are two forms of the receptor differing only by the length of the C-terminal which is shorter by 15 aminoacids in one of the two forms. This does not exclude the possibility of the existence of subtypes of these receptors. The pharmacology of these two forms, as far as it is known, is very similar. The distribution, however, differs in the sense that the short form is enriched in the caudate nucleus while the other form seems to be distributed throughout the brain, with the exception of the cerebellum [8]. The synthesis of selective high affinity antagonists has made possible the development of radioligands for these receptors, which have allowed for a detailed study of the characteristics and distribution of these receptors. Interestingly, the areas more enriched in this receptor are, again, the limbic system and the basal ganglia, where it co-localizes with other serotonin receptors. There are already on the market several compounds, particularly gastrokinetic agents such as cisapride, cinitapride or clebopride, which are known to be 5-HT4 receptor agonists. The presence of 5-HT4 receptors in the gastrointestinal system is very well documented. 5-HT4 receptors have also been described in the human heart. Indications for 5-HT4 agents are quite extensive, particularly it has been proposed that they could be useful in the treatment of irritable bowel syndrome, although no clinical information is available at the present time. With regard to central indications for 5-HT4 agents, it has been proposed that they could have beneficial effects in memory although, again, clinical data supporting these claims are still to be generated. The distribution in the brain indicates that, again, these receptors could be involved in many other functions, but the clarification of the possible therapeutic potential of 5-HT4 receptors remains to be established.

The New 5-HT Receptors

As mentioned above, many other receptors have now been identified by molecular cloning. Nothing is known about the possible functions of these receptors in the organism. What is known from them is the distribution of messenger RNA, and some attempts have been made to examine the distribution of binding sites using non-selective ligands. More interesting is the drug profile of these receptors when expressed in transfected cells. They all present high affinity for LSD and other ergots and, interestingly, the atypical neuroleptic clozapine has been found to have high affinity for both the 5-ht6 and 5-ht7. That, together with the localization of the messenger RNA coding for these receptors in areas such as the caudate nucleus for the 5-ht6, or some hypothalamic and limbic areas for the 5-ht7, had led to speculations of possible roles of these receptors in the mechanism of action of the atypical neuroleptics (for review see [5]). It has been suggested that serotonin 5-ht7 could be the 5-ht receptor mediating the role of this neurotransmitter in the control of circadian rhythms, an area where pharmacological development could be of interest. Thus, the new generation of recently cloned 5-ht receptors make an attractive target for further exploring the role of the serotoninergic system, not only at the brain level, but probably also in other organic systems.

What is Next in 5-HT Research?

The rapid pace at which 5-HT research has proceeded within the last few years will most probably continue in the future. To summarize:

(1) The molecular structure of the major types of 5-HT receptors is now known, and that of other unexpected receptors such as 5-ht5A, 5B, 5-ht6 and 5-ht7 are also described. Although it is now widely accepted that the newer generation of antidepressants such as fluoxetine (prozac) exert their clinical effects via the inhibition of serotonin-reuptake, it still remains to be determined via which of the numerous subtypes mentioned here, these elevated synaptic concentrations act.
(2) Selective and less selective ligands for the old and new receptors have been described, in particular new 5-HT1D/1B, 5-HT2, and 5-HT4 agents are now available.
(3) New experimental tools together with DNA recombinant techniques that are now beginning to be used in 5-HT research will widen our understanding of the biology and pharmacology of the 5-HT system.
(4) More importantly, clinical research oriented to the discovery of therapeutic utilities for the new 5-HT agents should result in important information on the role of the 5-HT mechanisms in man in health and disease.

Thus, it will come as no surprise if new 5-HT receptors are described, in particular subtypes of the newly described 5-HT receptor classes. Medicinal chemistry will soon provide us with new and more selective and potent agonists and antagonists of the old and new 5-IIT receptors. Results of the use of these molecules together with antisense oligonucleotide technologies and more "knock-out" mutants, and special clinical investigations will expand our understanding of the biology of the 5-HT systems and direct its use in therapeutics.

References

1. Peroutka SJ, Snyder SH. Multiple serotonin receptors: differential binding of [^{3}H]5-hydroxytryptamine, [^{3}H]lysergic acid diethylamide and [^{3}H]spiroperidol. Molec. Pharmac. 1979; 16: 687–699.
2. Leysen JE, Niemegeers CJE, Tollenaere JP, Laduron PM. Serotoninergic component of neuroleptic receptors. Nature 1978; 272: 168–171.
3. Fargin A, Raymond JR, Lohse MJ, Kobilka BK, Caron MG, Lefkowitz RJ. The genomic clone G-21 which resembles a β-adrenergic receptor sequence encodes the 5-HT1A receptor. Nature 1988; 335: 358–360.
4. Julius D, MacDermott AB, Axel R, Jessell TM. Molecular characterization of a functional cDNA encoding the serotonin 1C receptor. Science 1988; 241: 558–564.
5. Hoyer D, Clarke DE, Fozard JR, Hartig PR, Martin GR, Mylecharane EJ, et al. VII. International Union of Pharmacology Classification of Receptors for 5-Hydroxytryptamine (Serotonin). Pharmacological Reviews 1994; 46: 158–193.

6. Hartig PR, Branchek TA, Weinshank RL. A subfamily of 5-HT1D receptor genes. Trends Pharmac. Sci. 1992; 13: 152–159.
7. Maricq AV, Peterson AS, Brake AJ, Myers RM, Julius D. Primary structure and functional expression of the 5-HT3 receptor, a serotonin-gated ion channel. Science 1991a; 254: 432–437.
8. Gerald C, Adham N, Kao HT, Schechter LE, Olsen MA, Bard JA, et al. The 5-HT4 receptor: Molecular cloning of two splice variants. Poster Session 2, "5-HT Third IUPHAR Satellite Meeting on Serotonin", Chicago, Illinois, USA, August, 1994.
9. Palacios JM, Waeber C, Hoyer D, Mengod G. Distribution of serotonin receptors. Ann. N.Y. Acad. Sci. 1991a; 600: 36–51.
10. Palacios JM, Waeber C, Mengod G, Hoyer D. Autoradiography of 5-HT receptors: a critical appraisal. Neurochem. Int. 1991b; 18: 17–25.
11. Skingle M, Skopes DIC, Feniuk W, Connor HE, Carter MC, Clitherow MC. GR 127935: A potent orally active 5-HT1D receptor antagonist. Br. J. Pharmacol. 1993; 110: 9P.
12. Saudou F, Aït Amara D, Dierich A, LeMeur M, Ramboz S, Segu L, et al. Enhanced aggressive behaviour in mice lacking 5-HT1B receptor. Science 1994; 265: 1875–1878.
13. Richardson BP, Engel G. The pharmacology and function of 5-HT3 receptors. Trends Neurosci. 1986; 9: 424–428.
14. Bockaert J, Fozard JR, Dumuis A, Clarke DE. The 5-HT4 receptor: a place in the sun. Trends Pharmac. Sci. 1992; 13: 141–145.
15. Grossman CJ, Kilpatrick GJ, Bunce KT. Development of a radiogland binding assay for 5-HT4 receptors in guinea-pig and rat brain. Br. J. Pharmacol. 1993; 109: 618–624.

Pharmacological Sciences: Perspectives for
Research and Therapy in the Late 1990s
ed. by A.C. Cuello and B. Collier

Peptide Receptor Antagonists*

William Rostene[1] and Gérard Le Fur[2]

[1]INSERM U.339, Hôpital St. Antoine, 75571 Paris Cedex 12 France;
[2]SANOFI Recherche, 75008 Paris, France

Introduction

In the past 25 years, a large number of biologically active peptides has been identified in both brain and peripheral tissues. However, our understanding of their role in physiology and disease has been often undermined by the lack of potent and selective antagonists.

During the last 2 to 3 years, such compounds have become available. They are mainly non-peptide substances with no structural similarities with the endogenous peptide and represent a new series of chemical molecules.

The present review will focus on recent data obtained by various groups from industry and academic laboratories with such non-peptide antagonists for cholecystokinin (CCK), tachykinins and neurotensin (NT) receptors.

CCK Antagonists

CCK has been shown to play a role in various physiological processes including neural pathways mediating secretion, motility, analgesia, and satiety. The identification of at least two subtypes of CCK receptors (CCK-A, CCK-B/gastrin) which have a distinct regional distribution and pharmacological specificity has led to interest in the development of non-peptide antagonists selective for either receptor subtype.

Correspondence to: Dr. William Rostene, (address above).
*This review represents a summary of the work presented during the IUPHAR 94 in the symposium on "Peptide receptor antagonists". The section on cholecystokinin antagonists was presented by L.L. Iverson (Merck Sharp and Dohme, Harlow, UK) and J.D. Gardner (St. Louis University, MO, USA). The section on tachykinin antagonists was presented by R.M. Snider (Pfizer, Groton, CT, USA and Burroughs Wellcome Company Research Triangle Park, NC, USA) and J.M. Lundberg (Karolinska Institute, Stockholm, Sweden). The section on neurotensin antagonists was presented by D.Gully (Sanofi Recherche, Toulouse, France) and P. Kitabgi (IPMC CNRS, Sophia-Antipolis, Valbonne, France).

In 1985, Chang et al. [1] showed that asperlicin, a non-peptide molecule, isolated from *Aspergillus alliaceus*, was able to block CCK8 sulfate binding. This work led to the subsequent development of benzodiazepine CCK-A antagonists such as devazepide (L-364,718) and the selective CCK-B antagonist L-365,260.

In the last 2 years, a number of other non-peptide CCK receptor antagonists has been described, including the dipeptoid series based on the parent peptide, e.g., CI-988 and the pyrazolidinone series, e.g., LY-262,691. Although these compounds have provided valuable information, they represent first-generation compounds which have some limitations for their use as potential therapeutic agents such as limited aqueous solubility, weak affinity and relatively poor oral bioavailability and brain penetration. Such compounds as L-365,260, an orally active CCK-B antagonist with an IC_{50} of 8.5 nM was shown to potentiate morphine analgesia in rats and also prevent morphine tolerance, suggesting that CCK was able to antagonize the antinociceptive effect of opiates. It was recently reported that CCK4 and pentagastrin were able to induce panic attacks in normal human volunteers. L-365,260 (50 mg) can completely block this effect. However, this CCK-B antagonist was shown to be inactive in blocking endogenous panic attacks not stimulated by CCK4 in humans. One possibility of this lack of effect may be due to the low efficacy of such first-generation antagonists.

Freedman et al. [2] recently described a series of 1,4 benzodiazepine derivatives such as the tetrazole derivative L-368,935 and the basic amidine derivative L-740,093. Both compounds are much more soluble than L-365,260 and have 50–100 times higher affinities at the CCK-B/gastrin receptor *in vitro*. *In vivo*, L-740,093 was 100 times more potent than L-365,260 in inhibiting pentagastrin-induced gastric acid secretion in the rat (Table 1). L-740,093 also antagonized pentagastrin-induced excitation of rat ventromedial hypothalamic neurons and was 500 times more active than L-365,260 [2]. Such a compound may be a useful tool to discriminate between subtypes of CCK-B and gastrin receptors recently cloned and to investigate possible roles of CCK-B receptors in brain functions.

The various CCK-A and CCK-B non-peptide antagonists were recently used as radiolabeled ligands to characterize the different affinity

Table 1. *In vitro* binding properties of benzodiazepine amidine CCK-B antagonists

Compound	Guinea pig brain CCK-B IC_{50} (nM)	Rat pancreas CCK-A IC_{50} (nM)	ID_{50} mg/kg iv anti-Gastrin
L-365,260	8.5	740	0.83
L-740,093	0.1	1,600	0.01

Adapted with the permission of LL Iversen from [2].

states for CCK8 binding to transfected COS cells with cDNA for rat CCK-A and different CCK-B receptors [3]. Results of the group of J.D. Gardner [3] revealed that both rat and canine CCK-B receptors, like pancreatic CCK-A receptors, exist in three different affinity states for CCK8 (high (nM), low (100 nM) and very low affinity (μM or more). In rat, CCK-A and probably CCK-B receptors were most in the very low affinity state, whereas with canine CCK-B, most were in the low affinity state. At each state of the rat CCK-A receptor, CCK8 had a higher affinity than gastrin and the CCK-A preferring receptor antago-nist, L-364,718 had a higher affinity than the CCK-B antagonist, L-365,260. The opposite was observed for each state of the CCK-B receptor. Each antagonist had the same affinity for the low affinity state as for the very low state with each receptor. This ability to exist in multiple affinity states seems to be an intrinsic property of the CCK receptor molecule itself which may play important roles for the message induced by the signal transduction such as selective activation or inhibi-tion of specific cell functions and minimization of undesirable effects of pharmacological agents. Selective non-peptide antagonists may thus be useful for the elucidation of previously unrecognized regulatory mecha-nisms of target cell functions.

Tachykinin Antagonist

Substance P (SP) is a member of a large family of peptides, tachykinins (TK), sharing C-terminal amino acids such as NKA, NKB, NK and Ng. SP binds preferentially to a subtype of binding sites called NK_1, whereas NKA and NKB bind to NK_2 and NK_3 binding sites respec-tively. NK_1 sites are found in the spinal cord, brain dopaminergic areas, glial cells and in the periphery in the lung, lymphocytes and smooth muscles. Such distribution suggests that SP may play a major role in several diseases including various types of inflammation, pain, migraine, psychosis, anxiety and asthma.

Similarly, NK_2 receptors are found in lung, smooth muscles, bladder and in central neurons, whereas NK_3 were only reported until now in brain and spinal cord.

It was thus of great interest to develop potent, orally active, specific SP antagonists. Empirical screening program strategies including selec-tion of target tissues, discovery of a lead compound, synthesis, *in vitro* and *in vivo* assays and, finally, the development of a potent substance resulted during the last 5 years in several classes of SP antagonists such as the 2,3-substituted quinuclinidines and piperidines (CP 96345 and CP99994 from Pfizer), the perhydroisoindoles (RP 67580 from Rhône Poulenc Rorer) and some peptide-like compounds as FK 224 and FK 888 from Fujisawa. A selective non-peptide antagonist for NKA

(NK$_2$ receptors) has been also recently described (SR 48968 from Sanofi).

Thus the Pfizer antagonists were shown to displace ^{3}HSP binding on bovine caudate tissue with IC$_{50}$ values in the nM range, to block, following intravenous injection, SP-induced salivation in the rat and smooth muscle relaxation. Both CP 96345 and CP99994 were more potent in human tissues than in rodents.

Several models of pain such as formalin response or dorsal horn response to noxious stimulation were blocked by NK 1 antagonists, suggesting an important role of SP in coordinating pain.

It has long been postulated that endogeneous TK such as SP and NKA serve as neurotransmitters, especially of local axon reflex mediated responses. Although experiments using the initial TK antagonists of peptide nature suggested that TK's released from sensory nerves mediated both plasma protein extravasation, vasodilatation and bronchoconstriction, it was not until the recent development of potent selective TK receptor antagonists of non-peptide type that the transmitter status of TK1s was established.

Neurogenic plasma protein extravasation in, for example, airway mucosa caused by inhalation of irritants, such as cigarette smoke and lactic acid is mainly due to NK$_1$ receptor activation as revealed by experiments using CP 96345, RP 67580 and corresponding enantiomers lacking NK$_1$ blocking activity [4, 5]. Similar data are obtained also in other organs or when using other ways to activate sensory nerves, such as electrical antidromic stimulation or exposure to capsaicin. SP is most likely the main transmitter of neurogenic plasma protein extravasation.

In inflammation, there is an upregulation of NK$_1$ receptors and the above NK1 antagonists can blunt allergen-evoked plasma protein extravasation and nociceptive responses to sustained noxions stimuli.

The skin vasodilatation due to local mediator release from peripheral sensory nerves in response to low-frequency stimulation (antidromic vasodilatation) is, in contrast to the effects of exogenous TK's, not reduced by the recently developed NK$_1$ or NK$_2$ antagonists, but is reduced by the CGRP antagonist CGRP [8–37], suggesting that TK's are not likely to be involved in this response in spite of earlier suggestions using peptidic TK antagonists. Most likely, TK's released from perivascular sensory nerves do not reach the endothelial NK$_1$ receptors in arterioles mediating vasodilatation.

Bronchoconstriction in the guinea pig due to local sensory activation is markedly inhibited by the NK$_2$ antagonist SR 48968 [6], suggesting that NKA is most likely the main endogenous transmitter of sensory bronchoconstriction in the guinea pig. The non-peptide antagonist SR 48968 is also antitussive. NK$_1$ antagonists, on the other hand, have only marginal effects on the sensory bronchoconstriction *in vivo* or *in vitro*. It is clear that NK1 receptors are also present on guinea-pig bronchial smooth muscle, but evidently endogenously released TK's have little influence on these receptors. It can be concluded that with the use of these new TK receptor antagonists, the transmitter mechanisms for vascular and bronchomotor reactions upon sensory nerve activation have now been clarified. This will also increase the usefulness of these common models for future drug development.

Neurotensin Antagonists

Since the discovery and isolation of neurotensin (NT) in 1973, a great deal of information has accumulated regarding its distribution in the brain and peripheral tissues and its various central effects. It is now well established that the tridecapeptide NT is closely associated with dopamine (DA) transmission. NT-like immunoreactivity and NT receptors have been detected in all brain structures containing DA cell bodies and terminals such as the substantia nigra (pars compacta), the ventral tegmental area (VTA), the striatum, the nucleus accumbens and the prefrontal cortex [7].

Furthermore, NT has been speculated to play a role in DA-mediated neurological (e.g., Parkinson) and psychiatric (schizophrenia) diseases. Indeed, altered cerebrospinal fluid NT in schizophrenic patients and normalization after neuroleptic treatment has been reported. Moreover, when injected in the central nervous system of animals, NT exerts a variety of behavioral effects, such as hypolocomotion and hypothermia, resembling those of clinically used antipsychotic drugs. It has thus been proposed that NT may be an endogenous neuroleptic-like compound [7]. However, NY may produce opposite effects depending on its site of injection in the brain.

A better understanding of such complex effects may result from the use of highly potent, selective, orally-active antagonists of NT receptor: Such a compound, SR 48692, has been recently developed [8]. It competitively inhibits the binding of ^{125}I-NT to guinea pig, rat, mouse and human brain homogenates, to primary cultures of rat mesencephalic neurons, and to the cloned human and rat receptors transfected in different cell lines with K_i values ranging from 1 to 30 nM. SR 48692 has revealed potent antagonistic properties in numerous *in vitro* and *in vivo* models. It thus antagonizes *in vitro* NT-induced DA release, NT-stimulated Ca^{2+} mobilization in human adenocarcinoma cells HT29 and blocked *in vivo* the turning behavior induced by unilateral intrastriatal injection of NT in the mouse [8]. As it was recently reported that the stimulation of DA receptors with DA agonists could enhance NT release in the rat prefrontal cortex, the possibility of SR 48692 to affect the expression of behavioral responses associated with DA receptor stimulation was investigated. SR 48692 antagonizes turning behavior induced by intrastriatal administration of a D1 receptor agonist (+) SKF 38393 (0.1 mg), of a D2 receptor agonist bromocriptine (0.01 ng), of a mixed D1/D2 receptor agonist apomorphine (0.25 μg), and by amphetamine (10 μg), with respective ED_{50} of 0.02, 0.4, 0.03 and 0.06 mg/kg. The antagonism of SR 48692 in these models is specific since neither pilocarpine nor 3PPP-induced turning were affected by SR 48692. The injection of NT into the rat VTA induces contralateral rotations which are partially counteracted by SR 48692, at doses very close to those found to reduce NT-induced turning in mice. However, SR 48692 was unable to inhibit both the DA release in the nucleus accumbens evoked by NT injection into the VTA and DOPAC/DA ratios.

Finally, SR 48692 can counteract the hypomotility induced by i.c.v. injection of NT in rats. It also dose-dependently reverses the antagonist effect of intra-nucleus accumbens injection of NT on the amphetamine hyperlocomotion.

Taken together, these data provide evidence that NT specifically modulates the mesolimbic pathways, with different effects on cell bodies (VTA) compared to terminals (nucleus accumbens) and confirm that SR 48692 acts as a NT receptor antagonist able to reverse NT-induced behavioral manifestations associated with an altered function of the mesolimbic system, but not the changes in DA transmission.

Among all the behavioral effects induced by DA stimulation in rodents, only turning and yawning are inhibited by SR 48692, suggesting the participation of NT in these behaviors. However, not all the effects of exogenous NT are antagonized by SR 48692. Indeed, SR 48692 does not reduce the hypothermic effect obtained after i.c.v. administration of a high dose of NT in rats. SR 48692 is also unable to counteract the antinociception induced by i.c.v. injection of 100 ng of NT in rats as assessed by the reduced number of writhings induced by PBQ administration [9].

With its neuropharmacological profile, SR 48692 confirms the neuro-modulator role of NT in the brain and may open the way for new therapeutics in psychiatric diseases. The effects of SR 48692 on NT responses in peripheral tissues were further described. NT contracts the guinea-pig ileum through a neurogenic process that is partly mediated by acetylcholine and partly by substance P and relaxes the guinea-pig colon through a direct action on smooth muscle cells involving the opening of Ca^{2+}-dependent K^+ channels [10]. The non-peptide NT antagonist SR 48692 potently inhibited NT binding to membranes prepared from the guinea-pig ileum and colon with K_i values of approximately 3 nM. SR 48527 and SR 49711, two enantiomers structurally related to SR 48692 were respectively equipotent and 100-fold less potent than SR 48692 in inhibiting NT binding in both tissues. In both

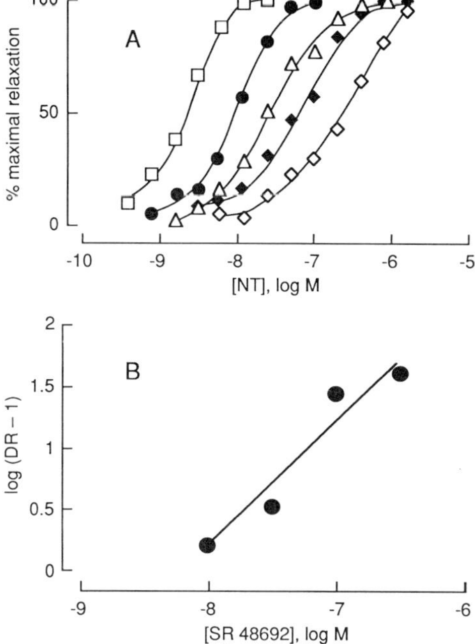

Fig. 1. Antagonism by SR 48692 of the NT-induced relaxation in guinea-pig colon preparations. A, cumulative concentration-response curves for NT-induced relaxation in the absence (□) or presence of SR 48692 at 10 (●), 30 (△), 100 (◆) and 30 nM (◇). B, Schild plot for the antagonist effect of SR 48692 (mean pA2 = 8.3 ± 0.1).

Adapted with the permission of P. Kitabgi from Labbé-Jullié et al. [11].

membrane preparations, NT binding was increased by Mg^{2+} and decreased by Na^+ and GTPS, whereas the binding potency of SR 48692 towards NT binding was not significantly affected by these agents. SR 48692 inhibited NT-induced contraction and relaxation (Fig. 1) in guinea pig ileum and colon preparations, respectively with K_i values between 4 and 5 nM. As in binding studies, SR 48527 was as potent as SR 48692, whereas SR 49711 was 100-fold less potent in antagonizing NT responses in both the guinea-pig ileum and colon. Altogether, these results show that NT receptors in the guinea pig ileum and colon, although functionally distinct, are coupled to G proteins and display similar biochemical and pharmacological properties, in particular with regard to their sensitivity and stereoselectivity toward non-peptide antagonists related to SR 48692. Because of their high potency to antagonize NT actions in intestinal preparations, such NT antagonists represent useful tools to study the physiological role of NT in the digestive tract.

References

1. Chang RS, Lotti VJ, Monaghan J, Birnbaum J, Stapley EO, Goetz MA et al. A potent non-peptide cholecystokinin antagonist selective for peripheral tissues isolated from *Aspergillus alliaceus.* Science 1985; 230: 177–179.
2. Patel S, Smith AJ, Chapman KL, Fletcher AE, Kemp JA, Marshall GR et al. Biological properties of the benzodiazepine amidine derivative L-740,093, a cholecystokinin-B gastrin receptor antagonist with high affinity *in vitro* and high potency *in vivo.* Mol. Pharmacol. 1994; 46: 943–948.
3. Talkad VD, Fortune KP, Pollo DA, Shah GN, Wank SA, Gardner JD. Direct demonstration of three different states of the pancreatic cholecystokinin receptor. Proc. Natl. Acad. Sci. USA 1994; 91: 1868–1872.
4. Delay-Goyet P, Franco-Cereceda A, Golsalves SF, Clingan CA, Lowe III JA, Lundberg JM. CP 96345 antagonism of NK_1 receptors and smoke-induced protein extravasation is unrelated to its cardiovascular effects. Eur. J. Pharmacol. 1992; 222: 213–218.
5. Delay-Goyet P, Satoh H, Lundberg JM. Differences in the potency of CP 96345 and RP 67580, two new non-peptide antagonists selective for NK_1 receptors in inhibiting responses evoked by stimulation of sensory nerves. Regul. Pept. 1993; 46: 304–306.
6. Satoh H, Lou YP, Lundberg JM. Inhibitory effects of capsazepine and SR 48968 on citric acid-induced bronchoconstriction in guinea-pigs. Eur. J. Pharmacol. 1993; 236: 367–372.
7. Kitabgi P, Nemeroff CB, editors. The neurobiology of neurotensin. Ann. New York Acad. Sci. Vol. 668, 1992, pp. 1–374.
8. Gully D, Canton M, Boigegrain R, Jeanjean F, Molimard JC, Poncelet M et al. Biochemical and pharmacological profile of a potent and selective non-peptide antagonist of neurotensin receptor. Proc. Natl. Acad. Sci. USA 1993; 90: 65–69.
9. Dubuc I, Costentin J, Terranova JP, Barnouin MC, Soubrié P, Le Fur G et al. Br. J. Pharmacol. 1994; 112: 352–354.
10. Kitabgi P. Effects of neurotensin on intestinal smooth muscle: application to the study of structure-activity relationships. Ann. N.Y. Acad. Sci. 1982; 400: 37–52.
11. Labbé-Jullié C, Deschaintres S, Gully D, Le Fur G, Kitabgi P. Effect of the non-peptide neurotensin antagonist, SR 48692, and two enantiomeric analogs, SR 48527 and SR 49711, on neurotensin binding and contractile responses in guinea-pig ileum and colon. J. Pharmacol. exp. Ther. 1994; 271: 267–276.

Pharmacological Sciences: Perspectives for
Research and Therapy in the Late 1990s
ed. by A.C. Cuello and B. Collier
© 1995 Birkhäuser Verlag Basel/Switzerland

Angiotensin Receptor Subtypes and their Pharmacology*

Pieter B.M.W.M. Timmermans[1], Tadishi Inagami[2], Juan M. Saavedra[3],
Raymond Ardaillou[4], Charles R. Rosenfeld[5],
and Frederick A.O. Mendelsohn[6]

[1]*Discovery Research, DuPont Merck Pharmaceutical Company, Experimental Station,
Wilmington, DE 19880-0400, USA;* [2]*Department of Biochemistry, Vanderbilt University
School of Medicine, Nashville, TN 37232, USA;* [3]*Department of Health and Human Services,
National Institutes of Health, Bethesda, MD 20892, USA;* [4]*Service d'Explorations
Fonctionnelles, INSERM 64, Hôpital Tenon, Paris, France;* [5]*Department of Pediatrics,
University of Texas, Southwestern Medical School, Dallas, TX 75235, USA;* [6]*Department of
Medicine, University of Melbourne, Austin Hospital, Heidelberg 3084,Victoria, Australia*

Summary. Angiotensin II (Ang II) is the principal mediator of the Renin-Angiotensin System,
but other peptides may have a role, including Ang IV and Ang 1–7. Our understanding of
Ang II receptor heterogeneity has been facilitated by the discovery of nonpeptide receptor
antagonists and the cloning of Ang II receptor subtypes. Losartan-sensitivity defines AT_1 and
PD123319 (and CGP42112)-sensitivity defines AT_2 receptor subtypes. AT_1 receptors predominate in most mammalian tissues and subserve the well-known effects of Ang II. The AT_2
receptors are abundant in fetal tissue, nonpregnant myometrium, discrete parts of brain,
adrenal cortex and medulla, and kidney. The functional significance of the AT_2 receptor is
largely unknown, but may involve ion channels, "growth", and cerebral autoregulation.
   Tremendous strides have been made in understanding the structure, function, and diversity
of Ang II receptors. These studies will continue to be facilitated by the advent of new receptor
antagonists. Additional research is needed, however, to clarify the role of the mammalian AT_2
site and the role of Ang 1–7 and Ang IV.

Nonpeptide Receptor Antagonists as Tools in Defining Angiotensin Receptor Heterogeneity

Although Ang II receptor heterogeneity had been previously suggested,
it was not until the discovery of specific nonpeptide receptor antagonists
that Ang II receptor subtypes were fully explored [1]. Three independent labs concurrently described unique structural classes of Ang II
receptor antagonists capable of inhibiting the binding of radiolabeled
Ang II to isolated membrane "receptors". Losartan, PD123319, and
CGP42112 are prototypes of these three series of compounds. Importantly, it was found that losartan inhibited binding in arterial tissue and

Correspondence to: Dr. Pieter B.M.W.M. Timmermans, DuPont Merck Pharmaceutical
Company, Experimental Station, P.O. Box 80400, Wilmington, DE 19880-0400, USA.
*Review based on Symposium "Angiotensin Receptor Subtypes and Their Pharmacology"
presented at IUPHAR '94, Montreal, Canada, July 26, 1994

adrenal cortex, but had little effect on binding in uterus and adrenal medulla. From these observations came the hypothesis that there are two distinct receptor subtypes for Ang II. With these new tools, many investigators characterized receptor subtypes in diverse species and tissue types. A standard nomenclature was proposed to designate losartan-sensitive binding sites as AT_1 and those sensitive to PD123319 or related compounds and CGP42112 as AT_2 [2]. If Ang II binding in a given tissue is not blocked by either class of antagonist, it is "atypical". It should be noted that the peptide Ang II antagonists such as saralasin are not selective and inhibit both binding sites.

Nonpeptide Ang II receptor antagonists have played an important role in the cloning of the Ang II receptors [3]. The amino acid sequences of the AT_1 receptor for the rat, mouse, cow, pig, and man have now been determined and shown to have high sequence homology (> 90%). Further, AT_1 receptor subtypes have been cloned from rat, mouse, and human genomic libraries and designated AT_{1A} and AT_{1B}. Controversy exists about the existence of these additional subtypes in humans [4, 5]. Both subtypes are inhibited by losartan and not by PD123319. AT_1 receptor subtypes have also been suggested by binding data from rat kidney (also designated AT_{1A} and AT_{1B}, but do not correspond to the cloned receptors) [6]. As will be discussed in more detail, the AT_2 receptor has also been isolated from rat and human cDNA plasmid libraries by expression cloning.

Ang II is the principal "angiotensin peptide" [7] but other peptide fragments originating from angiotensinogen such as Ang III, Ang IV, and Ang 1–7 may have important roles [8, 9] (Fig. 1). Losartan and PD123319 or CGP42112 inhibit binding to Ang III essentially like Ang II and do not inhibit Ang IV binding. A specific binding site to Ang 1–7 has not been identified, but interesting interactions with angiotensin receptors have been reported. Some of the functional responses to Ang IV and Ang 1–7 are blocked by losartan, suggesting an interaction of these peptides with the AT_1 receptor. The abundance of the Ang IV sites in rat brain encourages future research in this area and in the search for specific antagonists of this site.

The distribution of Ang II receptor subtypes has been studied with the peptide and nonpeptide antagonists used to displace radiolabeled ligands (^{125}I-Ang II, ^{125}I-Sar^1-Ang II, ^{125}I-[Sar^1, Ile^8]Ang II, ^3H-DuP 753 or ^{125}I-CGP42112). Caution should be used in interpreting data with the labeled antagonists because of their high nonspecfic protein binding and observations that the binding epitope for the antagonist is likely not identical to the natural ligand. Using Ang II as the ligand has also shown to give different results than using Sar^1,Ile^8-Ang II. Surprisingly, common cell culture contaminates such as mycoplasma also contain "specific binding sites for Ang II" [1].

AT_1 receptor ("losartan-sensitive") subtypes predominate in terms of density in most tissues although the ratio of the AT_1/AT_2 receptor subtypes varies significantly between species and between tissues within species. The AT_2 are the predominate subtypes in most fetal tissues, but in newborn and adult tissues the AT_1 represents the majority of sites except where localized in discrete parts of certain tissues such as kidney, brain or adrenal medulla. Ang II heterogeneity is also relevant to humans (Table 1). Human fetal tissue contains predominantly AT_2 sites, e.g., preglomerular vessels in human kidney and the cerebellum contain AT_2 sites. It is interesting to note that in the inferior olive of the rat, the AT_2 site predominates, whereas in humans, the AT_1 site predominates, so caution must be used in extrapolating findings in the rat to other species.

Atypical binding sites, those binding Ang II but which are not sensitive to inhibition by losartan or PD123319 or CGP42112, have been described in amphibians, in neuroblastoma cells, and in avian species. Whether these sites should be designated AT_3, AT_4...AT_n is under consideration by the Angiotensin Receptor Subcommittee of the IUPHAR Committee for Receptor Nomenclature and Classification [10]. The natural ligand for these sites may be different, e.g., Val^5-Ang II, for avians, and the functional response coupling is quite different from mammalian receptors, e.g., three receptors have been identified in the domestic fowl, one subserving vasodilation.

Binding sites are academically interesting, but do these binding sites have any physiological, pathophysiological function or therapeutic rele-

Table 1. Human angiotensin II receptor subtypes

Tissue	Receptor subtype	Identifying ligand
Kidney		
Glomeruli-cortex	AT_1 (AT_2, fetal)	Losartan; PD123177
Preglomerular vessels	AT_2	
Heart		
Atria	AT_1/AT_2	Losartan; CGP42112A
	AT_1 (functional response)	Losartan
Fibroblasts	AT_1	PD123177; ICI D8731
Vascular smooth muscle	AT_1	Losartan; PD123177
Brain		
Cerebellum	AT_2	PD123177; losartan
Paraventricular nucleus; mediuan eminence; substantia nigra	AT_1	Losartan; PD123177
Adrenal		
Glomerulosa	AT_1/AT_2	CGP42112A; EX89
Fetal cells	AT_2	Losartan
Uterus	AT_2	CGP42112A; EX89
Placenta	AT_1	CGP42112; losartan
Platelets	AT_1	Losartan
Adipocytes	AT_1	Losartan

vance? The use of antagonists to define "receptor function" requires that the antagonists have high specificity for the Ang II receptor *in vitro* and *in vivo*. Losartan, as the first of what is now a large number of AT_1 subtype-selective antagonists, has been the most widely studied and it has been shown to be highly specific for Ang II receptors both *in vitro* and *in vivo*. There are conflicting reports on whether losartan antagonizes thromboxane (U46,619) or releases prostaglandins. Interestingly, losartan and L-158,809 in high concentrations interfere with tachykinin NK_3 binding *in vitro*. In most settings, however, high specificity of losartan and of structurally-related AT_1-selective antagonists has been demonstrated. The second requirement of an antagonist used to define functional receptor subtypes is a lack of "intrinsic" or partial agonist activity. The peptide Ang II receptor antagonists such as saralasin demonstrated clear agonist (Ang II-like) activity in many experimental preparations and this was a serious limitation to its clinical use. Losartan has been shown to be devoid of intrinsic activity. For example, saralasin produces a dose-related pressor response in rats, whereas losartan has no effect on blood pressure and can block the pressor response to saralasin. *In vitro*, the majority of data also suggest that losartan has no "intrinsic" activity, but two reports suggest that intracellular calcium has been released in response to high concentrations of losartan. The data with PD123319 and CGP42112, although more limited, suggest that PD123319 also has affinity for the AT_1 site at higher concentrations and CGP42112 (a peptide) has significant intrinsic activity. Thus, these antagonists must be used with the understanding that high concentrations may not be specific and that concentration response studies are imperative.

Defining the functional subtypes of Ang II receptors with nonpeptide receptor antagonists can be accomplished by blocking the responses to exogenously administered or applied Ang II and/or by administering the antagonist and observing the resultant change. As will be described in each of the organ systems below, the evidence for Ang II subtypes based on inhibition of binding is well documented, but the evidence for receptor syubtypes based on inhibiting function is less so. Antagonizing the Renin-Angiotensin System (RAS) with inhibitors of Ang II synthesis (ACE inhibitors) clearly showed that nonselective blockade results in effects on vascular, adrenal, renal, cardiac, and cerebral functions. Are these effects Ang II-mediated or bradykinin-mediated? Since losartan and other AT_1-selective antagonists and ACE inhibitors are equivalent in most experimental preparations, Ang II is likely involved. Further, in Ang II-infusion studies in which cardiac and vascular toxicity is produced, only the receptor antagonists (not ACE inhibitors) prevent the Ang II pathology. Since nonpeptide receptor antagonists block the effects of Ang II wherever it is formed (other proteases can form Ang II from angiotensinogen or Ang I), differences are possible. In addition, although a clear role of the AT_2 site has not been established, differences could arise between AT_1-selective blockers and nonselective blockers such as ACE inhibitors or salarasin. This cannot be easily resolved by the use of subtype-selective antagonists in combination because of the differences in their pharmacokinetic profiles. For example, losartan and its metabolite, EXP3174, provide long-lasting Ang II receptor blockade in rat and humans. PD123319, by contrast, has a very short half-life *in vivo* (< 1 h). Future experiments will be aided by the recent introduction of nonpeptide receptor antagonists with high affinity for both receptor subtypes which have recently been described [11, 12]. These nonselective or "balanced" agents should provide additional tools to characterize the effects of Ang II and the functional role of receptor subtypes.

Diversity of Angiotensin Receptors: Cloning and Signaling Mechanism

Reflecting the diversity of physiological actions and signaling mechanisms of angiotensin, two major isoforms of Ang II receptors (AT_{1A}, AT_{1B}, AT_2) have been defined by pharmacological approaches using isoform-specific nonpeptide inhibitors. AT_1 was cloned from various mammalian and nonmammalian species. Human, rat, and mouse AT_1 was shown to consist of subtypes AT_{1A} and AT_{1B}; whereas, in other species like bovine, rabbit, and turkey, only one type of AT_1 has been

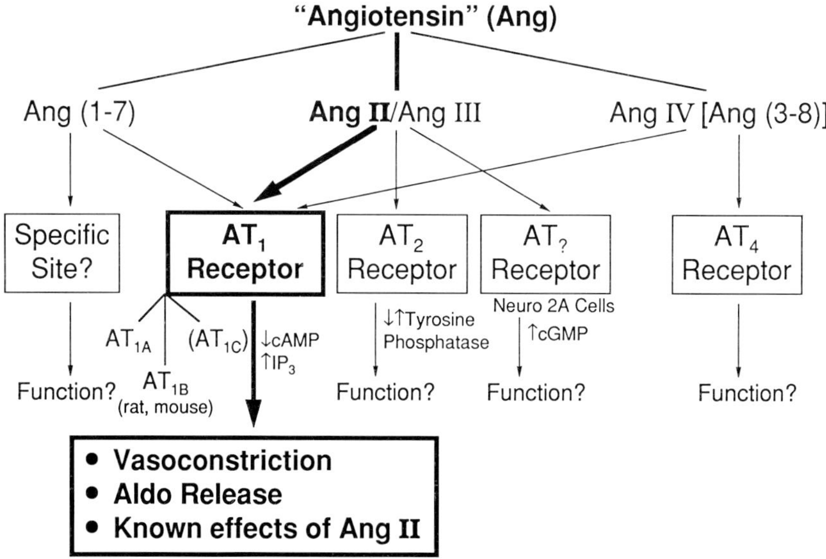

Figure 1. Angiotensin peptide receptor heterogeneity. Reprinted with permission from P.B.M.W.M. Timmermans and R.D. Smith, European Heart Journal, 1994; 15(D): 79–87.

found. AT_{1A} and AT_{1B} were highly homologous (96%) in their coding region sequence, but diverse in their noncoding region base sequence, suggesting differences in gene expression mechanisms. Whether human AT_{1B}, cloned recently from human placental cDNA library [5], is the counterpart of rodent AT_{1B} is yet to be clarified. AT_2 was cloned from rat pheochromocytoma cell line PC12W and rat fetal carcass. All of these Ang II receptors cloned to date turned out to have the seven-transmembrane domain structure and show G-protein coupled functions. However, between AT_1 receptors and AT_2, there are marked differences in their structure (only 32% amino acid sequence identity, as shown in Fig. 2) and functions.

AT_{1A} and AT_{1B} are coupled to the Gq-protein coupled-phospholipase C-β_1 to activate inositol *tris*phosphate production and transient increase in cytosolic Ca^{2+}, inhibition of adenylyl cyclase, and opening of a Ca-channel. More recently, this receptor was found to activate phospholipase $C\gamma_1$ by a protein tyrosine phosphorylation mechanism in several cells like vascular smooth muscle cells. Activation, rather than inhibition of adenylyl cyclase by AT_1 seems to mediate unique reactions in vascular smooth muscle. AT_2 did not show properties specific for the seven-transmembrane type receptors like ligand-induced internalization shift to a low affinity form stable GTP (guanosine 5'-triphosphate) analogs (GTP-γ-S). However, its structure deduced from the base sequence of cloned AT_2cDNA conformed with the structural characteristics of a seven-transmembrane domain receptor Interestingly, amino acid residues identified in AT_1 as essential for Ang II binding (Arg^{167}, Lys9, Trp^{253}, Thr^{260}, Asp^{263}) are well preserved in AT_{1A}, AT_{1B}, and AT_2. Activation or inhibition of phosphotyrosine phosphatase by AT_2 is one of them. This effect is abolished by pertussis toxin suggesting intermediacy by Gi or Go (GTP-activated protein subtype i or o). Interestingly, AT_1 and AT_2 are implicated in vascular smooth muscle cell growth in opposite directions. The AT_1 function supports cellular growth, whereas AT_2 is expressed in quiescent cells and the level of AT_2 expression is inversely correlated with the mitogenic activity of various cell types. Clarification of the exact mechanism of AT_2 function will show how the growth suppression takes place.

Functional Basis of Angiotensin Receptor Heterogeneity: The Blood Vessels and the Heart

Autoradiography and displacement with selective ligands have been used to quantify and characterize Ang II receptor subtypes in rat blood vessels and the heart.

Peripheral blood vessels, such as the rat aorta, express both receptor subtypes, AT_1 and AT_2, and the relative proportion of receptor sub-

AT_1		MALN	SSAEDGIKRI	QDDCPKAGRH	SYIFVMIPTL	
AT_2	MKDNFSFAAT	SRNITSSLPF	DNLNATGTNE	SAFNCSHKPA	DKHLEAIPVL	50
AT_1	YSIIFVVGIF	GNSLVVIVIY	FYMKLKTVAS	VFLLNLALAD	LCFLLTLPLW	
AT_2	YYMIFVIGFA	VNIVVVSLFC	CQKGPKKVSS	IYIFNLAVAD	LLLLATLPLW	100
	TM-1				TM-2	
AT_1	AVYTAMEYRW	PFGNHLCKIA	SASVTFNLYA	SVFLLTCLSI	DRYLAIVHPM	
AT_2	ATYYSYRYDW	LFGPVMCKVF	GSFLTLNMFA	SIFFITCMSV	DRYQSVIYPF	150
			TM-3			
AT_1	KSRLRRTMLV	AKVTCIIIWL	MAGLASLPAV	IHRNVYFIEN	TNITVCAFHY	
AT_2	LSQRRNP-WQ	ASYVVPLVWC	MACLSSLPTF	YFRDVRTIEY	LGVNACIMAF	200
		TM-4				
AT_1	ESRNSTLPIG	LGLT-KNILG	FLFPFLIILT	SYTLIWKALK	KAYEIQKNKP	
AT_2	PPEKYAQWSA	GIALMKNILG	FIIPLIFIAT	CYFGIRKHLL	KTNSYGKNRI	250
		TM-5				
AT_1	RNDDIFRIIM	AIVLFFFFSW	VPHQIFTFLD	VLIQLGVIHD	CKISDIVDTA	
AT_2	TRDQVLKMAA	AVVLAFIICW	LPFHVLTFLD	ALTWMGIINS	CEVIAVIDLA	300
		TM-6				
AT_1	MPITICIAYF	NNCLNPLFYG	FLGKKFKKYF	LQLLKYIPPK	AKSHSSLSTK	
AT_2	LPFAILLGFT	NSCVNPFLYC	FVGNRFQQKL	RSVFRVPITW	LQGKRETMSC	350
	TM-7					
AT_1	MSTLSYRPSD	NMSSSAKKPA	SCFEVE			
AT_2	RKSSSLREMD	TFVS				377

Figure 2. Amino acid sequences of rat AT_{1A} and rat AT_2 compiled from Iwai et al. [56], and Kambayashi et al. [57].

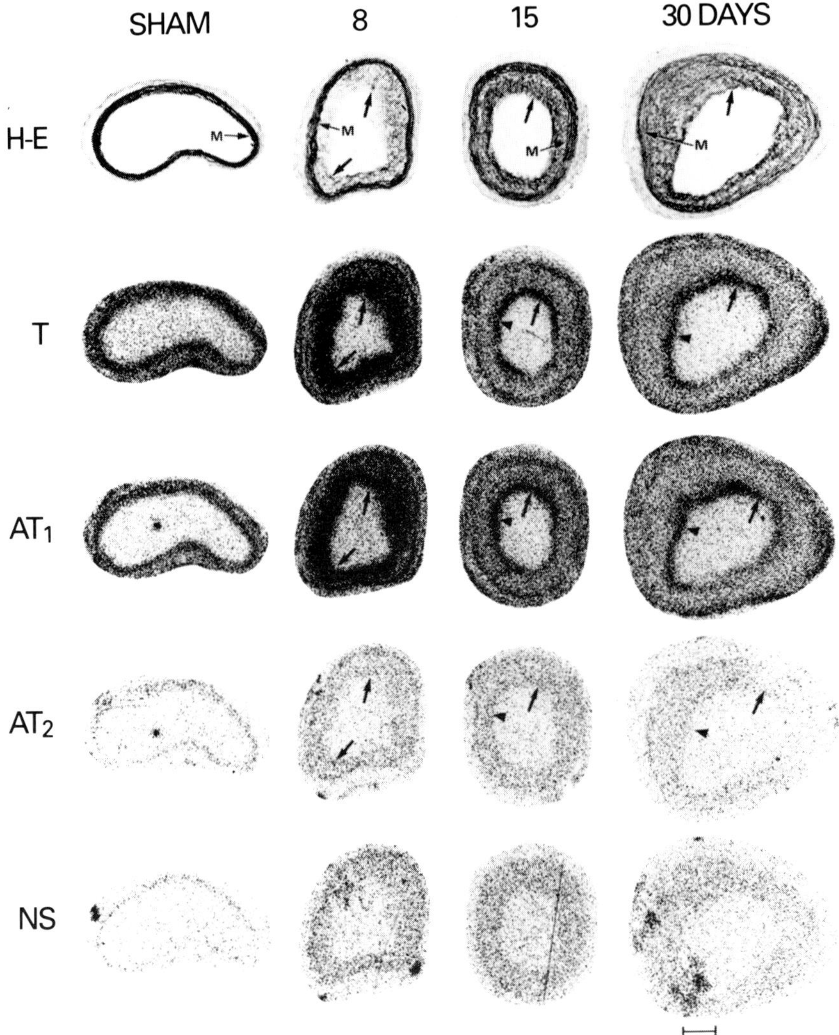

Figure 3. Autoradiography of Ang II receptors in rat carotid artery at 8 days after sham-surgery (SHAM) and at 8, 15, and 30 days after balloon injury. Consecutive sections show histology (H-E; hematoxylin-eosin stain), binding after incubation with 0.5 nM ^{125}I-[Sar^1]Ang II (T, total binding), binding as in T in the presence of 10 μM losartan (AT$_2$), and binding as in T in the presence of 5 μM Ang II (NS, nonspecific binding). M, media. Arrows point to neointima and arrowheads to binding in the subpopulation of proliferating neointimal cells. Bar = 0.2 mm. Reprinted with permission of Elsevier Science Ltd from Peptides, 1994; 15: 1205–1212.

types is developmentally regulated [13]. At embryonic day 18, and 2 weeks after birth, AT$_2$ receptors predominate. In adult animals, however, AT$_1$ receptors predominate (about 70% of the total) but the AT$_2$ subtype is also present [13].

Because of the relatively large number and higher proportion of AT_2 receptors in the immature, growing blood vessels, we asked the question of whether this receptor subtype was involved in mechanisms of blood vessel growth and repair. We used a model of balloon angioplasty to assess the expression of angiotensin II receptor subtypes during neointima formation [14]. To our surprise, AT_1, but not AT_2, receptors were highly expressed in actively growing neointima cells (Fig. 3) [14]. The above results indicate that, in peripheral blood vessels, both mechanisms of vasoconstriction and blood vessel repair are under the control of AT_1 receptors. The function and possible relevance of AT_2 receptors in blood vessels remain to be determined.

In addition to blood vessels, AT_1 receptors are present at all cardiovascular regulatory levels. For example, AT_1 receptors are highly expressed in brain areas related to cardiovascular control, such as the nucleus of the solitary tract, the peripheral sympathetic ganglia, and the conduction system of the heart [15–18]. In the myocardium, the number of Ang II receptors is not high, and the relative proportion of receptor subtypes cannot be clearly identified.

Autoradiography revealed that angiotensin receptors were expressed not only in peripheral but also in brain arteries as well [19]. Characterization of these receptors indicated that the predominant, and perhaps the exclusive, angiotensin receptor subtype in cerebral arteries of the rat was the AT_2 subtype. *In vivo* studies of cerebral blood flow indicated that stimulation of AT_2 receptors by infusion of Ang II under complete AT_1 blockade with losartan resulted in a shift of the upper limit of cerebral blood flow autoregulation toward higher blood pressures [20]. A similar effect was obtained by administration of two selective AT_2 ligands, PD123177 and CGP42112 [21, 22]. These data indicate the cerebrovascular AT_2 receptors may play a role in the control of cerebral blood flow and cerebrovascular autoregulation, and that, at least in this model, AT_2-selective ligands should be considered as AT_2 receptor agonists.

In conclusion, a predominant role of AT_1 receptors can be postulated for the central and sympathetic regulation of cardiovascular function, as well as for the regulation of the heart rhythm and the regulation of vascular tone and blood vessel repair. AT_2 receptors, on the other hand, seem to play a role in cerebrovascular regulation. Selective manipulation of angiotensin receptor subtypes will help in the clarification of the exact role of this peptide in cardiovascular control and in the development of future therapeutic strategies.

Functional Basis of Angiotensin Receptor Heterogeneity: The Kidneys and the Adrenals

AT_1 receptors are present in the renal vessels including the juxtaglomerular apparatus, the cortical arterioles and the outer medulla

Table 2. Renal effects of Ang II and distribution of the renal receptors for Ang II

	Distribution	Effects
AT_1 receptors	Renal vessels	
	Cells of the juxtaglomerular apparatus	Inhibition of renin secretion
	Cortical arterioles	Increase of arteriole resistance - Vascular
	Outer medulla vessels	Increase of resistance smooth muscle cell hypertrophy
	Glomerular mesangial cells	Contraction - Decrease of Kf
	Proximal tubule cells	Stimulation of Na^+/H^+ exchange - Tubule cell hypertrophy
	Interstitial cells of the medulla	PGE_2 production
	Fetal kidney	Development
AT_2 receptors	Fetal kidney	Development
	Renal capsule	?
	Arcuate and interlobular cortical arterioles	?
	Proximal tubule cells	Regulation of fluid reabsorption

vessels, in the glomerular mesangial cells and in the tubule all along the nephron but essentially in the proximal tubule on both the basolateral and the apical sides (Table 2). There are also AT_1 receptors in the interstitial cells of the renal medulla and in the fetal kidney. Two isoforms of AT_1 receptors, AT_{1A} and AT_{1B}, have been described in the rat. Both have been found in the kidney with a predominance for AT_{1A} [23, 24].

AT_2 receptors are essentially present in the fetal kidney [25], but have also been observed in different parts of the adult kidney including the renal capsule and the arcuate and interlobular cortical arterioles [26]. Their presence in the renal tubule is suggested by the described effects of AT_2 antagonists on urinary sodium and water excretion.

AT_1 receptors mediate most of the renal effects of Ang II [27, 28]. Their role has been demonstrated in Ang II-dependent vasoconstriction of the renal vessels, particularly of the glomerular afferent and efferent arterioles resulting in decrease of the cortical renal plasma flow. Theyalso mediate the contraction of glomerular mesangial cells resulting in decrease of Kf and of the glomerular filtration rate. In the proximal tubule cells, AT_1 receptors are responsible for the stimulatory effect of Ang II on the Na^+/H^+ exchanger activity with increase in Na^+ reabsorption and in H^+ secretion. AT_1 receptors in the glomerular afferent arteriole mediate the inhibitory effect of Ang II on renin secretion. AT_1 receptors also play a role in the trophic effects of Ang II both in the fetal and in the adult kidney, particularly in hypertrophy and cell proliferation of the vascular smooth muscle cells, of the mesangial cells, and of the proximal tubule cells. It has not yet been possible to identify specific roles for AT_{1A} and AT_{1B} receptor in the rat kidney.

AT_2 receptors are likely to be implicated in the development of the fetal kidney and, perhaps, in renal compensatory hypertrophy in the adult. Their role in water and sodium excretion is possible, but still hypothetical.

AT_2 receptors and, to a lesser extent, AT_1 receptors are present in rat adrenal chromaffin cells, but AT_2 receptors are only present in the rat pheochromocytoma-derived cell line PC12W [29, 30].

In the adrenal cortex, distribution depends on the zone of the cortex and on the species studied. In the adrenal zona glomerulosa, AT_1, mainly AT_{1B} receptors in the rat and AT_2 receptors have been found with major species differences in their relative distribution and the affinity of AT_1 antagonists for AT_1 receptor sites. For example, AT_1/AT_2 distribution is 60%/40% in rats and 80%/20% in humans. Concentrations inhibiting 50% of binding (IC_{50}) vary over a large range: 5 nM in the rat, 51 nM in the monkey, 260 nM in the dog, and 730 nM in the beef. AT_1 and AT_2 receptors are also differently distributed in the zona fasciculata according to the species. Both receptor types were found in bovines and ovines, only AT_1 receptors are present in humans and none of them could be found in rats [30, 31].

In the adrenal medulla, Ang II stimulates the release of epinephrine and norepinephrine via AT_1 receptor activation. The role of AT_2 receptors is still unknown. In the adrenal zona glomerulosa, Ang II acutely stimulates aldosterone synthesis and release in all species. Chronically, Ang II stimulates expression of the genes encoding for the steroidogenic enzymes. In the adrenal zona fasciculata, Ang II stimulates cortisol secretion by bovine and human cells. It upregulates expression of adrenocorticotrophic hormone (ACTH)-induced cyclic adenosine monophosphate (cAMP) production in bovine cells. All these Ang II effects on the adrenal cortex depend on AT_1 activation.

Ang II Receptors and Function During Pregnancy

Studies of the AT receptors during pregnancy are relatively few in number. Although AT receptors are expressed in several reproductive tissues, including the myometrium and placenta [32–36], their function and the mechanisms responsible for the control of their expression remain unclear. In the nonpregnant rat uterus there is evidence for receptor down-regulation in the presence of elevated plasma Ang II levels [32], a response similar to that seen in vascular smooth muscle. When the AT receptor subtypes were examined, the nonpregnant rat uterus (which may have included both myometrium and endometrium) was reported to express both AT_1 and AT_2 receptors in a 40:60 ratio [33]. However, it was not noted if the expression of the receptor subtypes was modified during pregnancy or in the presence of elevated plasma Ang II levels in the nonpregnant animal. These investigators also suggested that only AT_2 receptors were present in human "uterine" tissues, presumably obtained from nonpregnant women. In the placenta, the subtype predominantly expressed appears to be species and tissue related [34, 35]. For example, in the human, only AT_1 receptors are found in the umbilical cord artery and vein, villous tissue, and chorionic plate [34]. In the guinea pig and rabbit, there is a predominance of AT_1 receptors in the latter two tissues; however, there also is modest expression of the AT_2 receptor, which accounts for 3–25% of total binding [35]. The function and mechanisms responsible for the expression of these receptors has not been thoroughly determined, but the placental hormones may play an important role in determining both. Because of these deficits in our understanding of the AT receptor, we have examined several of these questions.

Normal pregnancy is associated with enhanced activation of the RAS in several species [37]. This may in large part reflect, not only the effects of increased estrogen synthesis on renin substrate production, but also the fall in systemic vascular resistance and thus an increase in renal renin. In several species, pregnancy also is associated with a decrease in

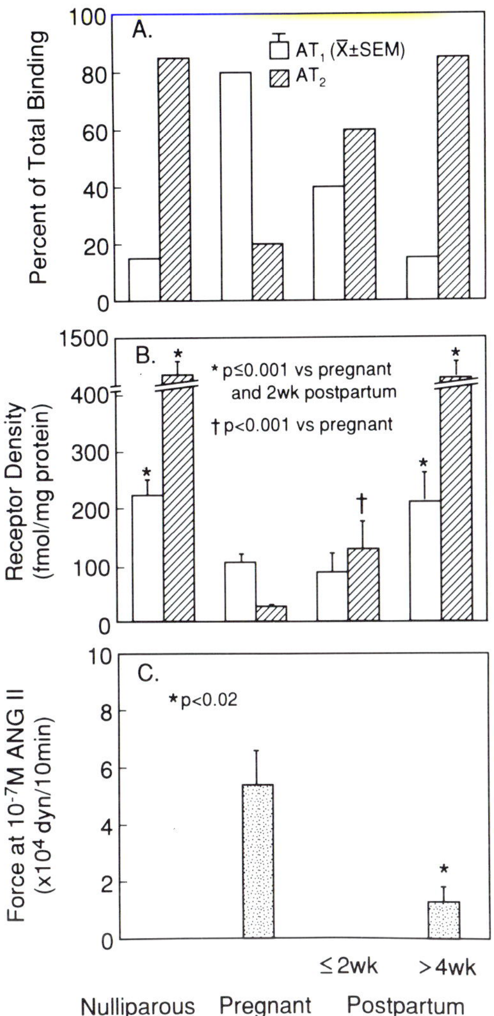

Figure 4. Alterations in (A) relative and (B) absolute binding density of myometrial AT receptor subtypes and (C) the stress generated by myometrial strips from near-term pregnant and postpartum sheep in the presence of 10^{-7} M Ang II. Reprinted with permission from [36]

the systemic pressor responses to infused Ang II, which is no longer seen in women with pregnancy-induced hypertension [38–40]. Furthermore, there is evidence in both women and sheep that the uteroplacental vasculature is even more refractory to the vasoconstrictor effects of Ang II than the systemic vasculature as a whole [41, 42]. This refractoriness to infused Ang II is not due to several factors, including decreases in total AT receptor binding density or affinity in vascular smooth muscle, despite a four- to five-fold rise in plasma Ang II levels, since neither are altered in the mesenteric artery, aorta, and uterine artery during normal ovine pregnancy [43]. In contrast, there is a marked down-regulation of the myometrial AT receptor during ovine pregnancy (Fig. 4B), total binding density falling about 90% in tissues studied near-term [36]. When the distribution of the receptor subtypes was examined, the AT_2 receptor was the predominant receptor expressed in the myometrium from nulliparous and nonpregnant sheep, accounting for 85% of receptor binding density (Fig. 4A), whereas the AT_1 receptor predominated near term, accounting for 80% of receptors present [36]. Although both the AT_1 and AT_2 receptors were down-regulated in pregnancy, the former had decreased only 51% as compared to 98% for the latter. Thus, there is differential regulation of the AT receptor in the myometrium as compared to vascular smooth muscle during pregnancy. Moreover, this reflects in part differences in the expression of the AT receptor subtypes.

Since myometrial function is also altered during pregnancy, the stress responses of myometrial strips to Ang II were studied *in vitro*. Stress responses (Fig. 4C) to Ang II were greater in tissues from near-term pregnant ewes as compared to that seen in tissues from postpartum and nonpregnant animals [36]. These responses were inhibited by losartan, the AT_1 receptor antagonist, but unaffected by PD123177, the AT_2 receptor antagonist, demonstrating that the AT_1 receptor is responsible for Ang II-induced myometrial stresses and implying that the predominance of the AT_2 receptor in the nonpregnant and postpartum myometrium did not modify stress responses to Ang II. Thus the functional role of the myometrial AT_2 receptor is unclear at present.

These data demonstrate that the AT_2 receptor can be the predominant receptor subtype expressed in an adult tissue and that this expression can be regulated during a normal physiologic process, i.e., pregnancy. Although the data also suggest an inhibitory role of the AT_2

Figure 5. Ang II receptor subtypes in rat brain coronal sections at the level of the rostral forebrain (A), hypothalamus (B), midbrain (C), and medulla oblongata (D). Blue represents background levels and red high receptor density. At each level there are four conditions – total binding (Total, $AT_1 + AT_2$), binding in the presence of an excess (10 µM) of losartan (AT_2 receptors), or an excess (10 µM) of PD123177 (AT_1 receptors), or in the presence of diothiothreitol (DTT, 5 mM) which inhibits binding to AT_1 and enhances binding to AT_2 receptors. Ang II receptors in the rostral forebrain are all AT_1 including SFO, PaAP, SCh, LOT (panel A), as are those in the hypothalamus: PaAP, PaMP, SCh (panel B), whereas at the same level those in the thalamus (MD, VPL/VPM) are AT_2 (panel B). In the midbrain, receptors in ZO of SC and MG are AT_2, whereas those in the CA1 and CA3 are AT_1 (panel C). In the dorsal medulla (SOL, DM 10) receptors are AT_1, whereas those in the inferior olivary nucleus (IO) are AT_2 (panel D).
Abbreviations are: CA1, CA3 – Fields CA1 and CA3 of Ammon's horn; DM10 – Dorsal motor nucleus of vagus; DTT – dithiothreitol; IO – inferior olivary n.; LOT – nucleus of lateral olfactory tract; MD – mediodorsal thalamic n.; MG – medial geniculate; PAAP-paraventricular hypothalamic n., anterior parvocellular part; PAMP – paraventricular hypothalamic, medial parvocellular part; SFO – subfornical organ; SOL – nucleus of the solitary tract; VPL/VPM-ventroposteromedial/ventroposterolateral thalamic nuclei; ZO of SC – zonal layer of superior colliculus.

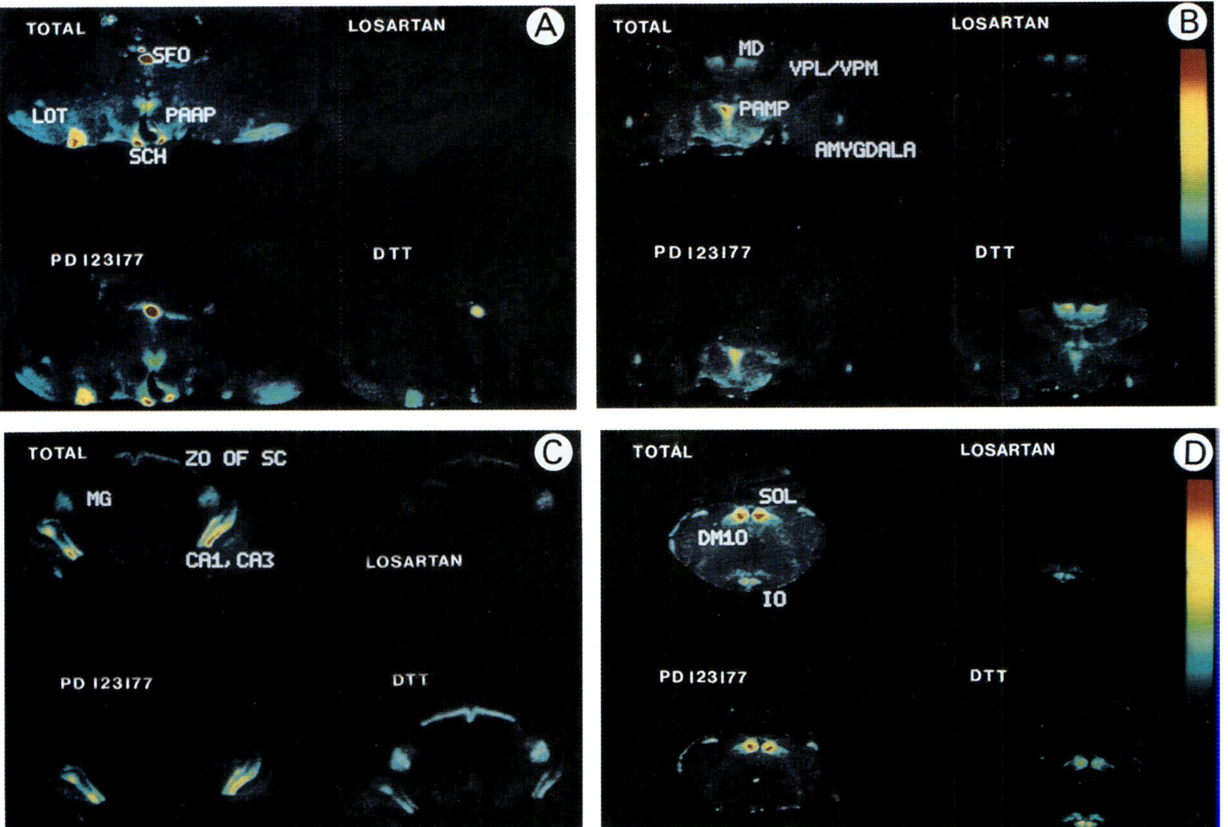

A
TOTAL
LOSARTAN
SFO
LOT
PAAP
SCH
PD 123177
DTT
B
TOTAL
LOSARTAN
MD
VPL/VPM
PAMP
AMYGDALA
PD 123177
DTT
C
TOTAL
ZO OF SC
MG
CA1, CA3
LOSARTAN
PD 123177
DTT
D
TOTAL
LOSARTAN
SOL
DM10
IO
PD 123177
DTT

order to evaluate likely sites where these actions are mediated, Ang II receptor binding was localized in the brains of rats, rabbits, and humans by *in vitro* autoradiography by using the antagonist analog, ^{125}I-[Sar^1,Ile^8]Ang II, which is subtype nonselective. Brain Ang II receptor subtypes were differentiated into AT$_1$ and AT$_2$ subtypes by using unlabeled nonpeptide antagonists specific for the two subtypes: AT$_1$ binding was determined as that inhibited by losartan (10 µM) and AT$_2$ binding as that inhibited by PD123177 (10 µM). The reducing agent, dithiothreitol (DTT), decreased binding to AT$_1$ receptors and enhanced binding to AT$_2$ receptors.

In the rat, many brain structures, such as the vascular organ of the lamina terminalis, subfornical organ (OVLT), median preoptic nucleus, hypothalamic paraventricular nucleus, area postrema, nucleus of the solitary tract, dorsal motor nucleus of the vagus, which are known to be related to the central actions of the Ang II, contain exclusively AT$_1$ Ang II receptors [44–46].

By contrast, the locus coeruleus, ventral and dorsal parts of lateral septum, superior colliculus and subthalamic nucleus, many nuclei of the thalamus, and the nuclei of the inferior olive contain predominantly AT$_2$ Ang II receptors (Fig. 5).

The detailed binding characteristics of each subtype were determined by competition studies with a series of analogs of angiotensin and antagonists. The pharmacological specificity obtained in rat superior colliculus and the nucleus of the solitary tract agreed well with published data on AT$_1$ and AT$_2$ receptors, respectively.

In the rabbit brain, AT$_1$ receptors are found in very high concentrations in the forebrain circumventricular organs–the subfornical organ, organum vasculosum of the lamina terminalis, and the median eminence, as observed in the rat, but there is very little labeling in the area postrema [47]. In the paraventricular nucleus, median preoptic nucleus, and supraoptic nucleus, there were high levels of predominantly AT$_1$ receptors. High densities of AT$_1$ receptors were also found in the nucleus of the solitary tract and the rostral and caudal ventrolateral medulla. All of these regions have putative roles in the regulation of blood pressure and fluid and electrolyte balance.

In the rabbit brain, there is less AT$_2$ receptor binding than in the rat, with most found in the molecular layer of the cerebellum and in the septohypothalamic nuclei. In the subthalamic nucleus, the mediodorsal and ventroposterior nuclei of the thalamus, locus coeruleus and inferior olivary nuclei, areas containing mostly AT$_2$ receptors in the rat, no binding was detected in the rabbit except in the locus coeruleus which contains moderate levels of AT$_1$ receptors [47].

The human brain is remarkable in that AT$_1$ receptors predominate in all regions except the cerebellum where AT$_2$ receptors have a preponderance (60:40%) over AT$_1$ (Fig. 6). In humans, a high density of AT$_1$

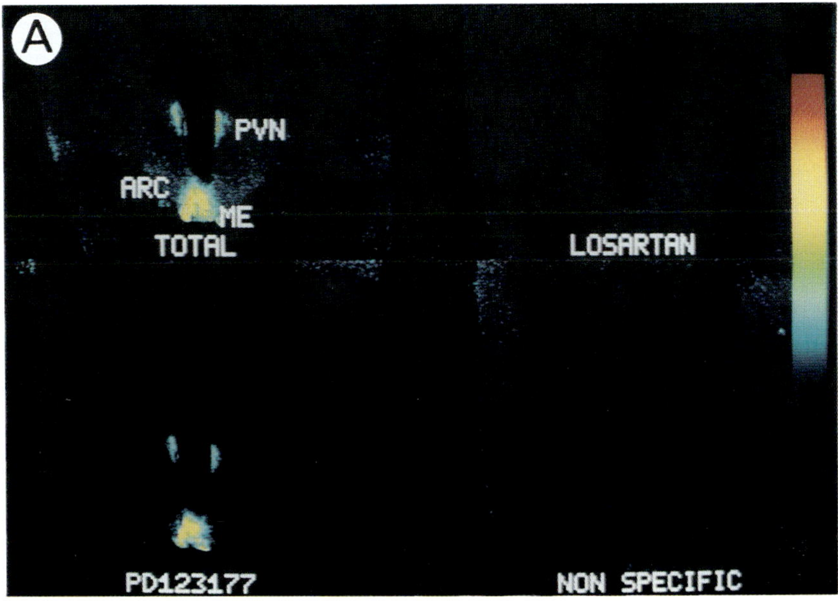

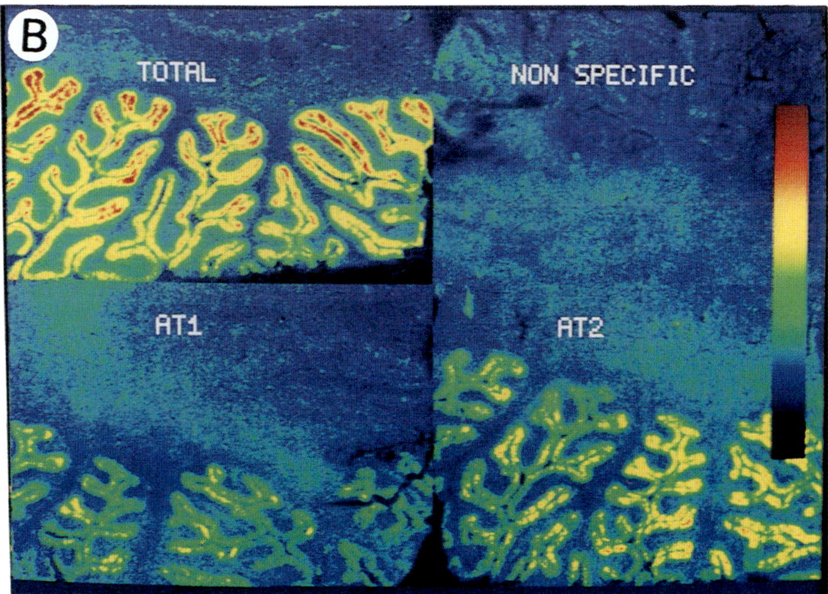

Figure 6. Ang II recptor subtypes in human hypothalamus (panel A) and cerebellar cortex (panel B). At each site, the four panels are as for Fig. 3. In the hypothalamus (PVN, Arc, ME), the receptors are exclusively AT_1, whereas in the molecular layer of the cerebellar cortex (B), both AT_1 and AT_2 receptors occur.

Abbreviations are: PVN – paracentricular n.; Arc – arcuate n.; Me – median eminence.

receptors occurs in the circumventricular organs, including the subfornical organ, vascular organ of the lamina terminalis, and median eminence, but not the area postrema or pineal, in the hypothalamic
paraventricular and arcuate nuclei and in the substantia nigra pars
compacta, the interpeduncular nucleus and some of the raphe nuclei.
Moderate densities occur in the lateral septum, preoptic periventricular
nucleus, caudate nucleus, putamen, bed nucleus of the stria terminalis,
rostral linear nucleus, caudal linear nucleus, dorsal and paramedian
raphe nuclei, locus coeruleus, and A5/periolivary region. A fuller account of Ang II receptors in the human brain has been published
[48–50]. The high density of Ang II receptors in the substantia nigra
occurs over pigmented, presumably dopaminergic, neurons [51]. The
binding in this site, and in the striatum, is not observed in any of the
other species we have studied. It displays similar pharmacological
characteristics to AT_1 receptors in other regions. In the cerebellum, both
AT_1 and AT_2 receptors are found in the molecular layer and have a
striped appearance.

Further studies were undertaken to determine possible actions of Ang
II in some brain regions bearing AT_1 receptors.

In the rostral ventrolateral medulla (RVLM), an area known to be
critically involved in blood pressure regulation by virtue of tonically
active sympathetic premotor neurones, we evaluated the effect of microinjection of Ang II in the anesthetized cat, and found a neurogeni-
cally-mediated rise in blood pressure [52, 53]. This region is innervated
by Ang II immunoreactive terminals in the rat, suggesting that neurally-
released angiotensin might modulate the excitability of these cardiovascular C1 neurones.

An adjacent area, the caudal ventrolateral medulla (CVLM) is the
site of noradrenaline-containing A1 neurones which are known to
project to innervate hypothalamic magnocellular neurosecretory neurones and thereby regulate vasopressin secretion in response to barore-
ceptor input in the medulla oblongata, and is also rich in AT_1 receptor.
Microinjection of Ang II into the CVLM produced a large increase in
vasopressin secretion, and more strikingly, injection of the antagonist,
Sarile, virtually abolished vasopressin secretion in response to hemor-
rhage [54]. These observations suggest that neuronal Ang II may exert
a tonic facilitation of the A1 neurones, or alternatively, the peptide may
be a transmitter at this synapse in the baroreflex-mediated vasopressin
pathway.

It is of interest that both of these medullary regions appear to involve
actions of Ang II on catecholamine-containing neurones. A further site
where we identified AT_1 receptors in association with catecholamine
neurones, was the dopaminergic cells of the substantia nigra in the
human brain and their terminals in the striatum. Since Ang II is well
known to prejunctionally facilitate noradrenaline release from sympa-

thetic nerve terminals, we evaluated the effect of Ang II on dopamine turnover from the terminals of the nigrostriatal neurones using microdialysis in the striatum of conscious, freely-moving rats. Dopamine and its metabolites, DOPAC and HVA, were measured in the microdialysis effluent by HPLC with electrochemical detection. During perfusion with Ang II (1 µM), DOPAC output increased markedly and returned to control with vehicle perfusion during the recovery period. This increase in DOPAC output with Ang II was completely blocked by co-administration of the AT_1-selective antagonist, losartan. Administration of losartan alone led to a significant depression of DOPAC ouput relative to vehicle, suggesting that dopamine release is under a tonic facilitatory influence of Ang II via the AT_1 receptor subtype. Parallel but smaller changes were seen with HVA output. In the presence of the dopamine uptake blocker, nomifensine, dopamine levels also rose markedly with Ang II, confirming an effect on release of dopamine, rather than inhibition of reuptake. The findings of increased dopamine, DOPAC and HVA output following Ang II infusion, which are blocked by losartan, together with the decreased basal output of these dopamine metabolites following losartan alone, is strong evidence for an *in vivo* role of Ang II in modulating dopaminergic transmission, probably via presynaptic AT_1 Ang II receptors which facilitate dopamine release [55].

These results reveal that AT_1 receptors predominate in rostral forebrain, hypothalamus, and autonomic control centers of the medulla oblongata in all three species. In these sites, animal experiments reveal multiple actions of Ang II associated with central regulation of body fluid volume and blood pressure. However, the distribution and density of AT_2 bearing sites in regions such as the septum, thalamus, subthalamic nuclei, locus coeruleus, cerebellum and inferior olivary nuclei, where Ang II actions are not yet defined, show marked species differences. Physiological experiments in three regions where AT_1 receptors are associated with catecholamine neurones has revealed new actions of angiotensin in central regulation of blood pressure (adrenaline cell group in the RVLM), vasopressin release (noradrenaline cell group in CVLM), and dopamine release (terminals of nigrostriatal neurones).

References

1. Timmermans PBMWM, Wong PC, Chiu AT, Herblin WF, Benfield P, Carini DJ, et al. Angiotensin II receptors and angiotensin II receptor antagonists. Pharmacol. Rev. 1993; 45(2): 205–251.
2. Bumpus FM, Catt KJ, Chiu AT, DeGasparo M, Goodfriend T, Husain A, et al. Nomenclature for angiotensin receptors. Hypertension 1991; 17(5): 720–721.
3. Inagami T. Diversity of angiotensin receptors: Cloning and signaling mechanism. Can. J. Physiol. Pharmacol. 1994; 72(Suppl 1): 31 (Abstract).

4. Aiyar N, Baker E, Wu HL, Nambi P, Edwards RM, Trill JJ, et al. Human AT1 receptor is a single copy gene: Characterization in a stable cell line. Mol. Cell. Biochem. 1994; 131(1): 75–86.
5. Konishi H, Kuroda S, Inada Y, Fujisawa Y. Novel subtype on human angiotensin II type 1 receptor: cDNA cloning and expression. Biochem. Biophys. Res. Commun. 1994; 199: 467–474.
6. Douglas JG, Hopfer U. Novel aspect of angiotensin receptors and signal transduction in the kidney. Annu. Rev. Physiol. 1994; 56(0): 649–669.
7. Smith RD, Timmermans PBMWM. Human angiotensin receptor subtypes. Curr. Opin. Nephrol. Hypertens. 1994; 3: 112–122.
8. Hanesworth JM, Sardinia MF, Krebs LT, Hall KL, Harding JW. Elucidation of a specific binding site for angiotensin II (3–8), angiotensin IV, in mammalian heart membranes. J. Pharmacol. Exp. Ther. 1993; 266(2): 1036–1053.
9. Benter LF, Ferrario CM, Morris M, Diz DI. Chronic intravenous angiotensin (1–7) infusions activate antihypertensive mechanisms in spontaneously hypertensive rats. Am. J. Hypertens. 1994; 7(4): H22 (Abstract).
10. DeGasparo M, Husain A, Alexander W, Catt KJ, Chiu AT, Drew M, et al. Update of the nomenclature of angiotensin receptors. Hypertension 1995; 25(5): 924–927.
11. Olson RE, Liu J, Lalka GK, Duncia JV, Wexler RR, Chiu AT, et al. Nonpeptide angiotensin II antagonists with affinity for AT1 and AT2 receptors. Unpublished.
12. Mantlo NB, Kim D, Ondeyka D, Chang RSL, Kivlighn SD, Siegl PKS, et al. Potent imidazo[4,5-b]pyridine based AT1/AT2 angiotensin II receptor antagonists. Bioorg. Med. Chem. Letter 1994; 4(1): 17–22.
13. Viswanathan M, Tsutsumi K, Correa FMA, Saavedra JM. Changes in expression of angiotensin receptor subtypes in the rat aorta during development. Biochem. Biophys. Res. Commun. 1991; 179(3): 1361–1367.
14. Viswanathan M, Strömberg C, Seltzer A, Saavedra JM. Balloon angioplasty enhances the expression of angiotensin II AT1 receptors in neointima of rat aorta. J. Clin. Invest. 1992; 90: 1707–1712.
15. Saavedra JM, Viswanathan M, Shigematsu K, Localization of angiotensin II AT1 receptors in the rat heart conduction system. Eur. J. Pharmacol. 1993; 235: 301–303.
16. Strömberg C, Tsutsumi K, Viswanathan M, Saavedra JM. Angiotensin II AT1 receptors in rat superior cervical ganglia – characterization and stimulation of phosphoinositide hydrolysis. Eur. J. Pharmacol. 1991; 208(4): 331–336.
17. Tsutsumi K, Saavedra JM. Differential development of angiotensin II receptor subtypes in the rat brain. Endocrinology 1991; 128: 630–632.
18. Tsutsumi K, Saavedra JM. Characterization and development of type-1 and type-2 angiotensin II receptors in rat brain. Am. J. Physiol. 1991; 261(1 30-1): R209–R216.
19. Tsutsumi K, Saavedra JM. Characterization of AT2 angiotensin II receptors in rat anterior cerebral arteries. Am. J. Physiol. 1991; 261: H667–H670.
20. Strömberg C, Naveri L, Saavedra JM. Angiotensin AT2 receptors regulate cerebral blood flow in rats. NeuroReport 1992; 3(8): 703–704.
21. Strömberg C, Naveri L, Saavedra JM. Nonpeptide angiotensin AT1 and AT2 receptor ligands modulate the upper limit of cerebral blood flow autoregulation in rats. J. Cerebral Blood Flow Metab. 1993; 13(2): 298–303.
22. Naveri L, Strömberg C, Saavedra JM. Angiotensin II AT2 receptor stimulation extends the upper limit of cerebral blood flow autoregulation: Agonist effects of CGP 42112 and PD123319. J. Cerebral Blood Flow Metab. 1994; 14(1): 38–44.
23. Kakar SS, Sellers JC, Devor DC, Musgrove LC, Neill JD. Angiotensin II type 1 receptor subtype cDNAs: Differential tissue expression and hormonal regulation. Biochem. Biophys. Res. Commun. 1992; 183: 1090–1096.
24. Ye MQ, Healy DP. Characterization of an angiotensin Type 1 receptor partial cDNA from rat kidney – Evidnece for a novel AT1B receptor subtype. Biochem. Biophys. Res. Commun. 1992; 185(1): 204–210.
25. Grone HJ, Simon M, Fuchs E. Autoradiographic characterization of angiotensin receptor subtypes in fetal and adult human kidney. Am. J. Physiol. 1992; 262(2): F326–F331.
26. Gibson RE, Thorpe HH, Cartwright ME, Frank JD, Schorn TW, Bunting PB et al. Angiotensin II receptor subtypes in the renal cortex of rat and rhesus monkeys. Am. J. Physiol. 1991; 261(3 Pt 2): F512–F518.

27. Chansel D, Czekalski S, Pham P, Ardaillou R. Characterization of angiotensin II receptor subtypes in human glomeruli and mesangial cells. Am. J. Physiol. 1992; 262(3): F432–F441.

28. Edwards RM, Aiyar N. Angiotensin II receptor subtypes in the kidney. J. Am. Soc. Nephrol. 1993; 3(10): 1643–1652.

29. Leung KH, Roscoe WA, Smith RD, Timmermans PBMWM, Chiu AT. Characterization of biochemical responses of angiotensin II (AT2) binding sites in the rat pheochromocytoma PC12W cells. Eur. J. Pharmacol. 1992; 227: 63–70.

30. Balla T, Baukal AJ, Eng S, Catt KJ. Angiotensin II receptor subtypes and biological responses in the adrenal cortex and medulla. Mol. Pharmacol. 1991; 40: 401–406.

31. Ouali R, Poulette S, Penhoat A, Saez JM. Characterization and coupling of angiotensin II receptor subtypes in cultured bovine adrenal fasciculata cells. J. Steroid Biochem. Molec. Biol. 1992; 43(4): 271–280.

32. Devynck MA, Rouzaire-Dubois B, Chevilotte E, Meyer P. Variations in the number of uterine angiotensin receptors following changes in plasma angiotensin levels. Eur. J. Pharmacol. 1976; 40: 27–37.

33. Whitebread S, Mele M, Kamber B, DeGasparo M. Preliminary biochemical characterization of two angiotensin II receptor subtypes. Biochem. Biophys. Res. Commun. 1989; 163: 284–291.

34. Kalenga MK, DeGasparo M, Dehertogh R, Whitebread S, VanKrieken L, Thomas K. Angiotensin II receptors in the human placenta are type AT1. Reprod. Nutr. Dev. 1991; 31(3): 257–267.

35. Kalenga MK, DeGasparo M, Whitebread S, VanKrieken L, Thomas K, Dehertogh R. Identification du recepteur a l'angiotensine II et determination de ses sous-types dans les annexes foetales du lapin et du cobaye. Reprod. Nutr. Dev. 1992; 32: 47–54.

36. Cox BE, Ipson MA, Shaul PW, Kamm KE, Rosenfeld CR. Myometrial angiotensin II receptor subtypes change during ovine pregnancy. J. Clin. Invest. 1993; 92: 2240–2248.

37. Wilson M, Morganti AA, Zevoudakis I, Letcher RL, Romney BM, Von Deyon, P et al. Blood pressure, the renin aldosterone system and sex steroids throughout normal pregnancy. Am. J. Med. 1980; 68: 97–104.

38. Gant Jr. NF, Daley GL, Chand S et al. A study of angiotensin II pressor responses throughout primigravid pregnancy. J. Clin. Invest. 1973; 52: 2682–2689.

39. Rosenfeld CR, Gant Jr. NF. The chronically instrumented ewe. A model for studying vascular reactivity to angiotensin II in pregnancy. J. Clin. Invest. 1981; 67: 486–492.

40. Paller MS. Mechanism of decreased pressor responsiveness to Ang II, NE, and vasopressin in pregnant rats. Am. J. Physiol. 1984; 247: H100–H108.

41. Naden RP, Rosenfeld CR. Effect of angiotensin II on uterine and systemic vasculature in pregnant sheep. J. Clin. Invest. 1981; 68: 468–474.

42. Erkkola RU, Pirhonen JP. Flow velocity waveforms in uterine and umbilical arteries during the angiotensin II sensitivity test. Am. J. Obstet. Gynecol. 1992; 162: 1193–1197.

43. Mackanjee HR, Shaul PW, Magness RR, Rosenfeld CR. Angiotensin II vascular smooth muscle receptors are not down-regulated in near-term pregnant sheep. Am. J. Obstet. Gynecol. 1991; 165: 1641–1648.

44. Allen AM, Paxinos G, Mendelsohn FAO. Localization of angiotensin II receptor binding sites in the rat brain. In: Kuhar MJ, editor. Handbook of chemical neuroanatomy Amsterdam: Elsevier, 1992, pp. 1–37.

45. Song K, Zhuo J, Allen AM, Paxinos G, Mendelsohn FAO. Angiotensin II receptor subtypes in rat brain and peripheral tissues. Cardiology 1991; 79(Suppl 1): 45–54.

46. Song K, Allen AM, Paxinos G, Mendelsohn FAO. Mapping of angiotensin II receptor subtype heterogeneity in rat brain. J. Comp. Neurol. 1992; 316(4): 467–484.

47. Aldred GP, Chai SY, Song K, Zhou J, MacGregor DP, Mendelsohn FAO. Distribution of angiotensin II receptor subtypes in the rabbit brain. Regul. Pept. 1993; 44(2): 119–130.

48. Allen AM, Chai SY, Clevers J, McKinley MJ, Paxinos G, Mendelsohn FAO. Localization and characterization of angiotensin II receptor binding and angiotensin converting enzyme in the human medulla oblongata. J. Comp. Neurol. 1988; 269: 249–264.

49. Allen AM, Paxinos G, McKinley MJ, Chai SY, Mendelsohn FAO. Localization and characterization of angiotensin II receptor binding sites in the human basal ganglia, thalamus, midbrain, pons and cerebellum. J. Comp. Neurol. 1991; 312: 291–298.

50. Allen AM, McKinley MJ, Paxinos G, Oldfield BJ, Mendelsohn FAO. Angiotensin II receptors in the human central nervous system. In: F.A.O. Mendelsohn and G. Paxinos, (editors). Receptors in the Human Nervous System, Academic Press 1991; pp. 123–142.
51. Allen AM, MacGregor DP, Chai SY, Donnan GA, Kaczmarczyk S, Richardson K, et al. Angiotensin II receptor binding associated with nigrostriatal dopaminergic neurones in human basal ganalia. Ann. Neurol. 1992; 32(3): 339–344.
52. Allen AM, Dampney RAL, Mendelsohn FAO, Angiotensin receptor binding and pressor effects in the cat subretrofacial nucleus. Am. J. Physiol. 1988; 255: H1011–H1017.
53. Allen AM, Sasaki S, Dampney RAL, Mendelsohn FAO. Actions of angiotensin II in the ventrolateral medulla oblongata. Central Neural Mech. Cardiovasc. Regul. 1991; 1: 95–103.
54. Allen AM, Mendelsohn FAO, Gieroba ZJ, Blessing WW. Vasopressin release following microinjection of angiotensin II into the caudal ventrolateral medulla oblongata in the anaesthetized rabbit. J. Neuroendocrinol. 1991; 2: 867–873.
55. Mendelsohn FAO, Jenkins TA, Berkovic SF. Effects of angiotensin II on dopamine and serotonin turnover in the striatrum of conscious rats. Brain Res. 1993; 613(2): 221–229.
56. Iwai N, Yamano Y, Chaki S, Konishi F, Bardhan S, Tibbetts C, et al. Rat angiotensin II receptor: cDNA sequence and regulation of the gene expression. Biochem. Biophys. Res. Commun. 1991; 177, No. 1: 299–304.
57. Kambayashi Y, Bardhan S, Takahashi K, Tsuzuki S, Inui H, Hamakubo T, et al. Molecular cloning of a novel angiotensin II receptor isoform involved in phosphotyrosine phosphatase inhibition. J. Biol. Chem. 1993; 268(33): 24543–24546.

Pharmacological Sciences: Perspectives for
Research and Therapy in the Late 1990s
ed. by A.C. Cuello and B. Collier
© 1995 Birkhäuser Verlag Basel/Switzerland

Foundations and Future of Molecular Pharmacology

Palmer Taylor[1], Elliott M. Ross[2], Paul B. Sigler[3], and Brian D. Sykes[4]

[1]*Department of Pharmacology, University of California, San Diego, La Jolla, CA 92093-0636 USA;* [2]*Department of Pharmacology, University of Texas, Southwestern, Dallas, TX 75235-9041 USA;* [3]*Howard Hughes Medical Institute, Yale University, New Haven, CT 06510 USA;* [4]*Department of Biochemistry, University of Alberta, Edmonton, Alberta, Canada T6G 2H7*

A symposium designed to highlight some of the recent advances in molecular pharmacology was organized to honor Sir Arnold Burgen, a founder and important contributor to the development of this discipline. Sir Arnold Burgen's career is marked by notable achievements on both sides of the Atlantic. Upon returning from McGill University to the United Kingdom in 1962 as Sheild Professor of Pharmacology at the University of Cambridge, Professor Burgen began plans for a molecular pharmacology program. The group was supported by the Medical Research Council as a formal Research Unit in 1965. As director of the Medical Research Council from 1972–1982, the Unit remained under his direction at the Mill Hill laboratories in London. The discipline of molecular pharmacology as conceived by him involved the use of physicochemical techniques to study the true working macromolecules of pharmacology which encompass receptors, sites of antimetabolite action, membrane lipids, membrane lipid-protein interfaces and other proteins and nucleic acids critical to drug action and disposition [1]. These initial developments in molecular pharmacology actually preceded the application of recombinant DNA technology in biology by nearly a decade. The field of molecular biology has popularized the modifier "molecular" and perhaps diverted a meaning originally intended by investigators in several disciplines to encompass the study of protein and macromolecular structure in relation to function.

Early research in the Molecular Pharmacology Unit pioneered the use of nuclear magnetic resonance spectrometry to study anesthetic perturbation of membranes and ligand interactions with proteins. These approaches were coupled with spectroscopic and fast reaction techniques to provide a comprehensive analysis of ligand-macromolecule interactions. Over the years experimental endeavors expanded with the sophistication of instumentation and the increased structural resolution of macromolecules. Research in this Unit was not only pioneering in the application of magnetic resonance techniques in pharmacology

SIR ARNOLD BURGEN

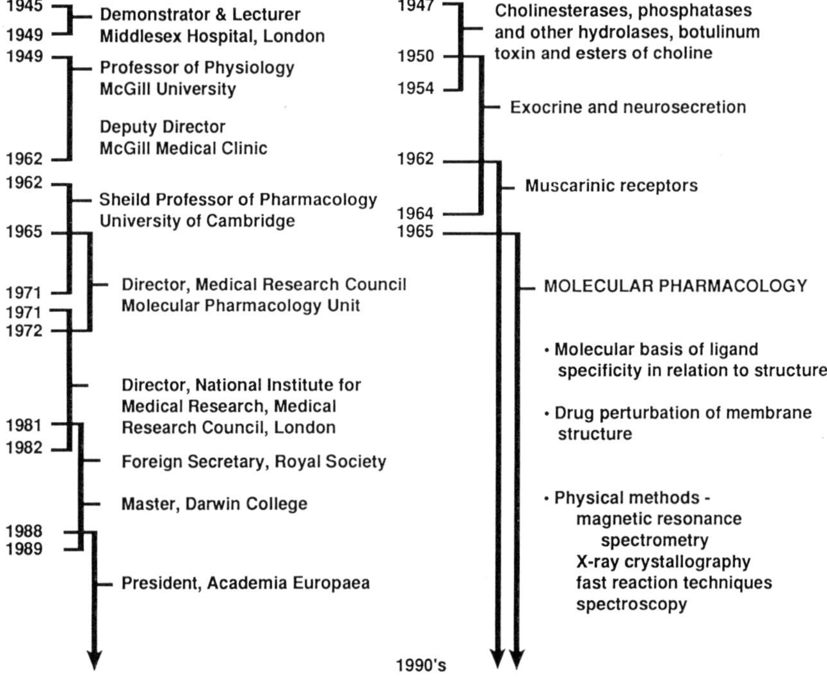

Fig. 1. Positions and research interests of Sir Arnold Burgen.

but provided cornerstone developments in the understanding of biological structures and macromolecular conformation through the use of magnetic resonance.

The four speakers in the symposium were selected to represent various phases of Sir Arnold Burgen's research career outlined in Fig. 1. His early research was devoted to the study of cholinesterases and phosphatases and, in particular, modes of enzyme inhibition by organophosphates and carbamoylating agents. This field has experienced evolution into the molecular arena with the cloning of the cholinesterase genes, analysis of sequence relationships and the subsequent analysis of crystal structures of the enzyme and several of its complexes [1–3].

Dr. Palmer Taylor reviewed recent studies on cholinesterase gene structure, the bases for structural diversity of the cholinesterases and the structural features that contribute to the capacity of certain ligands to discriminate between acetylcholinesterase and butyrylcholinesterase. In particular, he described how site-specific mutagenesis, formation of chimeric proteins and analysis of kinetic parameters can delineate side chains responsible for dictating dimensions of the acyl pocket and choline subsite in the active center of the enzyme as well as contribute to the force fields in forming complexes. In addition, he described the structural characteristics of the peripheral site on acetylcholinesterase and the nature of the interaction of the peptide, fasciculin with that site [4].

When Sir Arnold Burgen returned to Cambridge, he also developed a program on the muscarinic receptor to examine coupling between receptor occupation and response. Dr. Elliott Ross described the features in the muscarinic receptor and related receptors whose sequence spans the membrane seven times and are responsible for interaction with G proteins. Particular domains in the third intracellular loop with contributions from the second loop and cytoplasmic carboxyl-terminus were highlighted. Often the formation of chimeric receptors and site-specific mutants relax the specificity constraints for receptor coupling to G proteins, thereby producing a more promiscuous receptor [6–8]. A small peptide, mastoparan, has been studied in depth by Ross and colleagues [8] and has been shown to be a mimic of the receptor activating sequence in stimulating the G-protein. Nuclear magnetic resonance studies reveal unique features of mastoparan conformation which will be critical to understanding the molecular interactions at the interface between the peptide and the cytoplasmic loop of the G-protein. Dr. Ross also pointed out the importance of studying receptor-G protein interactions with multiple end-points with and without effector molecules present. Recent data clearly show that effector molecules can have GTPase stimulatory activity and can alter the kinetics of receptor-G protein interactions and turnover [7].

Dr. Paul Sigler of Yale University continued with the subject of G-proteins by extending considerations of structure to atomic resolution. He discussed the first three-dimensional structure of a hetero-trimeric G-protein, that of transducin-α, the G protein involved in coupling visual phototransduction. Structural resolution is at 1.8Å and only the amino-terminal 25 amino acids are missing from the structure. Dr. Sigler's group now has high-resolution structures of the GTP-γ-S, GDP and AlF_4^- complexes, [10, 11], and a comparison of structures provides unusual insight into the molecular aspects of activation. The bound GTP-γ-S molecule is occluded deep within a cleft between a domain homologous to the small GTPases such as Ras [12] and a helical domain unique to the heterotrimeric G proteins. A significant conformation change is required to open the cleft for nucleotide exchange. The terminal gamma phosphate on GTP through association with Mg^{++} interacts at three switch positions affecting the juxta-position of three loops which interact with the effector molecule. Difference density maps clearly reveal how the terminal phosphate positioned near the center of the molecule, by converting a bent helix to a linear helix and by increasing the tension on the chains, positions the loops at the outer surface of the molecule. The AlF_4^- complex of the GDP-transducin-α-complex shows that AlF_4^- is positioned where the γ-phosphate in GTP resides and reveals the existence of a glutamine containing loop which functions as a tautomer with the phosphate and AlF_4^-.

Dr. Brian Sykes presented recent data using nuclear magnetic resonance relaxation to study peptide-receptor interactions [13]. To this end his group has been examining the interaction of transforming growth factor-α (TGF-α) with a 85 kDa extracellular domain of the epidermal growth factor receptor. The former is a relatively small peptide while the latter can be produced as a 80 kDa soluble polypeptide. Initial studies through nuclear Overhauser measurements have established the preferred conformation of TGF-α. A second phase of the study was directed to the examination of methyl proton relaxation rates which reflect TGF-α association with the receptor. Immobilization can be measured by the relaxation parameters for the methyl protons on the amino acid side chains of TGF-α (i.e., alanine, valine, leucine and isoleucine). The studies using different ligand-receptor ratios clearly reveal differential stabilization of the associated molecule with the amino-terminal residues remaining far more mobile in the complex. Further data should continue to yield invaluable information on the bound conformation of the peptide and thereby prove useful in the design of non-peptide analogues of TGF-α. By combining the analysis of relaxation rates for the rapid exchange processes with the slower overall binding constants measured on a Biocore instrument by refractive index changes, an overall kinetic profile for TGF-α association can be ascertained.

Sir Arnold Burgen reviewed some of the developments in molecular pharmacology over the past 30 years since the MRC Unit in Cambridge was started. He pointed out that the objectives of molecular pharamcology are straightforward and encompass ascertaining the molecular determinants of macromolecular recognition or drug specificity and the mechanism of transduction of ligand-elicited cellular signals. Application, however, is often limited by sensitivity or resolution of the signals and by the necessity of isolating and purifying the system under study. Nevertheless, he pointed out that the great advances made in instrumentation over the last three decades. In NMR, for example, early studies were done on 60 MHz instruments, whereas at present 600–700 MHz instruments with capacity for rapid pulse measurements of multiple nuclei are available. With new area detectors and advances in growing crystals, novel X-ray crystal structures appear far more frequently, and investigators are completing structures which were thought to be too complex or intractable to crystallization. Moreover, receptors can now be readily purified, and recombinant DNA techniques have yielded comparative sequences of receptors and provided the capacity to study site-specific mutant and recombinant receptors. Though mutagenesis techniques are powerful, single amino acid substitutions only yield a limited perspective on overall molecular structure, and this approach is most informative when combined with other techniques.

Although the symposium could only touch on a few of the pharmacological systems in which molecular approaches are being applied, the systems amenable to study and the sophistication of the instruments and technology continue to expand. Hence, ideas conceived over 30 years ago have provided the cornerstone for continually expanding endeavors within pharmacology.

References

1. Burgen ASV. The drug-receptor complex. J. Pharm. & Pharmacol. 1966; 18: 137–175.
2. Taylor P, Radic' Z. The cholinesterases: from genes to proteins. Ann. Rev. Pharmacol. Toxicol. 1994; 34: 281–320.
3. Massoulie' J, Pezzementi L, Bon S, Krejei T, Vallette F-M. Molecular and cellular biology of cholinesterases. Prog. Neurobiol. 1993; 41: 31–91.
4. Sussman JL, Harel M, Frolow F, Oefner C, Goldman A, Toker L et al. Atomic structure of acetylcholinesterase from *Torpedo california*: a prototopic acetylcholine binding protein. Science 1991; 253: 872–879.
5. Radic' Z, Duran R, Vellom DC, Li Y, Cervenanski C, Taylor P. Site of fasiculin interaction with acetylcholinesterase. J. Biol. Chem. 1994; 269: 11233–11239.
6. Ross EM, Burstein G. Regulation of the M1 muscarinic receptor-Gα phospholipase C-β pathway by nucleotide exchange and GTP hydrolysis. Life Science 1993; 52: 413–419.
7. Wong SK, Ross EM. Chimeric muscarinic cholinergic: beta adrenergic receptors that are functionally promiscuous among G proteins. J. Biol. Chem. 1994; 269: 18968–18976.
8. Bernstein G, Blank JL, Jhon DY, Exton JM, Rhee SG, Ross EM. Phospholipase C-beta 1 is a GTPase-activating protein for Gq/11 its physiologic regulator. Cell 1992; 70: 411–418.
9. Higashijimi T, Ross EM. Mapping of the mastoparan-binding site on G proteins. J. Biol. Chem. 1991; 266: 12655–12662.
10. Noel JP, Hamm HE, Sigler PB. The 2.2Å crystal structure of transducin-α complexed with GTPγS. Nature 1993; 366: 654–663.
11. Lambright DG, Noel JP, Hamm HE, Sigler PB. Structural determinants for activation of the α-subunit of a heterotrimeric G protein. Nature 1994; 368: 654–663.
12. Pai EF, Kreugel U, Petsko GA, Good RS, Kabiel W, Hiughofer A. Refined crystal structure of the triphosphate conformation of M-ras p21 at 1.35 s resolution. EMBJ 1990; 9: 2351–2359.
13. Campbell AP, Sykes BD. The two-dimensional transferred nuclear Overhauser effect: theory and practice. Ann. Rev. Biophysics & Biomolecular Structure 1993; 22: 99–122.

Signal Transduction

Pharmacological Sciences: Perspectives for
Research and Therapy in the Late 1990s
ed. by A.C. Cuello and B. Collier
© 1995 Birkhäuser Verlag Basel/Switzerland

Lipid Messengers and Protein Kinase C for Intracellular Signalling

S. Nakamura[1], Y. Asaoka[2], F. Hullin[2] and Y. Nishizuka[1,2]

[1]*Department of Biochemistry, Kobe University School of Medicine, Kobe 650;*
[2]*Biosignal Research Center, Kobe Univeristy, Kobe 659, Japan*

Summary. Inositol phospholipid hydrolysis by phospholipase C was once thought to be the sole mechanism to produce diacyglycerol which relays information of extracellular signals to intracellular processes through activation of protein kinase C. It is now becoming clearer that stimulation of cell surface receptors initiates degradation of various membrane phospholipids. Many of their metabolites have potentials to induce, intensify, and prolong the activation of protein kinase C that is needed for sustained cellular responses.

Signals are transduced across the cell membrane through a highly elaborate network of molecular events which are initiated by stimulation of receptors or by opening of ion channels. It was shown first that hydrolysis of inositol phospholipid (PI) by phospholipase C produces DAG which is necessary for activation of protein in kinase C (PKC) [1]. Later, it became clear that stimulation of receptors elicits also a degradation cascade of various membrane phospholipids, particularly phosphatidylcholine (PC), by activation of phospholipase D and phospholipase A_2 [2, 3]. Many of the lipid metabolites appear to act as mediators for intracellular signalling (Figure 1).

Signal-induced PI hydrolysis causes a variety of cellular responses, but the formation of DAG from PI is normally transient. The DAG level in membranes often increases again with a slow onset, especially when cells are stimulated by long-acting signals such as some growth factors. Sustained activation of PKC is shown to be essential to cell proliferation and differentiation [1]. PC hydrolysis by phospholipase D activation is now thought to play a role in this second phase of DAG elevation [2, 3]. Phosphatidic acid, a product of PC hydrolysis by phospholipase D, is converted to DAG by removal of its phosphate.

Phospholipase D is found in particulate fractions of many tissues and cell types. The particulate fractions *per se* are inactive, and cytosol and GTP-γ-S are necessary for the enzymatic activity [4, 5]. This cytosolic factor appears to be a small GTP-binding protein, which was recently identified as ADP-ribosylation factor [6, 7]. The activation of phospholipase D by GTP-γ-S in cell-free systems is further enhanced by ATP,

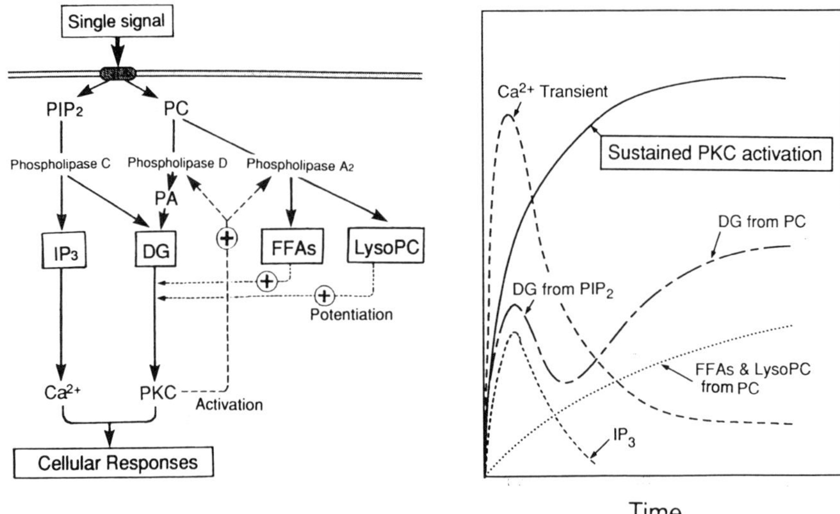

Fig. 1. Schematic representation of signal-induced membrane phospholipid degradation cascade for sustained PKC activation. PIP_2, phosphatidylinositol 4,5-bisphosphate; FFAs, free *cis*-unsaturated fatty acids.

and this enhancement is blocked by genistein, implying that tyrosine phosphorylation is involved in this phospholipase D activation. In permeabilized cell preparations, the enzyme is activated further by the addition of ATP and phorbol ester or a membrane-permeant DAG, and this activation is sensitive to PKC inhibitors such as staurosporin and calphostin C. It seems likely, then, that phospholipase D may be regulated by multiple mechanisms. PKC, once activated by DAG that is initially derived from PI hydrolysis, plays a role in the subsequent PC hydrolysis by phospholipase D to maintain the DAG level for sustained PKC activation.

On the other hand, phospholipase A_2 is abundant in mammalian tissues, and the receptor-mediated activation of this enzyme has been proposed [8]. The hydrolysis of PC by phospholipase A_2 generates two lipid mediators, free *cis*-unsaturated fatty acids and lysophosphatidyl-choline (lysoPC). In the presence of DAG, *cis*-unsaturated fatty acid causes nearly full activation of PKC at the basal level of Ca^{2+} concentrations [9]. Neither saturated fatty acid nor *trans*-unsaturated fatty acid is capable of enhancing the PKC activation. It is also shown that several *cis*-unsaturated fatty acids greatly enhanced the DAG-dependent cellular responses such as platelet release reaction [10]. For the release of serotonin, a membrane-permeant DAG and Ca^{2+}-ionophore are both essential, and fatty acid alone is inactive. Kinetic analysis with the Ca^{2+}-sensitive fluorescent dye fura-2 indicates that *cis*-unsaturated fatty acids markedly increase an apparent affinity of PKC activation to Ca^{2+}, thereby causing release reaction without a large increase in the

Ca^{2+} concentration. It is possible that PKC, once activated initially by PI hydrolysis, may intensify its own enzymatic activity even after the Ca^{2+} concentration returns to the basal level, when DAG and *cis*-unsaturated fatty acids become available (Fig. 1).

LysoPC, the other product of PC hydrolysis, shows a detergent-like activity, and is toxic to the cell. However, in the presence of a membrane-permeant DAG or phorbol ester, lysoPC greatly enhances cellular responses, paprticularly those in the long-term such as cell proliferation and differentiation. For instance, lysoPC significantly enhances the activation of human T-lymphocytes that is induced by a membrane-permeant DAG and ionomycin [11]. Similarly, lysoPC exerts a profound stimulatory effect on differentiation of HL-60 cells to macrophages as measured by CD11b expression and appearance of phagocytic activity [12]. In both cases, lysoPC is active only when both DAG and ionomycin are present, indicating that this lysophospholipid interacts with the PKC pathway. Other lysophospholipids are ineffective except for lysophosphatidylethanolamine which is, however, far less active than lysoPC. The biochemical mechanism of this lysoPC action remains unclear, but it is plausible that signal-induced phospholipase A_2 activation may also be involved in cell signalling.

PKC comprises a large family of proteins with multiple isoforms. To date, at least 11 structurally related enzymes have been identified in mammalian tissues by molecular cloning and enzymological analysis (Table 1). These isoforms may be divided into three subgroups. Classical PKC (cPKC) isoforms (α, β1, β2, γ) show typical characteristics of PKC which are activated by Ca^{2+} and DAG or phorbol ester in the presence of phosphatidylserine (PS). This subgroup of enzymes are

Table 1. PKC isoforms in mammalian tissues

	Subspecies	Amino acid residues	Ca^{2+} and lipid activators*	Tissue expression
cPKC	α	672	Ca^{2+}, DAG, PS, FFAs, lysoPC	Universal
	βI	671	"	Some tissues
	βII	671	"	Many tissues
	γ	697	"	Brain only
nPKC	δ	673	DAG, PS	Universal
	ε	737	DAG, PS, FFA, PIP_3	Brain and others
	η(L)	683	DAG, PS, PIP_3, Cholesterol sulfate	Skin, lung, heart
	θ	707	?	Muscle, T-cell, etc.
	μ	912	?	NRK cells
aPKC	ζ	592	PS, FFA, PIP_3?	Universal
	λ(ι)	587	?	Many tissues

*The activators for each isoform are determined with calf thymus H1 histone and bovine myelin basic protein as model phosphate acceptors. FFAs, *cis*-unsaturated fatty acids; for other abbreviations, see text.

further activated by *cis*-unsaturated fatty acid and lysoPC. New PKC (nPKC) isoforms (δ, ε, η, θ, μ) do not possess a Ca^{2+}-sensitive domain in their molecules, but are activated by DAG or phorbol ester in the presence of PS. Some of these isoforms are activated also by PI-trisphosphate which is a product of PI-3-kinase, or cholesterol sulfate, or by *cis*-unsaturated fatty acids. On the other hand, atypical PKC (aPKC) isoforms (ζ, λ (ι)) require PS for their activation, but do not respond to DAG and phorbol ester, nor to Ca^{2+}. At present, the lipid mediators for the activation of this subgroup of enzymes remain unknown.

Once the cell is disrupted, most protein kinases phosphorylate physiological and non-physiological substrate proteins non-selectively. Biochemical and histochemical analysis indicates that the isoforms so far examined show subtly different kinetics, different mode of activation, specific patterns of tissue distribution and intracellular localization, and distinct behaviours during cellular responses [1].

Extensive studies have been carried out to identify the specific role of each isoform. Some isoforms may play a crucial role in regulating membrane functions, such as downregulation of receptors, modulation of ion channels and pumps, and hydrolysis of membrane phospholipids that generate lipid mediators [1]. Release and exocytosis are well known to be under the control of PKC. For signalling to the nucleus, tyrosine kinases associated with growth factor receptors activate MAP kinases through a series of provocative protein-protein interactions involving Ras and Raf-1 (Fig. 2). It is almost certain that stimulation of G-protein-coupled receptors also activates MAP kinases through PKC pathway [13]. The α-isoform has been proposed to phosphorylate and activate Raf-1, which in turn activates MAP kinase through a protein kinase cascade [14]. However, the precise mechanism of the PKC signalling pathway or network to cause the MAP kinase activation has not been unequivocally established. It has been proposed that, in T- and B-lymphocytes, PKC links with MAP kinases at the point of Ras activation [15, 16].

When PKC was first found to be activated by DAG, it was thought that this DAG resulted solely from the hydrolysis of PI following induction by hormones and neurotransmitters. The pertinent question that followed was obviously how PKC activation could evoke cellular responses. Extensive studies in numerous laboratories have answered a part of this question. Today, it is becoming increasingly clear that several other membrane phospholipids may also be involved in transmitting information from a variety of extracellular signals across the membrane. However, it remains to be explored whether most of the phospholipid metabolites discussed herein indeed act as the signal mediators or messengers in physiological processes, because there is still a wide chasm between observations made *in vitro* and those derived from biological systems.

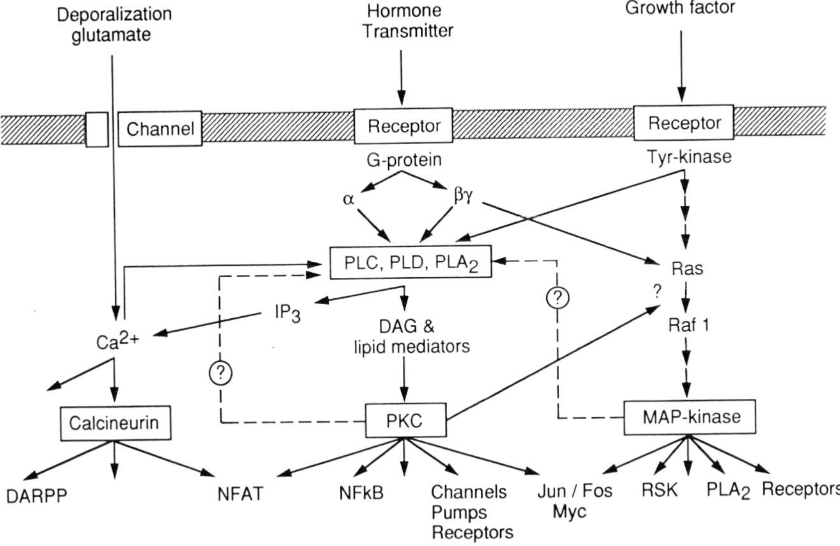

Fig. 2. Possible intracellular signalling network involving PKC. RSK, cell cycle-regulated S6 protein kinase; DARPP, dopamine and cAMP-regulated phosphoprotein; for other abbreviations, see text.

The main features of signalling pathways are, nevertheless, beginning to emerge. Interactions among various pathways together with the multiplicity of the PKC family produce enormous variation of cellular responses. Presumably, the function of each member of the PKC family may well coordinate with dynamic intracellular membrane lipid metabolism. Cellular responses often persist for a long period of time, and the sustained activation of PKC may be essential in these processes. However, unusually persistent activation of PKC such as that caused by phorbol esters, stable overexpression, or by alteration of lipid metabolism may all result in pathological responses, such as tumorigenesis. Spatiotemporal aspects and switch-off mechanism of the PKC activation must be explored for full understanding of physiological cellular regulation.

References

1. Nishizuka Y. Intracellular signaling by hydrolysis of phospholipids and activation of protein kinase C. Science 1992; 258: 607–614.
2. Billah MM, Anthes JC. The regulation and cellular functions of phosphatidylcholine hydrolysis. Biochem. J. 1990; 269: 281–291.
3. Exton JH. Signaling through phosphatidylcholine breakdown. J. Biol. Chem. 1990; 265: 1–4.
4. Olson SC, Bowman EP, Lambeth JD. Phospholipase D activation in a cell-free system from human neutrophils by phorbol 12-myristate 13-acetate and guanosine 5'-O-(3-thio triphosphate). J. Biol. Chem. 1991; 266: 17236–17242.

5. Anthes JC, Wang P, Siegel MI, Egan RW, Billah MM. Granulocyte phospholipase D is activated by a guanine nucleotide dependent protein factor. Biochem. Biophys. Res. Commun. 1991; 175: 236–243.
6. Brown HA, Gutowski S, Moomaw CR, Slaughter C, Sternweis PC. ADP-ribosylation factor, a small GTP-dependent regulatory protein, stimulates phospholipase D activity. Cell 1993; 75: 1137–1144.
7. Cockcroft S, Thomas GMH, Fensome A, Geny B, Cunningham E, Gout I, Hiles I, Totty NF, Truong O, Hsuan JJ. Phospholipase D: a downstream effector of ARF in granulocytes. Science 1994; 263: 523–526.
8. Dennis EA, Rhee SG, Billah MM, Hannun YA. Role of phospholipases in generating lipid second messengers in signal transduction. FASEB J. 1991; 5: 2068–2077.
9. Shinomura T, Asaoka Y, Oka M, Yoshida K, Nishizuka Y. Synergistic action of diacylglycerol and unsaturated fatty acid for protein kinase C activation: Its possible implications. Proc. Natl. Acad. Sci. USA 1991; 88: 5149–5153.
10. Yoshida K, Asaoka Y, Nishizuka Y. Platelet activation by simultaneous actions of diacylglycerol and unsaturated fatty acids. Proc. Natl. Acad. Sci. USA 1992; 89: 6443–6446.
11. Asaoka Y, Oka M, Yoshida K, Sasaki Y, Nishizuka Y. Role of lysophosphatidylcholine in T-lymphocyte activation: Involvement of phospholipase A_2 in signal transduction through protein kinase C. Proc. Natl. Acad. Sci. USA 1992; 89: 6447–6451.
12. Asaoka Y, Yoshida K, Sasaki Y, Nishizuka Y. Potential role of phospholipase A_2 in HL-60 cell differentiation to macrophages induced by protein kinase C activation. Proc. Natl. Acad. Sci. USA 1993; 90: 4917–4921.
13. Lange-Carter CA, Pleiman CM, Gardner AM, Blumer KJ, Johnson GL. A divergence in the MAP kinase regulatory network defined by MEK kinase and Raf. Science 1993; 260: 315–319.
14. Kolch W, Heidecker G, Kochs G, Hummel R, Vahidi H, Mischak H, Finkenzeller G, Marmé D, Rapp UR. Protein kinase Cα activates Raf-1 by direct phosphorylation. Nature (London) 1993; 364: 249–252.
15. Downward J, Graves JD, Warne PH, Rayter S, Cantrell DA. Stimulation of p21[ras] upon T-cell activation. Nature (London) 1990; 346: 719–723.
16. Harwood AE, Cambier J. B cell antigen receptor cross-linking triggers rapid protein kinase C independent activation of p21[rasl]. J. Immunol. 1993; 151: 4513–4522.

Pharmacological Sciences: Perspectives for
Research and Therapy in the Late 1990s
ed. by A.C. Cuello and B. Collier

Mechanisms for the Regulation of Inositol Lipid Signaling and Calcium Homeostasis

Stephen K. Fisher

Neuroscience Laboratory, Mental Health Research Institute and Department of Pharmacology, University of Michigan, Ann Arbor, MI 48104-1687, USA

Summary. Receptor activation of phosphoinositide-specific phospholipase C (PLC) represents a major mechanism for signal transduction in eukaryotic cells. In recent years, considerable progress has been made in the identification and characterization of the individual components of this pathway. Much less information, however, is available regarding mechanisms that may underlie the regulation of this pathway. In this review, some potential mechanisms for the regulation of receptor-stimulated phosphoinositide hydrolysis are discussed. Evidence is presented that regulation of this pathway can occur at the level of either receptor, G_q or PLC.

Introduction

Of the mechanisms used by cells for transmembrane signaling, the enhanced phosphodiesteratic breakdown of phosphatidylinositol 4,5-bisphosphate (PIP_2), a reaction catalyzed by a phosphoinositide (PPI)-specific phospholipase C (PLC), is one of the most widely encountered. For example, in neural tissues alone, more than 30 pharmacologically distinct receptors are now known to operate via this pathway, in which two second messengers, namely inositol 1,4,5-triphosphate [$I(1,4,5)P_3$] and diacylglycerol (DAG) are formed [1]. Although discovered more than 40 years ago [2], the significance of receptor stimulated PPI hydrolysis was not fully appreciated until the second messenger functions of $I(1,4,5)P_3$ and DAG had been established. Because of the central role that Ca^{2+} homeostasis and protein phosphorylation events play in cellular physiology, an understanding of the regulation of receptor-stimulated PPI hydrolysis assumes major importance. There is now increasing evidence that such a regulation can occur at multiple levels of the inositol lipid signaling pathway, some examples of which are discussed below, based upon the presentations made at this conference by the indicated participants.

Measurement of Receptor Stimulated Phosphoinositide Hydrolysis

Although an ability to accurately measure receptor-stimulated phospholipase C activity is a prerequisite for determination of both the occurrence and rate of desensitization of PPI hydrolysis, no single methodological approach is without its drawbacks [3]. Ideally, PLC activity would be monitored as the rate of $I(1,4,5)P_3$ formation in intact cells in the absence of any significant breakdown of the second messenger molecule (i.e., in a manner analogous to that employed for monitoring cyclic AMP accumulation). However, two issues are problematic in this regard. The first is that $I(1,4,5)P_3$ is metabolized very rapidly in most cells ($t_{1/2} \sim 5$ s), and the second is the absence of specific cell permeant inhibitors of $I(1,4,5)P_3$ metabolism. The most commonly used technique is the measurement of formation of a total inositol phosphate fraction in the presence of Li^+. Although this technique monitors the *accumulation* of PLC derived products and provides a sensitive measure of PLC activity, a disadvantage is that only an indirect estimate of $I(1,4,5)P_3$ production is provided. Measurement of the rate of formation of $I(1,4,5)P_3$ alone by means of either its mass or radiolabel, can overcome the latter problem. However, this approach suffers from the disadvantage that changes in the rate of metabolism of $I(1,4,5)P_3$ may complicate interpretation of the results. Thus, although the *net* formation of $I(1,4,5)P_3$ may decrease to zero, PLC activity, when monitored as the accumulation of a total inositol phosphate fraction in the presence of Li^+, may remain constant [4]. For these reasons, the application of both approaches is desirable to verify that changes in the rate of receptor-stimulated PPI hydrolysis have indeed occurred. Measurement of DAG production as an index of PLC activity can be even more problematic since lipids other than the phosphoinositides can be the source of DAG (M.J.O. Wakelam, University of Birmingham, UK). For example, in bombesin-stimulated Swiss 3T3 fibroblasts, both the phosphoinositides and phosphatidylcholine serve as precursor lipids for DAG formed upon receptor activation [5]. Similar results have been obtained for human neuroblastoma cells [6]. Thus, the release of inositol phosphates, rather than of DAG, is the most reliable index of receptor-stimulated PPI hydrolysis.

Receptor-stimulated PPI Hydrolysis Rapidly Desensitizes

The occurence of both a reduction of receptor number and functional response in response to chronic agonist occupancy of PPI-linked receptors has been well established. Largely due to the methodological difficulties outlined above, however, the occurence of a rapid attenuation of receptor-stimulated PPI hydrolysis is less well documented. In contrast to previous speculation, it is now evident that receptor-mediated PPI hydrolysis can rapidly desensitize following agonist occupancy of a number of PPI-linked receptors. Thus, in AR4-2J pancreatoma cells, substance P, bombesin, and cholecystokinin receptors all undergo a rapid attenuation of responsiveness, with the onset occurring within 15–30 s of agonist addition [7]. Similarly, in both neuroblastoma cells and in CHO cells transfected with the cDNA for the m_3 muscarinic cholinergic receptor (mAChR) the initial high rate of PPI hydrolysis observed following agonist addition (10–20% of the inositol lipid pool/min), rapidly attenuates. In most systems studied thus far, a number of characteristics are apparent. First, desensitization of stimulated PPI hydrolysis is often only partial and may precede a secondary sustained phase of inositol lipid turnover. Second, the rate of desensitization may be receptor specific for a given cell type, and third, the same receptor may exhibit different rates of desensitization dependent upon the cell type. Although the mechanism underlying this rapid attenuation of inositol lipid hydrolysis is not yet established, certain possibilities can be considered (S.R. Nahorski, University of Leicester, UK). For example, receptor internalization is unlikely to be responsible, due to its relatively slow onset (> 5 min). In addition, because attenuation of responsiveness is evident regardless of whether the net formation of $I(1,4,5)P_3$ or the accumulation of a total inositol phosphate fraction is monitored, an increased rate of metabolism of $I(1,4,5)P_3$ cannot account for the reduction in production of PPI-derived metabolites. One possible explanation for the attenuation of PLC activity is that changes in the phosphorylation status of the receptor alter its ability to couple to the effector enzyme. In this context, a number of PPI-linked receptors are known to undergo phosphorylation in response to agonist addition and in a time-frame similar to that observed for the desensitization of stimulated PPI hydrolysis. However, an issue that remains unresolved is that of the identity of the kinase(s) involved. In the most comprehensively studied example to date, phosphorylation of the m_3 mAChR in CHO cells was shown to be mediated by a kinase that is distinct from that of either protein kinases A or C, Ca^{2+}-calmodulin dependent protein kinase, or a β-adrenergic receptor kinase [8, 9]. Thus, although receptor phosphorylation constitutes a potential means whereby PLC activity can be modulated, a mechanistic link between the two parameters is yet to be established, and the identity of the kinase(s) is still to be determined.

Phospholipase C is Regulated by Guanine Nucleotide Independent and Dependent Mechanisms

Multiple forms of PLC, each regulated by distinct mechanisms are known to exist. To date, three major classes of PLC have been identified, β, γ and δ, with MWs of 130–155, 145 and 85 KDa, respectively. Each of these classes of enzyme is an immunologically distinct entity and the product of separate genes. In addition, a number of isoforms of each subtype of PLC has been isolated, such that nine distinct isozymes are now identified ($\beta_1-\beta_4$, $\delta_1-\delta_2$, $\gamma_1-\gamma_3$). Unexpectedly, only a very limited sequence homology exists between the isozymes, including two regions of 50–70 amino acids (X and Y, ref. [10]) that exhibit 40–50% homology between isozymes but are differentially localized within each enzyme. Also, it should be noted that in PLCγ, there is a region between the X and Y domains that is homologous to those found in non-receptor tyrosine kinases of the src family and of GTPase-activating proteins (SH2 binding domain). PLC may be activated by at least three separate mechanisms. Receptors which possess intrinsic tyrosine kinase activity, such as those for the platelet derived and epidermal growth factors, can activate PLC independently of a G-protein. This is accomplished via the phosphorylation of cytoplasmic domains of the receptor which in turn, via the SH2 binding domain, permits the recruitment of cytoplasmic PLCγ into a membrane locus, with the subsequent phosphorylation and activation of the enzyme. A variation on this theme is evident for platelets in which thrombin addition does not induce tyrosine phosphorylation of PLCγ, but does result in translocation of the cytosolic enzyme to the plasma membrane (E.G. Lapetina, Burroughs Wellcome, North Carolina, USA). In this instance, it appears that thrombin induces the association of a p21^{ras}-GTPase activating protein, rasGAP, with rap1B, a small MW GTP binding protein [11]. Given the constitutive association of the rasGAP with PLCγ, it is proposed that by binding to activated rap1B, rasGAP promotes the translocation of PLCγ to the membrane. Such a model also accounts for the inhibition of enzyme activity observed in the presence of elevated tissue concentrations of cAMP, since upon its phosphorylation by protein kinase A, rap1B translocates to the cytosol and can no longer facilitate translocation of the PLCγ.

For other receptors, such as the histamine, bradykinin or mAChR, PLC activation occurs through an intervening guanine nucleotide binding protein which may be either pertussis toxin insensitive or sensitive. Activation of the former category of receptors involves the mediation of a protein of the G_q class (G_q, G_{11}, G_{14}, G_{15}, or G_{16}), none of which are subject to pertussis toxin modification. The ability of α_q to regulate PLC is isozyme specific (PLCβ_1 > PLCβ_3 $\gg$ PLCβ_2), while α_q does not regulate either PLCγ or PLCδ. Intracellular concentrations of Gα_q are themselves subject to regulation following continuous agonist occupancy of cellular surface receptors by means of a mechanism which involves a transiently enhanced proteolytic degradation of the G protein [12]. The loss of G_q could conceivably result in a heterologous regulation of PLC, given that distinct receptors may share the same G protein pool.

A third mechanism for the activation of PLC is linked to the availability of $\beta\gamma$ subunits derived from G-proteins (T.K. Harden, University of N. Carolina, USA). For receptors that operate via a pertussis toxin sensitive mechanism, there is no evidence that the α subunit derived from either G_i or G_o can mediate PLC activation. This enigma was clarified by the discovery that the addition of $\beta\gamma$ subunits results in an isoform-specific activation of mammalian PLC (i.e., $PLC\beta_3 \gtrsim PLC\beta_2 \gg PLC\beta_1$), or turkey erythrocyte $PLC\beta$, a pattern distinct from that observed for α_q activation of PLC [13]. Little specificity is observed in terms of the ability of $\beta\gamma$ dimers of defined composition to activate turkey erythrocyte $PLC\beta$ (which most closely resembles mammalian $PLC\beta_2$). The possibility that there exists in cells a dual regulation of PLC by G protein subunits raises the issue of relative prevalence of these two modes of PLC activation. In this context, it should be noted (a) that ten-fold lower concentrations of α_q are required for activation of PLC than are needed for $\beta\gamma$ stimulation, and (b) that $PLC\beta_3$, the isoform responsive to both α_q and $\beta\gamma$ subunits is also the most widespread in distribution. A further consideration is that due to the relative abundance of G_s, G_i and G_o (relative to G_q), the former G proteins are more likely to serve as the source of the $\beta\gamma$ subunits.

Regulation of PLC Activity by Intra- and Extracellular Ca^{2+}

In addition to a reliance of all PLC isoforms on nanomolar concentrations of Ca^{2+}, the activity of the enzyme may be further regulated by changes in the concentration of cytosolic Ca^{2+} that accompany receptor activation. Four lines of evidence support this contention. First, an influx of Ca^{2+}, mediated by agents other than receptor agonists, can result in an enhanced PPI hydrolysis. Second, in permeabilized cells, physiologically relevant concentrations of Ca^{2+} can modulate the extent of PLC activation. Third, following either the chelation of extracellular Ca^{2+}, or inclusion of Ni^{2+} (an agent which blocks Ca^{2+} entry), the biphasic generation of $I(1,4,5)P_3$ observed following mAChR activation of SH-SY-5Y neuroblastoma cells becomes monophasic. Fourth, depletion of intracellular Ca^{2+} stores with thapsigargin results in a diminution of agonist-stimulated $I(1,4,5)P_3$ production. Thus, both intra- and extracellular Ca^{2+} may play a role in PLC regulation.

The mechanism whereby activation of PPI-linked receptor results in an enhanced Ca^{2+} influx is yet to be established. According to the capacitative model for Ca^{2+} entry into cells [14], depletion of the $I(1,4,5)P_3$-sensitive intracellular Ca^{2+} pool (by means of either agonist or thapsigargin) provides the signal that facilitates the entry of Ca^{2+} across the plasma membrane. It has been proposed that a small negatively charged molecule termed CIF (Ca^{2+} influx factor) is responsible for this increased permeability of the plasma membrane [15]. The evidence in favor of this possibility was critically reviewed by J.W. Putney (NIEHS, North Carolina, USA). Although many of the original observations have been successfully repeated, problems remain. First, a CIF has been identified in some, but not all cells. Second, administration of CIF to certain cells does not elicit the expected response and third, the issue of whether CIF is a single compound or a mixture of components is not yet resolved. Thus, the search for this intracellular mediator of Ca^{2+} entry continues.

Acknowledgements

The author wishes to thank Ms. Jo Ann Kelsch for preparation of the manuscript and Dr. A.M. Heacock for helpful comments. This work was supported by NIH NS 23831 and NIMH MH 46252.

Abbreviations

CIF:	calcium influx factor
DAG:	diacylglycerol
G-protein:	guanine nucleotide binding protein
$I(1,4,5)P_3$:	inositol 1,4,5,-triphosphate
mAChR:	muscarinic acetylcholine receptor
PIP_2:	phosphatidylinositol 4,5-bisphosphate
PLC:	phosphoinositide-specific phospholipase C
PPI:	Phosphoinositide
rasGAP:	p21^{ras}-GTPase activating protein

References

1. Fisher SK, Heacock AM, Agranoff BW. Inositol lipids and signal transduction in the nervous system: an update. J. Neurochem. 1992; 58: 18–38.
2. Hokin LE, Hokin MR. Enzyme secretion and the incorporation of P^{32} into phospholipides of pancreas slices. J. Biol. Chem. 1953; 203: 967–977.
3. Wojcikiewicz RJH, Tobin AB, Nahorski SR. Desensitization of cell signalling mediated by phosphoinositidase C. Trends in Pharmacol. Sci. 1993; 14: 279–285.
4. Plevin R, Wakelam MJO. Rapid desensitization of vasopressin-stimulated phosphatidylinositol 4,5,-bisphosphate and phosphatidylcholine hydrolysis questions the role of these pathways in sustained diacylglycerol formation in A10 vascular-smooth-muscle cells. Biochem. J. 1992; 285: 759–766.
5. Pettitt TR, Wakelam MJO. Bombesin stimulates distinct time-dependent changes in the sn-1,2-diradylglycerol molecular species profile from Swiss 3T3 fibroblasts and analysed by 3,5-dinitrobenzoyl derivatization and h.p.l.c. separation. Biochem. J. 1993; 289: 487–495.
6. Lee C-H, Fisher SK, Agranoff BW, Hajra AK. Quantitative analysis of molecular species of diacylglycerol and phosphatidate formed upon muscarinic receptor activation of human SK-N-SH neuroblastoma cells. J. Biol. Chem. 1991; 266: 22837–22846.
7. Menniti FS, Takemura H, Oliver KG, Putney JW, Jr. Different modes of regulation for receptors activating phospholipase C in the rat pancreatoma cell line AR4-2J. Mol. Pharmacol. 1991; 40: 727–733.
8. Tobin AB, Nahorski SR. Rapid agonist-mediated phosphorylation of m3-muscarinic receptors revealed by immunoprecipitation. J. Biol. Chem. 1993; 268(13): 9817–9823.
9. Tobin AB, Keys B, Nahorski SR. Phosphorylation of a phosphoinositidase C-linked muscarinic receptor by a novel kinase distinct from β-adrenergic receptor kinase, FEBS Letts. 1993; 335: 353–357.

10. Rhee SG, Suh P-G, Ryu S-H, Lee SY. Studies of inositol phospholipid-specific phospholipase C. Science 1989; 255, 546–550.
11. Torti M, Lapetina EG. Role of rap1B and p21^{ras}GTPase-activating protein in the regulation of phospholipase C-γ1 in human platelets. Proc. Natl. Acad. Sci. USA 1992; 89: 7796–7800.
12. Mitchell FM, Buckley NJ, Milligan G. Enhanced degradation of the phosphoinositidase C-linked guanine-nucleotide-binding protein $G_{q\alpha}/G_{11\alpha}$ following activation of the human M1 muscarinic acetylcholine receptor expressed in CHO cells. Biochem. J. 1993; 293: 495–499.
13. Boyer JL, Waldo GL, Harden TK. $\beta\gamma$-Subunit activation of G-protein-regulated phospholipase C. J. Biol. Chem. 1992; 267: 25451–25456.
14. Putney JW, Jr. Capacitative calcium entry revisited. Cell Calcium 1990; 11: 611–624.
15. Randriamampita C, Tsien RY. Emptying of intracellular Ca^{2+} stores releases a novel small messenger that stimulates Ca^{2+} influx. Nature 1993; 364: 809–814.

Pharmacological Sciences: Perspectives for
Research and Therapy in the Late 1990s
ed. by A.C. Cuello and B. Collier
© 1995 Birkhäuser Verlag Basel/Switzerland

G-Protein-Linked Receptors and Tyrosine Kinase-Mediated Signal Transduction Pathways: A Mid-1990s Perspective, with Working Hypotheses

Morley D. Hollenberg

*Endocrine Research Group, Departments of Pharmacology & Therapeutics and Medicine,
The University of Calgary, Faculty of Medicine, Calgary, AB Canada T2N 4N1, Canada*

Summary. G-protein signalling paradigms are reviewed, and data coming from four laboratories are summarized, indicating that G-protein-coupled agonists may act via tyrosine kinase pathways. Approaches to the analysis of these putative G-protein-regulated tyrosine kinase pathways are outlined and working hypotheses are put forward to explain the link between serpentine receptors such as the ones for angiotensin-II, vasopressin and bombesin and the activation of tyrosine kinases such as P125^{FAK}.

Introduction

G-Protein Signal Paradigms

The landmark studies of Rodbell and co-workers [1] demonstrated the crucial role that guanosine triphosphate (GTP) plays in the regulation of adenylate cyclase by hormones, such as glucagon and epinephrine. This requirement for GTP is now understood in terms of the central function of an enlarging family of heterotrimeric ($\alpha_x\beta_y\gamma_z$) guanine nucleotide-binding G-proteins, which serve as essential signal transduction elements for a vast array of tissue agonists ranging from neuropeptides to odorants and light [2–5]. A central mechanism for the regulation of cell processes by G-proteins comprises the exchange of GDP for GTP on the α-subunit of the heterotrimeric G-protein complex. This GDP-GTP exchange, catalysed by the hormone-occupied receptor [2, 6–8], leads to the dissociation of the heterotrimeric G-protein complex into its α- and $\beta\gamma$-subunits. Separately, the α_{GTP} and $\beta\gamma$ subunits are able to amplify the receptor-triggered signal by interacting with and regulating a number of "effector" moieties, including adenylate cyclase, phospholipase C and a number of ion channels [9, 10]. The ability of the agonist-occupied receptor to act as a GDP-GTP exchange factor for the heterotrimeric G-protein complex has elements in common with the modulation of other so-called "small" G-proteins, such as ras, bacterial elongation factor (EF) tu, or rho, wherein a GDP-GTP exchange reaction is central to the activation of these cellular regulators [11]. The structural requirements in the protein domains on receptors and on their cognate α-subunits, responsible for the interactions that promote the GDP-GTP exchange reactions are beginning to be understood. Whether some receptors are able to interact with and regulate G-proteins other than those of the α-superfamily [9] remains an open question. It is the multiplicity of targets of the activated G-protein oligomers that provide for the diversity of cellular responses regulated by G-protein-coupled receptors.

G-Protein-Mediated Signal Amplification

In large part, G-protein-coupled signal amplification results from phosphorylation-dephosphorylation cascades, modulated by the interactions of the α_{GTP}- and $\beta\gamma$-subunits with their various effector targets. Perhaps the longest understood of these cascades comprises the stimulation of adenylyl cyclase by α_s-GTP, followed by the activation of cyclic AMP-dependent protein kinase (PKA) that catalyses the phosphorylation of serine and threonine residues. More recently clarified are the mechanisms whereby either the α_q-GTP isoforms or free $\beta\gamma$ dimers can activate selected β-isoforms of phospholipase C. The second messengers, diacylglycerol (DAG) and inositol 1,4,5-triphosphate (IP$_3$) released by phospholipase C action, coordinately initiate phosphorylation cascades either via the DAG/Ca^{2+}-stimulation of kinase C (PKC) or, as a result of an IP$_3$-mediated elevation of intracellular calcium with a consequent stimulation of calcium-calmodulin-regulated protein kinases [12]. As with PKA, serine and threonine residues have been found to be targets for the PKC isoforms and the calcium-calmodulin-regulated protein kinases. Since the mid-1970s, it has thus become generally accepted that the diverse actions of a plethora of G-protein-coupled agonists can be rationalized in terms of phosphorylation-dephosphorylation cascades involving serine and threonine-containing motifs in a variety of regulatory proteins. The question that served as a catalyst for the symposium outlined in this report was: might tyrosine phosphorylation cascades also be involved in the response of tissues to G-protein-coupled agonists?

Tyrosine Phosphorylation, G-Protein-Coupled Agonists and Cross-Talk with other Receptor Systems

In view of the above synopsis, it came perhaps as a surprise that, in addition to stimulating serine/threonine phosphorylation, a number of G-protein-coupled agonists, including vasopressin, angiotensin-II, bradykinin, bombesin and endothelin have also been observed to stimulate tyrosine phosphorylation of multiple substrates in a variety of cultured cell systems, including murine Swiss 3T3 fibroblasts, rat liver-derived WB epithelial cells and rat aorta-derived smooth muscle cells [13–18]. The intense interest in tyrosine phosphorylation reactions as key elements of signal transduction pathways was kindled in the 1970s as a result of the discovery that the transforming gene of the tumorigenic Rous sarcoma virus (p60[v-src]) encodes a protein kinase that phosphorylates tyrosine residues in a wide variety of substrates [19, 20]. It soon became apparent that tyrosine kinase pathways were essential not only for the effects of tumorigenic viruses but also for the actions of agents such as epidermal growth factor-urogastrone (EGF-URO) and platelet-derived growth factor (PDGF) that act via receptors possessing intrinsic ligand-regulated tyrosine kinase activity [21, 22]. Thus, tyrosine phosphorylation reactions have become the *sine qua non* for the actions of many so-called growth factors, including not only those such as EGF-URO and PDGF that act via tyrosine kinase receptors, but also those cytokines, such as interferon-α/γ, interleukin-4 and prolactin, which act via kinase-deficient receptor subunits, capable nonetheless of recruiting the tyrosine kinases JAK-1 and -2 into the signal process [23–28]. It is important to note that the G-protein-coupled agonists listed above in connection with their ability to augment cellular tyrosine phosphorylation are also recognized as "growth factors" that stimulate the multiplication and/or hypertrophy of cultured cells. More widely, these G-protein-coupled agonists are also known for their ability to stimulate rapid tissue responses such as the contraction of smooth muscle or the synthesis of adrenal steroids.

Just as G-protein-coupled agonists can cause both immediate (e.g. contraction) and delayed (e.g. cell growth) responses in their target tissues, so can (it is now recognized) growth factors such as EGF-URO and PDGF cause rapid tissue responses (muscle contraction, ion transport) as well as their more widely recognized effects on gene transcription and cell replication [29, 30]. Further, it is widely appreciated that a hormone such as insulin, acting via a tyrosine kinase receptor, can stimulate a series of serine/threonine protein kinase cascades involved in the regulation of glucose metabolism [31]. Thus, it has become clear that there is an intricate cross-talk, at the level of protein tyrosine kinase and protein serine/threonine kinase pathways (as well as the appropriate targeted phosphatases), between G-protein-coupled agonists and those agonists such as EGF-URO and PDGF that act *via* receptors with intrinsic tyrosine kinase activity.

Symposium Presentations

It was with the above cross-talk in mind that it was decided to draw together, in the context of the 1994 IUPHAR symposium on G-protein-linked receptors and tyrosine kinase signal transduction pathways [32], the research of four groups focused on the regulation of cell function by tyrosine kinase pathways. The goal was to present data relevant not only to the behaviour of cultured cells [33, 34] but also to the behaviour of either acutely isolated cell systems [35] or intact tissues [36]. In each of the four types of systems discussed (cultured murine Swiss 3T3 fibroblasts, cultured rat WB hepatic epithelial cells, acutely dissociated bovine adrenal glomerulosa cells, or isolated guinea pig or rat gastric smooth muscle), three main approaches have been used: (1) substrate identification, (2) kinase identification and (3) use of tyrosine kinase inhibitors.

Phosphotyrosyl Proteins

A number of attempts has been made to identify proteins that become rapidly phosphorylated on tyrosine in step with agonist stimulation [21]; western blot analysis and immunoprecipitate employing antiphosphotyrosine monoclonals have been of considerable value in this regard. A variety of proteins ranging in molecular weight from about 35 to 200 kDa have been observed to be reproducibly phosphorylated within minutes of agonist stimulation. In particular, the focal adhesion-associated proteins, paxillin and the tyrosine kinase, p125[FAK], have been positively identified as two of the 3T3 fibroblast proteins that become rapidly phosphorylated on tyrosine after the addition of G-protein-coupled agonists to Swiss 3T3 cells [37–39]. In other cell systems, such as rat WB epithelial cells, it would appear that a set of proteins distinct from those affected in 3T3 cells become tyrosine phosphorylated in response to G-protein-coupled agonist such as angiotensin-II [34]. In the WB cells, p125[FAK] appears to represent a minor substrate; the other WB cell phosphoproteins have yet to be identified. What is clear from the work described by Huckle, Rozengurt and their colleagues [33, 34] is that in each cell system, a distinct set of tyrosine phosphorylated substrates may be found to play a functional role.

Agonist-Activated Tyrosine Kinases

Apart from the observation that p125[FAK] is one of the tyrosine kinases that can become tyrosine phosphorylated in the course of bombesin or vasopressin action, the identity of the enzyme or enzymes responsible for substrate phosphorylation in response to the stimulation of 3T3 or WB cells with G-protein-coupled agonists remains an open question. Of note are the observations of Earp and colleagues, who described tyrosine kinase activity that could be recovered from antiphosphotyrosine immunoprecipitates prepared using angiotensin-II-treated WB cells. Further, Earp and colleagues mentioned that over 20 distinct tyrosine kinase enzymes can be detected in the WB cells. The identification of the signal-associated tyrosine kinase that becomes activated by G-protein-coupled agonists would thus appear to represent a challenging task.

Tyrosine Kinase Inhibitors and Analysis of Cell Signalling

The use of tyrosine kinase inhibitors has figured prominently in analysing the signal transduction pathways activated by G-protein-coupled agonists in the different test systems discussed in the course of the symposium. For instance, in the 3T3 cell system, tyrphostin-23 was able to attenuate bombesin-stimulated tyrosine phosphorylation [40]. In another context, De Léan [41] reported on work from his laboratory, indicating that both genistein and tyrphostin-23 were able to abrogate angiotensin-II-stimulated glomerulosa cell steroidogenesis. Genistein and tyrphostin-47 were also observed by Hollenberg and colleagues to block the contractile actions of angiotensin-II in a gastric smooth muscle bioassay system [42]. Whereas the data obtained with the tyrosine kinase inhibitors must be interpreted with caution [43], these agents, exhibiting considerable specificity for enzymes of the tyrosine kinase family [44] point to an important role for non-receptor tyrosine kinases in G-protein-stimulated processes ranging from the contraction of smooth muscle [30] to the above-mentioned regulation of adrenal steroidogenesis.

Working Hypotheses for Tyrosine Kinase Activation by G-Protein-Coupled Agonists

In view of the activation of phospholipase C by G-protein-coupled processes, as discussed above, an important question to ask is: how might the released second messengers, DAG and IP_3 lead to the stimulation of tyrosine kinase activity? One possible mechanism clearly stems from the activation of protein kinase C. Nonetheless, even though the activation of kinase C by the combined action of DAG and IP_3-augmented intracellular Ca^{2+} can lead to increased cellular tyrosine phosphorylation, it is also evident that G-protein-coupled tyrosine phosphorylation (e.g. stimulated by bombesin or vasopressin) can also occur via a kinase C-independent mechanism [38]. As an alternative, the work reported by Earp and colleagues focused on elevated intracellular Ca^{2+} as a stimulus for increased tyrosine phosphorylation. What became evident in comparing the data coming from the Rozengurt group working with cultured murine Swiss 3T3 cells and from the Earp group working with cultured rat WB liver cells, is that the role of elevated intracellular calcium in stimulating protein tyrosine phosphorylation may vary dramatically from one cell type to another. For instance, elevated intracellular Ca^{2+} does not appear to play a role in agonist stimulation of p125^{FAK} phosphorylation in Swiss 3T3 cells [38], whereas elevated intracellular calcium appears to be a key factor in stimulation WB cell substrate tyrosine phosphorylation [13, 34]; further, in the WB cell, in contrast with other substrates, p125^{FAK} does not appear to be phosphorylated in response to elevated cellular calcium. The nature of the Ca^{2+}-stimulated tyrosine kinase in WB cells remains to be determined.

An interesting aspect of G-protein-coupled agonist stimulation of p125^{FAK} phosphorylation in Swiss 3T3 cells relates to the role of cytoskeletal elements [38]. Cytochalasin-D has been found to abrogate bombesin-induced p125^{FAK} phosphorylation without affecting other rapid cellular responses such as Ca^{2+} mobilization and PKC activation [38]. In a similar vein, Rozengurt reported on recently acquired data indicating that botulinum toxin C_3, which can interfere with rho p21, so as to disrupt the actin filament network [45], can also attenuate bombesin-induced p125^{FAK} phosphorylation. The role of the actin cytoskeleton in the ability of G-protein-coupled agonists to stimulate tyrosine phosphorylation would thus appear to be an area well worth exploring.

Given the general paradigm of G-protein-linked signalling outlined above, it is possible to set forth a number of working hypotheses that may account for the stimulation of substrate tyrosine phosphorylation by G-protein-coupled agonists. First, tyrosine phosphorylation may result in many instances from the activation of kinase C. The kinase C activation would be due to the elevations in cellular DAG and Ca^{2+} resulting from $\alpha_q/\beta\gamma$-mediated activation of phospholipase C [9, 46, 47]. Once activated, kinase C may regulate cellular tyrosine kinases either directly *via* ser/thr phosphorylation of regulatory kinase domains, or indirectly, *via* the activation of cellular tyrosine phosphatases. In other instances, one can hypothesize that a receptor G-protein interaction, via released subunits, might directly regulate ion-channel-mediated Ca^{2+} influx [10], so as to influence a Ca^{2+}-regulated tyrosine kinase. Alternatively, one might hypothesize that released $\beta\gamma$ subunits on their own might regulate tyrosine kinase targets involved in the maintenance of actin cytoskeletal structures. As another possibility one could postulate that the receptor-activated G-protein α-subunits (α_i or α_q) might be able to interact directly with as yet unidentified non-receptor tyrosine kinases, or tyrosine phosphatases, so as to modulate their function and thereby trigger increased cellular tyrosine phosphorylation. A final working hypothesis that can be put forward relates to the potential ability of serpentine receptor intracellular domains to interact with and directly regulate membrane effectors other than G-proteins. In keeping with "mobile" or "floating" receptor hypothesis put forward some time ago (summarized in [48]), there is no limit on the types of membrane effectors that could be modulated by an agonist-occupied receptor. Thus, a final possibility to be considered is that the receptors for those agonists such as bombesin, angiotensin-II, vasopressin and thrombin, which can stimulate cellular tyrosine phosphorylation, may act via the physical recruitment of cytoplasmic tyrosine kinases akin to Jak 1 or Jak 2. Such "recruited" tyrosine kinases have been found to play key roles in signal transduction mediated by cytokine receptors [27, 28]. Overall, it would thus be prudent to anticipate that mechanisms involving both activated G-protein subunits and novel serpentine receptor-effector interactions might account for the stimulation of protein tyrosine phosphorylation by agonists such as bombesin, vasopressin, endothelin, thrombin and angiotensin-II.

Acknowledgements

Work in the author's laboratory has been made possible by grants from the Canadian Medical Research Council and the Heart and Stroke Foundation of Alberta.

References

1. Rodbell M. The role of hormone receptors and GTP-regulatory proteins in membrane transduction. Nature 1980; 284: 17–22.
2. Hepler JR, Gilman AG. "G-proteins". Trends in Biol. Sci. 1992; 17: 383–387.
3. DeVivo M, Iengar R. G-protein pathways: signal processing by effectors. Mol. Cell. Endocrinol. 1994; 100: 65–70.
4. Milligan G. Mechanisms of multifunctional signalling by G protein-linked receptors. Trends in Pharmacol. Sci. 1993; 14: 239–244.
5. Savarese TM, Fraser CM. *In vitro* mutagenesis and the search for structure-function relationships among G protein-coupled receptors. Biochem. J. 1992; 283: 1–19.
6. Levitzki A, Bar-Sinai A. The regulation of adenylyl cyclase by receptor-operated G proteins. Pharmacol. Ther. 1991; 50: 271–283.
7. Levitzki A, Marbach I, Bar-Sinai A. The signal transduction between β-receptors and adenylyl cyclase. Life Sciences 1993; 52: 2093–2100.
8. Birnbaumer L. G proteins in signal transduction. Annu. Rev. Pharmacol. Toxicol. 1990; 30: 675–705.
9. Simon MI, Strathmann MP, Gautam N. Diversity of G proteins in signal transduction. Science 1991; 252: 802–808.
10. Clapham DE, Neer EJ. New roles for G-protein $\beta\gamma$-dimers in transmembrane signalling. Nature 1993; 365: 403–406.
11. Bourne HR, Sanders DA, McCormick F. The GTPase superfamily: conserved structure and molecular mechanism. Nature 1991; 349: 117–126.
12. Berridge M. Inositol trisphosphate and calcium signalling. Nature 1993; 361: 315–325.
13. Huckle WR, Prokop CA, Dy RC, Herman B, Earp S. Angiotensin II stimulates protein-tyrosine phosphorylation in a calcium-dependent manner. Molec. Cell. Biol. 1990; 10: 6290–6298.
14. Force T, Kyriakis JA, Bonventre JV. Endothelin, vasopressin, and angiotensin II enhance tyrosine phosphorylation by protein kinase C-dependent and -independent pathways in glomerular mesangial cells. J. Biol. Chem. 1991; 266: 6650–6656.

15. Tsuda T, Kawahara T, Shii K, Koide M, Ishida Y, Yokoyama M. Vasoconstrictor-induced protein-tyrosine phosphorylation in cultured vascular smooth muscle cells. FEBS Lett. 1991; 285: 44–48.
16. Zachary I, Gil J, Lehmann W, Sinnett-Smith J, Rozengurt E. Bombesin, vaspressin, and endothelin rapidly stimulate tyrosine phosphorylation in intact Swiss 3T3 cells. Proc. Natl. Acad. Sci. USA 1991; 88: 4577–4581.
17. Leeb-Lundberg LMF, Song X-H. Bradykinin and bombesin rapidly stimulate tyrosine phosphorylation of a 120-kDa group of proteins in Swiss 3T3 cells. J. Biol. Chem. 1991; 266: 7746–7749.
18. Molloy CJ, Taylor DS, Weber H. Angiotensin II stimulation of rapid protein tyrosine phosphorylation and protein kinase activation in rat aortic smooth muscle cells. J. Biol. Chem. 1993; 268: 7338–7345.
19. Collett MS, Erikson RL. Protein kinase activity associated with the avian sarcoma virus *src* gene product. Proc. Natl. Acad. Sci. USA 1978; 75: 2021–2024.
20. Hunter T, Sefton BM. Transforming gene product of Rous sarcoma virus phosphorylates tyrosine. Proc. Natl. Acad. Sci. USA 1980; 77: 1311–1315.
21. Glenney JR, Jr. Tyrosine-phosphorylated proteins: mediators of signal transduction from the tyrosine kinases. Biochim. Biophys. Acta 1992; 1134: 113–127.
22. Schlessinger J, Ullrich A. Growth factor signalling by receptor tyrosine kinases. Neuron 1992; 9: 383–391.
23. Velazquez L, Fellous M, Stark GR, Pellegrini S. A protein tyrosine kinase in the interferon α/β signalling pathway. Cell 1992; 70: 313–322.
24. Silvennolnen O, Ihle JN, Schlessinger J, Levy DE. Interferon-induced nuclear signalling by Jak protein tyrosine kinases. Nature 1993; 366: 583–585.
25. Shual K, Zlemlecki A, Wilks AF, Harpur AG, Sadowski HB, Gilman MZ, et al. Polypeptide signalling to the nucleus through tyrosine phosphorylation of Jak and Stat proteins. Nature 1993; 366: 580–583.
26. Wang L-M, Keegan AD, Li W, Lienhard GE, Pacini S, Gutkind JS, et al. Common elements in interleukin 4 and insulin signalling pathways in factor-dependent hematopoietic cells. Proc. Natl. Acad. Sci. USA 1993; 90: 4032–4036.
27. Miyajima A, Kitamura T, Harada N, Yokota T, Arai K-I. Cytokine receptors and signal transduction. Ann. Rev. Immunol. 1992; 10: 295–331.
28. Kishimoto T, Taga T, Akira S. Cytokine signal transduction. Cell 1994; 76: 253–262.
29. Hollenberg MD. The acute actions of growth factors in smooth muscle systems. Life Sci. 1994; 54: 223–235.
30. Hollenberg MD. Tyrosine kinase pathways and the regulation of smooth muscle contractility. Trends in Pharmacol. Sci. 1994; 15: 108–114.
31. Cohen P. Dissection of the protein phosphorylation cascades involved in insulin and growth factor action. Biochem. Soc. Trans. 1993; 214: 555–567.
32. Rozengurt E, Earp HS, Hollenberg MD, De Léan A. G-protein-linked receptors and tyrosine kinase signal transduction pathways. Can. J. Physiol. Pharmacol. 1994; 72, Suppl. 1, 39.
33. Rozengurt E, Sinnett-Smith J, Zachary I, Seckl M, Rankin S. Agonist stimulation of tyrosine phosphorylation in cultured cell systems. Can. J. Physiol. Pharmacol. 1994; 72, Suppl. 1, 39.
34. Earp HS, Huckle WR. Intracellular calcium and the regulation of agonist-stimulated tyrosine activity. Can. J. Physiol. Pharmacol. 1994; 72, Suppl. 1, 39.
35. De Léan A. Involvement of protein tyrosine kinase in secretagogue-induced steroid production. Can. J. Physiol. Pharmacol. 1994; 72, Suppl. 1, 39.
36. Hollenberg MD. Tyrosine kinase pathways and the actions of agonists in gastric and vascular smooth muscle systems. Can. J. Physiol. Pharmacol. 1994; 72, Suppl. 1, 39.
37. Zachary I, Rozengurt E. Focal Adhesion kinase (p125[FAK]): a point of convergence in the action of neuropeptides, integrins, and oncogenes. Cell 1992; 71: 891–894.
38. Sinnett-Smith J, Zachary I, Valverde AM, Rozengurt E. Bombesin stimulation of p125 focal adhesion kinase tyrosine phosphorylation. J. Biol. Chem. 1993; 268: 14261–14268.
39. Zachary I, Sinnett-Smith J, Turner CE, Rozengurt E. Bombesin, vasopressin, and endothelin rapidly stimulate tyrosine phosphorylation of the focal adhesion-associated protein paxillin in Swiss 3T3 cells. J. Biol. Chem. 1993; 268: 22060–22065.

40. Seckl M, Rozengurt E. Tyrphostin inhibits bombesin stimulation of tyrosine phosphorylation, c-*fos* expression, and DNA synthesis in Swiss 3T3 cells. J. Biol. Chem. 1993; 268: 9548–9 54.
41. Bodart V, Ong H, De Léan A. A role for protein tyrosine kinases in the steroidogenic pathway of angiotensin II in bovine zona glomerulosa cells. Can. J. Physiol. Pharmacol. 1994; 72, Suppl. 1, 555.
42. Yang S-G, Saifeddine M, Laniyonu AA, Hollenberg MD. Distinct signal transduction pathways for angiotensin-II in guinea pig gastric smooth muscle: Differential blockade by indomethacin and tyrosine kinase inhibitors. J. Pharmacol. Exp. Ther. 1993; 264: 958–66.
43. Wolbring G, Hollenberg MD, Schnetkamp PPM. Inhibition of GTP-utilizing enzymes by tyrphostins. J. Biol. Chem. 1994; 269: 22470–22472.
44. Levitzki A. Tyrphostins: tyrosine kinase blockers as novel antiproliferative agents and dissectors of signal transduction. FASEB J. 1992; 3275–3282.
45. Miura T, Kikuchi A, Musha T, Kuroda S, Yaku H, Sasaki T, et al. Regulation of morphology by *rho* p21 and its inhibitory GDP/GTP exchange protein (*rho* GDI) in Swiss 3T3 cells. J. Biol. Chem. 1993; 268: 510–505.
46. Exton JH. Phosphatidylcholine breakdown and signal transduction. Biochim. Biophys. Acta 1994; 1212: 26–42.
47. Sternweis PC. The active role of beta gamma in signal transduction. Cell Biol. 1994; 6: 198–203.
48. Cuatrescasas P, Hollenberg MD. Membrane receptors and hormone action. Adv. Protein Chem. 1976; 30: 251–451.

Pharmacological Sciences: Perspectives for
Research and Therapy in the Late 1990s
ed. by A.C. Cuello and B. Collier

Receptor-G Protein-Effector Coupling: Coding and Regulation of the Signal Transduction Process

Mark M. Rasenick[1], Marc G. Caron[2], Annette C. Dolphin[3], Brian K. Kobilka[4] and Gunter Schultz[5]

[1]*Department of Physiology & Biophysics, University of Illinois College of Medicine, Chicago, IL, USA;* [2]*Department of Physiology and Howard Hughes Medical Institute, Duke University Medical School, Durham, NC, USA;* [3]*Department of Pharmacology, Royal Free Hospital Medical School, London, UK;* [4]*Division of Cardiovascular Medicine and Howard Hughes Medical Institute, Stanford University Medical School, Stanford, CA, USA;* [5]*Department of Pharmacology, Free University of Berlin, Berlin, Germany*

Summary. Regulation of G protein mediated signal transduction is thought to occur primarily as the result of the occupancy of a receptor by the appropriate agonist. There is much regulation of this process, however, which occurs independently of straightforward agonist occupancy. It is demonstrated here that receptors themselves might have intrinsic activity and the expression of such intrinsically activated receptors might alter the activated state of a G protein and its effector within a given cell. The regulation of receptor display on the cell surface is also discussed and it is seen that the ability to sequester, internalize and restore receptors to the plasma membrane is quite different for different types of G protein coupled receptors. The suggestion that a given receptor uses specific G protein α, β, and γ subunits in order to couple receptor to effector is also explored. Such coding would allow a cell to shape its response to a given agonist. It further proposed that elements of the cytoskeleton (tubulin) form complexes with Gs, Gil or Gq and activate those proteins via the direct transfer of GTP. In this way, changes in cell shape could be conveyed into the intracellular mileu. Further, these cytoskeletal elements might act as the contact point for interplay among various signal transduction systems within the cell. Finally, it is observed that voltage-dependent calcium channels undergo a complex regulation in which G protein β subunits compete with one of the polypeptides making up the channel. Such a process allows for the neurotransmitter modulation of neurotransmitter release. Such modulation returns full circle back to the receptor-mediated activation of G proteins and the possibility that this process is modulated in the cellular interior.

Receptors for a variety of biogenic amines, acetylcholine, amino acid neurotransmitters, peptide hormones and purines are thought to contain seven membrane-spanning domains. These heptaspan receptors couple to G proteins so that the effect of a given agonist can be communicated to the cell interior (Fig. 1). The multitude of identified and cloned receptors is sufficiently large and ongoing that any attempt to list them is fruitless, any such list being immediately incomplete.

G-protein-mediated signaling systems include stimulation (G_s) and inhibition (G_i) of adenylyl cyclase, the gating of K^+ (G_i) and Ca^{2+} (Go)

Correspondence to: Mark M. Rasenick at: Department of Physiology & Biophysics, 901 S. Wolcott, Chicago, IL 60612-7342, USA.

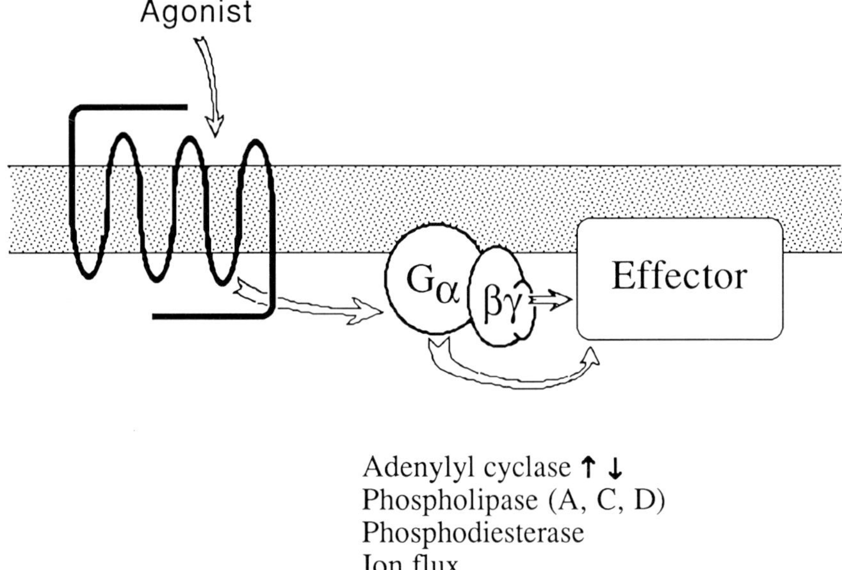

Fig. 1. Coupling between heptaspan receptors, G proteins and effector molecules. Sequential information flow from receptor to G protein to effector is depicted. Activation of effectors by G protein α and βγ subunits is shown. Some G protein-regulated biological processes are listed.

channels, the activation of phosphoinositide phospholipase c (G_q/G_{11}) and Na/H^+ exchange (G_{13}) [1, 2]. G proteins (herinafter known as G) are heterotrimeric in structure, and consist of α, β, and γ subunits, α separating from βγ (in solution) subsequent to its binding of GTP. Molecular weights for the α subunits range from 39–52 KDa, the β's are about 36 KDa, and the γ subunits range from 8–14 KDa. Although, the regulation of effectors by G proteins is generally mediated by the activated α subunit, recent evidence suggests that βγ may also serve some regulatory function [3–5].

While the activation of receptor by agonist and the subsequent activation of G protein by receptor and effector by G protein (whether α or βγ subunit) has been reviewed extensively, it is often depicted as a linear sequence of event involving agonist binding, G protein activation, and subsequent effector engagement. The actual scenario of events is likely to be far more complex. In this chapter some of the factors which help to shape the response to a hormone or neurotransmitter subsequent to the binding of that molecule to its assigned receptor are examined. This work can do no more than offer a few examples which provide an initial sketch of the elaborate canvas upon which the events of cellular signal transduction are portrayed.

Agonist Independent Activation of Dopamine D1B Receptors

While precise changes occurring in heptaspan receptors subsequent to agonist activation have yet to be elucidated, it has become clear that alterations of certain amino acids of the carboxy end of the third cytolasmic loop are sufficient to yield a constitutive activation of α_1 adrenergic receptors and β adrenergic receptors [6, 7]. Such receptors promote activation of Gq or Gs in the absence of agonist. More recently, it has been observed that certain adrenergic receptors may display such a constitutive activation in their native conformation [8].

It had been observed that D1B receptors showed a higher affinity for dopamine than D1A receptors and that cells which normally express high levels of D1B receptors register elevated activity of adenylyl cyclase over similar cells expressing D1A [9]. Expression of D1A and D1B receptors in human embryonic kidney cells showed that basal levels of cAMP accumulation were two to three times greater when D1B receptors were expressed. This phenomenon was seen with the expression of either rat or human receptor types. Curiously, cAMP accumulation in the presence of dopamine ($10 \, \mu$M) was about 1.5 times greater in the cells expressing D1A receptors. Thus, dopamine caused a 10–15 fold incease in cAMP accumulation in cells expressing D1A but only a two to five fold increase in cells expressing D1B [10]. Despite the greater stimulation of cAMP accumulation by $1 \, \mu$M dopamine, most D1 agonists were about 10 fold more potent at D1B receptors than D1A receptors in the transfected cells (Table 1).

Thus, it seems possible that changes in expression of closely related receptor subtypes might alter the "basal" level of cAMP within a cell. Such an alteration could alter the strength or frequency of firing of rhythmic neurons. Further, changes in the intrinsic level of cAMP might

Table 1. Comparison of human DM1A and DM1B receptors expressed in HEK 293 cells: Agonist affinity and adenylyl cyclase activation

	D1A	D1B
K_D Dopamine nM (n = 5)	4900 ± 282	1080 ± 36
Intracellular cAMP (CA/TU)	0.0045	0.12
Intracellular cAMP + dopamine ($10 \, \mu$M)	0.084	0.029

Human dopamine D1A and D1B receptors were expressed transiently using $5 \, \mu$g of DNA/2.5×10^2 cells. Expression levels in transfected 293 cells were about 1700 fmol/mg of membrane protein for both human receptors. Competition binding curves were performed and results are expressed as geometric means ± S.E. for five independent experiments. All curves displayed a slope factor near unity and were fitted best to a one-site model using LIGAND. K_D indicates equilibrium dissociation binding constants. cAMP levels were determined under basal conditions (no agonist) and under maximal stimulation using $10 \, \mu$M dopamine (DA). The results shown are the mean ± S.E. of triplicate determinations of single wells from a six-well dish. The data are a representative example of experiments repeated at least five times. CA/TU = [^{3}H]cAMP formed divided by the total ^{3}H adenine uptake. Data from Tiberi and Caron (10).

allow for a more elaborate modulation of neuronal function by the variety of signals linked to the inhibition of adenylyl cyclase.

G Protein-Coupled Receptor Trafficking

The appearance of receptors on the cell surface is regulated not only by the synthesis of those receptors, but also by shuttling between cell surface and cell interior, often in response to agonist. Various heptaspan receptors have been found to exhibit a variety of different intracellular targeting and trafficking behaviors [11–13]. These behaviors include the subcellular location of the receptor at steady-state and agonist mediated changes in the distribution of receptors. The three subtypes of alpha 2 adrenergic receptors are highly homologous (50–60% identity between the different subtypes) and have similar pharmacological and G protein-coupling properties, yet they have distinct trafficking properties [13]. The α 2b receptor behaves like the β 2 adrenergic receptor in that it resides in the plasma membrane, but is rapidly and reversibly internalized to endosomes following agonist activation. The α 2a receptor also resides on the plasma membrane, but is not internalized with agonist activation. The α 2c subtype is found predominantly in intracellular vesicles, even in cells which have not been exposed to agonists [13]. These differences in intracellular trafficking may reflect special distribution of receptors in the plasma membrane of differentiated cells *in vivo* such as at a synapse between neurons or between sympathetic nerves and smooth muscle cells. Receptor targeting to specific plasma membrane microdomains may also contribute to the specificity of receptor, G protein and effector interactions [14].

The process of agonist mediated internalization (also called sequestration) of the beta 2 receptor was examined using immunocytochemical methods [13, 15, 16]. These studies demonstrate that, in the presence of agonists, beta 2 receptors employ the same endosomal sorting system used by transferrin receptors. Furthermore, the process of sequestration is a dynamic one, with beta 2 receptors continuously cycling between the plasma membrane and endosomes in the presence of agonist [15, 16]. Sequestration can be resolved into two steps: an agonist-dependent redistribution of the receptor within the plasma membrane to coated invaginations, followed by an agonist independent internalization of the receptor [16]. The agonist-dependent step can occur at reduced temperature (16°C), while the agonist independent internalization from the coated invagination is blocked at 16°C.

A possible role for receptor trafficking in receptor function was recently demonstrated for the thrombin receptor. Thrombin plays a major role in the regulation of hemostasis through activation of fibrinogen. In addition, thrombin elicits a complex pattern of biological responses in endothelial cells, smooth muscle cells, platelets, and mononuclear cells by activating a receptor. The thrombin receptor is a G protein-coupled receptor but is activated by a unique mechanism. Thrombin cleaves the receptor's amino terminal exodomain to unmask a new amino terminus [17]. This new amino terminus then serves as a tethered peptide ligand, binding to sites within the body of the receptor to effect receptor activation. This irreversible proteolytic activation mechanism stands in contrast to the reversible liganding mechanisms utilized by adrenergic receptors, and leads to unique problems not associated with other G protein-coupled receptors: inactivation of receptors and recovery of responsiveness following cleavage of cell surface thrombin receptors by thrombin.

When the cellular trafficking of thrombin receptors is compared to that of beta 2-adrenergic receptors expressed in the same cell line, two differences are observed (Fig. 2): [1] In contrast to the reversible internalization of the beta 2 adrenergic receptor after agonist activation, most thrombin receptors are internalized and targeted to lysosomes after activation. [2] Almost all cellular beta receptor visualized by immunofluorescence is located on the plasma membrane. In contrast, approximately half of a cell's thrombin receptors are found in an intracellular membrane compartment. Upon activation and internalization of cell surface thrombin receptors, the intracellular pool of thrombin receptors is translocated to the plasma membrane (Fig. 2).

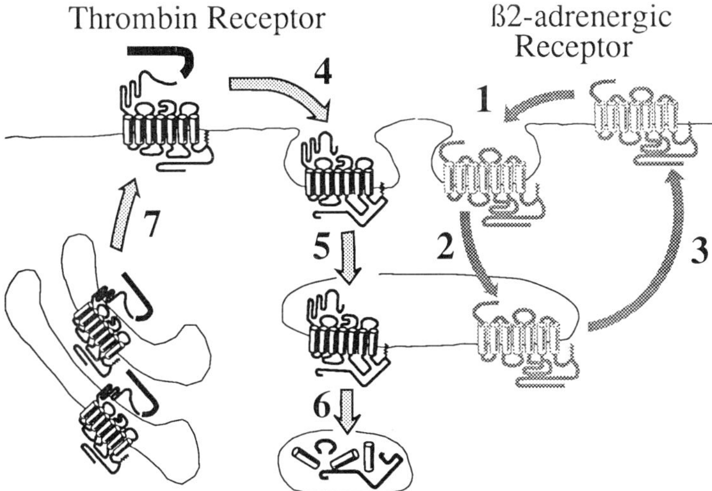

Fig. 2. Schematic representation of thrombin and β2 adrenergic receptor trafficking. Thrombin receptors and β2-adrenergic receptors undergo agonist-dependent endocytosis and can be co-localized in the same endosomes. At this point intracellular trafficking pathways diverge, with the β2 receptor recycling to the plasma membrane and the activated thrombin receptors being delivered to lysosomes. Following thrombin stimulation, a pool of naive thrombin receptors in an intracellular compartment is translocated to the cell surface.

The replenishment of plasma membrane thrombin receptors is associated with a recovery of responsiveness to thrombin [11].

Intracellular Regulation of G Protein Activation by Elements of the Cytoskeleton

Regulation of neurotransmitter responsiveness can occur at the level of receptor sorting, but it can also occur directly at the level of the G protein. It has been observed that certain elements of the cytoskeleton (specifically tubulin) are able to form complexes with certain G proteins subsequently activating those same proteins.

Many of the studies which have shaped understanding of G protein-mediated signal transduction have been done with purified components in reconstituted systems. While such studies have been of enormous benefit in understanding possible interactions between serpentine receptors, G proteins, and effector molecules such as Phospholipase C or adenylyl cyclase, clearly, there is a relationship between the ordered environment of the synaptic membrane and G protein mediated signal transduction systems. In particular, it appears that elements of the cytoskeleton alter the coupling among receptors and G proteins involved in the stimulation or inhibition of adenylyl cyclase. Given the increasing number of processes attributed to G proteins, some mechanism which channels individual receptors, G proteins, and effectors is likely to exist. This might be particularly true in the nervous system, where rapid and discrete response is a hallmark of synaptic transmission.

Recently, it has been observed that α subunits of G proteins may exist in complexes with synaptic membrane tubulin and undergo a directed transfer of nucleotide from the latter [18]. This appears to be a highly specific process, as tubulin has been shown to bind, with high affinity ($Kd \approx 130$ nM), to only three G proteins, αs and αi1 and αq. Even though several other G proteins (αi2, αi3, αo, and transducin (αr) are quite closely related to αi1, their affinity for tubulin is much lower [19]. It appears that distinct domains on tubulin are responsible for binding to G protein and for regulating the process of nucleotide transfer. There is also distinct commonalty between the receptor-G protein interface and the tubulin-G protein interface. Tubulin is capable of transferring GTP to a recombinant $G\alpha$ under conditions where that $G\alpha$ is incapable of binding nucleotide from the medium [20]. In fact, tubulin is capable of bypassing a "tightly coupled" receptor to activate a G protein [21] (Fig. 3).

It is suggested that tubulin, perhaps in response to a neurotransmitter not normally coupled to a G protein, can be engaged to activate Gs or Gil and their attendant intracellular effectors. This could be especially

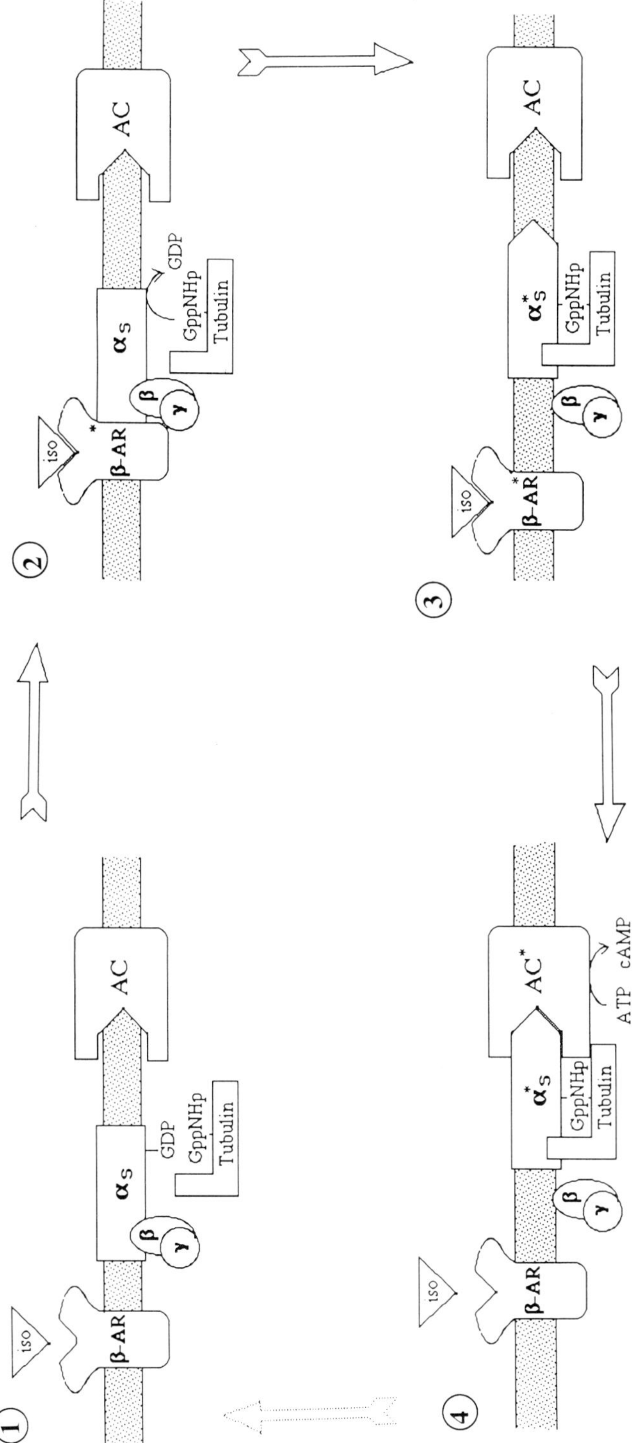

Fig. 3. Tubulin activation of Gsα in the absence of receptor agonist. In this schematic, tubulin with GTP in both exchangeable and non-exchangeable binding sites forms a complex with Gα and activates that protein by "sharing" its GTP. Tubulin is able to activate G proteins under conditions where those proteins normally are inactive. In the membrane, where an agonist-occupied receptor is normally required for activation of Gα, tubulin bypasses the receptor.

important with respect to the possibility that adenylyl cyclase can be activated (depending on subtype) by $\beta\gamma$ subunits, liberated, perhaps from a Gα coupled to a receptor not normally associated with the stimulation of adenylyl cyclase [22]. Such stimulation cannot occur without the prior activation of Gsα. The mechanism proposed here would allow tubulin, liberated, perhaps from the bonds of MAP2 on the post-synaptic membrane in response to an increase in intracellular Ca^{2+}, to activate Gs, Gi1 or Gq in a receptor-independent fashion. Thus, tubulin may provide a conduit for the interactions among neurotransmitters.

Specific complexes of Gα and $\beta\gamma$ Dictate Receptor and Effector Specificity

Studies of the hormonal control of voltage-dependent calcium channels by the whole-cell patch clamp technique allow the exact identification of the G proteins involved in stimulation and inhibition of endocrine and neuronal calcium currents. Whereas stimulation of calcium channels in endocrine and possibly in vascular smooth muscle involves Gi2 and protein kinase C (stimulated within the concurrent PI response), Go is involved in inhibition of endocrine and neuronal calcium currents [23–25].

Two approaches were used to identify the Gα subtypes involved in these phenomena. In one set of experiments, the expression of G-protein subunits was suppressed by intranuclear injection of unprotected antisense oligonucleotides annealing to α, β- or γ-subunit RNA, or by injection of antisense oligonucleotides protected against degradation by phosphorothioate end groups, with electrophysiological determinations performed 2 days later. In another set of experiments, $G_{\alpha o}$ subunits purified from bovine brain were infused through the patch pipette into pertussis toxin-pretreated cells, in which hormonal modulations of calcium currents were uncoupled, and reconstitution of the hormonal response was attempted.

In the rat pituitary cell line GH3, Gαo1 β3 γ4 was identified as the Go subform interacting with the inhibitory M4 muscarinic receptor, whereas the galanin effect on calcium currents appeared to involve an Gαo2 β3 γ2 heterotrimer; the Gαo1 β3 γ4 subform being used less efficiently by this receptor. In contrast, the inhibitory somatostatin effect involved the Gαo2 β1 γ3 heterotrimer [23, 26, 27]. The assumption of specific roles of Go1 and Go2 for the functional coupling of different receptors to calcium channels was supported by the ability of Gαo1 to reconstitute the muscarinic and the galanin effects and of Gαo2 to reconstitute the somatostatin effects in pertussis toxin- pretreated cells [28]. Similar data were obtained in the rat insulinoma cell line RINm5F, which in contrast to GH3 and other cell lines expressed much more Gαo1 than Gαo2 [29]. While the endocrine cell lines GH3 and RINm5F express L-type calcium channels, the human neuroblastoma cell line SH-SY5Y largely shows N-type calcium currents. Reconstitution in the pertussis toxin- pretreated cells by purified Gαo subunits showed that the muscarinic M2 receptor expressed in these cells functionally coupled to calcium channels after infusion of Gαo1 or Gαo2. In contrast, the effects of somatostatin, dopamine (through D2 receptors), and of μ- and δ-opioid agonists were reconstituted only by 4–8 nM Gαo2, but not by Gαo1 at this concentration [28].

These data clearly show that Go is the G protein involved in inhibition of L- and N-type calcium channels. Go1 and Go2 play different roles by interacting with different sets of receptors and it appears that heptaspan receptors interact with specific $(\alpha\beta\gamma)$G-protein heterotrimers.

Activation of Effectors by G Proteins

The role of the voltage-dependent calcium channel (VDCC) β-subunit has been examined, with respect to the biophysical properties of voltage-activated calcium currents (I_{Ba}), and the mechanism of action of the GABA$_B$ agonist $(-)$-baclofen to inhibit I_{Ba} in cultured rat dorsal root ganglion (DRG) neurons. Following depletion of β-subunit immunoreactivity in DRGs by the microinjection of an antisense oligonucleotide, there was a marked reduction in I_{Ba} compared to nonsense-injected cells, and a depolarizing shift in the voltage for activation of the current. Bay K8644 (1 μM) was no longer an effective agonist following β-subunit depletion [30]. These results are consistent with studies in which β-subunit has been co-expressed with the VDCC α1 subunit (for review see [31]).

$(-)$- Baclofen inhibits I_{Ba} in these neurons [32], by a mechanism that is partially voltage-dependent, and involves the G protein Go [33, 34]. In the present study, a maximal concentration of $(-)$-baclofen (50 μM) produced a greater inhibition of the residual current in β-subunit depleted cells, of $44.4 \pm 5.0\%$ $(n = 14)$ compared to $26.9 \pm 5.3\%$ $(n = 12)$ in nonsence-injected and $27.9 \pm 5.10\%$ $(n = 12)$ in controls cells. In the presence of internal GTPγS the effect of $(-)$-baclofen remained enhanced in cells in which β-subunit immunoreactivity had been depleted (Fig. 4).

There was no difference in the ability of the N channel blocker and ω-conotoxin GVIA to inhibit I_{Ba} following β-antisense oligonucleotide injection. It is thus unlikely that the enhanced response to $(-)$-baclofen is due to a selective sparing of the β-subunit associated with N-type VDCCs.

From these results, the hypothesis is refuted that agonist-induced inhibition of VDCCs occurs solely via a reduction in the interaction

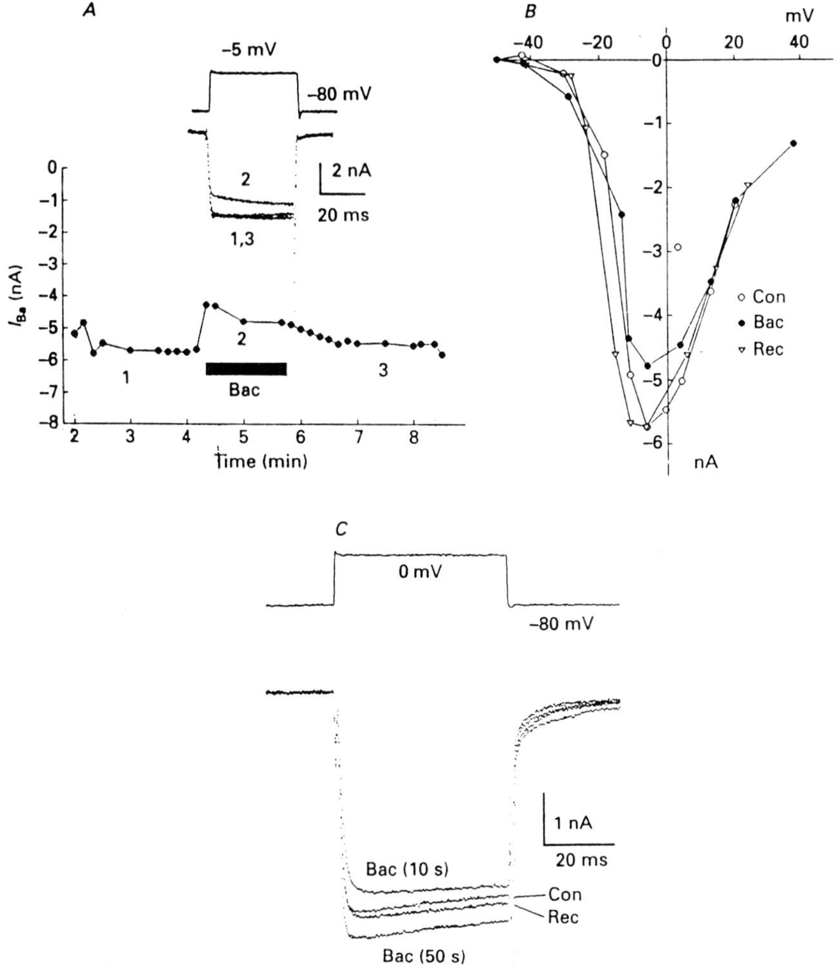

Fig. 4. Effect of baclofen (50 μm) on I_{Ba} in a cell injected with antisense DNA complementary to G_0. A, time course of effect on 50 μm ($-$)-baclofen (bar) on the maximum I_{Ba}. The inset traces show the currents at the times labelled 1, before; 2, during; and 3, after ($-$)-baclofen application in a cell injected 26 h previously with antisense DNA against $G\alpha_0$. B, current-voltage relationships from the same cell before, during and after ($-$)-baclofen application. C, the transient inhibitory effect of ($-$)-baclofen (50 μm) in another $G\alpha_0$ antisense-injected cell. Baclofen produced a small inhibition at 10 s after the start of its application, followed by a small reversible enhancement after 50 s. Reprinted by permission of the Physiological Society from [33].

between the VDCC α1- and β-subunits. A model is proposed whereby activated Gαo competes with the VDCC β-subunit, for a binding site on the VDCC α1-subunit, and is thus more effective when the stoichiometry between the VDCC α1- and β-subunits is altered by depletion of the β-subunit.

Conclusion

Early models for signal transduction through G protein-coupled receptors allowed for a straightforward "linear" activation scheme. Clearly, the process can be regulated at the level of the receptor by altering the efficiency to which it couples to its cognate G protein or by altering its ability to withdraw from the reappear on the cell surface. Further, a complex code of α, β, and γ subunits might dictate information flow from specific receptors to their designated receptors. This way, cells could respond distinctly to two different hormones or neurotransmitters which acted on the same intracellular effector molecule. It also appears that cell structure and function are intimately related. Specific elements of the cytoskeleton are able to modify signal transduction by complexing with specific G protein α subunits. Such a mechanism could allow for second messenger "crosstalk" as well as for the intracellular modulation of extracellular signals. Finally, it has been seen that a channel thought to be sensitive to membrane voltage is also modulated by G protein such that Ca^{2+} influx and, subsequently, neurotransmitter release might be modified in response to a signal initiated at the $GABA_B$ receptor. Clearly, the complexity of G protein-mediated signaling is far more extensive than originally imagined. It is entirely possible that we have not yet begun to appreciate the extent of the elegant complexity with which G protein-mediated signal transduction systems are integrated into a myriad of cellular processes.

Acknowledgements

This work was supported by grants from the U.S. Public Health Service, the U.S. National Science Foundation, the Medical Research Council (U.K.), the Wellcome Trust and the Deutsche Forschungsgemeinschaft. We would like to thank Jiang Chen for his assistance with the artwork.

References

1. Simon M, Strathman M, Gautam N. Diversity of G proteins in signal transduction. Science 1991; 252: 802–808.
2. Rasenick MM. G's (a poem). Trends in Biochemical Science 1992; 17: 71.
3. Taussig R, Iniguez-Lluhi J, Gilman A. Inhibition of adenylyl cyclase by $G_{i\alpha}$. Science 1993; 261: 218.
4. Federman AD, Conklin BR, Schrader KA, Reed RR, Bourne HR. Hormonal stimulation of adenylyl cyclase through Gi-protein $\beta\gamma$ subunits. Nature 1992; 356: 159–161.
5. Iyenger R. Molecular and functional diversity of mammalian Gs-stimulated adenylyl cyclases. FASEB Journal 1993; 7: 768–775.
6. Cotecchia S, Exum S, Caron MG, Lefkowitz RJ. Regions of the alpha 1-adrenergic receptor involved in coupling to phosphatidylinositol hydrolysis and enhanced sensitivity of biological function. Proc. Natl. Acad. Sci. USA 1990; 87: 2896–2900.
7. Samama P, Cotecchia P, Costa T, Lefkowitz RJ. A mutation-induced activated state of the $\beta 2$ adrenergic receptor: extending the ternary complex theory. J. Biol. Chem. 1993; 268: 4625–4636.

8. Tiberi M, Jarvis KR, Silva C, Falardeau P, Gingrich JA et al. Cloning, molecular characterization, and chromosomal assignment of a gene encoding a second D1 dopamine receptor subtype: Differential expression pattern in rat brain compared with rat D1A receptor. Proc. Natl. Acad. Sci. USA 1991; 88: 7491–7495.

9. Lefkowitz RJ, Cottechia S, Samama P, Costa T. Constitutive activity of receptors coupled to guanine nucleotide regulatory proteins. Trends in Pharm. Sci. 1993; 14: 303–307.

10. Tiberi M, Caron MG. High agonist-independent activity is a distinguishing feature of the dopamine D1B receptor subtype. J. Biol. Chem. 1994; 269: 27925–27931.

11. Hein LK, Ishii SR, Coughlain SR, Kobilka BK. Intracellular targeting and trafficking of thrombin receptors: a novel mechanism for resensitization of a G protein-coupled receptor. J. Biol. Chem. 1994; 269: 27719–27726.

12. Keefer JR, Limbird LE. The alpha 2A-adrenergic receptor is targeted directly to the basolateral membrane domain of Madin-Darby canine kidney cells independent of coupling to pertussis toxin-sensitive GTP-binding proteins. J. Biol. Chem. 1993; 268: 11340–11347.

13. vonZastrow M, Link MR, Daunt D, Barsh G, Kobilka BK. Subtype-specific differences in the intracellular sorting of G protein-coupled receptors. J. Biol. Chem. 1993; 268: 763–766.

14. Neer EJ, Clapham DE. Roles of G protein subunits in transmembrane signalling. Nature 1988; 333: 129–134.

15. vonZastrow M, Kobilka BK. Ligand-regulated internalization and recycling of human beta 2-adrenergic receptors between the plasma membrane and endosomes containing transferrin receptors. J. Biol. Chem. 1992; 267: 3530–3538.

16. vonZastrow M, Kobilka BK. Antagonist-dependent and -independent steps in the mechanism of adrenergic receptor internalization. J. Biol. Chem. 1994; 269: 18448–18452.

17. Vu TK, Hung DT, Wheaton VI, Coughlin SR. Molecular cloning of a functional thrombin receptor reveals a novel proteolytic mechanism of receptor activation. Cell 1991; 64: 1059–1068.

18. Roychowdhury S, Wang N, Rasenick MM. G protein binding and G protein activation by nucleotide transfer involve distinct domains on tubulin: regulation of signal transduction by cytoskeletal elements. Biochemistry 1993; 32: 4955–4961.

19. Wang N, Yan K, Rasenick MM. Tubulin binds specifically to the signal-transducing proteins, Gsa and Gia1. J. Biol. Chem. 1990; 265: 1239–1242.

20. Roychowdhury S, Rasenick MM. Tubulin-G protein association stabilizes GTP binding and activates GTPase: Cytoskeletal participation in neuronal signal transduction. Biochemistry 1994; 33: 9800–9805.

21. Popova JS, Johnson GL, Rasenick MM. Chimeric Gαs/Gαi2 proteins define domains on Gαs which interact with tubulin for the β adrenergic activation of adenylyl cyclase. J. Biol. Chem. 1994; 269: 21748–21754.

22. Andrade R. Enhancement of β-adrenergic responses by Gi-linked receptors in rat hippocampus. Neuron 1993; 10: 83–88.

23. Kleuss C, Heschler J, Ewel C, Rosenthal W, Schultz G, Wittig B. Assignment of G-protein subtypes to specific receptors inducing inhibition of calcium currents. Nature 1991; 353: 43–48.

24. Gollasch M, Kleuss C, Hescheler J, Wittig B, Schultz G. G12 and protein kinase C are required for thyrotrophin-releasing hormone induced stimulation of voltage-dependent Ca^{2+} channels in rat pituitary GH3 cells. Proc. Natl. Acad. Sci. USA 1993; 90: 6265–6269.

25. Hescheler J, Schultz G. G-protein involved in the calcium channel signaling system. Curr. Opinion Neurobiol. 1993; 3: 360–367.

26. Kleuss C, Scherubl H, Hescheler J, Schultz G, Wittig B. Different β-subunits determine G-protein interaction with transmembrane receptors. Nature 1992; 358: 424–426.

27. Kleuss C, Scherubl H, Hescheler J, Schultz G, Wittig B. Selectivity in signal transduction determined by gamma subunits of heterotrimeric G proteins. Science 1993; 259: 832–834.

28. Nurnberg B, Friedrich P, Hescheler J. Distinct properties of three α-subtypes of the G-protein Go purified from mammalian brains. Naunyn-Schmiedeberg's Arch. Pharmacol. 1993; 347: R60.

29. Nurnberg B, Degtiar VE, Harhammer R, Uhde M, Hescheler J, Schultz G. Hormone-induced $Go\alpha$ subtype-specific inhibition of calcium currents. Naunyn-Schmiedeberg's Arch. Pharmacol. 1994; 349: R13.
30. Berrow N, Campbell V, Fitzgerald E, Brickley K, Dolphin AC. Antisense depletion of β-subunits modulates the biophysical and pharmacological properties of neuronal calcium channels. J. Physiol. 1995; 482: 481–491.
31. Hofmann F, Biel M, Flockerzi VI. Molecular basis for Ca^{2+} channel diversity. Ann. Rev. Neurosci. 1994; 17: 399–418.
32. Dolphin AC, Scott RH. Calcium channel currents and their inhibition by (−)-baclofen rat sensory neurones: modulation by guanine nucleotides. J. Physiol. 1987; 386: 1–17.
33. Campbell V, Berrow NS, Dolphin AC. $GABA_B$ receptor modulation of Ca^{2+} currents in rat sensory neurones by the G protein Go: Antisense oligonucleotide studies. J. Physiol. 1993; 470: 1–11.
34. Menon-Johansson AS, Berrow NS, Dolphin AC. Go transduces $GABA_B$ receptor modulation of N-type calcium channels in cultured dorsal root ganglion neurones. Pflugers Arch. 1993; 425: 325–333.

Pharmacology of Ion Channels

Pharmacological Sciences: Perspectives for
Research and Therapy in the Late 1990s
ed. by A.C. Cuello and B. Collier

Novel Aspects of the Pharmacology of Calcium Channel Modulators

Théophile Godfraind[1] and Jeffrey Atkinson[2]

[1]*Laboratoire de Pharmacologie, Université Catholique de Louvain, FARL 5410, 1200 Bruxelles, Belgium;* [2]*Laboratoire de Pharmacologie, Faculté de Pharmacie de l'Université Henri Poincaré Nancy 1, 54000 Nancy, France*

Progress in the Molecular Structure of Calcium Channels in Different Tissues

Although many electrophysiological properties of the cardiac and smooth muscle L-type calcium channels (CC) are identical, they differ in their hormonal regulation. β-adrenergic stimulation increases the calcium inward current by cAMP-dependent phosphorylation of the CC or a closely associated protein in cardiac myocytes [1], but not in smooth muscle cells [2]. Cloning of the cDNA of both channels showed that they are composed of at least three proteins, the channel containing α_1 subunit and two auxillary subunits, the β and α_2/δ [3]. The α_1 subunit, a product of the class C gene, has several splice variants, containing identical putative phosphorylation sites. The cardiac and smooth muscle CC complexes presumably contain different β subunits [4] and an identical α_2/δ protein. The cloned α_1 subunit has been expressed stably or transiently, alone or in combination with the α_2/δ subunit and different β subunits, in CHO and HEK 293 cells. In each system, the α_1 subunit codes for a functional L-type CC which has many properties of the native cardiac or smooth muscle channel [5]. Recent reports show that barium currents (I_{Ba}) of the stable expressed α_{1C} subunit increased following treatment with dBcAMP [6], the catalytic subunit of cAMP kinase [7] or forskolin [8]. Dialysis of the CHO cells stably expressing the cardiac (α_{1Ca}) or smooth muscle (α_{1Cb}) α_1 subunit with the PKI inhibitor peptide 5–25 (1 mM), the pure catalytic subunit of cAMP kinase (cAMP kinase, 25 μM), or a combination of okadaic acid (1 μM) and cAMP kinase, had no significant effect on I_{Ba}. Predepolarization of the membrane potential to $+50$ mV increased I_{Ba} produced by a test pulse to $+20$ mV immediately afterwards. This facilitation was not affected by dialysis with cAMP kinase, PKI peptide or GDPβS [9]. Superfusion of HEK 293 cells stably expressing the α_{1Cb} subunit

---

Correspondence to: J. Atkinson, address as above.

with the cAMP kinase inhibitor H 89 had no significant effect on I_{Ba}. Furthermore, the I_{Ba} of these cells was not increased by forskolin (5 μM in the presence of 20 μM IBMX).

Transient expression of the α_{1Ca} subunit together with the $\beta 2a$ and the α_2/δ subunits induced large I_{Ba} ($\sim$ 40 pA/pF) in HEK 293 cells. Dialysis of HEK 293 cells with PKI peptide (1 mM), the cAMP kinase inhibitor H 89 or cAMP kinase (1–25 µM) did not modify I_{Ba}. Furthermore, the transiently expressed channel showed no facilitation (conditions: 7). I_{Ba} was not affected by dialysis with cAMP kinase under facilitation conditions [7]. Forskolin (plus IBMX) had no effect on I_{Ba}. These results suggest that the expressed α_{1Ca} or α_{1Cb} subunit of the L-type CC is not affected by cAMP kinase under the conditions used. The co-expression of the $\beta 2a$ and the α_2/δ subunits increased the current density, but did not affect the lack of cAMP-dependent modulation of the expressed channel.

Calcium Channel Modulators: Structure-Activity Relationships

First-generation CC blockers (verapamil, nifedipine, and diltiazem) have achieved major prominence for their therapeutic applications in cardiovascular disorders and their potential roles in other disorders ranging from achalasia to vertigo [10]. These agents and their second generation analogs owe their use to an ability to interact selectively with the L-type, voltage-gated CC. The voltage-gated CCs are a homologous subset of a "super-family" of voltage-gated ion channels, including the L, T, N, P, and at least two other classes, each electrophysiologically and pharmacologically characterized [11].

The CC may be regarded as a receptor with specific binding sites for activator and antagonist ligands, coupled to the gating machinery of the channel, and endowed with specific structure-activity relationships including stereoselectivity [10]. Structure-activity relationships have been best defined for the 1–4 dihydropyridines (DHP) which include both activators and antagonists, but there are probably not less than 7 or 8 specific drug binding sites associated with the L-type CC, including that for the investigational agent Ro 40-5967 [12].

Three components of the structure-activity relationships of the DHP are of major interest: the binding site topography and location, and the binding site properties relative to channel state. The structural and conformational requirements for DHP interaction have been reviewed in detail [13]. Biochemical studies have located the DHP binding site on the α_1 subunit of the heteromeric channel complex, in the extracellular regions of domains III and IV [14]. A series of vectorial DHP probes orients the binding site on the extracellular membrane, some 12–14 A deep [15].

The state-dependent interactions of DHP are well established electrophysiologically, antagonist activity increasing with the level of depolarization. This voltage-dependent binding contributes to the potency and selectivity of the DHP antagonists [16]. Activators show little voltage-dependence regardless of potency [17]. DHP with different 3,5-ester substitution patterns demonstrate different levels of voltage-dependent binding [16, 17]. This probably underscores the different patterns of vascular selectivity observed between nifedipine and second-generation DHPs [18].

Although the DHP structure is most active at L-type CCs, it is associated with molecules active at other channels and receptors [12], including T-type CC, sodium channels, several potassium channels, and a number of G-protein-coupled receptors such as α-adrenoceptors and PAF receptors [19, 20]. This suggests that the DHP nucleus is a general pharmacophore.

Multiple Types of Calcium Channels, Diversity of Form and Function

Understanding the molecular diversity of CCs and their varied contributions to physiological functions represents a formidable challenge. Some of the most powerful agents for discrimination among CCs are ω-conotoxins, small peptides derived from venomous marine snails [21], which selectively inhibit specific types of voltage-dependent CC [22]. ω-conotoxins have been essential for uncovering the functional diversity of CC within the nervous system. The best-known example, ω-CTx-GVIA, is representative of other ω-conotoxins in sharing a common disulfide-linked structural backbone. ω-CTx-GVIA potently blocks N-type channels, with no effect on L, T, P, Q and R-type channels. The selective interaction of ω-CTx-GVIA with N-type channels enabled their purification, cellular localization, and discovery of their roles in controlling Ca^{2+} entry, neurotransmitter release and neuronal migration. In contrast to the selective neurotoxins that block Na^+ or K^+ channels such as TTX or charybdotoxin, little is known about the structural basis of how CC are blocked by ω-CTx-GVIA or other peptide toxins. We have characterized structural determinants of the N-type CC's susceptibility to toxin block.

N-type channel activity in *Xenopus* oocytes was obtained with a cDNA construct comprised almost entirely of the α_1 subunit of the N-type channel [α_{1B}; 23–25]. To achieve robust expression, the 5' untranslated region and the amino terminus of the coding region were derived from α_{1A}, a subunit which expresses at high levels in oocytes. The construct expressed at levels comparable to α_{1A} while displaying N-type channel characteristics. Inward currents were high-voltage activated and typically $\geq 2\,\mu A$ with 5 mM external Ba^{2+}. Currents were susceptible to inactivation with steady depolarizations and were rapidly blocked by ω-CTx-GVIA (5 μM; τblock = 30.8 $\pm$ 0.9). The current was unresponsive to FPL 64176 (1 μM; n = 3), a potent agonist of L-type channels, or ω-Aga-IVA (100 nM; n = 3), a blocker of P and Q-type channels. Thus, the expressed current behaves like an N-type CC.

Unlike N-type CCs, CCs encoded by α_{1A} are unresponsive to ω-CTx-GVIA. We constructed chimeras comprising all possible combinations of motifs from toxin-sensitive (N) and toxin-insensitive (A) α_1 subunits, designated by four letters giving the identity of motifs I–IV. All motif chimeras formed functional channels, but the time-course and degree of block by ω-CTx-GVIA varied. Among the chimeras that contain three motifs from N and 1 from A, NNAN showed the greatest alteration in the onset of block relative to the parental N-type channel, a nine-fold slowing, while ANNN, NANN, and NNNA displayed more moderate slowing relative to wild-type. Among the chimeras consisting of one motif from N and three from A, only AANA was capable of a detectable degree of ω-CTx-GVIA-responsiveness. Toxin block was slow but significant, in contrast to behavior of wild-type α_{1A} channels (AAAA).

When chimeras containing two motifs from each parent were grouped as complementary pairs, toxin block developed much more rapidly for the chimera that contained motif III, and more slowly for the one that did not. Thus, striking changes in toxin interactions results in every case where replacements were made in motif III, for N->A substitution against an N-type background, A->N substitution against a class A background, and with pairwise combinations of motifs in the N_2A_2 chimeras.

ω-CTx-GVIA acts at the external face of the channel, possibly by occluding the pore. Accordingly, we introduced changes in the putative external loops of the N-type channel, on either side of the pore-lining segment known as H5, switching sets of residues from those in the N-type channel to those in α_{1A}. In either motifs II and III, changes in the H5-H6 loop had little effect on block kinetics. In contrast, substitutions in IIS5-H5 caused appreciable slowing of block onset. The largest contribution came from residues in the putative external loop between IIIS5 and IIIH5, where N-type and α_{1A} channels differ at 11 positions. At nine of these the respective amino acid side chains bear a different charge. Swapping of five amino acids in this loop (mutant III-2) caused a dramatic slowing exceeding that produced by replacing motif III in entirety. For both mutant III-2 and wild-type currents, blocking rate was linearly related to ω-CTx-GVIA concentration; association rate coefficients (K_{on}) were 0.49 μM^{-1} min^{-1} (wild-type) and 0.016 μM^{-1} min^{-1} (mutant III-2), a 30-fold difference.

These experiments reveal molecular features of a voltage-gated CC that are important for its high affinity interaction with a peptide toxin.

The heterotetrameric architecture of the CCs allowed us to examine the contributions of various motifs, singly and in combination. Further analysis indicated the importance of external loops of individual motifs. Taken together, the results support the idea that the ω-conotoxin molecule interacts with extracellular aspects of each motif, presumably by lodging within the outer vestibule of the channel. In light of close structural similarities among ω-conotoxins, our conclusions for N-type channels and ω-CTx-GVIA may find general applicability to other CC and ω-toxins.

Factors Responsible for the Tissue Selectivity of Calcium Channel Blockers

The question posed in the 1960s as to how first-generation CC blockers of such diverse structure could have similar pharmacological properties has been answered by the demonstration of distinct sites on the α_1 subunit of the L-type CC with which the various structures interact (see above). The question of the late 1990s is: "are there any pharmacological differences amongst the second-generation CC blockers?" Such selectivity could be related to drug, tissue, and/or stimulus characteristics [26].

The high potency of the DHP, nisoldipine, in coronary arteries provides a good example of drug selectivity. *In vitro* mechanical studies on ratios of IC_{50} values in human arterial and cardiac preparations show that these are low for first-generation CC blockers (1 for diltiazem and verapamil, 14 for nifedipine) but much higher for nisoldipine (1555), whereas the binding affinity of nisoldipine for vascular and cardiac membranes is similar [27]. Furthermore the inhibitory effect of nisoldipine – and other second-generation DHPs such as lacidipine and isradipine, but not the first-generation CC blockers – is markedly time-dependent, increasing slowly following depolarization [27–30]. Preincubation with a depolarizing solution increased the nisoldipine inhibition of arterial contraction (but not that of first-generation CC blockers) producing kinetics similar to those of nisoldipine binding [28, 30]. This combination of voltage- and time-dependency contributes to the coronary artery selectivity of nisoldipine and may involve a differential distribution of L-type CC isoforms.

The observation that neonatal hearts are more sensitive to the negative inotropic effects of CC blockers than adult hearts provides a first example of tissue selectivity. This is correlated to the CC distribution which is on peripheral sarcolemma in neonatal hearts but on junctional areas of T-tubules in close vicinity to the terminal cisternae of the sarcoplasmic reticulum in adult hearts [31].

A second example of tissue selectivity stems from the observation that CC blockers lower blood pressure more effectively in hypertensive than in normotensive animals and humans. *In vitro*, depolarized arteries from SHR relax slower than those from normotensive WKY rats upon removal of the depolarizing solution [32]. Such postcontraction tone is a general feature of the vasculature being present in compliance (aorta) and resistance vessels (mesenteric artery). As it is suppressed by chronic administration of antihypertensive doses of nisoldipine and by *in vitro* pretreatment of arteries with nisoldipine, it appears to be due to an abnormally prolonged activation of CCs after transfer of arteries from the depolarizing to the physiological solution. The antihypertensive – as opposed to the hypotensive – action of nisoldipine could be related to the mechanisms involved in the suppression of postcontraction tone. Thus the sensitivity of a given tissue to the action of a given CC may be influenced by vascular pathology such as hypertension.

Calcium Channel Blockers and Tissue Remodeling

Vessel wall remodeling in aging and age-linked pathologies such as atherosclerosis and hypertension, is a programmed structural change in the balance between (i) mitosis, cell hypertrophy and migration, secretion of extracellular matrix, etc., and (ii) necrosis, apoptosis, digestion of extracellular matrix, etc. Calcium ions are involved in all of these processes and there is a multitude of reports on potential protective mechanisms based on interference with one or more of these events. But we do not know whether such hypotheses translate into chronic *in vivo* reality. Part of the difficulty arises from the question as to whether the primary change in vascular calcium handling followed by an increase in the wall calcium content, *viz.*, vascular calcium overload, is intra- or extracellular. During aging, for example, the total calcium content of the human aorta increases some 30- to 50-fold [33], as it does in animal models of this phenomenon such as the vitamin D_3 plus nicotine rat model [34]. However, in the latter model, as in hypertension and aging in rats, increases in intracellular calcium levels in viable smooth muscle cells are far less than in the total wall calcium content (Capdeville-Atkinson et al., unpublished results), suggesting that vascular calcium overload is a more extracellular phenomenon. Thus CC blockers may interfere with the early, initial stages of vascular calcium overload by preventing the massive calcium influx into smooth muscle cells *via* CCs, and the chain of events leading to cell death and extracellular deposition of calcium. Secondly they could interfere in a non-specific fashion, at higher doses, with the deposition of extracellular calcium directly on matrix components.

In atherosclerosis, animal data illustrate these two principles: CC blockers act often only at high doses and are "preventive", retarding the progression of early lesions but incapable of inducing regression [35].

Data in man [36, 37] confirm this – the main effect of CCs being on the development of new lesions. Amongst the many mechanisms proposed for the anti-atherogenic action of CC blockers [35] endothelial protection merits further investigation. Endothelial injury – central to atherosclerosis and aging [38] – could involve calcium overload as CC blockers restore endothelial structure and function in animals [39].

In essential hypertension (*viz.* an increase in diastolic and systolic pressures) the main effect of CC blockers on structure is probably pressure-dependent, the case for a supplementary, pressure-independent effect of CC blockers on vessel (and heart) structure is still controversial [40]. One possible pressure- (and flow-) independent effect of CCs merits further investigation. In the future the treatment of isolated systolic hypertension of the elderly will gather importance. The age-associated increase in the calcium content of the medial extracellular matrix, especially the elastic component (see above) may be central to the development of characteristic structural (dilatation and increased arterial rigidity) and hemodynamic complications (decreased compliance and isolated systolic hypertension). Prevention of age-linked medial elastocalcinosis and, therefore, isolated systolic hypertension could form a new target for CC blockers. Our vitamin D_3 plus nicotine rat model develops many of these features: a reduction in the medial elastin content and an increase in arterial diameter accompanied by a decrease in arterial compliance and isolated systolic hypertension [34].

References

1. Kameyama M, Hofmann F, Trautwein W. On the mechanism of the β-adrenergic regulation of Ca-channel in the guinea-pig heart. Pflügers Arch. 1985; 405: 285–293.
2. Welling A, Felbel J, Peper K, Hofmann F. Hormonal regulation of the calcium current of airway smooth muscle. Am. J. Physiol. 1992; 262: L351–L359.
3. Hofmann F, Biel M, Flockerzi V. Molecular basis for Ca^{2+} channel diversity. Ann. Rev. Neurosci. 1994; 17: 399–418.
4. Hullin R, Singer-Lahat D, Freichel M, Biel M, Dascal N, Hofmann F et al. Calcium channel β subunit heterogeneity: functional expression of the cloned cDNA from heart, aorta and brain. EMBO J. 1992; 11: 885–890.
5. Welling A, Bosse E, Cavalié A, Bottlender R, Ludwig A, Nastainczyk W et al. Stable coexpression of calcium channel α_1, β and α_2/δ subunits in a somatic cell line. J. Physiol. 1993; 471: 749–765.
6. Yoshida A, Takahashi M, Nishimura S, Takeshima H, Kokubun S. Cyclic AMP-dependent phosphorylation and regulation of the cardiac dihydropyridine-sensitive Ca channel. FEBS 1992; 390: 343–349.
7. Sculptoreanu A, Rotman E, Takahashi M, Scheuer T, Catterall WA. Voltage-dependent potentiation of the activity of cardiac L-type calcium channel α_1 subunits due to phosphorylation by cAMP-dependent protein kinase. Proc. Natl. Acad. Sci. USA 1993; 90: 10135–10139.
8. Perez-Reyes E, Yaun W, Wei X, Bers DM. Regulation of the cloned L-type cardiac calcium channel by cyclic-AMP-dependent protein kinase. FEBS 1994; 342: 119–123.
9. Kleppisch T, Pedersen K, Bosse E, Flockerzi V, Hofmann F, Hescheler J. Double-pulse facilitation of smooth muscle α_1 subunit Ca^{2+} channels expressed in CHO cells. EMBO J. 1994; 13: 2502–2507.

10. Janis RA, Triggle DJ. Drugs acting on calcium channels. In: Hurwitz L, Partridge LD, Leach JF, editors. Calcium channels. Their properties, function, regulation and clinical relevance. Boca Raton, Florida: CRC Press, 1991: 195–249.
11. Catterall WA. Excitation-contraction coupling in vertebrate skeletal muscle: a fate of two calcium channels. Cell 1991; 64: 871–874.
12. Rampe D, Triggle DJ. New synthetic ligands for L-type calcium channels. Prog. Drug. Res. 1993; 40: 191–238.
13. Triggle DJ, Langs DA, Janis RA. Ca^{2+} channel ligands: structure-function relationships of the 1,4-dihydropyridines. Med. Res. Revs. 1989; 9: 123–180.
14. Catterall WA, Striessnig J. Receptor sites for Ca^{2+} channel antagonists. Trends Pharmacol. Sci. 1992; 13: 256–262.
15. Baindur N, Rutledge A, Triggle DJ. A homologous series of permanently charged 1,4-dihydropyridines: novel probes designed to localize drug binding sites on ion channels. J. Med. Chem. 1993; 36: 3743–3745.
16. Triggle DJ. Structure-function correlations of 1,4-dihydropyridines calcium channel antagonists and activators. In: Hondeghem L, editor. Molecular and cellular mechanisms of antiarrhythmic agents. Mt. Kiscoe, New York: Futura, 1989: 269–291.
17. Zheng W, Stoltefuss J, Goldmann S, Triggle DJ. Pharmacologic and radiogland binding studies of 1,4-dihydropyridines in rat cardiac and vascular preparations: stereoselectivity and voltage-dependence of antagonist and activator interactions. Mol. Pharmacol. 1992; 41: 535–591.
18. Triggle DJ. Calcium channel antagonists: mechanisms of action, vascular selectivities and clinical relevance. Clev. Clin. J. Med. 1992; 59: 617–627.
19. Frank CA, Forst JM, Grant T, Harris RJ, Kau ST, Lii JH et al. Dihydropyridine K_{ATP} potassium channel openers. Bioorg. Med. Chem. Lett. 1993; 3: 2725–2726.
20. Cooper K, Fray MJ, Parry J, Richardson K, Steele J. 1,4-dihydropyridines as antagonists of platelet activating factor. I. Synthesis and structure activity relationships of 2-(4-heterocyclyl) phenyl derivatives. J. Med. Chem. 1992; 35: 3115–3129.
21. Olivera BM, McIntosh JM, Cruz LJ, Luque FA, Gray WR. Purification and sequence of a presynaptic peptide toxin from *Conus. geographus* venom. Biochem. 1984; 23: 5087–5090.
22. Olivera BM, Miljanich G, Ramachandran J, Adams ME. Calcium channel diversity and neurotransmitter release. Ann. Rev. Biochem. 1994; 63: 823–867.
23. Williams ME, Brust PF, Feldman DH, Saraswathi P, Simerson S, Maroufi A et al. Structure and functional expression of an ω-conotoxin-sensitive human N-type calcium channel. Science 1992; 257: 389–395.
24. Dubel SJ, Starr TVB, Hell J, Ahlijanian MK, Enyeart JJ, Catterall WA et al. Molecular cloning of the α_1 subunit of an ω-conotoxin-sensitive calcium channel. Proc. Natl. Acad. Sci. USA 1992; 89: 5058–5062.
25. Fujita Y, Mynlieff M, Dirksen RT, Kim MS, Niidome T, Nakai J et al. Primary structure and functional expression of the ω-conotoxin-sensitive N-type channel from rabbit brain. Neuron 1993; 10: 585–598.
26. Godfraind T. Analysis of factors involved in the tissue selectivity of calcium antagonists. In: Frank GB, Bianchi CP, Terkevrs H, editors. Excitation-contraction coupling in skeletal, cardiac and smooth muscle. New York: Plenum, 1992: 103–120.
27. Godfraind T, Salomone S, Dessy C, Verhelst B, Dion R, Schoevaerts J-C. Selectivity scale of calcium antagonists in the human cardiovascular system (based on *in vitro* studies). J. Cardiovasc. Pharmacol. 1992; 20 (suppl. 5): S34–S39.
28. Wibo M, De Roth L, Godfraind T. Pharmacologic relevance of dihydropyridine binding sites in membranes from rat aorta: kinetic and equilibrium studies. Circ. Res. 1988; 62: 91–96.
29. Godfraind T, Dessy C, Salomone S. A comparison of the potency of selective L-calcium channel inhibitors in human coronary and internal mammary arteries exposed to serotonin. J. Pharmacol. Exp. Ther. 1992; 263: 112–122.
30. Salomone S, Godfraind T. Radiogland and functional estimates of the interaction of the 1,4-dihydropyridines, isradipine and lacidipine, with calcium channels in smooth muscle. Br. J. Pharmacol. 1993; 109: 100–106.
31. Wibo M, Bravo G, Godfraind T. Postnatal maturation of excitation-contraction coupling in rat ventricle in relation to the subcellular localization and surface density of 1,4-dihydropyridine and ryanodine receptors. Circ. Res. 1991; 68: 662–673.

32. Godfraind T, Kazda S, Wibo M. Effects of chronic treatment by nisoldipine, a calcium antagonistic dihydropyridine, on arteries of spontaneously hypertensive rats. Circ. Res. 1991; 68: 974–982.
33. Fleckenstein A, Frey M, Zorn J, Fleckenstein-Grün G. Calcium, a neglected key factor in hypertension and arteriosclerosis. Experimental vasoprotection with calcium antagonists or ACE inhibitors. In: Laragh JH, Brenner BM, editors. Hypertension pathophysiology, diagnosis and management. New York: Raven Press, 1990: 471–509.
34. Atkinson J, Poitevin P, Chillon JM, Lartaud I, Levy B. Vascular calcium overload produced by vitamin D_3 plus nicotine treatment diminishes arterial distensibility in rats. Am. J. Physiol. 1994; 266: H540–H547.
35. Weinstein DB, Heider JG. Antiatherogenic properties of calcium antagonists. Am. J. Med. 1989; 86 (suppl. 4A): 27–32.
36. Lichtlen PR, Hugenholtz PG, Rafflenbeul W, Hecker H, Jost S, Deckers JW. Retardation of angiographic progression of coronary heart disease by nifedipine. Lancet 1990; 335: 1109–1113.
37. Waters D, Lespérance J, Francetich M, Causey D, Théroux P, Chiang YK et al. A controlled clinical trial to assess the effect of a calcium channel blocker on the progression of coronary atherosclerosis. Circulation 1990; 82: 1940–1953.
38. Atkinson J, Tatchum-Talom R, Corman B. Effect of chronic angiotensin I converting enzyme inhibition on the aging processes. III. Endothelial function of the mesenteric arterial bed. Am. J. Physiol. 1994; 267: R136–R143.
39. Henrion D, Chillon JM, Capdeville-Atkinson C, Atkinson J. Effect of chronic treatment with the calcium entry blocker, isradipine, on vascular calcium overload produced by vitamin D_3 and nicotine in rats. J. Pharmacol. Exp. Ther. 1992; 260: 1–8.
40. Mulvany MJ. The development of vascular hypertrophy. J. Cardiovasc. Pharmacol. 1992; 19 (suppl. 2): S22–S27.

Pharmacological Sciences: Perspectives for
Research and Therapy in the Late 1990s
ed. by A.C. Cuello and B. Collier
© 1995 Birkhäuser Verlag Basel/Switzerland

Calcium Channels, Calcium Channel Antagonists and the Functioning of the Gastrointestinal Tract

Jan D. Huizinga[1], Casey Van Breemen[2], Kenton M. Sanders[3],
Ryuji Inoue[4], Paul A. Cann[5], Theophile Godfraind[6], and
Marie Odile Christen[7]

[1]*Department of Biomedical Sciences, McMaster University, Hamilton, ON L8N 3Z5, Canada;*
[2]*Department of Pharmacology, University of British Columbia, Vancouver, BC V6T 1Z3,
Canada;* [3]*University of Nevada School of Medicine, Reno, Nevada 89557, USA;* [4]*Department of
Pharmacology, Kyushu University, Fukuoka 812, Japan;* [5]*South Tees Health/Middlesbrough
Hospital, Middlesbrough, Cleveland TS5 5AZ, U.K.;* [6]*Laboratoire de Pharmacologie,
Université Catholique de Louvain, B-1200 Brussels, Belgium;* [7]*Solvay Pharma, Laboratoires de
thérapeutique moderne L.T.M., 92151 Suresnes Cedex, France*

Summary. Calcium ions play an essential role in the generation of motor activity of the
gastrointestinal tract. Calcium ions enter gastrointestinal smooth muscle cells to activate
contractile proteins through voltage-activated ion channels, through ion channels opened by
activation of plasma membrane receptors and through release from internal stores. The
pharmacology of these pathways is under intense investigation. However, at the moment only
drugs that affect voltage dependent calcium channels have reached clinical significance.

The Voltage Activated L-Type Calcium Channel

Smooth muscle cells of the gastrointestinal (GI) tract are activated by at
least three types of electrical behavior. In tonic regions of the GI tract,
contractions result from slow changes in membrane potential induced
by neural or hormonal inputs [1]. In most regions of the GI tract,
however, contractions must be followed by nearly full relaxation to
allow refilling between segmental contractions or peristaltic sweeps of
contraction. This is accomplished by coupling contractions to rhythmi-
cally occurring action potentials which include slow waves and superimposed spiking activity [2, 3] (Fig. 1). Part of this electrical activity is
generated through voltage-activated L-type calcium channels. Release
of Ca^{2+} from internal stores plays a regulatory and complementary role
in the contractile response; the majority of Ca^{2+} needed for activation
in most gastrointestinal smooth muscle tissue is provided by entry
through L-type Ca^{2+} channels. Patch clamp studies of many types of GI
smooth muscle cells have shown the expression of these channels [4–6];
when they are blocked, the mechanical performance of GI muscles is

Correspondence to: Dr. J. Huizinga, McMaster University, HSC-3N5C, 1200 Main Street
West, Hamilton, ON L8N 3Z5, Canada.

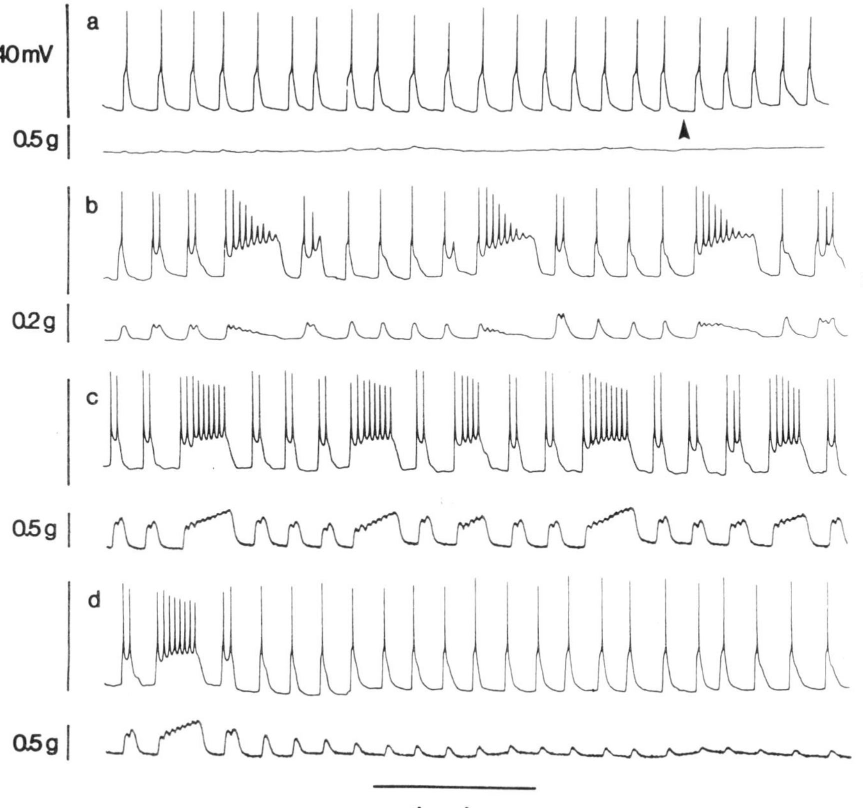

1 min

Fig. 1. Action of carbachol on spontaneously active muscle of dog colon. Upper tracings, electrical activity; lower tracings, mechanical activity. (a) spontaneous activity with slow waves and superimposed spikes. Carbachol (2×10^{-8} M), added at arrow, depolarized the membrane. (b) 5 min after addition of carbachol: a regular pattern emerged, prolonged slow waves alternated with slow waves of normal duration. On top of the slow waves spikes emerge. Note that the amplitude of the plateau potentials of the prolonged slow waves reached higher levels than the plateau potentials of the slow waves of unchanged duration. (c) 5 min after carbachol (5×10^{-8} M). Notice the increased incidence of prolonged slow waves and increased spiking activity. (d) return to Krebs' solution at the beginning of the tracing. Reprinted from Huizinga et al. J. Pharm. Exp. Ther. 1984; 231: 692–699.

severely decreased or blocked [7]. Phasic contractile activity depends upon periodic increases in the open probability of L-type Ca^{2+} channels. In whole-cell voltage clamp experiments, current through L-type Ca^{2+} channels can usually be resolved with test potentials positive to -50 mV. These Ca^{2+} channels first activate, and then undergo voltage- and Ca^{2+}-dependent inactivation. The activation and inactivation properties of Ca^{2+} channels predict that inactivation is incomplete over a range of potentials between -50 and -20 mV. This observation predicts that a small Ca^{2+} current continues to flow during the plateau phase of electrical slow waves in GI muscles. Measurements of this current and resulting changes in intracellular Ca^{2+} concentration show that the small Ca^{2+} current is capable of providing enough $[Ca^{2+}]_i$ to explain excitation-contraction coupling during slow waves [5]. Small increases in the degree and duration of depolarization dramatically change the accumulation of $[Ca^{2+}]_i$ during depolarizing steps, thus explaining the steep dependence of contractile amplitude on slow wave amplitude and duration. In addition, wherever spikes occur superimposed on the slow waves, further marked increases in force development are noted (Fig. 1). Blockers of L-type calcium channels inhibit the increase in $[Ca^{2+}]_i$ caused by such depolarizations.

In gastrointestinal smooth muscle that depends upon electrical slow waves for activation, excitatory agonists produce a small depolarization by influencing the open probabilities of K channels [8] or by activating non-selective cation channels [9]. As a consequence, an increase in the amplitude and duration of slow waves and in many tissues increased spiking activity superimposed on the slow waves occurs. These electrical phenomena reflect Ca^{2+} entry, resulting in increased force of contraction.

Classical L-type calcium channel blockers form a heterogeneous group of agents that differ from a chemical, pharmacological and therapeutic point of view [10–12]. They can be divided into three major groups: phenylalkylamines, dihydropyridines and benzothiazepines, whose prototype compounds are verapamil, nifedipine and diltiazem respectively. These agents exert their action at different receptor domains carried on the alpha-1-unit of L-type calcium channels. Recent studies by Godfraind et al. indicate molecular heterogeneity of calcium channels. Using rabbit small intestine, the α-1 subunit was found to be different from corresponding units in the aorta, lung and heart [13]. In the extracellular loop, between the third and fourth segments of domain IV and the transmembrane loop, primary sequence variations were found corresponding to alternative splicing phenomena. Specificity of pharmacological targeting of intestinal smooth muscle is also sought through pharmacokinetic means. Pinaverium for example is a quaternary ammonium compound that, taken orally, is not absorbed [41] and thereby action on the cardiovascular system is avoided. Pinaverium bromide reduced the voltage dependent inward current through L-type calcium channels in single cells of the rabbit small intestine [15]. Pinaverium was found to interact with the dihydropyridine binding site, at the external surface of the plasma membrane, in a competitive manner [14].

Release of Calcium from Internal Stores

The sarcoplasmic reticulum in gastrointestinal smooth muscle cells stores calcium. This calcium complements calcium entry through the plasma membrane in generating contractions. It further plays a regulatory role in calcium homeostasis and smooth muscle electrical activity [9]. The physical relationships between the sarcoplasmic reticulum, the cytosol and the plasma membrane are complex. There is increasing morphological and physiological evidence for a "buffer barrier" between the superficial SR and the plasma membrane, as originally proposed by Van Breemen and co-workers [16]. Since this buffer barrier is far more extensive in certain types such as the interstitial cells of Cajal, this suggests that in such cells specific cell functions are associated with SR-plasma membrane interactions [17]. One such function may be the activation of plasma membrane ion channels by calcium released from the SR [18]. In intestinal smooth muscle, IP_3 and ryanodine receptors are not distributed equally, supporting the hypothesis of a specialized function of superficial SR [19].

The role of SR calcium and the buffer barrier in regulating calcium entry from the plasma membrane ion channels has thus far been studied mainly in vascular smooth muscle. Depletion of the SR slows down depolarization induced contraction without decreasing the rate of Ca^{2+} influx. Force did not develop until the SR was refilled to close to physiological levels [20]. Prevention of SR Ca^{2+} accumulation with caffeine abolished a delay between the rise in $[Ca^{2+}]_i$ and force development and changed the nature of the $[Ca^{2+}]_i$ increase to monophasic [21]. These and other data led to the hypothesis that elevation of $[Ca^{2+}]_i$ can be restricted to a sub-plasmalemmal space and that the Ca^{2+} pump of the peripheral SR is responsible for local Ca^{2+} gradients. Ganitkevich and Isenberg have shown that in urinary bladder smooth muscle, $[Ca^{2+}]_i$ transients induced by depolarizing pulses have a component of calcium induced calcium release [22]. If the rate of Ca^{2+} entry is slowed down by gradual depolarization, which allows for inactivation of Ca^{2+} channels during the ramp, the SR buffering function becomes dominant. Inhibition of this buffering action increases the ramp induced $[Ca^{2+}]_i$ transient, while a decline in Sr Ca^{2+} content diminishes the pulse-induced $[Ca^{2+}]_i$ transient.

Not only is SR calcium involved in regulation of calcium entry, the SR is also involved in calcium extrusion. This pathway involves efflux from the SR into a restricted space where Na/Ca exchange effectively removes calcium from the cell. In fact, in vascular smooth muscle it is suggested that steady-state maintenance of the buffer barrier function requires continuous SR Ca^{2+} unloading to the extracellular space [23], implying that Ca^{2+} extrusion from the cytoplasm proceeds partly via the SR-Ca^{2+} pump [24, 25]. Recently, it was found that abolition of SR Ca^{2+} accumulation by caffeine, thapsigargin or ryanodine caused a 60% inhibition of the rate of Ca^{2+} extrusion measured as the rate of decline of $[Ca^{2+}]_i$ in the absence of extracellular Ca^{2+} (Chen, Haynes and van Breemen, unpublished results). In the *vena cava* smooth muscle preparation, removal of external Na^+ caused a similar inhibition, which was not additive to the inhibition induced by thapsigargin. These data suggest that in this smooth muscle preparation Ca^{2+} extrusion proceeds via two separate pathways: 40% is mediated by the plasmalemmal Ca^{2+}, Mg^{2+}-ATPase and 60% is first taken up into the SR from where it is released into a specialized SR-plasmallemmal junctional space to be extruded by the Na^+/Ca^{2+}-exchanger.

The pharmacology of SR calcium release is investigated mostly at the single cell level at the moment, and the lack of specificity of drugs used has precluded meaningful tissue or whole animal evaluation, nevertheless therapeutic approaches using this pathway are being discussed [26]. The pharmacology of the refilling pathways into the SR calcium store is also investigated. This is clearly not identical in all smooth muscle but it is of interest that in some smooth muscle cells the refilling is inhibited by L-type calcium channel blockers [27] so that calcium channel antagonists may block contractions also in smooth muscle cells that depend on intracellular calcium release for contraction.

Calcium Entry via Muscarinic Receptor Activation

Muscarinic receptor-activated nonselective cation channels ($I_{ns,Ach}$) are ubiquitously found in gastrointestinal smooth muscle, and are thought to play a central role in the excitatory control of gut motility [9]. Two main characteristics of these channels make them suitable for receptor mediated calcium entry, they open near the resting membrane potential and they show a high Ca permeability.

Prolonged exposure of single smooth muscle cells from the guinea-pig ileum to acetylcholine generated a sustained atropine sensitive inward current with a concomitant contraction, when the cells were voltage clamped at $-60\,mV$ (Fig. 2) [28, 29]. When the cell was pretreated with Ca-free solution over several minutes, the Ach-induced contraction was greatly attenuated and correspondingly, the inward current was also reduced. This inward current evoked by Ach is mediated mainly by nonselective cationic channels of $20-25\,pS$, which appear to be more permeable to divalent than to monovalent cations: the selectivity sequence estimated from the reversal potential is $Ba^2 \geq Ca^{2+} > Na^+ = Li^+ \geq Cs^+ \geq K^+ \gg Mg^{2+}$ [30]. Nevertheless, rapid removal of external Na reduced the amplitude of the Ach-induced current by about 90%, suggesting that about 10% of the current is carried by Ca^{2+} ions near the resting membrane potential. The $[Ca^{2+}]_i$ rise due to this source of Ca entry is about $10\,nM$ [31]. It is likely that the mechanisms which reduce $[Ca^{2+}]_i$ such as Ca sequestration and extrusion are activated simultaneously upon muscarinic receptor stimulation. The extent of calcium entry through the $I_{ns,Ach}$ channel will remain an intriguing

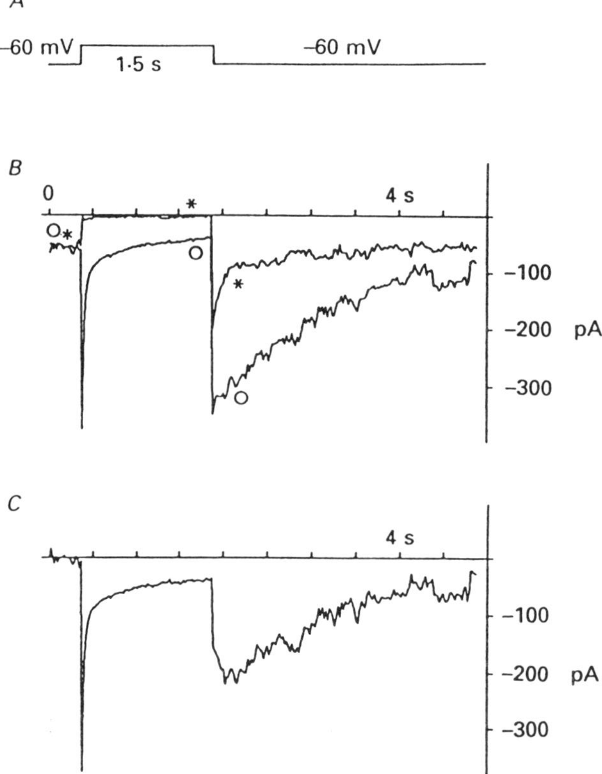

Fig. 2. Ca influx through L-type calcium channels markedly influences the Ach-induced, non-specific inward current. (A) Pulse protocol in the presence of 300 μM Ach in the absence (O) or presence of 300 μM nitrendipine (*). (B) Inward currents during depolarizing pulse, and tail currents reflecting $I_{ns,Ach}$. (C) Depiction of the nitrendipine sensitive current as difference between (O) and (*). Note: $I_{ns,Ach}$ before the pulse is insensitive to nitrendipine. During the pulse, at 0 mV, $I_{ns,ACh}$ does not contribute to the record because it is set at the reversal potential, thus the nitrendipine-sensitive current is I_{Ca}. Tail $I_{ns,Ach}$ after the pulse is strongly diminished by nitrendipine. Reprinted with permission from [43].

question, because this concerns the most essential property of the receptor-operated Ca entry hypothesis [32].

The pharmacological properties of the $I_{ns,Ach}$ channel have recently been investigated. Surprisingly, many of the agents which are frequently used in pharmacological research for other channels inhibit the Ach-induced cationic current [33]. For example, K channel blockers such as tetraethylammonium, 4-aminopyridine, quinine and procaine are all able to block the current. Furthermore, even the agents that have been believed to selectively block the voltage-dependent Ca channels such as D-600 [34] and nicardipine blocked the Ach-induced current; nicardipine at 10 μM about 70% reduction occurs (Inoue, Waniishi and Ito, unpublished data). These results suggest that there may be few drugs that selectively block the receptor-operated nonselective cation channels. However, several drugs derived from nonsteroidal anti-inflammatory drugs have recently been introduced as relatively selective blockers of nonselective cation channels in several tissues [35]. Diphenylamine-2-carboxylate (DPC) derivatives such as 3,'5'-dichloro-DPC, flufenamic acid and mefenamic acid have been proved potent in inhibiting the nonselective cation channels. Further refinement of these derivatives might lead to the discovery of selective blockers for the receptor-operated nonselective cation channels in smooth muscle.

Although most studies of non-selective cation channels have been performed via activation of the conductance with cholinergic agonists, recent studies have shown that a variety of excitatory neurotransmitters and hormones activate these channels. These observations suggest an important convergence of excitatory agonists at the non-selective cation conductance, making the role of these channels extremely important in excitation-contraction coupling in the GI tract. Since L-type calcium channel blockers inhibit Ach-induced calcium entry through L-type calcium channels as well as through non-specific calcium channels, these drugs will be effective in cholinergically mediated contractions. It may well be that it is of critical importance to inhibit cholinergically induced spasmogenic contractions in GI motility disorders such as IBS.

The Role of Calcium Channel Blockers in Gastrointestinal Disease

Since contractile activity of gastrointestinal smooth muscle is dominantly mediated by L-type calcium channels (Fig. 3), drugs affecting these channels have been evaluated for GI motility disorders [12]. There is a considerable body of evidence to support the use of nifedipine in certain conditions of esophageal motor disfunction [36]. It reduces pressure in the lower esophageal sphincter and the amplitude of esophageal body contractions. It also decreases the frequency of non-peristaltic waves. This would all be of little clinical value unless symptoms were improved but this has been well documented too. Esophageal dysmotility can produce unpleasant, frightening and disabling symptoms of chest pain that may resemble angina, or cause difficulty in swallowing. Achalasia is the failure of the tonic, lower esophageal sphincter to relax with swallowing and allow a food bolus to pass into the stomach. "Diffuse esophageal spasm" refers to periods of non-peristaltic and powerful contraction and "Nutcracker" esophagus describes peristaltic contractions of very high amplitude. These abnormalities can respond well to calcium channel antagonists, in particular nifedipine. Clinical value in esophageal problems can be limited by unwanted cardiovascular actions and other adverse effects such as headache, flushing and peripheral edema.

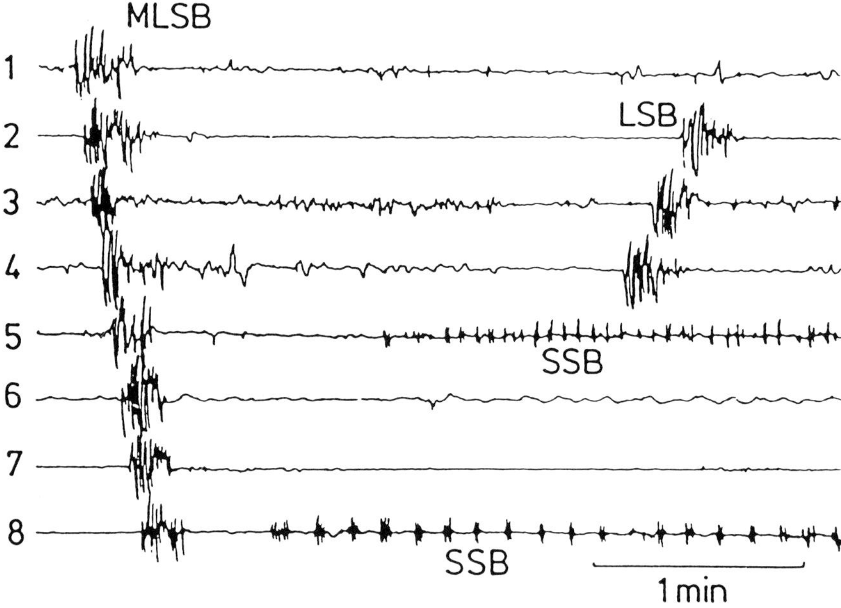

Fig. 3. Fioramonti et al. [44] measured electrical activities generated by the smooth muscle layers of the colon of an IBS patient. Various patterns of electrical activity were noted. These activities are all mediated by L-type calcium channels [45, 46]. MLSB = migrating long spike burst. LSB = long spike burst. SSB = short spike burst. The numbers refer to different intraluminal electrodes positioned 3 cm apart. Reprinted with permission from [44].

Irritable bowel syndrome describes patients with abdominal pain, associated with a perceived disturbance of bowel function. Currently, no common underlying organic disease has been identified [37]. IBS is very common and makes up between 20% and 50% of the general gastroenterological workload, affecting ~5 million people in the United States [38]. The etiology of IBS is not well understood, but diet, emotional state and hypersensitivity of the gut to a wide range of stimuli are thought to be important factors. Disorders of motility and transit have been described and related to symptoms [37]. Unlike esophageal dysmotility, there is not yet a recognized characteristic, pathognomonic abnormal pattern of motility for IBS. The classical L-type calcium channel antagonists have not been intensively investigated in IBS; few double-blind, controlled trials have been carried out [39]. This may be because of their adverse effects and the difficulty of performing such clinical trials under clinical conditions where the evaluation is based on symptoms and the placebo effect is high [40]. Pinaverium bromide is a calcium channel antagonist that has selective activity for gastrointestinal smooth muscle. It achieves this because of low absorption, rapid hepatic metabolism and hepatobiliary excretion [41]. This means that most of an oral dose remains in the gut. Oral doses that have clear and significant effects on gut physiology and symptoms have no effect on the cardiovascular system. Moreover, it does not cause headache, flushing or peripheral edema. Pinaverium inhibits the colonic motor response to eating [42]. It accelerates transit in the distal colon in patients with constipation. It appears that it can reduce "spasm" type motor activity without inhibiting propulsive patterns of motility. It therefore seems appropriate for patients with IBS.

In many laboratories, the pharmacology of the various pathways leading to increased cytosolic calcium is investigated. The different systems discussed are clearly interrelated and the pharmacology overlapping. In the laboratory we are still working with rather non-specific drugs to unravel the control systems of intracellular calcium regulation. In some cases we can use pharmacokinetic properties of drugs to attain at least some degree of specificity *in vivo* such as with pinaverium. The use of L-type calcium channel blockers in clinical practice is already extensive and clinical trials are being conducted to better understand the *in vivo* actions of the drugs. Current research is directed towards elucidating molecular specificity of intestinal L-type calcium channels as well as using pharmacokinetic methods to target the GI tract.

Acknowledgements

This work was supported by the Medical Research Council, Canada.

References

1. Szurszewski JH. Electrophysiological basis for gastrointestinal motility. In: Johnson LR, (editor), Physiology of the gastrointestinal tract, Vol. 2. New York: Raven Press, 1981: 1435–1466.
2. Sanders KM. Ionic mechanisms of electrical rhythmicity in gastrointestinal smooth muscles. Annu. Rev. Physiol. 1992; 54: 439–453.
3. Huizinga JD. Action potentials in gastrointestinal smooth muscle. Can. J. Physiol. Pharmacol. 1991; 69: 1133–1142.
4. Langton PD, Burke EP, Sanders KM. Participation of Ca currents in colonic electrical activity. Am. J. Physiol. 1989; 257: C451–C460
5. Vogalis F, Publicover NG, Hume JR, Sanders KM. Relationship between calcium current and cytosolic calcium in canine gastric smooth muscle cells. Am. J. Physiol. 1991; 260: C1012–C1018.
6. Molleman A, Thuneberg L, Huizinga JD. Characterization of the outward rectifying potassium channel in a novel intestinal smooth muscle preparation. J. Physiol. 1993; 470: 211–229.
7. Barajas-López C, Huizinga JD. Different mechanisms of contraction generation in circular muscle of canine colon. Am. J. Physiol. 1989; 256: G570–G580.
8. Sims SM, Singer JJ, Walsh JV, Jr. Cholinergic agonists suppress a potassium current in freshly dissociated smooth muscle cells of the toad. J. Physiol. (Lond) 1985; 367: 503–529.
9. Sims SM, Janssen LJ. Cholinergic excitation of smooth muscle. NIPS 1993; 8: 207–212.
10. Godfraind T, Miler R, Wibo M. Calcium antagonism and calcium entry blockade. Pharmacol. Rev. 1986; 38: 321–416.

11. Huizinga JD, Liu LWC. Role of calcium channels in pharmacological modulation of gastrointestinal motility. In: Christen MO, Paoletti R (editors), Calcium antagonists in gastroenterology research and perspectives. Boston: Kluwer Academic Publishers, 1993: 31–39.

12. De Ponti F, Giaroni C, Cosentino M, Lecchini S, Frigo G. Calcium-channel blockers and gastrointestinal motility: basic and clinical aspects [review]. Pharmacol. & Ther. 1993; 60: 121–148.

13. Feron O, Octave JN, Christen MO, Godfraind T. Quantification of two splicing events in the L-type calcium channel alpha-1 subunit of intestinal smooth muscle and other tissues. Eur. J. Biochem. 1994; 222: 195–202.

14. Feron O, Wibo M, Christen MO, Godfraind T. Interaction of pinaverium (a quaternary ammonium compound) with 1,4-dihydropyridine binding sites in rat ileum smooth muscle. Br. J. Pharmacol. 1992; 105: 480–484.

15. Beech DJ, MacKenzie I, Bolton TB, Christen MO. Effects of pinaverium on voltage-activated calcium channel currents of single smooth muscle cells isolated from the longitudinal muscle of the rabbit jejunum. Br. J. Pharmacol. 1990; 99: 374–378.

16. van Breemen C. Calcium requirement for activation of intact aortic smooth muscle. J. Physiol. (Lond) 1977; 272: 317–329.

17. Liu LWC, Thuneberg L, Huizinga JD. Regulation of colonic pacemaker frequency by intracellular calcium in sacroplasmic reticulum [abstract]. J. Gastroint. Motil. 1993; 5: 201.

18. Huizinga JD, Farraway L, Den Hertog A. Generation of slow-wave-type action potentials in canine colon smooth muscle involves a non-L-type Ca^{2+} conductance. J. Physiol (Lond) 1991; 442: 15–29.

19. Wibo M, Godfraind T. Comparative localization of inositol 1,4,5-triphosphate and ryanodine receptors in intestinal smooth muscle: an analytical subfractionation study. Biochem. J. 1994; 297: 415–423.

20. Loutzenhiser R, Leyten P, Saida K, van Breemen C. Ca^{2+} compartments and Ca^{2+} mobilization during contraction of smooth muscle. In: Grover AK, Daniel EE (editors), Calcium and smooth muscle contractility. Clifton NJ: Humana Press Inc. 1985.

21. Chen Q, Cannell M, van Breemen C. The superficial buffer barrier in vascular smooth muscle. Can. J. Physiol. Pharmacol. 1992; 70: 509–514.

22. Ganitkevich VY, Isenberg G. Caffeine-induced release and reuptake of Ca^{2+} by Ca^{2+} stores in myocytes from guinea-pig urinary bladder. J. Physiol. (Lond) 1992; 458: 99–117.

23. van Breemen C, Cauvin C, Johns A, Leijten P, Yamamoto H. Ca^{2+} regulation of vascular smooth muscle [review]. Federation Proceedings 1986; 45: 2746–2751.

24. Nishimura J, Khalil RA, van Breemen C. Agonist-induced vascular tone [review]. Hypertension 1989; 13: 835–844.

25. Chen Q, van Breemen C. The superficial buffer barrier in venous smooth muscle;. sarcocplasmic reticulum refilling and unloading. Br. J. Pharmacol. 1993; 109: 336–343.

26. Ehrlich BE, Kaftan E, Bezprozvannaya S, Bezprozvanny I. The pharmacology of intracellular Ca^{2+}-release channels. Trends in Pharmacol. Sci. 1994; 15: 145–149.

27. Godfraind T, Christen MO, Dessy C, Feron O, Morel N, Octave JN et al. Characterization of receptors for calcium channel blockers on intestinal smooth muscle [abstract]. Can. J. Physiol. Pharmacol. 1994; 72 (Suppl. 1): 19.

28. Inoue R, Kitamura K, Kuriyama H. Acetylcholine activates single sodium channels in smooth muscle cells. Pflugers Archiv – Eur. J. Physiol. 1987; 410: 69–74.

29. Vogalis F, Sanders KM. Cholinergic stimulation activates a non-selective cation current in canine pyloric circular muscle cells. J. Physiol. (Lond) 1990; 429: 223–236.

30. Inoue R, Isenberg G. Acetylcholine activates nonselective cation channels in guinea pig ileum through a G protein. Am. J. Physiol. 1990; 258: C1173–C1178.

31. Pacaud P, Bolton TB. Relation between muscarinic receptor cationic current and internal calcium in guinea-pig jejunal smooth muscle cells. J. Physiol. (Lond) 1991; 441: 477–499.

32. Cousins HM, Edwards FR, Hirst GD, Wendt IR. Cholinergic neuromuscular transmission in the longitudinal muscle of the guinea-pig ileum. J. Physiol. (Lond) 1993; 471: 61–86.

33. Chen S, Inoue R, Ito Y. Pharmacological characterization of muscarinic receptor-activated cation channels in guinea-pig ileum. Br. J. Pharmacol. 1993; 109: 793–801.

34. Inoue R, Isenberg G. Effect of membrane potential on acetylcholine-induced inward current in guinea-pig ileum. J. Physiol. (Lond) 1990; 424: 57–71.
35. Siemen D, Hescheler J, editors. Nonselective cation channels: pharmacology, physiology and biophysics. Basel: Birkhäuser Verlag 1993.
36. Hongo M, Traube M, McAllister RG, Jr., McCallum RW. Effects of nifedipine on esophageal motor function in humans: correlation with plasma nifedipine concentration. Gastroenterology 1984; 86: 8–12.
37. Collins SM. The irritable bowel syndrome. Can. Med. Assoc. J. 1988; 138: 309–316.
38. Heaton KW. Epidemiology of irritable bowel syndrome. European Journal of Gastroenterology and Hepatology 1994; 6: 465–469.
39. Godfraind T, Govoni S, Paoletti R, Vanhoutte PM, editors. Calcium antagonists. Pharmacology and Clinical Research. Dordrecht, The Netherlands: Kluwer Academic Publishers 1993.
40. Cann PA, Read NW, Brown C, Hobson N, Holdsworth CD. Irritable bowel syndrome: relationship of disorders in the transit of a single solid meal to symptom patterns. Gut 1983; 24: 405–411.
41. Christen MO. Action of pinaverium bromide, a calcium-antagonist, on gastrointestinal motility disorders. Gen. Pharmac. 1990; 21 No. 6: 821–825.
42. Fioramonti J, Frexinos J, Staumont G, Bueno L. Inhibition of the colonic motor response to eating by pinaverium bromide in irritable bowel syndrome patients. Fundamental & Clinical Pharmacology 1988; 2: 19–27.
43. Inoue R, Isenberg G. Intracellular calcium ions modulate acetylcholine-induced inward current in guinea-pig ileum. J. Physiol. (Lond) 1990; 424: 73–92.
44. Frexinos J, Fioramonti J, Bueno L. Colonic myoelectrical activity in IBS painless diarrhoea. Gut 1987; 28: 1613–1618.
45. Huizinga JD, Waterfall WE. Electrical correlate of circumferential contractions in human colonic circular muscle. Gut 1988; 29: 10–16.
46. Huizinga JD. Electrophysiology of human colon motility in health and disease. Clin. Gastroenterol. 1986; 15: 879–901.

Pharmacological Sciences: Perspectives for
Research and Therapy in the Late 1990s
ed. by A.C. Cuello and B. Collier
© 1995 Birkhäuser Verlag Basel/Switzerland

Aspects of Potassium Channel Modulation

Gillian Edwards[1], Icilio Cavero[2], Gregory J. Kaczorowski[3],
Olaf Pongs[4], Uli Quast[5], Norio Taira[6] and Arthur H. Weston[1]

[1]*School of Biological Sciences, University of Manchester, G.38 Stopford Building, Oxford Road,
Manchester M13 9PT, UK;* [2]*Rhône-Poulenc Rorer, Centre de Recherche de Vitry-Alfortville,
13, Quai Jules Guesde, 94400 Vitry Sur Seine, France;* [3]*Merck Sharp & Dohme Research
Laboratories, P.O. Box 2000, Rahway, New Jersey 07065-0900, USA;* [4]*Zentrum für Molekuläre
Neurobiologie, Institut für Neurale Signalverarbeitung, Martinistraße 52 – Haus 42, D-20246,
Hamburg, Germany;* [5]*Pharmakologisches Institut, Eberhard-Karls-Universität, Wilhelmstraße
56, D-7400 Tübingen 1, Germany;* [6]*Tohoku University School of Medicine, Aoba-ku 980,
Sendai, Japan*

Introduction

Advances in molecular biology are giving unprecedented insights into
potassium (K)-channel structure and function. In parallel, the development of channel modulators and their investigation using electrophysiological techniques has revealed much, but posed many questions. What
is the importance of the recently-discovered channel β-subunits? What
is the nature of the so-called ATP-sensitive K-channel and how do
K-channel openers exert tissue-protective actions? Can the K-channels
in lymphocytes be exploited therapeutically? The purpose of this chapter is to give the reader an insight into this dynamic pharmacological
field and to provide a stimulus for further discussion and experimentation.

Implications of Subunit Composition for the Properties of Voltage-Gated K-Channels

Background

K-channels are ubiquitous membrane proteins which occur in most
excitable and non-excitable cells [1]. Many cDNAs which encode K-
channel-forming subunits have been cloned and characterized [2, 3]
using *in vitro* expression systems. From these studies the major finding
has been an unexpectedly close relationship not only between voltage-

Correspondence to: Gillian Edwards, address as above.

and ligand-gated K-channels, but also between K-, Na- and Ca-channels [4]. Furthermore, distinct structural domains control channel selectivity, conductance, gating and inactivation [5, 6].

The Role of α- and β-Subunits

(i) Basic Structural Features

A tetrameric assembly of α-subunits is apparently sufficient to express functional voltage-gated K-channels and to elicit outward currents in *in vitro* expression systems [7]. The derived primary sequences of α-sub-

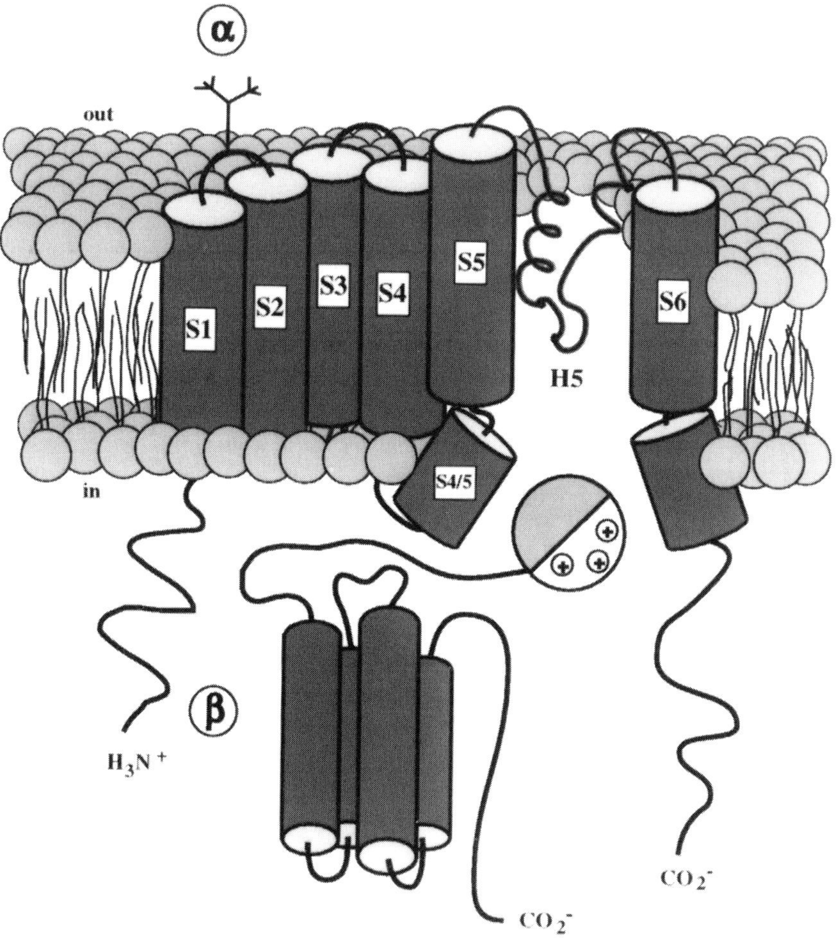

Fig. 1. Hypothetical model of the structures of α- and β-subunits comprising rat voltage-gated K channels. For further details, see text.

units are quite similar and six hydrophobic segments S1 to S3, S5, H5 and S6, together with a positively charged segment S4 interspersed between segments S3 and S5 can always be recognized (Fig. 1). The S1 to S6 segments probably traverse the cellular membrane such that sequences between S1 and S2, S3 and S4, S5 and S6, respectively, are facing the extracellular space. Accordingly, the amino terminal, carboxyterminal and the sequences between S2 and S3, and S4 and S5 are facing the cytoplasmic side of the membrane. The pore-forming H5 region is probably tucked into the membrane such that it enters and exits the lipid bilayer from the extracellular side.

Rapidly-inactivating A-type K-channels may have an amino terminal inactivating domain (Fig. 1) which is able to close the open channel from the inside at depolarized membrane potentials. This type of inactivation is often referred to as "N-type" inactivation [8].

(ii) The α-Subunit

Most of the vertebrate K-channel cDNAs which have been cloned to date encode K-channel α-subunits, which express slowly-inactivating K-channels in *in vitro* systems. This is somewhat unexpected since one could imagine the need for a great variety of A-type K-channels, whereas the requirement for many delayed rectifiers is not so obvious. It has been reported that the addition of β-subunits converts non-inactivating delayed-rectifier type K-channels (formed from α-subunits alone) into A-type K-channels [9]. This conversion might represent a basic cellular mechanism for the modulation of K-channel properties.

(iii) The β-Subunit

Certain K-channel β-subunits may harbour an N-terminal inactivating domain with structural features similar to those in the α-subunit-inactivating domain. Like the inactivating domains of α-subunits, the activity of the β-subunit-inactivating domain does not require a covalent bond to the β-subunit core sequence. A peptide corresponding to its amino terminus may be added to the bathing solution of inside-out patches which express slowly inactivating, delayed-rectifier type K-channels. This leads to a rapid inactivation of the elicited outward currents [9].

Conclusions

The assembly of α- and β-subunits in a hetero-oligomeric complex (Fig. 1) may considerably expand the capability of excitable cells to express diverse voltage-gated K-channels. One possible biological significance of such separation may be suggested from the observation that β-subunit activity regulates the conversion of delayed rectifier-type K-channels into those of the A-type, and vice-versa. Such plastic changes in electrical excitability may contribute to molecular mechanisms underlying learning and behaviour.

K-Channel Modulation and the Control of Human T-Lymphocyte Activation

Background

Lymphocytes use a variety of ionic pathways to regulate Ca^{2+} homeostasis and thereby modulate cellular signal transduction processes. During activation of lymphocytes by mitogens *in vitro*, many initial events are coupled to transmembrane ion fluxes and consequent changes in membrane potential, including the influx of Ca^{2+} and the efflux of K^+ [10]. Concurrently, hydrolysis of membrane phosphatidylinositides occurs which yields inositol 1,4,5 trisphosphate and diacylglycerol. The resulting sustained elevation in intracellular Ca^{2+} concentration, together with stimulation of protein kinase C, promotes lymphocyte activation, eventually leading to the expression of lymphokines, DNA synthesis and cell division.

Voltage-Gated K-Channels

Patch-clamp information from human and rodent T-cells shows that the predominant outward currents are due to voltage-gated K-channel activity. Three distinct types of K-channel (n- n'- and l-type) can be differentiated based on their biophysical and pharmacological properties [11–13]. The only voltage-gated K-channel present in human T-cells has been designated the n (normal)-type. This channel displays a unitary conductance of 12–16 pS, activates at potentials positive to -50 mV, and inactivates with a time constant of 100–200 ms. It is also sensitive to inhibition by 4-aminopyridine (4-AP), tetraethylammonium ion (TEA), L-type Ca^{2+} channel inhibitors and charybdotoxin (ChTX; see below). Based upon its biophysical properties, the n-type channel probably controls both cell volume [14] and resting potential [15] of human T-lymphocytes. The channel is the product of the *Shaker*-related K-channel gene, $K_V 1.3$ [16], expression of which in *Xenopus* oocytes produces currents that are very similar to those characteristic of n-type channels in lymphocytes.

Effects of Selective K-Channel Inhibitors

Using the lipophilic cation [^{3}H]tetraphenylphosphonium ion to monitor membrane potential, margatoxin (17; MgTX), noxiustoxin (NxTX) and charybdotoxin (ChTX) produce an equivalent depolarization of human T-cells from a value of -60 mV to -25 mV [18]. These results indicate that the resting membrane potential is controlled by $K_V1.3$ channel activity. In other studies [19], ChTX blocked T-lymphocyte proliferation and lymphokine production stimulated only by signal transduction pathways which elevated intracellular Ca^{2+}. Similarly, MgTX and NxTX also inhibited lymphokine production with a rank order of potency predicted from their ability to interact with $K_V1.3$ (inhibition of [125I]ChTX binding). Importantly, the three peptides also prevent the rise in intracellular Ca^{2+} that occurs following stimulation by anti-CD3, with MgTX being the most potent agent in this model.

Conclusions

If depolarization is important in preventing T-cell activation, then any means of inducing this phenomenon should have the same functional consequences as exposure to $K_V1.3$ inhibitors. Consistent with this idea, high K^+ (80 mM) depolarizes T-cells and blocks the ability of anti-CD3 to increase intracellular Ca^{2+} [19]. The magnitude of the effects are identical to, but not additive with, those produced by $K_V1.3$ inhibitors. Furthermore, depolarization by high K^+ blocks IL-2 production and T-lymphocyte proliferation [20]. Although the relationship between membrane potential and changes in cytosolic Ca^{2+} during lymphocyte activation is not completely understood, the present data demonstrate that inhibition of $K_V1.3$ channels depolarizes human T-lymphocytes and this alters Ca^{2+} homeostasis so that (mitogen-activated) Ca^{2+}-dependent pathways of cell activation are specifically blocked. These results suggest a strategy for the development of novel immunosuppressant drugs based on the discovery of selective $K_V1.3$ channel inhibitors.

K_{ATP} – a Nucleotide-Sensitive K-Channel

Background

Few K-channels have been so widely studied over the past few years as the so-called ATP-sensitive K-channel (K_{ATP}). This channel is inhibited by micromolar concentrations of $[ATP]_i$ (or non-hydrolysable analogs) and requires phosphorylation by MgATP for normal functioning. On run-down, it can be reactivated by nucleoside diphosphates; its open probability is enhanced by K-channel openers like levcromakalim, the action of which is inhibited by glibenclamide and other sulphonylureas [21]. Under symmetrical high K^+ conditions, its unitary conductance lies in the range 40–80 pS.

Effects of Nucleoside Diphosphates

ADP opposes the inhibitory effect of ATP and under physiological conditions, channel opening is probably enhanced by a fall in the $[ATP]_i:[ADP]_i$ ratio rather than by a decrease in $[ATP]_i$ alone [22]. When K_{ATP} is fully-phosphorylated, nucleoside diphosphates other than ADP are also capable of reducing the inhibitory effect of ATP on channel opening. In addition, in detached membrane patches, nucleoside diphosphates (of which ADP is the least efficacious) are capable of stimulating the opening of partially-dephosphorylated (run-down) channels which no longer open spontaneously [23, 24]. These re-activated channels retain their sensitivity to the inhibitory effect of $[ATP]_i$. Interestingly, although K-channel openers are incapable of stimulating the opening of K_{ATP} after run-down, these agents can enhance the opening effect of nucleoside diphosphates [23]. Indeed, the presence of nucleoside diphosphates may be essential for the actions of some K_{ATP}-channel openers, e.g. nicorandil [25]. Further dephosphorylation of K_{ATP} produces a totally quiescent channel, even in the presence of nucleoside diphosphates, although the opening effect of these nucleotides returns after exposure to MgATP [24].

Effects of K-Channel Openers

There is a good correlation between the ability of the structurally-diverse K-channel openers to inhibit the binding of [³H]P1075 (a pyridine) in rat aortic strips and to relax rat aorta precontracted with noradrenaline [26]. This may indicate that the K-channel openers share a common site of action. Nevertheless, this is unlikely to be directly associated with the channel since, under a variety of conditions, specific binding of radiolabelled K-channel openers cannot be demonstrated in membrane preparations from either rat aorta or an insulin-secreting cell line (RINm5F) [26, 27].

In addition to enhancing the opening of K_{ATP}, K-channel openers simultaneously inhibit the delayed rectifier channel (K_V) in both smooth muscle and insulinoma cells [21]. Both of these effects are also produced by conditions which are anticipated to be dephosphorylating [28]. Despite the marked increase in current associated with the opening of K_{ATP}, the peak outward current at positive potentials ($+40$ to $+50$ mV) is not enhanced by the presence of K-channel openers. Thus, under these conditions, the increase in the current carried by K_{ATP} almost exactly matches the decrease in the delayed rectifier current [28].

The basis for the dual effects of the K-channel openers is not certain. It is possible that these agents modify the phosphorylation status of the two channels (K_{ATP} and K_V) producing these opposite effects. However, an alternative explanation is that, at least in smooth muscle, K_{ATP} is not a distinct entity but is a partially dephosphorylated state of K_V [28]. Thus the K-channel openers could induce ATP-sensitivity and remove the voltage-sensitivity of K_V, allowing it to open at potentials more hyperpolarised than its normal activation potential (approximately -40 mV), and thus also modify inactivation.

It seems unlikely that the cardiac K_{ATP} could be a state of K_V, since there are at least two (and possibly three) different types of cardiac K_V channel which vary in proportion in different species. Furthermore, an ATP-sensitive K-channel (rcK_{ATP}-1) opened by pinacidil and with structural features typical of an inwardly-rectifying K-channel has just been cloned from cardiac muscle [29]. Nevertheless, it seems remarkable that whereas the cardiac K_{ATP} shows no voltage-sensitivity in isolated patches, the K-channel openers only appear to induce an outward current under whole-cell recording conditions since there is no evidence of any summation of inward current with that attributed to the inward rectifier (K_{IR}) [30]. One interpretation of this observation could be that the K-channel openers modify the voltage-sensitivity of K_{IR} and introduce ATP-sensitivity.

Conclusions

Initially it may appear unlikely that the K-channel openers could exert effects on two such structurally-different channels as K_V and K_{IR}, the α-subunits of which comprise six and two membrane-spanning regions, respectively. Nevertheless, these two channel types share structural homology in the H5 region which dips into the membrane and which lies between S5 and S6 of K_V (see Fig. 1) and between the two membrane-spanning regions of K_{IR} [31]. Furthermore, removal of four membrane-spanning regions of K_V1.1 (S1 to S4) produces an inwardly-rectifying channel [32]. Further studies to establish the relationship betwen 'K_{ATP}' and rectifying K-channels are clearly required.

Drug-Induced Modulation of Cardiac K_{ATP} During Ischemia

Background

The cardiac K_{ATP} is gated by changes in both $[ATP]_i$ and $[ADP]_i$ [24]. Under normoxic conditions when $[ATP]_i$ is approximately 3–4 mM, K_{ATP} has a very low open probability and thus does not contribute to normal cardiac function. However, when $[ATP]_i$ declines during metabolic stress, $[ADP]_i$ increases and pH_i falls. Under these conditions K_{ATP} fluctuates between closed and open states. The channels are virtually time-independent, their density on a single myocyte is very high and their unitary conductance is approximately 70 pS. Calculations show that the opening of less than 1% of the total K_{ATP} population can generate a current which can substantially shorten cardiac action potential duration.

Ischemic Preconditioning; An Endogenous Cardioprotective Strategy

If one or more 3–10 min periods of ischemic stress are interspersed by reperfusion (preconditioning), subsequent prolonged ischemia results in fewer abnormalities than in control hearts [33]. This so-called first "window" protective effect of preconditioning appears within a few minutes of the ischemic stimulus. It is, however, temporary and is lost if the reperfusion period between the end of the initial ischemic insult and subsequent prolonged metabolic stress exceeds 30 to 120 min or if the duration of the ischemic stress exceeds 30 to 90 min.

The mechanisms responsible for preconditioning appear to be biochemical in nature, since even a heart arrested with hyperkalaemia can be preconditioned to reduce energy consumption [34]. In certain species, blockade of adenosine receptors abolishes preconditioning-induced cardioprotection [33] which thus may be triggered by the release of adenosine. Furthermore, blockade of K_{ATP} with glibenclamide prevents mammalian cardiac myocytes from becoming preconditioned by brief ischemia [35]. It has thus been suggested that the adenosine receptor may be coupled to K_{ATP} [36].

Openers of K_{ATP} Afford Cardioprotection

In the *in vitro* perfused rat heart subjected to a transient (30 min) global ischemia, the K-channel openers nicorandil, pinacidil, cromakalim and aprikalim are cardioprotective, an action which is completely prevented by glibenclamide [37, 38]. In the perfused guinea-pig right ventricular wall, pinacidil and aprikalim attenuate the development of contracture during no-flow ischemia and reperfusion and they enhance the recovery of contractility during reperfusion. These effects are seen at concentrations that do not affect baseline contractility or accelerate the loss of inotropism at the early onset of ischemia [39].

K-channel openers administered to dogs before a stunning insult substantially accelerate the recovery of regional ventricular function on reperfusion. This effect is found at doses which are devoid of hemodynamic activity at systemic and coronary levels. Pretreatment of dogs with glibenclamide abolishes K-channel opener-induced cardioprotection while glibenclamide alone can worsen ischemic damage [37, 40]. In a model of irreversible myocardial infarction, K-channel openers are also cardioprotective at doses which are devoid of hemodynamic activity [36, 41].

Drug-Induced Cardioprotection; The Mechanism Remains Elusive

Initially, the cardioprotective effects of K-channel openers were attributed to shortening of action potential duration, which would accelerate the loss of contractility at the onset of ischemia. This would save high-energy phosphates, to be used subsequently by the ischemic myocyte to preserve its viability for a longer period during oxygen deficiency. However, bimakalim affords cardioprotection without affecting the duration of the ventricular action potential [41]. Furthermore, the cardioprotective effects of aprikalim can occur independently of an initial cardioplegic activity [39]. Similarly, the cardioprotection afforded by preconditioning does not require a depression of contractility, since it is possible to precondition non-contracting hearts [34].

Conclusions

None of the proposed theoretical mechanisms appears fully to account for the cardioprotective actions of the K-channel openers [42]. However, clarification of the mechanism underlying ischemic preconditioning will certainly help to understand the protective effects of the K-channel openers and provide new therapeutic approaches to protect the heart from an ischemic insult.

Clinical Experiences with Potassium Channel Openers

Background

Clinical experiences with K-channel openers are limited. To date, two agents (nicorandil and pinacidil) have been quite extensively studied with the majority of data relating to nicorandil, which was introduced into the Japanese market in 1984. Although nicorandil is structurally a nitrate (N) and indeed possesses actions characteristic of this drug class, its opening action on certain types of potassium channel (K action) in cardiac and vascular smooth muscle was proposed at the beginning of the 1980s. Thus, nicorandil can be characterized as an N-K hybrid [43] with nitrate-like actions prevailing in most conductance and capacitance vessels, whereas K-channel opening is dominant in resistance vessels. Both actions combine to dilate large coronary arteries and to increase coronary blood flow, to reduce both pre- and after-load, with a tendency to increase cardiac output. Nicorandil can shorten the action potential duration and reduce the force of contraction of cardiac muscle via K-channel opening but in vasodilator doses the drug exhibits virtually no cardiodepressant action.

Effects on Systemic Haemodynamics in Patients with Cardiovascular Disease

Effects on systemic haemodynamics of single doses of nicorandil administered intravenously or orally to patients with cardiac disease generally resemble those observed in animal experiments. Thus, systolic and diastolic blood pressures fall, as do left ventricular end-diastolic pressure and pulmonary capillary wedge pressure, indicating a reduction in pre-load [44]. Systemic vascular resistance also declines after nicorandil and this effect leads to a reduction in afterload. Nicorandil exerts various effects on cardiac output but an increase is usually observed and this is accompanied by a rise in heart rate of sympathetic reflex origin [44].

Antianginal Effects

(i) Effects on Vasospastic Angina
In patients with vasospastic angina, intravenous nicorandil promptly relieves spontaneous or ergonovine-induced coronary spasm and abolishes ST-segment elevation in the ECG and the anginal pain. The large coronary arteries are dilated [45] as observed with nitrates. In patients with variant angina, administration of nicorandil for 2 to 3 consecutive days prevents the coronary spasm induced by ergonovine [44] and significantly reduces the incidence of chest pains [47].

(ii) Effects on Stable Effort Angina
In patients with stable effort angina, nicorandil prolongs exercise duration and delays the time to onset of ischaemic symptoms [48]. In patients with coronary artery disease and in those with pacing-induced myocardial ischaemia, nicorandil improves left ventricular function and regional wall motion and abolishes pacing-induced angina [49]. The long-term efficacy of nicorandil in stable effort angina pectoris has been shown in patients who initially received nicorandil in individually-titrated regimens followed by maintenance therapy for 1 year [50].

Experiences with Pinacidil and Levcromakalim

Both pinacidil and levcromakalim lack a nitro-moiety and thus their effects in man are perhaps more representative of those which can be expected from a drug acting predominantly by a K_{ATP}-channel opening mechanism.

The effects of pinacidil in man have been comprehensively reviewed, and clinical trials have been directed towards its possible use in hypertension [51]. Clinical studies in comparison with placebo or against other antihypertensive agents showed good efficacy with the general profile of a peripheral vasodilator [51]. Key side-effects were oedema and headache, and approval for use in the USA was only given for concurrent administration with a thiazide.

Clinical data with levcromakalim are relatively sparse [52]. Like pinacidil, levcromakalim exhibits powerful antihypertensive actions and vasodilator headache is sometimes experienced. However, fluid retention does not seem to be a problem [52]. Interestingly, there is an improvement in blood lipid profiles [52] a phenomenon also reported for pinacidil [51].

Conclusions

Nicorandil can be considered as an effective anti-anginal agent. More potent K-channel openers such as pinacidil and levcromakalim exhibit powerful anti-hypertensive actions but their true place in the treatment of cardiovascular disease may lie in their use as cardioprotective agents (see earlier). The appropriate trials for this application have been initiated and the results of these are awaited with eager anticipation.

References

1. Rudy B. Diversity and ubiquity of K^+ channels. Neurosci. 1988; 25: 729–749.
2. Pongs O. Molecular biology of voltage-dependent potassium channels. Physiol. Rev. 1992; 72: S69–S88.
3. Gutman GA, Chandy KG. Nomenclature of mammalian voltage-dependent potassium channel genes. Neurosci. 1993; 5: 101–106.
4. Jan LY, Jan YN. Tracing the roots of ion channels. Cell 1992; 69: 715–718.
5. Jan LY, Jan YN. Structural elements involved in specific K^+ channel functions. Ann. Rev. Physiol. 1992; 54: 537–555.
6. Pongs O. Structure-function studies of the pore of potassium channels. J. Membrane Biol. 1993; 136: 1–8.
7. MacKinnon R. Determination of the subunit stoichiometry of a voltage-activated potassium channel. Nature 1991; 350: 232–235.
8. Hoshi T, Zagotta WN, Aldrich RW. Two types of inactivation in Shaker K^+ channels: effects of alterations in the carboxy-terminal region. Neuron 1991; 7: 547–556.
9. Rettig J, Heinemann S, Lorra C, Parcej DN, Dolly JO, Pongs O. Non-inactivating voltage-gated potassium channels are converted to A-type channels by association with a β-subunit. Nature 1994; 369: 289–294.
10. Grinstein S, Foskett JK. Ionic mechanisms of cell volume regulation in leukocytes. Ann. Rev. Physiol. 1990; 52: 599–614.
11. DeCoursey TE, Chandy KG, Gupta S, Cahalan MD. Voltage-gated K^+ channels in human T lymphocytes: a role in mitogenesis? Nature 1984; 307: 465–468.

12. Matteson DR, Deutsch C. K channels in T lymphocytes: a patch clamp study using monoclonal antibody adhesion. Nature 1984; 307: 468 471.
13. Lewis RS, Cahalan MD. Subset-specific expression of potassium channels in developing murine T lymphocytes. Science 1988; 239: 771 775.
14. Deutsch C, Chen L-Q. Heterologous expression of specific K^+ channels in T lymphocytes: functional consequences for volume regulation. Proc. Natl. Acad. Sci. USA 1993; 90: 10036 10040.
15. Cahalan MD, Chandy KG, DeCoursey TE, Gupta S. A voltage-gated potassium channel in human T lymphocytes. J. Physiol. 1985; 358: 197 237.
16. Grissmer S, Dethlets B, Wasmuth JJ, Goldin AL, Gutman GA, Cahalan MD et al. Expression and chromosomal localization of a lymphocyte K^+ channel gene. Proc. Natl. Acad. Sci. USA 1990; 87: 9411 9415.
17. Garcia-Calvo M, Leonard RJ, Novick J, Stevens SP, Schmalhofer W, Kaczorowski GJ et al. Purification, characterization, and biosynthesis of margatoxin, a component of Centruroides margaritatus venom that selectively inhibits voltage-dependent potassium channels. J. Biol. Chem. 1993; 268: 18866 18874.
18. Leonard RJ, Garcia ML, Slaughter RS, Reuben JP. Selective blockers of voltage-gated K^+ channels depolarize human T lymphocytes: mechanism of the antiproliferative effect of charybdotoxin. Proc. Natl. Acad. Sci. USA 1992; 89: 10094 10098.
19. Lin C, Boltz RC, Blake T, Nguyen M, Talento A, Fischer P et al. Voltage-gated potassium channels regulate calcium-dependent pathways involved in human T lymphocyte activation. J. Exp. Med. 1993; 177: 637 646.
20. Freedman BD, Price MA, Deutsch CJ. Evidence for voltage modulation of IL-2 production in mitogen-stimulated human peripheral blood lymphocytes. J. Immunol. 1992; 149: 3784 3794.
21. Edwards G, Weston AH. The pharmacology of ATP-sensitive potassium channels. Annu. Rev. Pharmacol. Toxicol. 1993; 33: 597 637.
22. Dunne MJ, Petersen OH. Intracellular ADP activates K^+ channels that are inhibited by ATP in an insulin-secreting cell line. FEBS Lett. 1986; 208: 59 66.
23. Tung RT, Kurachi Y. On the mechanism of nucleotide diphosphate activation of the ATP-sensitive K^+ channel in ventricular cell of guinea-pig. J. Physiol. 1991; 437: 239 256.
24. Terzic A, Findlay I, Hosoya Y, Kurachi Y. Dualistic behaviour of ATP-sensitive K^+ channels toward intracellular nucleoside diphosphates. Neuron 1994; 12: 1049 1058.
25. Shen WK, Tung RT, Machulda MM, Kurachi Y. Essential role of nucleotide diphosphates in nicorandil-mediated activation of cardiac ATP-sensitive K^+-channel – a comparison with pinacidil and lemakalim. Circ. Res. 1991; 69: 1152 1158.
26. Quast U, Bray KM, Andres H, Manley PW, Baumlin Y, Dosogne J. Binding of the K^+ channel opener $[H^3]P1075$ in rat isolated aorta – relationship to functional effects of openers and blockers. Mol. Pharmacol. 1993; 43: 474 481.
27. Hoffman FJ, Lenfers JB, Niemers E, Pleiss U, Scriabine A, Janis RA. High affinity binding of a potassium channel agonist to intact rat insulinoma cells. Biochem. Biophys. Res. Acta. 1993; 190: 551 558.
28. Edwards G, Ibbotson T, Weston AH. Levcromakalim may induce a voltage-independent K-current in rat portal veins by modifying the gating properties of the delayed rectifier. Br. J. Pharmacol. 1993; 110: 1037 1048.
29. Ashford MLJ, Bond CT, Blair TA, Adelman JP. Cloning and functional expression of a rat heart K_{ATP} channel. Nature 1994; 370: 456 459.
30. Arena JP, Kass RS. Enhancement of potassium-sensitive current in heart cells by pinacidil: Evidence for modulation of the ATP-sensitive potassium channel. Circ. Res. 1989; 65: 436 445.
31. Kubo Y, Baldwin TJ, Jan YN, Jan LY. Primary structure and functional expression of a mouse inward rectifier potassium channel. Nature 1993; 362: 127 133.
32. Tytgat J, Vereecke J, Carmeliet E. A possible structural link between voltage-gated and inward rectifier K^+ channels. Biophys. J. 1994; 66: A425.
33. Parratt JR, Kane KA. K_{ATP} channels in ischemic preconditioning. Cardiovasc. Res. 1994; 28: 783 785.
34. Jennings RB, Murry CE, Reimer KA. Energy metabolism in preconditioned and control myocardium: effect of total ischemia. J. Mol. Cell Cardiol. 1991; 33: 1449 1458.
35. Tomai F, Crea F, Gaspardone A, Versaci F, De Paulis R, de Peppo AP et al. Blockade of ATP-sensitive potassium channels prevents myocardial preconditioning in man. Circulation 1994; 90: 700 705.

36. Gross GJ, Auchampach JA. Ischemic preconditioning during coronary angioplasty is prevented by glibenclamide, a selective ATP-sensitive K^+ channel blocker. Circ. Res. 1992; 70: 223–233.
37. Escande D, Cavero I. Potassium channel openers in the heart. In: Escande D, Standen N, editors. K^+ channels in cardiovascular medicine. Paris: Springer-Verlag, 1993: 225–244.
38. McPherson CD, Pierce GN, Cole WC. Ischemic cardioprotection by ATP-sensitive K^+ channels involves high energy phosphate preservation. Am. J. Physiol. 1993; 34: H1809–H1818.
39. Djellas Y, Mestre M, Cavero I. Aprikalim protection against ischemic injury occurs with an accelerated decrease in action potential duration but not in myocardial contractility. Circulation 1993; 88: I-632.
40. Auchampach JA, Maruyama M, Cavero I, Gross GJ. Pharmacological evidence for a role of ATP-dependent potassium channels in myocardial stunning. Circulation 1992; 86: 311–319.
41. Yao Z, Gross GJ. Effects of the K_{ATP} channel opener bimakalim on coronary blood flow, monophasic action potential duration, and infarct size in dogs. Circulation 1994; 89: 1769–1775.
42. Cavero I, Premmereur J. ATP-sensitive potassium channel openers are of potential benefit in ischemic heart disease. Cardiovasc. Res. 1994; 28: 32–33.
43. Taira N. Nicorandil as a hybrid between nitrates and potassium channel activators. Am. J. Cardiol. 1989; 63: 18J–24J.
44. Frampton J, Buckley MM, Fitton A. Nicorandil: a review of its pharmacology and therapeutic efficacy in angina pectoris. Drugs 1992; 44: 625–655.
45. Aizawa T, Ogasawara K, Nakamura F, Hirosaka A, Sakuma T, Nagashima K et al. Effect of nicorandil on coronary spasm. Am. J. Cardiol. 1989; 63: 75J–79J.
46. Lablanche J-M. Bauters C, Leroy F, Bertrand ME. Prevention of coronary spasm by nicorandil: comparison with nifedipine. J. Cardiovasc. Pharmacol. 1992; 20 (Supplement 3): S82–S85.
47. Kishida H, Murao S. Effect of a new coronary vasodilator, nicorandil, on variant angina pectoris. Clin. Pharmacol. Ther. 1987; 42: 166–174.
48. Kinoshita M, Nishikawa S, Sawamura M, Yamaguchi S, Mitsunami K, Itoh M et al. Comparative efficacy of high-dose versus low-dose nicorandil therapy for chronic stable angina pectoris. Am. J. Cardiol. 1986; 58: 733–738.
49. Thormann J, Schlepper M, Kramer W, Gottwik M, Kindler M. Effectiveness of nicorandil (SG-75), a new long-acting drug with nitroglycerin effects in patients with coronary artery disease: improved left ventricular function and regional wall motion and abolition of pacing-induced angina. J. Cardiovasc. Pharmacol. 1983; 5: 371–377.
50. Wagner G. Selected issues from an overview on nicorandil: tolerance, duration of action, and long-term efficacy. J. Cardiovasc. Pharmacol. 1992; 20 (Supplement 3): S86–S92.
51. Friedel HA, Brogden RN. Pinacidil: A review of its pharmacodynamic and pharmacokinetic properties, and therapeutic potential in the treatment of hypertension. Drugs 1990; 39(6): 929–967.
52. Hamilton TC, Beerahee A, Moen JS, Price RK, Ramju JV, Clapham JC. Levcromakalim. Cardiovasc. Drugs Rev. 1993; 11: 199–222.

Drug Metabolism

Pharmacological Sciences: Perspectives for
Research and Therapy in the Late 1990s
ed. by A.C. Cuello and B. Collier

The Proliferating P450s: Providers of Polysubstrate Pharmacology

Ronald W. Estabrook

Department of Biochemistry, University of Texas Southwestern Medical Center at Dallas, Dallas, Texas 75235-9038, USA

Summary. Over 300 P450s have been cloned and sequenced. A gene superfamily composed of at least 36 gene families has been described in 31 eukaryotes (including 11 mammalian and three plant species). Twelve families, made up of 22 subfamilies, exist in all mammals examined to date including 36 P450s identified in humans. Substrates for P450s represent a wide range of endogeneous and xenobiotic chemicals – including a large number of pharmacologically active drugs. Current areas of research on P450 focus on differences in protein structure related to the specificity of substrate structure for metabolism, the identify and response of specific transcriptional regulatory elements influencing the induction of individual P450s, the frequency of polymorphisms and their impact on drug metabolism and susceptibility to cancer, and the heterologous expression of functional P450s in systems that can be applied to drug discovery and toxicity evaluation studies. The diversity of P450s and their versatility in the metabolism of drugs continues to challenge the pharmacologist.

Introduction

The P450s are remarkable hemoproteins – remarkable because there are so many of them and they catalyze so many different reactions. At the present time over 300 different P450s have been cloned, sequenced and classified in a gene superfamily [1] which includes at least 36 gene families. A large array of chemical reactions are catalyzed by the P450s (estimated to exceed 60 different types) including hydroxylations, reductions, dehalogenations, epoxidations, sulfoxidations, dehydrogenations, N- and C-dealkylations, etc., to name but a few. The variety and diversity of chemicals that serve as substrates for P450s include nearly every type of organic compound. In addition to their central role in drug metabolism and chemical carcinogenesis, specific P450s serve as catalysts responsible for the synthesis of cholesterol and its further metabolism to steroid hormones and bile acids, the oxidation of polyunsaturated fatty acids to generate reactive epoxides that can serve as "second messengers", the metabolism of alcohol and volatile organic solvents, the activation of vitamins to physiologically important mediators, and the omega hydroxylation of prostanoids for attenuation of their activities – and the list goes on and on. The major proportion of P450s now known participate in reactions of interest to pharmacology and toxicology (Fig. 1A – families 1, 2, 3, and 4). It has been

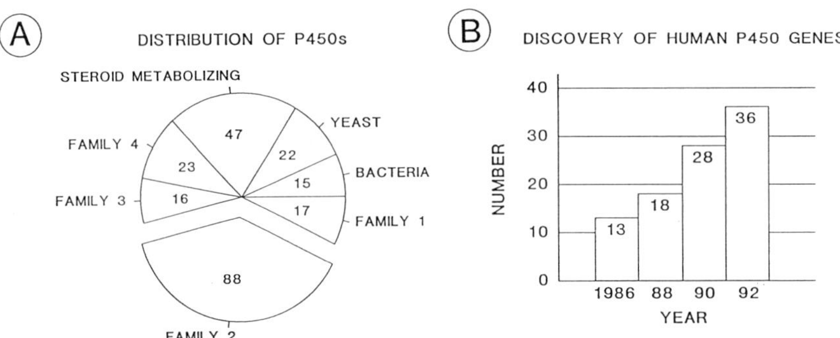

Fig. 1. (A) Distribution of P450s by gene family. Many drug metabolism reactions are catalyzed by P450s of family 2. (B) The increase in the number of human P450s cloned and sequenced from 1986 to 1992.

proposed that humans contain at least 60 different P450s (Fig. 1B) – although only 36 have been identified to date, but the number continues to expand.

The P450s play many additional roles in biolgy – in particular for the synthesis of biologically important secondary metabolites in plants. For example, P450s in plants are responsible for the synthesis of colors, flavors, alkaloids, suberins, flavonoids, and natural pesticides and fungicides; P450s in insects are central to the synthesis of hormones for developent (the ecdysones) and resistance to insecticides. Fish, reptiles, and birds have their own unique inventory of P450s. The full breadth of P450's role in biology is only now unfolding.

Structure/Function Relationships

Three P450s have been crystallized and their structures established by X-ray analysis [2, 3]. All three P450s have many structural features in common – yet they also have many significant differences. There are distinguishing areas of the molecule that are rich in β-pleated sheets (Domain I) or αhelical structures (Domain II) (Fig. 2). Each P450 has a heme buried in the protein surrounded on the distal face by a characteristic α-helix (I-helix) which borders the substrate-binding cavity, while the L-helix, containing the cysteine that contributes the thiolate ligand for interaction with the heme iron, lies on the proximal face of the heme. In spite of these similarities, a direct comparison of the three known structures reveals the uniqueness of each P450 as evidenced by the displacement of atoms by as much as 12–15 Angstroms when aligned to maximize similarities. The three P450s that have been crystallized are all soluble proteins isolated from bacteria. As yet the crystal structure of a membrane-associated P450, isolated from a mammalian

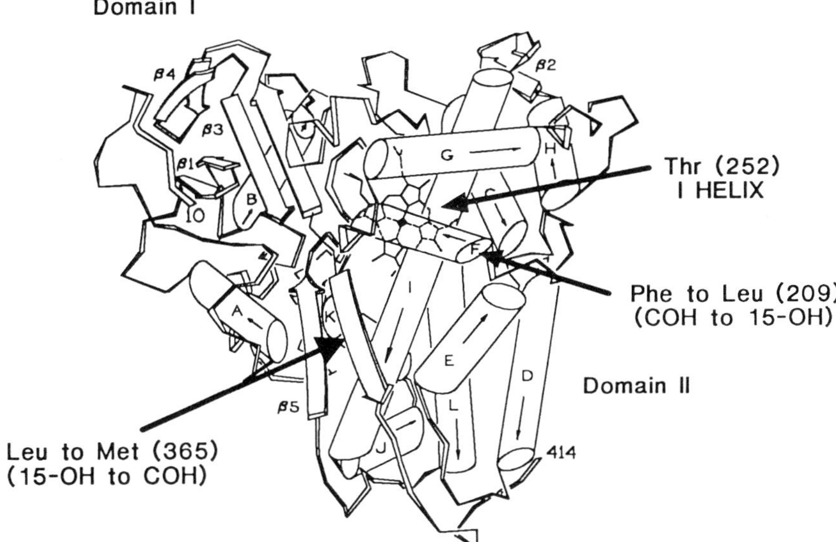

Fig. 2. The crystal structure of P450 101 showing the proposed locations of key amino acids studied by mutation analysis.

source, has not been achieved – leaving unanswered questions concerning differences imposed by the membrane-binding properties of these P450s.

What are the structural features that permit the discrimination of chemicals that serve as substrates for unique P450s and what are the properties of the protein which dictate the site of chemical change of these substrates, *i.e.* their orientation in the active site? What is the path of electron tunneling to the heme iron of a P450 from the donating reduced flavin moiety of the P450 reductase or cytochrome b_5 that interact at the surface of the molecule? What are the forces that direct the pattern of folding of each P450 protein thereby establishing surface properties that contain the topology which maximizs the "docking" of these companion proteins? These and many additional questions remain unanswered. Those investigators working on the biochemical and bio-physical properties of P450s anxiously await the crystallization and structural determination of a mammalian membrane-bound P450. This information is necessary as a guide on which to build the next level of the framework relating structure to function. This knowledge will be essential to the future development of drugs synthesized to contain specific pharmacologic properties.

Polymorphisms

Individuals differ in their response to drugs. Studies of drug metabolism have classified individuals into groups; such as "slow (poor) metabolizers" and "rapid (extensive) metabolizers". These differences are now recognized as due, in part, to the presence of mutant forms of P450. The metabolism of debrisoquine by P450 2D6 serves as the prototype for this type of P450 polymorphism [4]. Over 200 different natural mutations are now known which modify the enzymatic properties of this P450, resulting in profound changes in the pharmacokinetic disposition of debrisoquine. Clearly, many adverse responses to drugs may be attributable, in part, to the presence of similar modified forms of other P450s resulting from other mutations. In addition, numerous studies are in progress in an attempt to relate the presence and enzymatic properties of unique P450s to a genetic predisposition for cancer.

Most remarkable are experiments designed to express mutants of P450s [5] which show that the change of a single amino acid can markedly change the site and efficiency of metabolism, *e.g.* studies with mouse P450 2A4, which catalyzes the 7α-hydroxylation of coumarin, have shown that replacement of the phenylalaine located as amino acid 290 with a leucine by mutation of the cDNA converts this P450 to a form active for the 15α-hydroxylation of testosterone (Fig. 2). One must question why the structure of a P450 is so delicately poised that changing one amino acid out of 500 has such a profound influence on the enzymatic properties of the hemoprotein?

Regulation of Expression – The Induction of P450s

The association of a tolerance to drugs with changes in the content of P450 has been recognized for over 40 years [6, 7]. Introduction of the techniques of molecular biology have revealed the manner in which many chemicals serve as inducers of P450s and other proteins. It is generally accepted today that a series of reactions occurs whereby a chemical: (a) reacts with a specific receptor; (b) this complex of the receptor with the chemical ligand is then transported to the nucleus where the receptor complex initiates an increase in transcriptional expression by an interaction with specific regulatory elements located on the DNA of specific genes; (c) this increase in transcription results in an increase in the synthesis of specific RNA species (which also may be stabilized) thereby extending the efficiency of translation. In addition, some chemicals may activate processes of post-translational modification of the newly synthesized protein (P450) which can target proteins for specific cellular locations or signal them for degradation. Best understood [8–10] is the induction of P450s and associated Phase II enzymes by polycyclic aromatic hydrocarbons (such as TCDD) via the AhH receptor (Fig. 3). It is believed that similar receptor systems exist for the induction of specific P450s by phenobarbital (P450s of family 2), synthetic steroids (P450s of family 3), and peroxisome proliferators

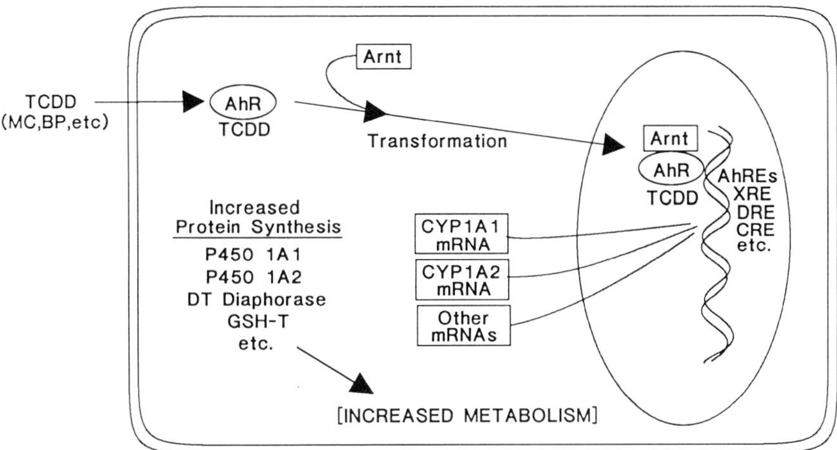

Fig. 3. Schematic representation of the AhH locus indicating the role of the Ah receptor (AhR) in regulation of transcription.

(P450s of family 4). A new language of acronyms has developed as the lexicon for identifying the ever increasing number of response elements that have been characterized.

Of major interest to those studying the regulation of synthesis of P450s is the question of the tissue-specific expression of a specific P450. What determines whether a P450 will be present in intestine, liver *or* lung? What dictates the time of appearance and level of expression of a P450 during fetal development? And, what similarities (differences) relate the presence and properties of orthologous proteins when comparing reactions in rodent species with those in humans? These and similar questions are central to our understanding of the pharmacologic and toxicologic properties of chemicals.

Heterologous Expression of P450s

The *in vitro* study of drug metabolism was limited for many years to the examination of reactions catalyzed by isolated microsomes or P450s purified from rodents. The availability of human tissue, and problems of contaminating infectious agents, restricted the detailed characterization of those human enzymes which catalyze the biotransformation of many drugs. Today, several different methods are available to study human P450s. Using PCR methods to clone a P450, combined with incorporating the specific cDNA into a number of useful plasmid vectors, one is able to transform bacteria for the high level expression of many human P450's that can be isolated and purified for biochemical and biophysical studies [11]. Also, one can transform mammalian cells for the stable expression of a P450 so that it resides in an environment that simulates the *in vivo* milieu of these membrane-bound proteins. In this way, one can couple studies of metabolism with an evaluation of the toxicity of metabolites formed during metabolism by human enzymes [12]. Clearly, many new opportunities now exist for the evaluation of drug metabolism by different human P450s – in particular, without the need to be encumbered by doubts questioning the validity of extrapolation from rodents to man.

Further, methodologies are in development for the application of recombinant P450s as catalysts in bioreactors for the synthesis of new specialized chemicals, in particular drugs.

Forecast of Opportunities for Drug Metabolism

Just prior to the recent International Congress of Pharmacology, the 10th International Symposium on Microsomes and Drug Oxidations was held at the University of Toronto. This was an impressive gathering of over 700 scientists from 36 countries. The 22 plenary lectures accompanying 60 symposium lectures and 388 poster presentations attest to the vitality of research on drug metabolism. A significant segment of the science presented was dedicated to genetic aspects affecting drug metabolism – in particular, the regulation of expression of the involved enzymes. Great interest focused on the relationship of nitric oxide synthase to P450. Research on drug metabolism continues to thrive. Opportunities to explore new vistas serve to attract young people interested in this fascinatiang arena of science that has so many practical applications useful to man.

Acknowledgements

This work was supported in part by grants from the USPHS, National Institutes of Health (GM-16488) and The Robert A. Welch Foundation (I-0959).

References

1. Nelson DR, Kamataki T, Waxman DJ, Guengerich FP, Estabrook RW, Feyereisen R et al. The P450 superfamily: update on new sequences, gene mapping, accession numbers, early trivial names of enzymes, and nomenclature. DNA and Cell Biology 1993; 12: 1–51.
2. Boddupalli SS, Hasermann CA, Ravichandran KG, Lu JY, Goldsmith EJ, Deisenhofer J et al. Crystallization and preliminary x-ray diffraction analysis of P450$_{terp}$ and the hemoprotein domain of P450$_{BM-3}$, enzymes belonging to two distinct classes of the cytochrome P450 superfamily. Proc. Natl. Acad. Sci. USA 1992; 89: 5567–5571.
3. Ravichandran KG, Boddupalli SS, Hasermann CA, Peterson JA, Deisenhofer J. Crystal structure of hemoprotein domain of P450$_{BM-3}$, a prototype for microsomal P450's. Science 1993; 261: 731–736.

4. Tyndale R, Aoyama T, Broly F, Matsunaga T, Inaba T, Kalow W et al. Identification of a new variant CYP2D6 allele lacking the phenotype. Pharmacogenetics 1991; 1: 26–32.
5. Lindberg RL, Negishi M. Alteration of mouse cytochrome P450coh substrate specificity by mutation of a single amino-acid residue. Nature 1989; 339: 632–634.
6. Conney AH. Pharmacological implications of microsomal enzyme induction. Pharmacological Reviews 1967; 19: 317–366.
7. Remmer H, Merker HJ. The effect of drugs on the formation of smooth endoplasmic reticulum and drug-metabolizing enzymes. Annals NY Acad. Sci. 1965; 123: 79–97.
8. Reye H, Reisz-Porszasz S, Hankinson O. Identification of the Ah receptor nuclear translocator protein (Arnt) as a component of the DNA binding form of the Ah receptor. Science 1992; 256: 1193–1195.
9. Burbach KM, Poland A, Bradfield CA. Cloning of the Ah-receptor cDNA reveals a distinctive ligand-activated transcription factor. Proc. Natl. Acad. Sci. USA 1992; 89: 8185–8189.
10. Harper PA, Giannone JV, Okey AB, Denison MS. *In vitro* transformation of the human Ah receptor and its binding to a dioxin response element. Molecular Pharmacology 1992; 42: 603–612.
11. Shet MS, Fisher CW, Holmans PL, Estabrook RW. Human cytochrome P450 3A4. Enzymatic properties of a purified recombinant fusion protein containing NADPH-P450 reductase. Proc. Natl. Acad. Sci. USA 1993; 90: 11748–11752.
12. Crespi CL, Gozalez FJ, Steimel DT, Turner TR, Gelboin HV, Penman BW et al. A metabolically competent human cell line expressing five cDNAs encoding procarcinogen-activating enzymes: application to mutagenicity testing. Chemical Research in Toxicology 1991; 4: 566–572.

Human Cytochromes P450: Regulation and Functional Variability*

Urs A. Meyer[1], Frank J. Gonzalez[2], F. Peter Guengerich[3],
Michael E. McManus[4], and Kyo-Ichiro Okuda[5]

[1]*Department of Pharmacology, Biozentrum of the University of Basel, CH-4056 Basel,
Switzerland;* [2]*National Cancer Institute, National Institutes of Health, Bethesda, MD 20892,
USA;* [3]*Department of Biochemistry and Center of Molecular Toxicology, Vanderbilt University,
School of Medicine, Nashville, Tennessee 37232-0146, USA;* [4]*Department of Physiology
and Pharmacology, University of Queensland, Queensland 4072, Australia;* [5]*Department
of Surgery I, Miyazaki Medical College, Kiyotake, Miyazaki, Japan 889-16*

Summary. Human cytochrome P450 enzymes involved in the metabolism of exogenous and
endogenous compounds are intensively studied. The regulation of cytochromes P450 by
transcriptional activation via the Ah receptor (CYP1A) and peroxisome proliferator activated
receptor (CYP4A) is increasingly understood at the molecular level, in contrast to induction
by phenobarbital and glucocorticoids. The tissue-specific and developmental regulation of
P450s by transcription factors such as HNF-Iα, DBP, C/EBP and Sp1 is complex. Individual
variation of the activity of these enzymes is markedly influenced by common genetic
polymorphisms, with numerous loss of function or decreased function and even increased
function alleles of P450 genes. There is evidence for an important role of P450 enzymes in the
activation of carcinogens. Variable extrahepatic expression of human P450s may contribute to
individual cancer risk. Progress has also been made in the understanding of the role of P4507α
(cholesterol 7α hydroxylase) in cholesterol homeostasis.

Human cytochrome P450 (P450) enzymes are the subject of intensive research because of
their obvious relevance to drug therapy and toxicity and their role in chemical carcinogenesis.
Moreover, the biotransformation of numerous endogenous compounds such as steroids,
retinoids, eicosanoids, and other endogenous compounds is catalyzed by P450s. The superfamily of P450 genes comprises over 300 genes in various species. Of these, over 40 human
P450 enzymes are known and classified within the 13 mammalian gene families.

Transcriptional and Post-Translational Regulation of P450 Expression

Xenobiotic-metabolizing cytochromes P450s are regulated at a variety
of different levels. P450 genes in the CYP1 family are induced by foreign
chemicals including dioxins, polycyclic aromatic hydrocarbons, and in
humans by drug omeprazole. This induction involves a heterodimeric
ligand receptor complex that interacts with regulatory elements upstream of target genes. CYP2A and CYP2B subfamily members exhibit
inducibilities by phenobarbital and related compounds in various species including humans. The CYP3A P450 genes and possibly members
of the CYP2B subfamily can be transcriptionally activated by high

Correspondence to: Urs A. Meyer at the Department of Pharmacology, Biozentrum of the
University of Basel, Klingelbergstr. 70, CH-4056 Basel/Switzerland
*From the Symposium "Human Cytochromes P450: Regulation and Functional Variability"

concentrations of synthetic steroids such as dexamethasone. CYP4A P450 genes are induced by the class of chemicals known as peroxisome proliferators which act through binding to a family of cytosolic receptors, the peroxisome proliferator activated receptors.

P450s can also be post-transcriptionally controlled. The CYP2E1 protein is regulated by a substrate-induced stabilization mechanism. Under certain conditions, the CYP2E1 mRNA can also be stabilized. The CYP3A P450s can be post-translationally activated by certain flavonoid compounds through a mechanism that involves dual occupancy of the enzyme active site by activator and substrate.

The rat has been extensively used as an experimental model to examine the tissue-specific and developmentally-programmed expression of P450s in the CYP2 family. Most P450 gene expression occurs under control of hepatocyte-enriched transcription factors that presently consist of four families distinguished by their DNA binding and dimerization motifs. The CYP2C6, CYP2E1 and CYP2D5 genes are activated during different stages of rat development. Indeed, experimental evidence suggests that they are activated by different transcription factors. Trans-activation transfection assays and *in vitro* DNA-binding experiments were used to determine that HNF-la controls liver-specific expression of the CYP2E1 gene [1]. The diurnally regulated DBP was found to be the principle factor involved in regulation of the CYP2C6 [2] and CYP7A1 [3] genes. The post-pubertal onset of expression of DBP is responsible for the adult-specific expression of these P450s. More recently the CYP2D5 gene was found to be under unique regulation by the factor C/EBPβ [4]. In contrast to other transcription factors studied to date, this liver-enriched factor only activates transcription in the presence of the ubiquitously-expressed factor Sp1. This occurs by a protein-protein interaction between C/EBPβ and Sp1. In the absence of Sp1, C/EBPβ is incapable of directly binding to the regulatory element upstream of the CYP2D5 gene. Developmental expression of C/EBPβ correlates with expression of the CYP2D5 gene providing evidence that this factor controls the gene in the intact animal. Cooperative protein-protein interaction by transcription factors represents another level of complexity and possible diversity to the processes that control tissue-specific gene expression.

Genetic Polymorphism of Human Drug Metabolizing Enzymes

Genetic polymorphisms of drug-metabolizing enzymes such as P450s give rise to subgroups in the population which differ in their ability to perform a certain drug biotransformation reaction. Polymorphisms are caused by mutations in the genes for these P450s which are maintained in the population at a high frequency ($> 1\%$) and cause decreased, increased or absent enzyme activity. Frequencies of these polymorphisms also show marked interethnic variation. The best studied of these genetic polymorphisms are the polymorphism of cytochromes P450 CYP2D6 and CYP2C19 (for review, see [5]). The significance for drug and carcinogen metabolism of genetic polymorphisms of CYP1A1, CYP1A2, CYP2A6, CYP2C9, CYP3A5, CYP2E1 and CYP4A11 genes awaits clarification.

Debrisoquine Polymorphism

The best studied example of a genetic variation in drug response is the debrisoquine polymorphism. Five to 10% of individuals in Caucasian and 1–2% in Asian populations are "poor metabolizers" and are homozygous for two recessive loss-of-function alleles of the gene encoding cytochrome P450 CYP2D6. These poor metabolizers of debrisoquine are inefficient in the metabolism of over 30 clincally used drugs. Over 90% of the mutations of the CYP2D6 gene that cause absence of the CYP2D6 protein and result in the debrisoquine poor metabolizer phenotype have been identified. The most common loss of function or null allele (CYP2D6-B, $\sim 75\%$ of poor metabolizer alleles) is characterized by multiple mutations including a point mutation at a splice-site recognition sequence that leads to a frameshift. Another mutant allele (CYP2D6-A, 5%) consists of a single basepair deletion in the coding sequence causing a frameshift, and yet another common loss of CYP2D6 activity is caused by the deletion of the entire CYP2D6 gene (CYP2D6-D 5–15%). Additional rare poor metabolizer alleles are continuing to appear in the literature. The variation among the individuals of the much larger group of so-called "extensive metabolizers" is in part due to CYP2D6 alleles which result in only slightly decreased enzyme activity [6] and to dominantly inherited gene amplification and gene duplication resulting in "ultarapid" metabolism [7]. All of the mutations can be identified by DNA analysis, using PCR amplification or by RFLP. Over 90% of phenotypes can be predicted in most populations. New mutations are efficiently identified with techniques such as PCR-SSCP followed by selected sequencing.

Mephenytoin Polymorphism

The mephenytoin polymorphism affects the metabolism of S-mephenytoin and several other drugs including diazepam, omeprazol and proguanyl. 3–5% of Caucasians and $\sim 20\%$ of Asians are of the recessive poor metabolizer phenotype. The predominant defect in poor metabolizers is a single bp mutation of CYP2C19 which causes aberrant splicing of the CYP2C19 mRNA and absence in the liver of the CYP2C19 protein. A PCR-based DNA test for this mutation identifies ~ 70% of loss of function of alleles of Caucasian and Japanese poor metabolizers [8]. A second mutant allele has also been identified and accounts for the remaining 30% of loss of function alleles in Japanese [9].

Considerable work remains to characterize variability in levels of other human P450s and in particular to clarify the role of low or high expression of P450s in the susceptibility to environmentally based diseases such as Parkinson's disease and cancer.

Metabolism of Carcinogens by Human P450s

There has been much in interest in the roles of P450 enzymes in the processing of carcinogens. Direct evidence for roles of human P450s in the etiology of cancer has been difficult to obtain, however. Considerable attention has been given to differences among individuals in the levels of expression of specific P450 enzymes and the roles these enzymes can have in the activation and detoxification of individual carcinogens [10].

Aflatoxin B₁ is a potent hepatocarcinogen in some experimental animals, and epidemiological studies suggest a role in human liver cancer, possibly synergistic with hepatitis B virus [11]. CYP3A4, the principal human liver P450 in many individuals, is a major enzyme involved in aflatoxin B_1 activation [12]. CYP3A4 and CYP1A2 catalyze the formation of aflatoxin Q1 and M1, respectively, both less toxic products than aflatoxin B_1. The two epoxide isomers of afltoxin B1 differ considerably in their reactivity with DNA and genotoxicity, with the *endo* isomer being at least 10^3 times more reactive than the *endo*. Some epoxide is known to be formed in human liver microsomes and can be trapped as the diastereomeric glutathione conjugate [13]. Studies with recombinant P450s reveal that CYP3A4 catalyzes the formation of the dangerous *exo* epoxide and that P4501A2 forms roughly equimolar mixture of *exo* and the detoxicated *endo*.

In other recent studies with recombinant human P450s CYP1A1 was most active in several oxidations of *benzo(a)pyrene* in contrast to previous beliefs.

An area of practical consideration is the oxidation of the industrial monomer *1,3-butadiene*, which is considerably more tumorigenic in mice than rats. The monoepoxide is formed in several animal species, including humans, by CYP2E1 (and possibly by CYP2A6). However, the overall pharmacokinetics predicted by *in vitro* and *in vivo* assays suggest strongly that humans are more similar to rats than mice in the metabolism of butadiene. Further, the genotoxic diepoxide seems to be formed only in mice. Conjugation of the diepoxide with glutathione by theta class transferases renders this compound *more* genotoxic in bacterial assays [14].

Extrahepatic P450s and Their Functions

Both for presystemic metabolism of drug substrates and the activation or inactivation of carcinogens, extrahepatic metabolism can be the decisive activity. The gastrointestinal tract represents a major portal of entry for many toxic chemicals and carcinogens. The expression of the CYP1A and CYP3A subfamilies, and CYP4B1 has recently been studied in the human gastrointestinal tract. RNA and immunoblotting as well as histological analyses did not reveal CYP1A1 or CYP1A2 mRNA or protein in cells of the human gut. However, when microsomes were prepared from human tissues for use as the activating source in a series of Ames *Salmonella* tests, activity toward the food-derived heterocyclic amine MeIQ was observed in a single colon microsomal preparation and this activity was completely abolished by the CYP1A inhibitor, a-naphthoflavone [15, 16].

In contrast to the CYP1A subfamily, CYP3A expression was demonstrated in epithelial cells lining the human oesophagus, small intestine, colon and rectum. CYP3A expression is particularly abundant in the small intestine where CYP3A4 represents the major form. From maximal levels in the duodenum, the level of CYP3A expression decreases down the intestine with only minimal expression in the rectum. RNA blot analysis of mRNA samples from 11 human colons revealed marked heterogeneity in CYP3A mRNA with at least two genes, CYP3A4 and CYP3A5, expressed in some colons and none in others. CYP3A5 mRNA was demonstrated in six of 11 colon samples, a higher proportion than routinely observed in liver [16].

Analysis of human tissues with probes for the major lung P450, CYP4B1, demonstrated low levels of expression only in the colon [17]. This expression was variable and observed in approximately half of the specimens studied. Interestingly, CYP4B1 is a major gastrointestinal P450 in the rabbit. Human colon microsomes were however unable to activate 2-aminofluorene, a rabbit CYP4B1 substrate, to a mutagen in the Ames Salmonella test. These studies serve to highlight species-specific differences in both CYP4B1 expression and function.

In summary, these studies demonstrate that whilst CYP3A enzymes are major gastrointestinal P450s, CYPIA1, CYP1A2 and CYP4B1 are minimally expressed. Of greater interest is the demonstrable heterogeneity in P450 complement between different sites within the gastrintestinal tract. This emphasizes the need to study each tissue as a distinct entity.

Human P450s Involved in Cholesterol Homeostasis

Cholesterol is converted into bile acid by the action of about 15 enzymes, of which three major P450s (P4507, P45012α and P45027) catalyze the important hydroxylations. Accumulation of cholesterol in the body causes hyperlipidemia, atherosclerosis and gallstone disease. The serum level of cholesterol is maintained at a constant level by homeostatic mechanisms. Two major enzymes, HMG-CoA reductase and cholesterol 7α-hydroxylase (P4507) are the most important enzymes, regulating cholesterol biosynthesis and cholesterol degradation, respectively. P4507 catalyzes a first step of cholesterol catabolism and constitutes the rate-limiting step of cholesterol degradation. The enzyme is regulated in different ways in the rat, i.e., by long-term (feed-back control by bile acids), mid-term (diurnal rhythm), and short-term regulation. To study the mechanism of these regulations at the molecular level, rat P4507 (CYP7) was purified to homogeneity, antibodies prepared and a P4507 cDNA was cloned and sequenced, using these antibodies. Immunoblotting and Northern blotting analyses revealed that long-term as well as mid-term regulation are both pretranslational [18]. The rat P4507 cDNA has been used to clone the human P4507 cDNA from a human liver cDNA library. This cDNA served to determine the gene structure of human P4507 [19] and to produce large amounts of protein by expression in *E. coli*. The antibodies raised against this protein inhibited cholesterol 7α-hydroxylase by 70% and recognized a protein of 58 kDa after heterologous expression of the human P4507 cDNA [20]. Cholesterol 7α-hydroxylase activity increased about six-fold in livers of cholestyramine-treated patients. However, the increase in immunoreactive P4507 protein was considerably less, i.e. only about two-fold. Although this finding is still preliminary, it suggests that there may be mechanisms other than pretranslational control in the feed-back regulation of cholesterol 7α-hydroxylase by bile acids in man [20]. The tools to study the regulation of this gene are now available.

References

1. Liu SY, Gonzalez FJ. Role of the liver-enriched transcription factor HNF-1α in expression of the CYP2E1 gene. DNA Cell Biol. 1995; 14: 285–293.
2. Yano M, Favey E, Gonzalez FJ. Role of the liver-enriched transcription factor DBP in expression of the cytochrome P450 CYP2C6 gene. Mol. Cell. Biol. 1992; 12: 2847–2854.
3. Lee YH, Alberta JA, Gonzalez FJ, Wasman DJ. Multiple, functional DBP-binding sites on the promoter of the cholesteriol 7α-hydroxylase P450 gene CYP7: proposed role in diurnal regulation of liver gene expression. J. Biol. Chem. 1994; 269: 14681–14689.
4. Lee YH, Yano M, Liu SY, Matsunaga E, Johnson PF, Gonzalez FJ. A novel *cis*-acting element controlling the rat CYP2D5 gene requiring cooperativity between C/EBPβ, and an Sp1 factor. Mol. Cell. Biol. 1994; 14: 1383–1394.

5. Meyer UA. Pharmacogenetics: The slow, the rapid, and the ultrarapid. Proc. Natl. Acad. Sci. USA 1994; 91: 1983–1984.
6. Broly F, Meyer UA. Debrisoquine oxidation polymorphism: phenotypic consequences of a 3-base-pair deletion in exon 5 of the CYP2D6 gene. Pharmacogenetics 1993; 3: 256–263.
7. Johansson I, Lundqvist E, Bertilsson L, Dahl M-L, Sjöqvist F, Ingelman-Sundberg M. Inherited amplification of an active gene in the cytochrome P450 CYP2D locus as a cause of ultrarapid metabolism of debrisoquine. Proc Natl Acad Sci USA 1993; 90: 11825–11829.
8. DeMorais SMF, Wilkinson GR, Blaisdell J, Nakamura K, Meyer UA, Goldstein JA. The major genetic defect responsible for the polymorphism of S-mephenytoin metabolism in humans. J. Biol. Chem. 1994; 269: 15419–15422.
9. DeMorais SMF, Wilkinson GR, Blaisel J, Meyer UA, Nakamura K, Goldstein JA. Identification of a new genetic defect responsible for the polymorphism of S-mephenytoin metabolism in Japanese. Mol. Pharmacol. 1994; 46: 594–598.
10. Guengerich FP, Shimada T. Oxidation of toxic and carcinogenic chemicals by human cytochrome P450 enzymes. Chem. Res. Toxicol. 1991; 4: 391–407.
11. Eaton DL, Gallagher EP. Mechanism of aflatoxin carcinogenesis. Annu. Rev. Pharmacol. Toxicol. 1994; 34: 135–172.
12. Raney KD, Shimada T, Kim DH, Groopman JD, Harris TM, Guengerich FP. Oxidation of aflatoxin B_1 and related dihydrofurans by human liver microsomes: significance of aflatoxin Q_1 as a detoxication product. Chem. Res. Toxicol. 1992; 5: 202–210.
13. Raney KD, Coles B, Guengerich FP, Harris TM. The *endo* 8,9-epoxide of aflatoxin B_1: a new metabolite. Chem. Res. Toxicol. 1992; 5: 333–335.
14. Thier R, Pemble SE, Taylor JB, Humphreys WG, Persmark M, Ketterer B, Guengerich FP. Expression of rat gluthathione S-transferase 5–5 in *Salmonella typhimurium* TA 1535 leads to base-pair mutations upon exposure to dihalomethanes. Proc. Natl. Acad. Sci. USA 1993; 90: 8576–8580.
15. McKinnon RA, Burgess WM, Gonzalez FJ, McManus ME. Metabolic differences in colon mucosal cells. Mutat Res. 1993; 290: 27–33.
16. McKinnon RA, Burgess WM, Hall P de la M, Roberts-Thomson SJ, Gonzalez FJ, McManus ME. Characterization of CYP3A gene subfamily expression in human gastrointestinal tissues. Gut. 1995; 36: 259–267.
17. McKinnon RA, Burgess WM, Gonzalez FJ, Gasser R, McManus ME. Species-specific expression of CYP4B1 in rabbit and human gastrointestinal tissues. Pharmacogenetics. 1994; 4: 260–270.
18. Noshiro M, Nishimoto M, Okuda K. Rat liver cholesterol 7α-hydroxylase J. Biol. Chem. 1990; 265: 10036–10041.
19. Nishimoto M, Noshiro M, Okuda K-I. Structure of the gene encoding human liver cholesterol 7α-hydroxylase. Biochem. Biophys. Acta 1993; 1172: 147–150.
20. Maeda Y-I, Eggertsen G, Nyberg B, Setoguchi T, Okuda K-I., Einarsson, et al. Immunological determination of human cholesterol 7α-hydroxylase. Eur. J. Biochem. 1995; 228: 144–148.

Pharmacological Sciences: Perspectives for
Research and Therapy in the Late 1990s
ed. by A.C. Cuello and B. Collier
© 1995 Birkhäuser Verlag Basel/Switzerland

Drug Conjugation: Diversity and Biological Significance

Peter I. Mackenzie[1], M.W. Anders[2], Yasushi Yamazoe[3],
Richard M. Weinshilboum[4], Kathleen M. Knights[1]
and John Caldwell[5]

[1]*Department of Clinical Pharmacology, School of Medicine, Flinders University of South
Australia, Bedford Park, Australia 5042;* [2]*Department of Pharmacology, University of
Rochester, Rochester, New York, USA;* [3]*Department of Pharmacology, School of Medicine,
Keio University, 35 Shinanomachi, Shinjuku-ku, Toyko 160, Japan;* [4]*Department of
Pharmacology, Mayo Medical School/Mayo Clinic, Rochester, MN 55905, USA;*
[5]*Department of Pharmacology and Toxicology, St Mary's Hospital Medical School,
Norfolk Place, London W2 1PG, UK*

Summary. The first conjugation reaction, the conversion of benzoic acid to hippuric acid, was
reported in 1842. Subsequent research lead to the discovery of major pathways involving
conjugation with glucuronic acid, sulfate, glutathione, methyl groups and coenzyme A, which
are mediated by families of enzymes in the cytoplasm and in intracellular membranes.
Conjugation often produces less biologically active products that are readily excreted. However, in some cases, the conjugate may be more active or may be further metabolized to
products that are toxic to the cell. The potential for predicting individuals that may be at risk
for drug- or xenobiotic-induced toxicities has been enhanced by the characterization of the
cDNAs and genes encoding conjugating enzymes and the discovery of mutations and
polymorphisms in these genes.

Conjugation of drugs and endogenous compounds with glucuronic acid,
glutathione, sulfate, methyl groups and CoA is mediated by multien-
zyme families of both membrane-bound and soluble proteins. Conjugation generally leads to a decrease in the biological activity of the compound and its excretion, although in some cases, a more reactive or
toxic metabolite is formed. Competition between drugs and/or endogenous substrates for the active sites of enzymes may also lead to adverse
effects. The potential for toxicity is dependent in part on the profile of
drug-conjugating enzymes in the cell or tissue and their substrate
selectivities. In this symposium review, selected enzymes involved in
conjugation will be briefly described and examples of potential toxicological importance presented.

Conjugation with Glucuronic Acid

The transfer of glucuronic acid from UDP glucuronic acid to a sulfur,
oxygen, nitrogen or carbon in many chemicals is mediated by mem-

brane-bound UDP glucuronosyltransferases in the endoplasmic reticulum [1]. More than 30 forms have been identified from human (eleven), rat (twelve), mouse (three), rabbit (three) and bovine (one) tissues. Based on sequence similarities, they can be divided into two families [2]. The members of family one are all derived from a single gene and hence have identical carboxy-terminal domains of 245 residues (responsible for UDP glucuronic acid binding) but different amino-terminal domains (280–290 residues involved in substrate selection) which are only about 37–49% similar in sequence. Members of this family glucuronidate bilirubin (UGT1*1 and 4), planar phenols (UGT1*6) and more bulky phenolic compounds (UGT1*7) and are found in the kidney, gastrointestinal tract, lung and brain as well as in the liver. The individual members of family 2 are encoded by separate genes and are mainly found in the liver, although some forms have also been detected in the kidney and lungs. Family 2 members glucuronidate steroids (e.g., UGT2B1, 2, 6, 7 and 10) and many foreign compounds including carboxylic-acid containing drugs (UGT2B1 and 7) and carcinogen/mutagen metabolites.

Significant sequence similarities exist between UDP glucuronosyltransferases and ceramide UDP galactosyltransferase, the first enzyme in the pathway leading to the synthesis of sphingolipids [3]. There all also similarities to plant, insect viral and bacterial UDP glycosyltransferases, suggesting that enzymes conjugating sugars (glucose, galactose, glucuronic acid) to small molecules are encoded by a superfamily of genes that evolved at about the same time as the cytochromes P450, about 3.5 billion years ago.

In general, the glucuronides of most chemicals are unreactive and excreted without further metabolism. However, the glucuronidation of some chemicals produces more biologically active or toxic products. Examples are morphine-6-glucuronide which is a more potent analgesic than morphine, the electrophilic acylglucuronides of several non steroidal antiinflammatory drugs that can bind irreversibly to sulfhydryl, amino and hydroxyl groups on proteins, and the steroid D-ring glucuronides that are cholestatic. Glucuronidation may also enable reactive metabolites such as N-hydroxyarylamines to be transported as their glucuronides to other organs via the blood or bile. Subsequent hydrolysis of the glucuronides would release the metabolites for further processing to DNA-binding or toxic species (for review, see ref. [4]).

Conjugation with Glutathione

The glutathione S-transferases (GST) are encoded by a family of genes that are present in bacteria, plants and mammals. Four classes of soluble enzymes, Alpha (GSTA1), Mu (GSTM1), Pi (GSTP1) and

Theta, as well as one microsomal enyzme have been identified [5]. The enzymes within a class are more than 50% identical in amino acid sequence.

Although primarily a detoxification pathway, conjugation with glutathione plays a key role in the toxicity of several haloalkenes, vicinal dihaloalkanes, dihalomethanes, and halogenated hydroquinones. For example, the nephrotoxicity of haloalkenes involves hepatic glutathione S-conjugate formation; excretion of glutathione S-conjugates in the bile; hydrolysis of glutathione S-conjugates to the corresponding cysteine S-conjugates; uptake of cysteine S-conjugates by amino acid transporters in the kidney; and bioactivation by renal cysteine conjugate β-lyase [6].

The formation of glutathione S-conjugates of haloalkenes is catalyzed preferentially by the microsomal glutathione S-transferase. 1,1-Dichloralkenes undergo an addition-elimination reaction to give S-(1-chloralkenyl) glutathione conjugates, whereas 1,1-difluoroalkenes undergo an addition reaction to give S-(1,1-difluoroalkyl)glutathione S-conjugates. The conjugates are hydrolyzed to cysteine S-conjugates by the sequential action of δ-glutamyltransferase and the dipeptidases, aminopeptidase M and cysteinylglycine dipeptidase in the intestine or on the basolateral or luminal membranes of the renal proximal tubules. The cysteine S-conjugates are transported by renal amino acid and anion transport systems into the kidney and become substrates for cysteine conjugate β-lyase, which is present in both cytosolic and mitochondrial fractions of renal proximal tubular cells. Cytosolic β-lyase is identical with glutamine transaminase K, but the mitochondrial lyase has not been well characterized.

Unstable thiols formed by the β-lyase-catalyzed bioactivation of cysteine S-conjugates undergo nonenzymatic elimination reactions to give electrophilic toxic products. 1,1-Dichloroalkene-derived cysteine S-conjugates, e.g., S-(1,2-dichlorovinyl)-L-cysteine, are nephrotoxic and mutagenic in bacterial and mammalian test systems. Bromine-lacking, 1,1-difluoroalkene-derived cysteine S-conjugates, e.g., S-(2-chloro-1,1,2-trifluoroethyl)-L-cysteine, are nephrotoxic, but not mutagenic, whereas bromine-containing, 1,1-difluoroalkene-derived cysteine S-conjugates, e.g., S-(2-bromo-2-chloro-1,1difluoroethyl)-L-cysteine, are both nephrotoxic and mutagenic. Recent studies show differences in the fate of the unstable thiols formed by the β-lyase-dependent bioactivation of the three types of cysteine conjugates [8]. Chloroalkene-derived S-(1-chloroalkenyl)-L-cysteine conjugates give 1-chloroalkenylthiolates as products; the thiolates eliminate chloride to give thioketenes that may react with tissue nucleophiles to give covalently bound adducts or with water to give haloacids as terminal products. Bromine-lacking, 1,1-difluoroalkene-derived S-(1,1-difluoroalkyl)-L-cysteine conjugates yield 1,1-difluoroalkylthiolates as products; these thiolates lose fluoride to give thioacyl fluorides, which acylate tissue nucleophiles or undergo hydrolysis to give haloacids as terminal products. Bromine-containing, 1,1-difluoroalkene-derived S-(2-bromo-1,1-difluoroalkyl)-L-cysteine conjugates yield 2-bromo-1,1-difluoroalkythiolates as metabolites. Loss of fluoride and hydrolysis of thioacyl fluorides gives 2-bromothiolacids as the favored tautomers, which may undergo a cyclization reaction to give α-thiolactones. These compounds may be hydrolyzed to give glyoxylic acid as the terminal product. Recent studies with a range of bromine-lacking and bromine-containing 1,1-difluoroalkene-derived cysteine conjugates shows that only bromine-containing conjugates that give glyoxylic acid as a terminal product are mutagenic, indicating that α-thiolactone formation is associated with mutagenicity.

Conjugation with Sulfate

Sulfotransferases, which use the activated sulfate compound PAPS as sulfate donor, are found in the cytosol. Mammalian forms have been divided into two families based on a comparison of 10 different sequences [9]. One of these families, sometimes referred to as the ST1 family, has four subfamilies which contain enzymes with the diagnostic substrates dopamine, triiodothyronine, N-hydroxy-2-acetylaminofluorene and estrone respectively. The other, sometimes referred to as the ST2 family, contains forms that sulfate alcohols including hydroxysteroids. Sequence comparisons between mammalian, plant and bacterial species and the presence of conserved regions near the carboxy-terminus that may be important in the binding of PAPS, suggest that sulfotransferase genes evolved from a common ancestor.

Sulfation is generally regarded as a detoxification pathway although reactive intermediates may be generated after sulfation of arylhydroxamic acids, hydroxylamines and benzyl alcohol. The first example of activation by sulfation was demonstrated with N-hydroxy-2-acetylaminofluorene which occurred through the formation of an N-O-sulfate ester. As demonstrated by cDNA expression in COS cells, this reaction is mediated by the rat enzyme ST1C1 [10]. Subsequently, sulfuric acid esters of N-hydroxy-4-aminobiphenyl and N-hydroxy-2-aminofluorene were shown to be major electrophilic metabolites responsible for liver DNA damage in rodents. O-Sulfation of N-hydroxyarylamines leads to the formation of DNA adducts. As O-sulfating activities vary considerably between different human liver samples, the potential for sulfation mediated-toxicity in some individuals may be quite high (reviewed in ref. [11]).

Conjugation with Methyl Groups

The methyltransferases are a group of enzymes that transfer methyl groups to drugs and neurotransmitters from the methyl donor, S-adenosyl-L-methionine. Polymorphisms in the genes encoding several methyltransferases in humans play a major role in determining the toxicity and therapeutic efficacy of many drugs.

Thiopurine methyltransferase (TPMT, EC 2.1.1.67) catalyzes the S-methylation of aromatic and heterocyclic compounds including the drug, 6-mercaptopurine [12]. Measurements of TPMT activities in red blood cells have demonstrated a trimodal frequency distribution which family studies demonstrated was due to a genetic polymorphism. The gene frequencies of the two alleles at the locus TPMT are such that 89% of the white population is homozygous for the trait of high enzyme activity, 11% is heterozygous and 0.3% is homozygous for very low or absent enzyme activity. Individuals with low activity who are exposed to standard therapeutic doses of 6-mercaptopurine or azathioprine often develop profound myelosuppression, whereas those homozygous for the trait of high enzyme activity may be at risk for undertreatment of diseases such as acute leukemia.

Genetic polymorphisms in the genes encoding other methyltransferases may also result in clinically-significant interindividual differences in the metabolism of endogenous molecules and drugs. For example, the metabolism of L-dopa and methyldopa may be affected by polymorphisms of catechol-O-methyltransferase (EC 2.1.1.6), a magnesium-dependent enzyme that catalyzes the methylation of catecholamine neurotransmitters. The clinical significance of variations in histamine N-methyltransferase (EC 2.1.1.8) and nicotinamide N-methyltransferase (EC 2.1.1.1) activities remains to be explained [13, 14]. The former enzyme provides the only mechanism for terminating histamine's neurotransmitter actions in the mammalian central nervous system whereas the latter catalyzes the N-methylation of pyridines to potentially toxic pyridinium ions. The recent cloning of cDNAs and genes encoding these enzymes will open the way for the application of molecular techniques to studies of methyltransferase genetic polymorphisms and their potential medical implications.

Conjugation with Coenzyme A

Numerous drugs and chemicals contain a carboxyl group which may be conjugated with glucuronic acid or an amino acid. The latter process involves activation of the carboxyl moiety to a high energy coenzyme A (CoA) intermediate with subsequent transfer of the acyl group to an amino acid such as glycine. Over the past few years it has become increasingly apparent that both the formation and fate of xenobiotic-CoAs may involve enzyme pathways additional to those associated with amino acid conjugation [15].

Xenobiotic-CoA formation is catalysed by ATP-dependent hepatic fatty acid-CoA ligases which are subdivided based on fatty acid specificity, i.e., short-chain (C2–C5), medium-chain (C6–C12) and long-chain (C8–C20). The mitochondrial medium-chain-CoA ligase is principally associated with formation of CoAs prior to amino acid conjugation, whilst the long-chain fatty acid-CoA ligase is involved in the formation of the acyl-CoA thioesters of a variety of hypolipidaemic peroxisome proliferators and the 2-arylpropionic acid NSAIDs [16]. This latter enzyme is located in the smooth endoplasmic reticulum and the outer membranes of both peroxisomes and mitochondria.

Studies using palmitic acid (C16) as a probe have demonstrated in rat hepatic peroxisomes and microsomes that formation of palmitoyl-CoA exhibits biphasic kinetics, corresponding to high affinity, low capacity and low affinity, high capacity components [17, 18]. Inhibition of the high affinity forms was observed with a variety of xenobiotics and evidence of widely varying Ki values and differing profiles of inhibition (competitive, non-competitive and mixed) suggested that multiple CoA ligases existed. Subsequent studies identified in addition to the long-chain ligases, kinetically distinct hepatic microsomal and peroxisomal nafenopin-CoA ligases [15]. Differential inhibition of the peroxisomal and microsomal forms of the nafenopin-CoA ligase was observed using clofibric acid and ciprofibrate. Additionally, the long-chain and nafenopin-CoA ligases were distinguished by antibody immunoreactivity and response to enzyme inducers such as clofibric acid, di-ethylhexylphthalate and phenobarbitone. In addition to the microsomal and peroxisomal studies, activity of a protein expressed in COS 7 cells from a cDNA coding for the rat liver long-chain ligase was studied and a Km of 8.5 μM was determined for palmitic acid. Activity of the expressed protein was expressed protein was examined in the presence of a variety of xenobiotics (e.g., clofibric acid, ciprofibrate, nafenopin, R(-) ibuprofen) and profiles of either activation or activation/inhibition were observed for a number of these xenobiotic probes. The latter observation may be explained by the xenobiotic binding at a site additional to the catayltic site.

These studies on rat hepatic mocrosomes, peroxisomes and a cDNA expressed protein indicate that palmitic acid is not an isoform specific substrate. Use of a variety of carboxylic acid xenobiotics as inhibitory probes has highlighted kinetic anomalies which may be interpreted as indicating the presence of discrete isoforms of "long-chain like" xenobiotic CoA ligases. These studies also reveal the potential for toxicity resulting from competition between xenobiotic and endogenous carboxylic acids for the same catalytic site.

Perspective

It is now some 160 years since the discovery of the first conjugation reaction: in 1842 the discovery of the conversion of benzoic acid to hippuric acid was not only the first metabolic reaction of a xenobiotic to be described but was the first demonstration of a biosynthesis of any type.

Following this observation, the major conjugation reactions of glucuronidation, sufation, methylation and mercapturic acid formation, all giving inactive, easily eliminated products, were revealed during the later years of the 19th Century. This led to the view that the conjugation reactions were inherently of little interest: these "detoxication mechanisms" with their polar, water soluble end products stood to one side as the pharmacological, toxicological and clinical consequences of the Phase I pathways were discovered during the 1960s and onwards. Indeed, as late as 1978, one of us was representing the conjugation reactions as the "poor relations" of drug metabolism [19] while pointing out the numerous areas where these rections were, in fact, determinants of the biological activities of drugs and xenobiotics of all sorts [19, 20]. The last 15 years have seen a critical reevaluation of the conjugation reactions and it is now generally recognized that they deserve the resurgence of interest which is reflected in the content of this symposium. This interest has three main strands: (i) the increasing appreciation of the role which "active conjugates" can have, including the interfaces which the conjugates represent between xenobiotic metabolism and endogenous biochemistry, (ii) the realization that conjugation reactions can be metabolic determinants of drug effect, and (iii) the more recent application of the tools of the "new biology" to these complex reactions. Much of current progress has revolved around the major reactions of glucuronic acid and glutathione conjugation, where all three of the above approaches have resulted in substantial growth in our understanding of both the fundamentals of these reactions and their consequences in the body, but it must be stressed that there are numerous other conjugation reactions deserving the same intensity of investigation as these processes. This is well exemplified by the very recent appreciation of the importance of acyl CoA formation in the biochemistry or xenobiotic carboxylic acids. For many years seen only in terms of their intermediacy in the amino acid conjugations, xenobiotic acyl CoAs are now realized to be key intermediates in a wide range of metabolic reactions, including participation in biosyntheses, and toxic manifestations of carboxylic acid drugs and environmental chemicals. It is evident that drug metabolism scientists have overcome their "myopic" view of only cytochrome P450, and now have extended their vision to the conjugation reactions in all their biochemical diversity and multitude of pharmacological and toxicological consequences. The future will see further substantial developments.

References

1. Miners JO, Mackenzie PI. Drug glucuronidation in humans. Pharmac. Ther. 1991; 51: 247–269.

2. Burchell BB, Nebert DW, Nelson DR, Bock KW, Iyanagi T, Jansen PLM et al. The UDP glucuronosyltransferase gene family. Suggested nomenclature based on evolutionary divergence. DNA Cell Biol. 1991; 10: 487–494.
3. Schulte S, Stoffel W. Ceramide UDP galactosyltransferase from myelinating rat brain: purification, cloning, and expression. Proc. Natl. Acad. Sci. USA 1993; 90: 10265–10269.
4. Mackenzie PI. Molecular aspects of UDP glucuronsyltransferases. In: Hodgson E, Philpot RM, Bend JR editors. Reviews in biochemical toxicology. In press.
5. Mannervik V, Awasthi YC, Board PG, Hayes JD, Di Illio C, Ketterer B et al. Nomenclature for human glutathione transferases. Biochem. J. 1992; 282: 305–308.
6. Dekant W, Anders MW, Monks TJ. Bioactivation of halogenated xenobiotics by S-conjugate formation. In: Andeers MW, Denkant W, Henschler D, Oberleithner H, Silbernagl S (editors. Renal disposition and nephrotoxicity of xenobiotics). San Diego: Academic Press, 187–215.
7. Dekant W, Vamvakas S, Anders MW. Formation and fate of nephrotoxic and cytotoxic glutathione S-conjugates: Cysteine conjugate β-lyase pathway. Adv. Pharmacol. 1994; 27: 117–164.
8. Finkelstein MB, Vamvakas S, Bittner D, Anders MW. Structure-mutagenicity and structure-cytotoxicity studies on bromine-containing cysteine S-conjugates and related compounds. Chem. Res. Toxicol. 1994; 7: 157–163.
9. Yamazoe Y, Nagata K, Ozawa S, Fato R. Structural similarity and diversity of sulfotransferases. Chem. Biol. Interactions 1994; 92: 107–117.
10. Nagata K, Ozawa S, Miyata M, Shimada M, Gong D-W, Yamazoe Y et al. Isolation and expression of a cDNA encoding a male-specific rat sulfotransferase that catalyzes activation of N-hydroxy-2-acetylaminofluorene. J. Biol. Chem. 1993; 268: 24720–24725.
11. Kato R, Yamazoe Y. Metabolic activation of N-hydroxylated metabolites of carcinogenic and mutagenic arylamines and arylamides by esterification. Drug Metab. Revs. 1994; 26: 413–430.
12. Weinshilboum RM. Methylation pharmacogenetics: thiopurine methyltransferase as a model system. Xenobiotica 1992; 22: 1055–1071.
13. Girard B, Otterness DM, Wood TC, Honchel R, Wieben ED, Weinshilboum RM. Human histamine N-methyltransferase pharmacogenetics: cloning and expression of kidney cDNA. Mol. Pharmacol. 1994; 45: 461–468.
14. Aksoy S, Szumlanski CL, Weinshilboum RM. Human liver nicotinamide methyltransferase: cDNA cloning, expression and biochemical characterization. J. Biol. Chem. 1994; 265: 14835–14840.
15. Knights KM, Roberts BJ. Xenobiotic acyl-CoA formation: evidence of kinetically distinct hepatic microsomal long-chain fatty acid and nafenopin-CoA ligases Chem. Biol. Interactions 1994; 90: 215–222.
16. Knights KM. Talbot UM, Baillie TA. Evidence of multiple forms of rat liver microsomal coenzyme A ligase catalysing the formation of 2-arylpropionyl-coenzyme A thioesters. Biochem. Pharmacol 1992; 44: 2415–2417.
17. Knights KM, Jones ME. Inhibition kinetics of hepatic microsomal long-chain fatty acid-CoA ligase by 2-arylpropionic acid non-steroidal anti-inflammatory drugs. Biochem Pharmacol 1992; 43: 1465–14715.
18. Roberts BJ, Knights KM. Inhibition of rat peroxisomal palmitoyl-CoA ligase by xenobiotic carboxylic acids. Biochem. Pharmacol. 1992; 44: 261–267.
19. Caldwell J. The conjugation reactions – the poor relations of drug metabolism? In: Aitio A, editor. Conjugation reactions in drug biotransformation. Amsterdam: Elsevier 477–485.
20. Caldwell J. Conjugation reactions in foreign compound metabolism: definition, consequences and species differences. Drug Metab. Rev. 1982; 13: 745–777.

Pharmacological Sciences: Perspectives for
Research and Therapy in the Late 1990s
ed. by A.C. Cuello and B. Collier
© 1995 Birkhäuser Verlag Basel/Switzerland

Interethnic Differences in Drug Metabolism and Pharmacogenetics

T. Inaba[1], G. Alvan[2], Y. Yamazoe[3], S.M.F. de Morais[4],
J.A. Goldstein[4], G.T. Tucker[5], and R. Kato[3]

[1]*Department of Pharmacology, University of Toronto, Toronto, Canada;* [2]*Karolinska Institute,
Huddinge University Hospital, Sweden;* [3]*Department of Pharmacology, Keio University, Japan;*
[4]*NIEHS, North Carolina, USA;* [5]*Department of Medicine and Therapeutics, The Royal
Hallamshire Hospital, University of Sheffield, Sheffield, U.K.*

The International Congress of Pharmacology was held in Canada for
the first time and one of the symposia was entitled "Interethnic Differences in Drug Metabolism and Pharmacogenetics". The purpose of the
symposium was to discuss genetic variation of drug metabolizing enzymes in different ethnic groups. Five speakers from Sweden, Japan,
USA, Canada and U.K. participated in this symposium and discussed
the pharmacogenetics of drug metabolism ranging from a molecular
basis to some clinical implications.

Pharmacodynamics, and Ethnicity

The variability observed between patients in therapeutic drug effects
and in the occurrence of adverse drug effects is certainly of fundamental
importance for clinical medicine and pharmacology [1]. For some time
now, the importance of genetic factors for all aspects of human life has
been more emphasized and genetics has taken back its prime role on the
scene [2]. Certainly, ethnicity is a factor that reflects to what extent
individuals within an ethnic group share genetic material of importance
for the interaction between genes and environment which will ultimately
govern biological phenomena such as drug responses [3]. In contrast to
the race concept which takes only distribution of genes into account, the
term "ethnic group" covers cultural and probably also environmental
aspects of human subpopulations. Thus purely genetic and cultural/environmental factors will combine to produce the possibilities of ethnic
influence on human drug response.

Although genetic variability of a character may be monogenic or
polygenic, the most thoroughly studied drug metabolic polymorphisms

Correspondence to: Dr. T. Inaba, Department of Pharmacology, Faculty of Medicine,
University of Toronto, Toronto M5 S1 A8, Canada

are the monogenic enzymes for N-acetylation, S-mephenytoin hydroxylation and debrisoquine/sparteine hydroxylation. Polymorphic patterns separate a patient population into at least two groups: slow and rapid, or differently named poor and extensive metabolizers. The discovery of the acetylation polymorphism in the late 1950s triggered a great interest in clinical application of the knowledge since some important drugs such as the antituberculotic agent isoniazid and the antihypertensive hydralazine as well as some sulpha compounds are metabolized by this enzyme. Both clinical efficacy and risk for adverse reactions are clearly correlated to plasma concentrations and acetylation status for these drugs and interethnic differences in the proportions of rapid and slow acetylators are well documented [2]. The most striking example is probably a severe reaction called the lupus-like syndrome which occurs almost exclusively in slow acetylators.

Before going into the evidence for ethnic influence on pharmacodynamics, there is one more point to be stressed which is the differences in frequencies of the polymorphisms between ethnic groups and some simple consequences of genetic laws. A key question is of course the relative influence of interethnic variation in comparison with the variability existing within a population. That question has to be addressed, for example, when new drugs are studied and developed with the aim to have them used world wide. Hardy-Weinberg's law tells us that the distribution of genotypes in a monogenic system. The difference in frequencies of poor metabolizers is obvious, but if we believe in the potential importance of a gene dose effect, we should particularly care about the marked differences in occurrence of heterozygotes formally classified as extensive metabolizers, but more likely to attain higher drug levels than the group of homozygous extensive metabolizers.

The importance of ethnicity for pharmacodynamics is a primary research topic where much less work has been done so far. The necessary prerequisities such as differences in gene pools and environments are certainly there, but large scale prospective studies showing clear cut implications are still lacking. The crucial question is if this pharmacogenetic and interethnic variability should influence the dosing within and between ethnic groups. This is of course a sensitive issue, both for the drug manufacturer and drug regulatory agencies. Such a decision has to be made after a careful review of the pharmacological, pharmacokinetic and pharmacodynamic information available. Interethnic differences seem substantial enough to have a potential influence on treatment outcome, and this is presently an area of investigation. Concerning the production of side-effects, one may conclude that ethnicity is a factor of proven importance.

Diazepam Metabolism and S-Mephenytoin Polymorphism

The metabolism of the anticonvulsant drug mephenytoin exhibits polymorphism in humans, with individuals being characterized as either extensive (EM) or poor (PM) metabolizers [4, 5]. Several drugs, including diazepam, have been investigated to determine whether their *in vivo* metabolisms are cosegregated with the polymorphic pharmacogenetic factor involved in S-mephenytoin 4′-hydroxylation. *In vivo* results suggested that the N-demthylation is a dominant metabolic pathway of diazepam and implicated it to polymorphic mephenytoin 4′-hydroxylation in Caucasian subjects [3]. In contrast to Caucasians, the plasma half-life of diazepam in Chinese subjects did not differ between EMs and PMs of mephenytoin.

The earlier studies *in vitro* indicated that diazepam was mainly metabolized by 3-hydroxylation, mediated by CYP3A form. It was reported that mephenytoin competitively inhibited diazepam N-demethylation, but not the 3-hydroxylation in rat liver microsomes [6]. In contrast, neither nordiazepam nor temazepam formation was inhibited by mephenytoin in human liver microsomes. These data appeared to suggest that mephenytoin hydroxylation and diazepam N-demethylation are catalyzed by different forms of P450. However, Yasumori et al. [7] showed that the rate of N-demethylation is about one-third that of 3-hydroxylation at a high substrate concentration of 200 μM, whereas the rate of N-demethylation increases with the decreasing substrate concentration. The kinetic parameters for diazepam N-demethylation were assessed in liver microsomes from EMs and PMs of mephenytoin. Involvement of at least two P450 forms in diazepam N-demethylation was suggested by a biphasic kinetic pattern in the EM, whereas a monophasic pattern was observed in the PM liver microsomes, particularly the lack of a low K_m component. The concentration-dependent N-demethylation of diazepam *in vitro* is consistent with these observations *in vivo*.

Identification of the Genetic Defects Responsible for the Mephenytoin Polymorphism in Humans

The frequency of the PM phenotype of mephenytoin 4′-hydroxylation is much higher in the Oriental (18–23%) than in the Caucasian population (3–5%). This polymorphism also affects the metabolism of a number of currently used drugs including omeprazole, proguanil, citalopram, and certain barbiturates. The metabolism of propranolol, certain antidepressants, and diazepam is also affected, albeit to a lesser extent.

Recently CYP2C19 was shown to be the principle S-mephenytoin 4′-hydroxylase in humans. Using a yeast cDNA expression system, Goldstein et al. showed that recombinant CYP2C19 stereospecifically hydroxylated S-mephenytoin at a rate about 100-fold higher than other members of the CYP2C subfamily and some 20-fold higher than human liver microsomes [8]. Moreover, S-mephenytoin 4′-hydroxylase activity correlated with CYP2C19 content in human livers. The report by de Morais et al. indicated that the principal defect in PMs of mephenytoin is a single base pair mutation producing an aberrant splice site in exon 5 of CYP2C19 [9]. The aberrantly spliced mRNA lacks the first 40 bp of exon 5, and the reading frame is altered to produce a premature stop codon. Only the aberrantly spliced mRNA was found in livers from individuals that were homozygous for this defect. A PCR-restriction enzyme genetic test was developed and used to genotype individuals who had been previously phenotyped *in vivo*. $CYP2C19_{m1}$ was found to account for 75–83% of the defective alleles in both Japanese and Caucasian PMs. Recently a second defect was identified and this mutation, $CYP2C19_{m2}$, accounted for the remainder of the PMs in Japanese subjects, but was rare or absent in Caucasian PMs [10]. PCR-restriction enzyme genetics tests showed that the two defects accounted for 100% of available Japanese poor metabolizers while $CYP2C19_{m1}$ accounted for 83% of Caucasian PMs. Moreover, in family studies of three Japanese PM probands, the co-inheritance of the $CYP2C19_{m1}$ and the $CYP2C19_{m1}$ alleles explained the autosomal recessive inheritance of the PM phenotype. The development of genetic tests for this polymorphism will be useful in investigating the clinical importance of this defect.

Inter-Individual Variability of Metabolite, Drug Ratio (Reduced Haloperidol/Haloperidol) in Plasma

Haloperidol (HAL) is a potent neuroleptic drug which is used very widely. There have been many attempts to establish optimal plasma levels of haloperidol. The known active metabolite of HAL is reduced haloperidol (RHAL) with an affinity to dopamine D2 receptor several orders of magnitude less than the parent compound. RHAL is present in human patient plasma and its concentration differs widely between subjects. For the sigmareceptor, however, RHAL has virtually the same affinity as HAL. The association of the sigma receptor with neurotoxicity would imply that RHAL concentrations would be important for the side-effects of HAL therapy.

Plasma data for both HAL and RHAL are abundant in the literature. After unsuccessful attempts by many laboratories to establish a therapeutic window for HAL, the use of ratios such as RHAL/HAL was proposed as an index for therapeutic drug monitoring [11]. Another aspect of this index is the fact that it is independent or sparsely correlated with HAL dosage, while HAL concentrations are linearly proportional to the dosage. The first report on the pharmacogenetics of RHAL/HAL demonstrated that its distribution in 45 patients in Japan showed a non-normal or likely bimodal distribution [12]. The antimode appeared to be 0.7 and most Japanese patients showed values below 0.7, this low ratio representing 82% of the population studied. If the same antimode is assumed to be applicable to other ethnic groups, the combined data for the five Caucasian studies (n = 72) showed the RHAL/HAL ratio to be 1.13, while those for the combined Oriental data (n = 80) were 0.58, about half of the Caucasian value. Besides the two major ethnic groups, Hispanic and African-American subjects were investigated and the data were not too different from the Caucasian data. In conclusion, plasma RHAL/HAL ratios were considerably higher in Caucasian patients than in Oriental patients. Because haloperidol is not suitable as a pharmacogenetic probe due to severe side-effects, the discovery of an alternative probe would be necessary for further understanding of the ethnic difference in RHAL/HAL ratios.

Determination of Drug Metabolism Status In Vivo: *Pharmacokinetic and Statistical Issues*

Probe substrates for various isoforms of cytochrome P450 and other drug metabolising enzymes are now widely used to assess genetic, environmental and ethnic differences in drug metabolism *in vivo*. The key issues in population phenotyping include: (a) separation of risk and exposure (with the aid of genotyping), (b) choice of the best experimental index to use, and selection and number of subjects for study, (c) objective criteria for assessing multimodality in frequency distributions. Failure to appreciate these issues has resulted in a lot of misinterpretation and overinterpretation of data in the pharmacogenetics literature.

Many groups are now using urinary caffeine metabolite ratios as indirect indices of both cytochrome P450 1A2 (CYP1A2) and N-acetyltransferase (NAT2) activity, particularly with a view to delineating susceptibility to various cancers. The fact that this might be done simply and noninvasively is attractive, but whether it can be done selectively, especially in the context of CYP1A2, is debatable. For example, a population of Arkansans was screened using a particular caffeine urinary metabolic ratio [(17U + 17X)/137X] claimed to mark CYP1A2 [13]. The fact that this ratio, although very sensitive to urine flow, discriminates between smokers and non-smokers was encouraging. The authors went on to suggest that the distribution in non-smokers (but not in smokers) is trimodal – suggestive of a major genetic polymorphism in CYP1A2. This diagnosis was based upon the use of probit plots – two break-points claimed in non-smokers, but none in smokers. This approach was then applied across several ethnic groups leading to the suggestion of markedly different proportions of the "slow", "intermediate", and "rapid" phenotypes according to racial origin. However, two major problems can be raised with these types of studies: (a) the assumption that the urinary metabolic ratio used is specific for marking

CYP1A2, and (b) the use of arbitrary break-points in probit plots to separate so-called "phenotypes". At least five different urinary metabolite ratios of caffeine have been proposed by different ivestigators as empirical probes for *in vivo* CYP1A2 activity, and claims for the frequency distribution of the activity of this enzyme in populations vary from "normal" to "trimodal". Recently, Notarianni et al. [14] have attempted to correlate these different ratios in a study of over 200 healthy subjects; Some of the ratios were correlated, but most were not. Clearly, they are not all marking the same things, and some may be better than others at indicating CYP1A2 activity.

Therefore, it would be useful to evaluate the underlying pharmacokinetic basis and limitations of these ratios. A computer simulation was carried out based on the quite extensive experimental data base for caffeine kinetics [15]. Systematic changes were applied to model parameters, individual enzyme intrinsic clearances and renal clearances. Finally, the sensitivity of each ratio to each of the variables was assessed by regression of log values. Reasonable parameter values were culled from the literature. The simulations conducted were consistent with experimental data in that they predicted all of Notarianni's correlations and lack of correlations. Also, this study underlines why the ratio (AFMU + 1X + 1U)/17U by Campbell et al. [16] turns out to be the best when compared with the gold standard (measurement of caffeine plasma clearance which depends 90% on 1A2) and the caffeine breath test.

The other problem that was raised with regard to data analysis was use of probit plots to define phenotypes. Simple simulations show that although such plots indicate non-normality the resulting S-shape does not uniquely indicate bimodality. For example, an S-shaped probit plot can be derived just as well from a unimodal distribution with low kurtosis as a truly bimodal distribution. Furthermore, identification of the antimode is not straightforward, and probit analysis does not provide an objective statistic as evidence of bimodality. In conclusion, one needs to be aware of the limitations of indirect *in vivo* indices of enzyme activity when making important conclusions about risk and susceptibility.

Acknowledgement

We thank Ms. Agnes Bleiwas for reading the manuscript.

References

1. Alvan G. Genetic polymorphism in drug metabolism. J. Int. Med. 1992; 231: 571–573.
2. Price Evans DA. Genetic factors in drug therapy – clinical and molecular pharmacogenetics. Cambridge: Cambridge University Press, 1993: 1–136.

3. Kalow W, Bertilsson L. Interethnic factors affecting drug response. In: Testa B, Meyer UA, editors. Advances in drug research. Vol. 25. London: Academic Press, 1994: 1–53.
4. Kupfer A, Preisig R. Pharmacogenetics of mephenytoin: a new hydroxylation polymorphism in man. Eur. J. Clin. Pharmacol. 1984; 26: 753–759.
5. Inaba T. Pharamcogenetic polymorphism: mephenytoin hydroxylation deficiency. In: Kato R, Estabrook RW, Cayen MN, (editors), Xenobiotic metabolism and disposition. London: Talor & Francis, 1989: 467–474.
6. Beischlag TV, Kalow W, Mahon WA, Inaba T. Diazepam metabolism by rat and human liver *in vitro*: inhibition by mephenytoin. Xenobiotica 1992; 22: 559–567.
7. Yasumori T, Li Q-H, Yamazoe Y, Ueda M, Tsuzuki T, Kato R. Lack of low K_m diazepam N-demethylase in livers of poor metabolizers for S-mephenytoin 4′-hydroxylation. Pharmacogenetics. 1994; 4: 323–331.
8. Goldstein JA, Faletto MB, Romkes-Sparks M, Sullivan T, Kitareewan S, Raucy JL, Lasker JM, Ghanayem BI. Evidence that CYP2C19 is the major S-mephenytoin 4′-hydroxylase in humans. Biochemistry 1994; 33: 1743–1752.
9. de Morais SMF, Wilkinson GR, Blasidell J, Nakamura K, Meyer UA, Goldstein JA. The major genetic defect responsible for the polymorphism of S-mephenytoin metabolism in humans. J. Biol. Chem. 1984; 269: 15419–15422.
10. de Morais SMF, Wilkinson GR, Blasidell J, Meyer UA, Nakamura K, Goldstein JA. Identification of a new genetic defect responsible for the polymorphism of S-mephenytoin metabolism in Japanese. Mol Pharmacol. 1994; 46: 594–598.
11. Inaba T, Someya T, Shibasaki M, Tang SW, Takahashi S. Influence of ethnicity on reduced haloperidol concentrations in blood. In: Lin KM et al. (editors), Psychopharmacology and psychobiology of ethnicity. Washington: Am. Psychiatric Press, 1993: 123–132.
12. Someya T, Takahashi S, Shibasaki M, Inaba T, Cheung SW, Tang SW. Reduced haloperidol/haloperidol ratios: polymorphism in Japanese psychiatric patients. Psychiatry Res. 1990; 31: 111–120.
13. Butler MA, Lang NP, Young JF, Caporaso NE, Vineis P, Hayes RB, Teitel CH, Massengi JP, Lawsen MF, Kadlubar FF. Determination of CYP1A2 and NAT2 phenotypes in human populations by analysis of caffeine urinary metabolites. Pharmacogenetics 1992; 2: 116–127.
14. Notarianni LJ, Oliver SE, Dobrocky P, Bennett PN, Silverman BW. Caffeine as a metabolic probe: A comparison of the metabolic ratios used to assess CYP1A2 activity. Brit. J. Clin. Pharmacol. In press.
15. Tucker GT, Rostami-Hodjegan A, Nurminen S, Jackson PR. (1994) In preparation.
16. Campbell ME, Spielberg SP, Kalow W. A urinary metabolite ratio that reflects systemic caffeine clearance. Clin. Pharmacol. Ther. 1987; 42: 157–165.

Pharmacological Sciences: Perspectives for
Research and Therapy in the Late 1990s
ed. by A.C. Cuello and B. Collier
© 1995 Birkhäuser Verlag Basel/Switzerland

Dietary Effects on Drug Metabolism*

Chung S. Yang[1], Peter G. Welling[2], Grant R. Wilkinson[3],
David G. Bailey[4] and Charles S. Lieber[5]

[1]*Laboratory for Cancer Research, College of Pharmacy, Rutgers University, Piscataway,
NJ 08855-0789, USA;* [2]*Warner-Lambert/Parke-Davis, 2800 Plymouth Road, Ann Arbor,
MI 48105-2430, USA;* [3]*Department of Pharmacology, Vanderbilt University School of
Medicine, Nashville, TN 37232-6600, USA;* [4]*Department of Medicine, Victoria Hospital,
375 South Street, London, Ontario, N6A 4G5, Canada;* [5]*The Alcohol Research and Treatment
Center and the Section of Liver Diseases and Nutrition, Bronx Veterans Administration Medical
Center and Mount Sinai School of Medicine, New York, NY 10468, USA*

Summary. In this chapter, we discuss the interactions of food and drugs. The effects of food
on the absorption of drugs and the molecular basis by which dietary chemicals affect drug
metabolism are discussed. The possible complication of applying information from laboratory
studies to humans is discussed together with a recent study on the effect of fasting on
chlorzoxazone metabolism. The significant effect of grapefruit juice ingestion on the disposition of dihydropyridine drugs illustrates the possible impact of dietary factors in clinical
pharmacology. The important interactions among alcohol, drugs, and nutrients in human
health are also discussed.

It is well recognized that the plasma clearances of drugs may vary
greatly among different individuals [1]. Genetic and environmental
factors may both contribute to this variation. The importance of environmental factors in affecting drug metabolism is illustrated by the
observed day-to-day variabilities in the clearance of certain drugs in
the same individuals. Diet is one of the most important environmental
factors which influence drug metabolism. Earlier studies have demonstrated that, in human volunteers, a low-protein diet caused a lower
metabolic clearance of antipyrine and theophylline than a high-protein
diet. Feeding cabbage- and brussels sprout-containing diets to volunteers decreased the half-life of antipyrine by 13% [1]. Food and dietary
components may affect the fate of a drug by the following mechanisms: (1) altering the rates of its absorption and systemic availability,
(2) reacting or tightly binding with the drug, (3) competing with the
drug for binding to plasma proteins, and (4) affecting phase I and
phase II metabolism. In this chapter, dietary factors which affect drug
absorption and metabolism are discussed with examples from recent
studies.

Correspondence to: Dr. Chung S. Yang, Laboratory for Cancer Research, College of
Pharmacy, Rutgers University, Piscataway, NJ 08855-0789, USA.
*From the Symposium *Dietary Effects on Drug Metabolism* at the XIIth International
Congress of Pharmacology, Montreal, Canada, July 28, 1994.

Effects of Food on the Absorption of Drugs

Increased Drug Absorption

Increased drug absorption due to the presence of food has been frequently reported [2]. Accumulated evidence suggests that more complete drug dissolution due to the presence of food itself, or as a result of food-induced gastrointestinal (GI) secretions or delayed gastric emptying, often has a significant positive effect on absorption, particularly for fat soluble compounds. This, together with other mechanisms brought about by food ingestion seems to outweigh the many negative effects that food may have on drug absorption. Drugs whose absorption is increased by food include alafosfalin, canrenone, chlorothiazide, diazepam, griseofulvin, mebendazole, nitrofurantoin, and riboflavin. A recent, dramatic example of increased drug absorption by food is provided by the experimental hypolipidemic agent CGP 43371 [3].

Reduced Drug Absorption

Several studies have reported drug absorption to be reduced in the presence of food. Such drugs include many penicillin and cephalosporin analogues, atenolol, hydrochlorothiazide, captopril, and ketoconazole. An excellent recent example of a specific food effect is provided by a study in which the absorption of the fluoroquinolone antibacterial agent, ciprofloxacin, was reduced by approximately 35% by coadministered milk or yogurt. The mechanism of inhibition may be related to calcium chelation. The extent of inhibition may give rise to therapeutic failure in the case of moderately susceptible organisms [4].

Delayed Drug Absorption

While the absorption of drugs in this category is delayed, the extent of absorption is generally not reduced when taken with food. Delayed absorption due to food has been demonstrated for a large number of drugs including cinoxacin, diclofenac, glipizide, piroxicam, tolmesoxide, and valproic acid. Delayed absorption may be beneficial in reducing peak-trough variability in circulating plasma drug concentrations when multiple doses are administered and in avoiding toxicity due to high peak levels. On the other hand, the delayed absorption may cause overall profiles to be prolonged and reduced to the point that circulating levels may fall below those required for therapeutic effect. An example of this is provided by the antibacterial cepharadine. The considerable reduction in peak serum cepharadine concentrations in nonfasted individuals may cause therapeutic failure with moderately susceptible organisms while, on the other hand, the prolonged serum profile may provide coverage for a more extended period for more susceptible organisms.

No Effect on Drug Absorption

Although food ingestion invariably has some effect on the rate or extent of drug absorption, the effect is often sufficiently small that it is statistically or clinically insignificant. A large number of drugs fall into this category, including chlorpropamide, ethambutol, pivampicillin, tolbutamide and tranexamic acid. Ampicillin suspension, aspirin enteric-coated granules, and digoxin elixir also fall into this category, while absorption of these compounds from less dispersed dosage forms is significantly altered by food in one way or another. These latter examples reflect the relative insensitivity of the absorption of dispersed dosage forms to changes in the GI environment.

The importance of food-drug interactions has been clearly recognized by the pharmaceutical industry and has also found expression in guidances regarding the testing of food-drug interactions from both conventional and controlled-release formulations. Because of the unpredictable nature and extent of food-drug interactions, each drug formulation needs to be examined on a case-by-case basis and such studies are a necessary component of a clinical drug development program. In the meantime, more studies are needed to determine the precise mechanisms of the many food-drug interactions affecting drug absorption.

Mechanisms by which Dietary Factors Affect Drug Metabolism

Dietary chemicals can influence drug metabolism by affecting the level of phase I and phase II drug-metabolizing enzymes, or the level of cofactors required for these enzymes, such as NADPH, glutathione, UDP-glucuronic acid, and 3'-phosphoadenosine-5'-phosphosulfate.

Modulation of Cytochrome P450 Enzyme Levels

The modulation of cytochrome P450 enzymes by dietary factors has received much attention because of wide involvement of P450 enzymes in drug oxidation. The action can be at the transcriptional level. For example, P450 2B1 is induced by diallyl sulfide (DAS), a compound derived from garlic. P450 1A1 is induced by dietary indole-3-carbinol, a compound derived from cruciferous vegetables. The induction is probably due to the binding of the *Ah*-receptor by indolo(3,2-b)carbozole, a condensation product of indole-3-carbinol formed under the acidic conditions of the stomach. Post-transcriptional mechanisms are also known to be involved. In the induction of P450 2E1 by fasting and diabetes in animal models, elevation of mRNA was observed, but transcriptional activation could not be convincingly demonstrated. Stabilization of P450 2E1 mRNA may play a role in this induction. In the induction of P450 2E1 by acetone, elevation of the mRNA was not observed, and protein stabilization was suggested to be a mechanism for the increased level of this enzyme. Ethanol, because of its frequent consumption by humans, is probably the most important agent which induces P450 2E1 [5]. A high fat diet has also been shown to increase the level of P450 2E1 in rodents [6].

Inactivation of P450 Enzymes

Dietary chemicals may decrease P450 enzyme levels by metabolism-dependent inactivation. For example, P450 2E1 is inactivated by DAS. DAS is metabolized to diallyl sulfoxide and then to diallyl sulfone which is a substrate and suicide inhibitor of P450 2E1 [7]. Phenethyl isothiocyanate (PEITC), a compound derived from cruciferous vegetables, also inactivates P450 2E1 by a suicide mechanism. These compounds can also inactivate other P450 enzymes; for example, when administered to mice, the ability of lung microsomes to activate 4-(methylnitrosamino)-1-(3-pyridyl)-1-butanone (NNK), a potent tobacco carcinogen, was markedly decreased [8]. This inhibitory action in the activation of NNK can account for the protective effects of these compounds against NNK-induced lung tumorigenesis in mice and rats.

Inhibition and Stimulation of P450 Activities

P450 2E1 is known to metabolize low molecular weight drugs such as enflurane, halothane, acetaminophen, and chlorzoxazone, as well as environmental chemicals such as acetone, ethanol, *N*-nitrosodimethylamine, benzene, alkanes, carbon tetrachloride, vinyl chloride, and other halogenated hydrocarbons. It is expected that each compound serves as a competitive inhibitor (competitive substrate) for the metabolism of another substrate by P450 2E1. Inhibition of P450-dependent reactions by a variety of dietary compounds has been discussed in a recent review [9]. Among these compounds, flavonoids, which occur widely in fruits, vegetables, and beverages, are most intriguing. In general, these compounds are more effective inhibitors toward P450 1A2-catalyzed reactions than those by P450s 3A4 and 2E1 [10]. Quercetin, kaempferol, and naringenin are inhibitors for P450 3A4-catalyzed oxidation of nifedipine, felodipine, and acetaminophen [10, 11]. Flavonoids with free phenolic groups seem to be more effective inhibitors than those without. The latter group of compounds, such as flavone, nobiletin, and tangeretin, are stimulators of P450 3A4-dependent metabolism of acetaminophen [10].

Application of Information from Laboratory Studies to Humans

In the application of information from laboratory studies to humans, several factors have to be considered. (1) Although similar mechanisms between humans and rodent models are observed in the metabolism of drugs, there are also marked differences. (2) The doses of dietary chemicals used in laboratory studies are usually higher than the amounts in human consumption; difficulties exist in the extrapolation of effects from high to low doses. (3) In laboratory studies, the conditions are well defined, which make the effects of dietary factors discernible, whereas in human studies, many factors are difficult to control and thus may confound the results. The following two examples illustrate the complications.

Effects of Fasting and Obesity on Chlorzoxazone Metabolism in Humans

The muscle relaxant, chlorzoxazone, has been suggested as a drug probe for studying the activity of P450 2E1. Prolonged (38 h) fasting of six healthy men (which produced a significant increase in circulating ketone bodies) was associated with a reduction in the oral clearance of chlorzoxazone [12]. The urinary recovery of 6-hydroxychlorzoxazone was extensive, and the reduced clearance reflected a lower 6-hydroxylation ability after fasting. The elimination half-life of the drug was increased, whereas its apparent volume of distribution was unaffected by fasting. This result is different from laboratory studies which predicted that an induction of P450 2E1 would increase clearance of chlorzoxazone. In a second study, the disposition of chlorzoxazone was studied in nine obese women and nine age-matched normal women. Obese women were found to have a significantly higher rate of oral clearance and distribution of chlorzoxazone on both absolute and weight-normalized bases [12].

Effects of Watercress Ingestion on Acetaminophen Metabolism

Based on the observation that PEITC inhibited P450 2E1 activity, it was predicted that ingestion of watercress, which has a rather high content of PEITC, will inhibit the oxidative metabolism of acetaminophen. This concept was tested in humans by administering 50 g of fresh watercress (in 100 ml homogenate) to each of 10 volunteers in the late evening, and followed by a dose (1 g) of APAP on the next morning. As predicted, watercress ingestion inhibited the oxidative metabolism, but not the conjugation reaction of APAP (Yang et al., unpublished results). However, the effects of other watercress components on APAP metabolism cannot be excluded.

Grapefruit Juice and Drug Interactions in Humans

The discovery that grapefruit juice can markedly augment drug bioavailability was deduced from findings in an ethanol interaction study with felodipine, a dihydropyridine calcium antagonist. Although ethanol did not change felodipine pharmacokinetics, plasma drug concentrations were several-fold higher than expected. Interestingly, double-strength grapefruit juice was used in this study to blind the taste of ethanol. Subsequently, double-strength grapefruit juice was shown to triple mean felodipine bioavailability compared to water by inhibition of presystemic drug elimination [13]. Double-strength orange juice did not produce an interaction. The clinical relevance of the grapefruit juice-felodipine interaction was supported by greater diastolic blood pressure reduction, heart rate increase and frequency of adverse drug experiences. A significant interaction also occurs with a normal amount (200 ml) of regular-strength grapefruit juice with either felodipine standard tablets or an extended release formulation. The magnitude of the interaction showed substantial inter-subject variability but was reproducible within individuals on retesting. Grapefruit juice inhibited the formation of the single primary metabolite, dehydrofelodipine [13, 14], a step mediated by P450 3A4. Grapefruit juice also inhibited secondary metabolism by reducing de-esterification of dehydrofelodipine to its M3 monocarboxylic acid derivative. Concomitant grapefruit juice administration has been shown to enhance plasma drug concentrations of the other dihydropyridines, nifedipine, nitrendipine and nisoldipine as well as cyclosporin and terfenadine, all of which are substrates for P450 3A4 [15]. In addition, grapefruit juice interacted with coumarin, a compound metabolized by P450 2A6 [15].

Identification of the active ingredient in grapefruit juice would allow prediction of other foods that may interact with drugs. Naringin, which is the most prevalent flavonoid in grapefruit juice, appeared to be a relatively weak inhibitor of felodipine oxidation by human liver microsomes. Its aglycone, naringenin, was more potent [11]. The same amount of naringin in solution produced a less but qualitatively identical effect compared to grapefruit juice in some individuals [14]. Naringenin was also a potent competitive inhibitor of caffeine 3-*N*-demethylation, a step mediated by P450 1A2 [16]. Quercetin, which is found in high concentrations in many vegetables and fruits also inhibited dihydropyridine oxidation.

The discovery that grapefruit juice can markedly augment drug bioavailability has provided fundamental new knowledge both to improve pharmacotherapy and to stimulate further research.

Interaction of Alcohol with Drugs and Nutrients

Whereas cytosolic alcohol dehydrogenase (ADH) of the liver represents the main pathway for ethanol metabolism, extrahepatic ADH is also contributory, especially in the stomach. The human stomach has both low and high K_m ADH isozymes, resulting in significant ethanol metabolism in gastric cells *in vitro*, and decreased bioavailability of ethanol *in vivo* [5]. Commonly used medications (such as H_2-blockers) may interact with ADH-mediated ethanol metabolism in the stomach and result in a significant elevation of blood alcohol levels. In addition, liver microsomes are the site for a distinct and adaptive system of ethanol oxidation by P450 2E1. This enzyme was induced 5–10 fold in biopsies of alcohol drinking subjects with selective perivenular location [17] and enhanced levels of hepatic P450 2E1 mRNA [18].

Chronic ethanol consumption increases the metabolism of (and tolerance to) various other drugs, including meprobamate, warfarin, phenytoin, tolbutamide, propranolol, and rifampin. Much of the medical significance of the ethanol-inducible P450 2E1 results from its unique capacity to activate many xenobiotic compounds to toxic metabolites [5]. Enhanced metabolism (and toxicity) pertains to a variety of drugs, such as isoniazid, phenylbutazone, and acetaminophen. In addition, there is an association between alcohol misuse and an increased incidence of upper alimentary and respiratory tract cancers. Many factors have been incriminated including the effect of ethanol on P450-dependent activation of carcinogens and the ethanol-induced depletion of retinoids and interaction with carotenoids. Indeed, new hepatic pathways of microsomal retinol and retinoic acid metabolism, inducible by either ethanol or drug administration, have been discovered [19]. Enhanced hepatic toxicity of β-carotene in the presence of ethanol has also been observed.

Ethanol interacts with a multitude of other dietary factors [20] including dietary fat, which potentiates the alcohol-induced hepatic steatosis and the induction of microsomal enzymes. Chronic ethanol consumption is also associated with an increase in P450 4A1, more pronounced in male than in female rats; the microsomal ω-hydroxylation of lauric acid is significantly increased. Products of ω-oxidation increase liver cytosolic fatty acid binding protein content and peroxisomal β-oxidation, an alternate pathway for fatty acid disposition. Peroxisomes also contain catalase, which can oxidize ethanol *in vitro* in the presence of an H_2O_2-generating system.

Ethanol is converted in equimolar amounts to acetaldehyde, which can form protein adducts, resulting in antibody production, enzyme inactivation, and decrease DNA repair. Moreover, acetaldehyde promotes GSH depletion, free radical-mediated toxicity, and lipid peroxidation. The decrease in GSH is associated with S-adenosylmethionine depletion and can be partly corrected by S-adenosylmethionine replenishment. In addition, acetaldehyde affects hepatic collagen synthesis in cultured lipocytes, and the effect is abolished by dilinoleoylphosphatidylcholine [5]. Furthermore, this phosphatidylcholine protects against fibrosis and cirrhosis in the baboon [5] and this compound is now being tested in man. Thus, interactions of ethanol with other drugs and nutrients are of great pharmacologic, toxicologic, and clinical significance.

Abbreviations

GI: gastrointestinal; DAS: diallyl sulfide; PEITC: phenethyl isothiocyanate; NNK: 4-(methylnitrosamino)-1-(3-pyridyl)-1-butanone; ADH: alcohol dehydrogenase; APAP: acetaminophen.

References

1. Conney AH. Induction of microsomal enzymes by foreign chemicals and carcinogenesis by polycyclic aromatic hydrocarbons: G. H. A. Clowes Memorial Lecture. Cancer Res. 1982; 42: 4875–4917.
2. Welling PG. Effects of food on drug absorption. Pharmac. Ther. 1989; 43: 425–41.
3. Sun JX, Cipriano A, Chan K, Klibaner M, John VA. Effect of food on the relative bioavailability of a hypolipidemic agent (CGP 43371) in healthy subjects. J. Pharm. Sci. 1994; 83: 264–266.
4. Neuvonen PJ, Kivistö KT, Lehto P. Interference of dairy products with the absorption of ciprofloxacin. Clin. Pharmacol. Ther. 1991; 50: 498–501.
5. Lieber CS. Alcohol and the liver: 1994 update. Gastroenterology 1994; 106: 1085–1105.
6. Yoo J-SH, Ning SM, Pantuck CB, Pantuck EJ, Yang CS. Regulation of hepatic microsomal cytochrome P450IIE1 level by dietary lipids and carbohydrates in rats. J. Nutrition 1991; 121: 959–965.
7. Brady JF, Ishizaki H, Fukuto JM, Lin MC, Fadel A, Gapac JM, et al. Inhibition of cytochrome P-450IIE1 by diallyl sulfide and its metabolites. Chem. Res. Toxicol. 1991; 4: 642–647.
8. Yang CS, Smith TJ, Hong J-Y. Cytochrome P450 enzymes as targets for chemoprevention against chemical carcinogenesis and toxicity: Opportunities and limitations. Cancer Res. 1994; 54: 1982s–1986s.
9. Smith TJ, Yang CS. Effects of food phytochemicals on xenobiotic metabolism and tumorigenesis. In: Huang M-T, Osawa T, Ho C-T, Rosen RT, editors. Food phytochemicals for cancer prevention I. Washington, D.C.: ACS Symposium, Series 546, 1994: 17–48.
10. Li Y, Wang E, Patten C, Chen L, Yang CS. Effects of flavonoids on cytochrome P450-dependent acetaminophen metabolism in rats and human liver microsomes. Drug Metab. Dispos. 1994; 22: 566–571.

11. Guengerich FP, Kim D-H. *In vitro* inhibition of dihydropyridine oxidation and aflatoxin B_1 activation in human liver microsomes by naringenin and other flavonoids. Carcinogenesis 1991; 11: 2275–2279.
12. O'Shea D, Davis SN, Kim RB, Wilkinson GR. Effect of fasting and obesity in humans on the 6-hydroxylation of chlorzoxazone: a putative probe of CYP 2E1 activity. Clin Pharmacol. Ther. 1994; 56: 359–367.
13. Bailey DG, Spence JD, Munoz C, Arnold JMO. Interaction of citrus juices with felodipine and nifedipine. Lancet 1991; 337: 268–269.
14. Bailey DG, Arnold JMO, Munoz C, Spence JD. Grapefruit juice-felodipine interaction: mechanism, predictability and effect of naringin. Clin. Pharmacol. Ther. 1993; 53: 637–42.
15. Bailey DG, Arnold JMO, Spence JD. Grapefruit juice and drugs: how significant is the interaction? Leading Article. Clin Pharmacokinet. 1994; 26: 91–98.
16. Fuhr U, Klittich K, Staib AH. Inhibitory effect of grapefruit juice and its bitter principal, naringenin, on CYP1A2 dependent metabolism of caffeine in man. Br. J. Clin. Pharmacol. 1993; 35: 431–436.
17. Tsutsumi M, Lasker JM, Shimizu M, Rosman AS, Lieber CS. The intralobular distribution of ethanol-inducible P450IIE1 in rat and human liver. Hepatology 1989; 10: 437–446.
18. Takahashi T, Lasker JM, Rosman AS, Lieber CS. Induction of cytochrome P450 2E1 in human liver by ethanol is due to a corresponding increase in encoding RNA. Hepatology 1993; 17: 236–245.
19. Leo MA, Kim CI, Lieber CS. NAD^+-dependent retinol dehydrogenase in liver microsomes. Arch Biochem. Biophys. 1987; 259: 241–249.
20. Lieber CS. A personal perspective on alcohol, nutrition, and the liver. Am. J. Clin. Nutr. 1993; 58: 430–442.

Neuropharmacology

Pharmacological Sciences: Perspectives for
Research and Therapy in the Late 1990s
ed. by A.C. Cuello and B. Collier

Neurotransmitter Functions of Mammalian Tachykinins: Substance P and Neurokinin A

Masanori Otsuka

Department of Pharmacology, Faculty of Medicine, Tokyo Medical and Dental University, Tokyo 113, Japan

Summary. The purpose of this chapter is to present some additional persuasive evidence obtained recently in our laboratory, that substance P (SP) and neurokinin A (NKA) act as neurotransmitters released from certain primary afferent C-fibers in the spinal cord and some sympathetic ganglia. In the isolated spinal cord preparations of neonatal rats several types of synaptic responses lasting 20–60 s to C-afferent stimulation were depressed by tachykinin NK_1 receptor antagonists, e.g., GR71251. The C-fiber responses were potentiated by a mixture of peptidase inhibitors, but not after the treatment with GR71251. Likewise, in the coeliac ganglion of the guinea pig, the slow excitatory postsynaptic potential (EPSP) evoked by nerve stimulation was depressed by GR71251 and potentiated by peptidase inhibitors. These results suggest that certain tachykinins, i.e., SP and NKA, and NK_1 receptors are involved in generation of slow EPSPs in the spinal cord of neonatal rat and prevertebral ganglia of guinea pig. The recent advent of nonpeptide tachykinin antagonists may open a new field in therapeutics of various diseases.

Historical Background of Substance P and Neurokinin A

von Euler and Gaddum in 1931 discovered that equine brain and intestine contained an unidentified substance which produced hypotensive and smooth muscle-contracting actions. This substance was named substance P (SP) and was soon found to be a peptide (or peptides, in retrospect). Around 1953, Lembeck and other investigators found that dorsal roots of mammalian spinal nerves contained a substance having an SP-like activity in a much larger amount than ventral roots, and based on this finding Lembeck proposed a foresightful hypothesis that SP may be a sensory transmitter. For many years following, however, further evidence supporting the hypothesis of Lembeck had not been obtained, so that the hypothesis was not generally accepted until the early 1970s. In 1971, Leeman and colleagues determined the structure of SP as an undecapeptide, and then we discovered that SP exerts a potent excitatory action on spinal neurons of the frog and the rat. In the 1970s and 1980s, abundant evidence was accumulated to support the concept that SP is a neurotransmitter (for review, see ref. [1]). But until recently

Correspondence to: M. Otsuka, Department of Pharmacology, Faculty of Medicine, Tokyo Medical and Dental University, Yushima 1-5-45, Bunkyo-ku, Tokyo 113, Japan

 M. Otsuka

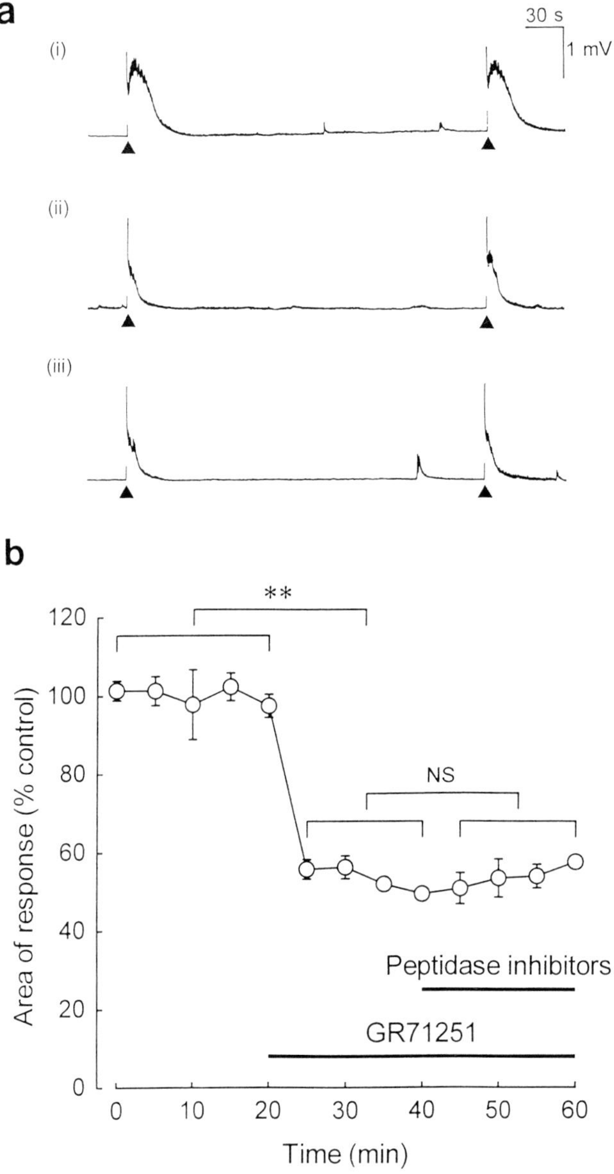

Fig. 1. Effects of GR71251 and the mixture of peptidase inhibitors on the saphenous nerve-evoked slow VRP in the isolated spinal cord-saphenous nerve preparations. Naloxone (0.5 μM) was present in perfusion solution throughout the experiments. (a) Sample records: (i) control responses in the presence of naloxone; (ii) responses after addition of GR71251 (5 μM); (iii) responses after addition of GR71251 and the mixture of peptidase inhibitors. The saphenous nerve was stimulated every 5 min with double square pulses of 100 μs duration and supramaximal intensity at 50 ms interval at ▲. (b) Time-course of the responses. Drugs were applied to spinal cords by perfusion during the periods indicated by the horizontal bars. Areas of the depolarizing responses in mV × s were measured and expressed as percentages of the

the concept was still received with some scepticism. One purpose of this chapter is to present some additional persuasive evidence obtained recently in our laboratory supporting the neurotransmitter role of SP. It is important to firmly establish the neurotransmitter status of SP, because SP is likely to serve as a model for many other putative peptide transmitters.

In 1983, two peptides having a C-terminal sequence in common with SP were discovered in mammalian CNS. These are neurokinin A (NKA) and neurokinin B (NKB), and together with many other peptides having the common C-terminal sequence, Phe-X-Gly-Leu-Met-NH_2, they are called tachykinins. SP, NKA and NKB are three main members of mammalian tachykinins. NKA is very often present with SP in the same neurons, and exerts effects similar to those of SP on neurons and peripheral tissues. Therefore, the present chapter focuses on the putative neurotransmitter functions of both SP and NKA.

Effects of Tachykinin NK$_1$ Receptor Antagonists on C-Afferent-Evoked Responses in the Isolated Spinal Cord of Neonatal Rat

For the identification of neurotransmitters, antagonists often play decisive roles. Since the introduction of tachykinin antagonists in 1981, great improvements have been achieved in their potencies, subtype specificities, etc., so that they now provide powerful experimental tools.

In the isolated neonatal rat spinal cord attached to peripheral nerves, stimulation of the saphenous nerve at a strength sufficient to activate C-fibers evoked a slow depolarizing response lasting 20–30 s in the ipsilateral L3 ventral root [2]. This response will be referred to as the saphenous nerve-evoked slow ventral root potential (VRP). The slow VRP was markedly depressed by tachykinin NK$_1$ receptor antagonists, spantide, GR71251 and GR82334 ([2, 3]; Fig. 1). A nonpeptide NK$_1$ receptor antagonist, RP67580, also depressed the saphenous nerve-evoked slow VRP, whereas its inactive enantiomer RP68651 did not have any detectable effect [4].

Since C-afferent fibers are known to be important for pain sensation, and both SP and NKA are contained in a subpopulation of C-afferent fibers, we examined the possible involvement of these tachykinins in the responses evoked by painful stimuli in the neonatal rat spinal cord. For this purpose, we developed two kinds of *in vitro* preparations, namely, the isolated spinal cord-tail preparation [5] and the isolated spinal

average value of control responses. Each point represents mean ± s.e.m. (n = 5). ** Represents a significant difference between the average values of the points under the horizontal bars (P < 0.01). NS: no significant difference. (Reproduced from ref. [9]).

cord-saphenous nerve-skin preparation of neonatal rats [6]. In both preparations, application to peripheral tissues of capsaicin, a specific stimulant of C-afferents, evoked a slow depolarizing response lasting 20–60 s in lumbar ventral roots, which was reversibly depressed by spantide [5, 6].

Another type of response evoked by C-afferent fibers is a long-lasting inhibition of monosynaptic reflex. In the isolated spinal cord attached to peripheral nerves of neonatal rat, a conditioning stimulation of the saphenous nerve produced a prolonged inhibition, lasting 20–30 s, of monosynaptic reflex. The duration of this inhibition was markedly shortened by spantide and GR71251 [3, 7, 8].

Our experiments with tachykinin NK_1 receptor antagonists support that tachykinins, i.e., SP and NKA, and NK_1 receptors are involved in C-afferent-evoked responses in the neonatal rat spinal cord.

Inactivation of Tachykinin Neurotransmitters

One of the important criteria for identification of neurotransmitters is the demonstration of the inactivation mechanism. What appears more important for the establishment of a neurotransmitter, however, is the demonstration that an inhibition of the inactivation mechanism results in a prolongation or a potentiation of synaptic responses. A typical example is the augmentation and prolongation of end-plate potential produced by cholinesterase inhibitors at the neuromuscular junction, which provided a convincing piece of evidence for the transmitter role of acetylcholine. We have therefore examined the effect of peptidase inhibitors on the saphenous nerve-evoked slow VRP. For this purpose, it was important to carry out the experiments in the presence of naloxone. Otherwise, the effect of the peptidase inhibitors on the enkephalinergic component in the slow VRP was more pronounced and masked the effect on the tachykininergic component.

In the presence of naloxone, the slow VRP was potentiated by a mixture of peptidase inhibitors, including thiorphan, actinonin and captopril (Fig. 2). However, when the tachykininergic component had been eliminated by adding an NK_1 receptor antagonist, GR71251, in addition to naloxone, the peptidase inhibitors no longer potentiated the slow VRP (Fig. 1). The latter result suggests that what is potentiated by the peptidase inhibitors is mainly the tachykininergic component [8, 9].

The saphenous nerve-induced long-lasting inhibition of monosynaptic reflex was similarly prolonged by the mixture of peptidase inhibitors. Furthermore, after the treatment with GR71251, the peptidase inhibitors no longer prolonged the inhibition of monosynaptic reflex [8].

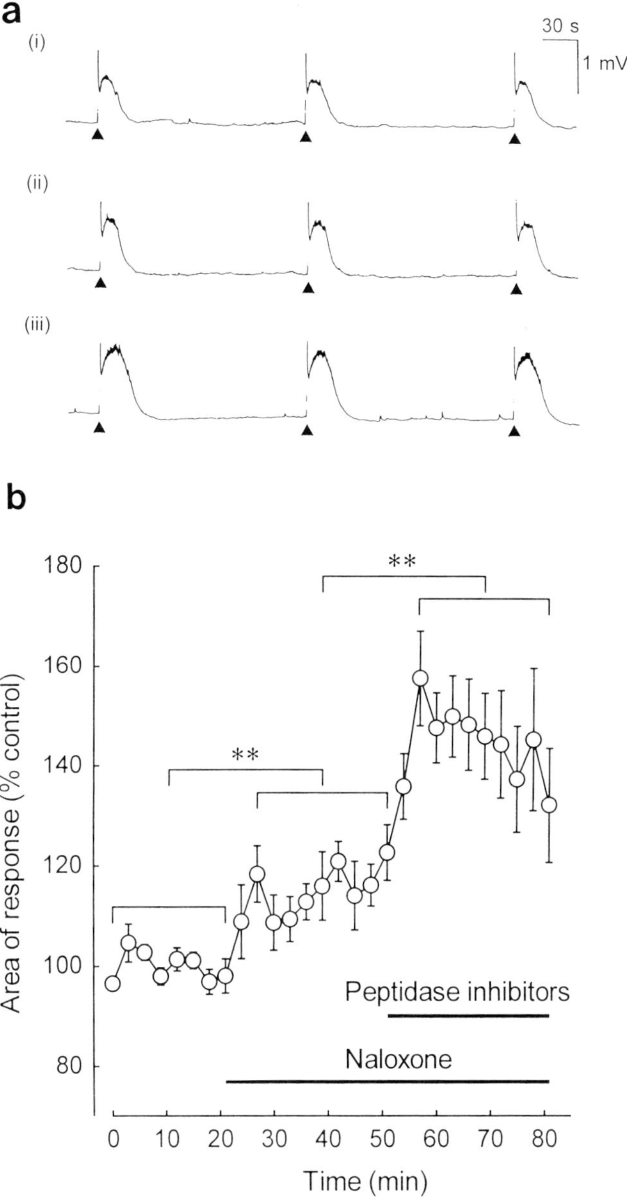

Fig. 2. Effects of naloxone and the mixture of peptidase inhibitors on the saphenous nerve-evoked slow VRP in isolated spinal cord-saphenous nerve preparations. (a) Sample records: (i) control responses; (ii) responses after addition of naloxone (0.5 μM); (iii) responses after addition of naloxone and the mixture of peptidase inhibitors. Experimental procedures were the same as in Fig. 1 except that the saphenous nerve was stimulated every 3 min at ▲. (b) Time-course of the responses. Naloxone (0.5 μM) and the mixture of peptidase inhibitors were applied during the periods indicated by the horizontal bars. Each point represents mean ± s.e.m. (n = 7). ** Represents a significant difference between the average values of the points under the horizontal bars (P < 0.01). (Reproduced from ref. [9]).

Descending Fiber-Induced Slow Depolarization of Motoneurons

The saphenous nerve-evoked slow VRP represents a polysynaptic response induced by primary afferents, in which tachykininergic slow excitatory postsynaptic potentials (EPSPs) in dorsal horn neurons are presumably involved. We wanted, however, to more directly study the synaptic potential which may be mediated by tachykinins. There is morphological evidence that bulbospinal fibers contain SP and NKA in addition to serotonin and TRH, and that some of these fibers form synapses with motoneurons. When we stimulated the cervical spinal cord electrically, and recorded extracellularly from a lumbar ventral root, a fast depolarizing potential followed by a slow depolarization was observed. A large part of this response was eliminated by the mixture of excitatory amino acid receptor antagonists, i.e., D(−)-2-amino-5-phosphonovaleric acid (D-APV) and 6-cyano-7-nitroquinoxaline-2,3-dione (CNQX), and a serotonin antagonist, ketanserin, but a small part of the slow depolarization remained, which could be recorded either extracellularly from lumbar ventral roots or intracellularly from motoneurons. This depolarization, lasting about 1 min, may partly represent slow EPSPs produced in motoneurons by tachykinins. In support of this notion, the slow depolarization that remained after the treatment with D-APV, CNQX and ketanserin was potentiated by a peptidase inhibitor, thiorphan, and depressed by GR71251 [10]. Furthermore, when we treated the isolated spinal cord with a serotonin neurotoxin, 5,7-dihydroxytryptamine, for several hours, the slow depolarization became smaller and was no longer potentiated by the peptidase inhibitor or depressed by GR71251. These results suggest that SP and NKA released from descending serotonergic fibers produce slow EPSPs in motoneurons.

Involvement of NK$_1$ Receptors in Slow EPSP in Coeliac Ganglion of Guinea Pig

Previous studies suggested that SP- and NKA-containing primary afferent axon collaterals form synapses with principal cells in the prevertebral ganglia of the guinea pig and that SP and NKA released from these axons produce slow EPSPs in the ganglion cells. We have examined the effects of NK$_1$ receptor antagonists on slow EPSPs in the coeliac ganglion of the guinea pig. In one group of ganglion cells classified as tonic cells, electrical stimulation of the mesenteric, i.e., postganglionic nerves induced a slow EPSP lasting a few minutes. The slow EPSP was partially or totally blocked by GR71251 or CP96345. In contrast, addition of peptidase inhibitors augmented and prolonged the slow EPSP.

In another group of ganglion cells, i.e., phasic cells, mesenteric nerve stimulation induced a shortening of the afterhyperpolarization following action potentials. This shortening of afterhyperpolarization lasted a few minutes, and was again partially blocked by GR71251 or CP96345 and prolonged by peptidase inhibitors. Thus, it appears that SP and NKA serve as neurotransmitters acting on NK_1 receptors in the coeliac ganglion and that they produce slow EPSPs in tonic cells and shortening of afterhyperpolarization in phasic cells [11, 12].

Blockade of NKA Action by NK_1 Receptor Antagonists

Although tachykinin antagonists provide important experimental tools for examining the possible transmitter roles of tachykinins, their profiles must be carefully examined for adequate interpretation of the experimental results. It is believed that SP acts mainly on NK_1 receptors. In support of this, we confirmed that the action of SP on spinal neurons of the neonatal rat was antagonized by NK_1 antagonists, spantide, GR71251, GR82334, and RP67580 [3–5, 13]. Furthermore the action of SP on prevertebral ganglion cells of the guinea pig was also antagonized by GR71251 and CP96345 [11, 12]. In addition, however, the actions of NKA on these neurons were also antagonized by these NK_1 antagonists, and there is evidence that NKA acts on these neurons via NK_1, but not NK_2, receptors [3, 4, 11–14]. Although the involvement of multiple subtypes of tachykinin receptors appears likely, our experimental results with NK_1 antagonists support the involvement of SP, NKA and NK_1 receptors in the synaptic transmission in the neonatal rat spinal cord and guinea pig prevertebral ganglia.

Tachykinins and Future Therapeutics

One important aspect of tachykinin research may be its possible relation to future therapeutics. In particular, the recent advent of nonpeptide tachykinin receptor antagonists may promote the development of new types of drugs. Tachykinins have been implicated in many diseases and disorders such as familial dysautonomia, Hirschsprung's disease, Crohn's disease, allergy and headache. Tachykinin-related drugs may be useful in therapy of some of these. Since SP and NKA are contained in a major subpopulation of primary afferent C-neurons, and C-neurons are important for pain sensation, there is little doubt that SP and NKA contribute to pain sensation. These tachykinins are released not only from central terminals of primary afferent neurons in spinal dorsal horn, but also from peripheral terminals to participate in inflammatory processes. Thus, tachykinin antagonists may serve as a new type of analgesic and anti-inflammatory drugs with entirely different mechanisms of action from those of aspirin-like drugs.

SP and NKA are widely distributed in mammalian bodies. Particularly high concentrations of SP and NKA are found in basal ganglia, substantia nigra, solitary nuclei and trigeminal nuclei in the brain and gastrointestinal tract and salivary glands in periphery. The physiological functions of the tachykinins are probably similar to those in the spinal dorsal horn and prevertebral ganglia as described in this chapter. Tachykinins, however, are not major transmitters compared with GABA or glutamate. In this respect, tachykinins may resemble opioid peptides. Injection of naloxone into normal subjects produces almost no effect, suggesting that the roles of opioid peptides in normal subjects are not important. Yet, opioid peptides are connected to a large field of narcotic analgesics. Similarly, tachykinins and their receptors may be connected to a large field of pharmacology and therapeutics of pain, inflammation, visceral and CNS functions.

References

1. Otsuka M, Yoshioka K. Neurotransmitter functions of mammalian tachykinins. Physiol. Rev. 1993; 73: 229–308.
2. Nussbaumer J-C, Yanagisawa M, Otsuka M. Pharmacological properties of a C-fibre response evoked by saphenous nerve stimulation in an isolated spinal cord-nerve preparation of the newborn rat. Br. J. Pharmacol. 1989; 98: 373–382.
3. Guo J-Z, Yoshioka K, Yanagisawa M, Hosoki R, Hagan RM, Otsuka M. Depression of primary afferent-evoked responses by GR71251 in the isolated spinal cord of the neonatal rat. Br. J. Pharmacol. 1993; 110: 1142–1148.
4. Hosoki R, Yanagisawa M, Guo J-Z, Yoshioka K, Maehara T, Otsuka M. Effects of RP67580, a tachykinin NK_1 receptor antagonist, on a primary afferent-evoked response of ventral roots in the neonatal rat spinal cord. Br. J. Pharmacol. 1994; 113: 1141–1146.
5. Otsuka M, Yanagisawa M. Effect of a tachykinin antagonist on a nociceptive reflex in the isolated spinal cord-tail preparation of the newborn rat. J. Physiol. (Lond.) 1988; 395: 255–270.
6. Yanagisawa M, Hosoki R, Otsuka M. The isolated spinal cord-skin preparation of the newborn rat and effects of some algogenic and analgesic substances. Eur. J. Pharmacol. 1992; 220: 111–117.
7. Yoshioka K, Sakuma M, Otsuka M. Cutaneous nerve-evoked cholinergic inhibition of monosynaptic reflex in the neonatal rat spinal cord: involvement of M_2 receptors and tachykininergic primary afferents. Neuroscience 1990; 38: 195–203.
8. Yanagisawa M, Yoshioka K, Kurihara T, Saito K, Seno N, Suzuki H, Hosoki R, Otsuka M. Enzymatic inactivation of tachykinin neurotransmitters in the isolated spinal cord of the newborn rat. Neurosci. Res. 1992; 15: 289–292.
9. Suzuki H, Yoshioka K, Yanagisawa M, Urayama O, Kurihara T, Hosoki R, Saito K, Otsuka M. Involvement of enzymatic degradation in the inactivation of tachykinin neurotransmitters in neonatal rat spinal cord. Br. J. Pharmacol. 1994; 113: 310–316.
10. Kurihara T, Yoshioka K, Otsuka M. Involvement of tachykinins in the slow depolarization of neonatal rat spinal motoneurones evoked by stimulation of descending pathway [abstract]. Jpn. J. Pharmacol. 1993; 61, Suppl. I: 243P.
11. Zhao F-Y, Saito K, Konishi S, Guo J-Z, Murakoshi T, Yoshioka K, Otsuka M. Involvement of NK_1 receptors in synaptic transmission in the guinea pig coeliac ganglion. Neurosci. Res. 1993; 18: 245–248.

12. Zhao F-Y, Saito K, Guo J-Z, Murakoshi T, Yoshioka K, Otsuka M. Involvement of NK$_1$ receptors in tachykininergic synaptic transmission in the coeliac ganglion of the guinea pig [abstract]. Neurosci. Res. 1993; Suppl. 18: S40.
13. Yanagisawa M, Otsuka M. Pharmacological profile of a tachykinin antagonist, spantide, as examined on rat spinal motoneurones. Br. J. Pharmacol 1990; 100: 711 716.
14. Suzuki H, Yoshioka K, Maehara T, Hagan RM, Nakanishi S, Otsuka M. Pharmacological characteristics of tachykinin receptors mediating acetylcholine release from neonatal rat spinal cord. Eur. J. Pharmacol. 1993; 241: 105 110.

Pharmacological Sciences: Perspectives for
Research and Therapy in the Late 1990s
ed. by A.C. Cuello and B. Collier

Molecular Events Underlying the Anti-Opioid Effect of Cholecystokinin Octapeptide (CCK-8) in the Central Nervous System

Ji-Sheng Han

Neuroscience Research Center, Beijing Medical Univeristy, Beijing 100083, China

Summary. Cholecystokinin octapeptide (CCK-8) has been known as the most potent neuropeptide possessing an anti-opioid activity. In the present study, *in vivo* and *in vitro* experiments were performed to explore the possible mechanisms of this action. The results indicate that CCK/opioid interactions take place at different levels of the transmembrane signal transduction pathways. The most prominent changes induced by CCK-8 are two fold: (1) the suppression of opioid receptor binding capability, and [2] an increase of the intracellular free calcium level that is opposite to the opioid effect of decreasing the intracellular calcium concentration. A dynamic balance between opioid and CCK activities may serve as a cardinal determinant for certain physiological reactions including the response to noxious stimulation.

Introduction

Cholecystokinin octapeptide (CCK-8) is widely and abundantly distributed in numerous brain regions and in the spinal cord. An array of physiological functions have been attributed to central CCK-8, among which the antagonistic effect on opioid analgesia has been clearly defined by behavioral and electrophysiological studies [1–5]. However, the molecular mechanisms whereby CCK-8 exerted its anti-opioid effect remain obscure. The aim of this study was to analyze the mechanisms underlying the anti-opioid effect of CCK-8 at different levels of the transmembrane signal transduction pathways, including the interactions taking place at the receptor level and G protein level, the second messenger cAMP and inositol phosphates, the possible involvement of protein kinase C (PKC), the intracellular calcium and the activities of voltage-gated calcium channels as well as the observation of changes in pain and analgesia as a result of CCK gene transfer in CNS.

CCK-8 Antagonizes Opioid Analgesia Mediated by μ- and κ- But Not δ-Receptors in the Spinal Cord of the Rat

Previously studies performed in our laboratory revealed that centrally administered CCK-8 antagonized the analgesic effect produced not only

Correspondence to: Prof. J. S. Han, Neuroscience Research Center, Beijing Medical University, 38 Xue Yuan Road, Beijing 100083, China.

by morphine but also by the endogenously released opioids elicited, e.g., by the electroacupuncture (EA) stimulation [4, 5]. Since opioid receptors were known to be divided into three different types, the μ-, δ- and κ-receptor, it would be interesting to characterize which of the three opioid receptors is most susceptible to CCK antagonism. Experiments were performed in rats using tail flick latency (TFL) as the endpoint of nociception. The analgesic effect produced by intrathecal (i.t.) injection of PL017, a μ specific opioid agonist, could be markedly antagonized by CCK-8 at a dose as small as 4 ng. Similar effect was observed when 66A-078, a κ-specific opioid agonist was used instead of the μ-agonist. In contrast, analgesia produced by i.t. injection of the δ-specific opioid agonist DPDPE could not be blocked by CCK-8 even at a dose as high as 40 ng [6]. Since the effect of CCK-8 could be totally reversed by the CCK receptor antagonist proglumide, this effect of CCK-8 is most likely mediated by the activation of CCK receptors rather than by the blockade of opioid receptors.

Modification by CCK-8 of the Binding of μ- and κ- But Not δ-Opioid Receptors

Following a pivotal study showing that CCK-8 suppressed the binding of opioid receptor to the universal opioid agonist [³H]etorphine [7], highly selective tritium labeled agonists for μ- (DAGO), δ- (DPDPE) and κ- (U69,593) opioid receptor respectively were used to clarify which type(s) of opioid receptor in rat brain membrane is suppressed by CCK-8. In the competition experiments, CCK-8 suppressed the binding of [³H]DAGO and [³H]U69,593 but not that of [³H]DPDPE to the respective opioid receptor. This effect was blocked by the CCK antagonist proglumide at 1 μM. In the saturation experiments CCK-8 at concentrations of 0.1 to 1.0 nM decreased the Bmax of [³H]DAGO binding sites without affecting the Kd; on the other hand, CCK-8 increased the Kd of [³H]U69,593 binding without changing the Bmax [8]. The results suggest that CCK-8 inhibits the binding of μ- and κ-, but not δ-opioid receptors via the activation of CCK receptors, which is in line with the findings mentioned above that CCK-8 suppressed the analgesia induced by opioid agonists acting on μ- and κ-receptors, but not that on δ-receptor.

Evidence Supporting a Direct Interaction Between Opioid Receptor and CCK Receptor

Two possibilities exist to explain the phenomena of CCK suppression of the opioid agonist binding: (a) receptor-receptor cross talk, (b) interaction via post-receptor or intracellular events. To substantiate the first possibility, we used the opioid antagonist [³H]naloxone which has no intrinsic activity, therefore would not induce post-receptor events. Radioreceptor assay with [³H]naloxone in rat brain membrane revealed two populations of [³H]naloxone binding sites, a high affinity site and a low affinity site. CCK-8 at 10, 100 and 1000 nM dose-dependently increased the Kd and decreased the Bmax of the high affinity site [9]. The results are in favor of a direct CCK-receptor/opioid-receptor interaction, although an indirect interaction via G proteins can not be ruled out.

Uncoupling of Opioid Receptors From Their Relevant G Proteins

The effect of opioids is known to be mediated by the G proteins, mainly the Gi protein [10]. If CCK produces an uncoupling of the opioid receptor with its relevant G protein, it would inevitably result in a decrease in post-receptor activities. This was tested using the universal opioid agonist [³H]etorphine ([³H]Et) for opioid binding assay in rat brain membranes. Scatchard analysis revealed that CCK-8 at 10 nM increased the Kd and decreased the Bmax of the high affinity Et binding sites. GTPγS, the hydrolysis-resistant GTP analog capable of dissociating the G protein from its receptor, increased Kd of opioid binding without affecting the Bmax. In the presence of both CCK-8 and GTPγS, the binding parameters of [³H]Et were essentially the same as they were in the presence of CCK-8 alone, suggesting that CCK-8 and GTPγS may act through one and the same mechanism. In another experiment it was found that GTPγS at $1-100\ \mu$M dose-dependently decreased the [³H]Et binding. In the presence of CCK-8 which produced a moderate decrease of opioid binding, the slope of the dose-response curve for GTPγS became flat. The two curves plateau at the same level, again suggesting that CCK-8 and GTPγS may work through the same mechanism, i.e., to uncouple opioid receptors from their relevant G protein, resulting in a decreased capability of opioid binding and a blockade of the post-receptor signal transduction [9].

The Influence of CCK-8 and Opioid Agonists on Spinal cAMP Content

It is well known that cAMP is involved in mediating opioid effects. It is therefore relevant to evaluate whether the effect of opioids on CNS cAMP content is affected by CCK-8. The spinal cAMP content was measured by radioimmunoassay. No significant changes in spinal cAMP content were found 10 min after i.t. injection of 5–40 ng CCK-8. In contrast, 25 ng PL017 (μ-agonist) or 20 μg DPDPE (δ-agonist) induced a remarkable decrease in spinal cAMP content. Injection of 300 ng of 66A-078, the κ-agonist, produced a slight decrease in cAMP content. The decrease in spinal cAMP content induced by the three opioid agonists was not reversed by CCK-8, even at a dose as high as 20 ng administered 10 min after the opioids. These results seem to indicate that central cAMP is not involved in the mechanisms of the anti-opioid effect of CCK-8 [11].

Activation by CCK-8 of the Phophoinositide (PI) Signaling System in CNS

Neonatal-rat brain (minus cerebella) cells were dispersed with trypsin. The intact cells were incubated with [³H]inositol for 3 hr. The labeled cells were then stimulated with the typical cholinomimetic carbachol in the presence of 10 nM LiCl. Carbachol at 1 mM stimulated an increase of IP_3 in brain cells in 10 min, peaked at 30 min, and approached the baseline at 45 min. A very similar effect was obtained when 10 nM CCK-8 was used instead of 1 mM carbachol. A bell-shaped dose-response curve was obtained showing that IP_3 formation increased when the concentration of CCK-8 was increased from 0.1 to 10 nM. A further increase of the CCK-8 concentration to 100–1000 nM resulted in a gradual decrease in IP_3 formation. The results provide a direct evidence for CCK-8 to stimulate PI turnover, and to increase IP_3 content in dissociated rat brain cells [12].

Effect of Opioid Ligands and CCK-8 on the Intracellular Free Calcium Concentration in Dissociated Rat Brain Cells [13]

In enzymatically dissociated brain cells prepared from neonatal rats, KCl at concentration of 25 and 50 mM produced a significant increase in intracellular free calcium concentration ($[Ca^{2+}]i$), and this increase could be prevented by verapamil or nifedipine (10 μM) known to block voltage-sensitive calcium channel.

Opioid receptor agonist ohmefentanyl (OMF, μ-specific), DPDPE (δ-specific) and 66A-078 (κ-specific) at concentration of 10 nM to 1 μM produced a marked suppression of the Ca^{2+} influx induced by high K^+ depolarization, without changing the $[Ca^{2+}]_i$ level in resting cells.

Specific opioid receptor antagonist β-FNA (μ-specific), ICI 174864 (δ-specific) and nor-BNI (κ-specific) exert no significant influence on the resting $[Ca^{2+}]_i$. However, the suppressive effect of OMF, DPDPE and 66A-078 on high K^+ depolarization-induced increase in $[Ca^{2+}]_i$ was markedly reversed by their respective antagonist β-FNA, ICI 174864 and nor-BNI.

CCK-8 at concentratons of 0.3, 3, and 30 nM dose-dependently mobilized Ca^{2+} from intracellular stores. While CCK-8 (30 nM) did not significantly affect the high K^+-induced increase of $[Ca^{2+}]_i$, it did reverse the opioid suppression of high K^+ induced increase of $[Ca^{2+}]_i$ caused by μ-agonist OMF and κ-agonist 66A-078, but not that caused by the δ-agonist DPDPE. This effect of CCK-8 remained even in the calcium-free medium. It is thus obvious that while opioid ligands suppress $[Ca^{2+}]_i$ by blocking voltage-operated Ca^{2+} influx, the anti-opioid effect of CCK-8 seems to be operated mainy via mobilization of Ca^{2+} from intracellular stores. That CCK-8 does not antagonize the $[Ca^{2+}]_i$ lowering effect of δ-agonist seems to fit in with the *in vivo* finding that δ agonist-induced analgesia was not antagonized by CCK-8, the underlying mechanisms of which deserve further investigation.

PKC Antagonism of Opioid Analgesia

Hydrolysis of PI produces both IP_3 and diacyl glycerol (DAG). Since phorbol ester TPA has been known to mimic DAG in stimulating PKC, TPA was used to simulate the effect of an increase in DAG following the activation of CCK receptor and to observe its effect on opioid analgesia. TPA injected intrathecally in five cumulative doses (6.25– 100 ng at 10 min intervals) produced no significant changes in the TFL. However, pretreatment with TPA markedly attenuated the analgesia elicited by 10 ng of the μ-agonist PL017. Analgesia induced by i.t. injection of 20 μg of DPDPE was also antagonized by TPA in a dose-dependent manner between 12.5 and 50 ng. Higher doses of TPA (50 and 100 ng) were needed to suppress the analgesia elicited by the κ-agonist 66A-078 (300 ng) (Zhang LJ, Han JS, to be published). The mechanisms whereby TPA suppresses opioid analgesia are not clear. It may be related to an activation of PKC which induces an increase in calcium conductance (resulting in a rise in $[Ca^{2+}]_i$ level) and a decrease in potassium conductance (leading to a tendency of depolarization) [14]. Both effects are opposite that of opioids.

Opioid Suppression of Voltage-Gated Calcium Current and its Reversal by CCK-8

Whole-cell patch-clamp technique was employed in acutely isolated rat dorsal root ganglion neurons to examine the effect of κ-opioid agonist, U-50488H on the voltage-gated calcium channels and to explore a possible interaction between U-50488H and CCK-8. The calcium current elicited in dorsal root ganglion neurons was significantly inhibited (20–25%) by the κ-agonist U-50488H, an effect readily reversed by the specific κ-antagonist Nor-BNI. This effect of U-50488H can also be reversed by CCK-8, whereas the effect of CCK-8 can be totally abolished by the CCK-B antagonist L-365260, suggesting that the effect of CCK-8 is mediated by the CCK-B receptors. Thus the phenomena observed in antinociceptive assays and in membrane binding assays can essentially be replicated at the channel level. Most of the voltage-gated calcium channels in the dorsal root ganglion neurons seem to belong to the N type, since they can be blocked over 70% by ω-conotoxin. In fact, the effect of U-50488H could only be barely observed in ω-conotoxin treated neurons [15].

Interesting enough was the finding that while the calcium current in dorsal root ganglion neurons can be inhibited by either μ-, δ- or κ-agonist, it was only the μ- effect (Liu NJ, Xu T, Xu C, Li CQ, Yu YX, Kang HG, Han JS, to be published) and the κ-effect [15], but not the δ-effect could be antagonized by CCK-8.

Intracerebral Transfer of CCK cDNA Vector in the Rat Decreased the Effectiveness of EA Analgesia

If CCK-8 in the CNS indeed plays an antagonistic role on opioid effects, one would anticipate that rats with a low level of CCK-8 content in the CNS would show a high responsiveness to EA stimulation and *vice versa*. A breed of rat with audiogenic seizure was developed in Peking Medical College in 1977 by Pei and his colleagues (hence the name P77PMC rat); the content of immunoreactive CCK-8 in the cerebral cortex, hippocampus and periaqueductal gray is only half that in Wistar rats from which the P77PMC rat was derived [16]. An interesting finding was that P77PMC rats happened to be very good responders for EA-induced analgesia. Regression analysis revealed a positive correlation between the susceptibility of audiogenic seizure and the effectiveness of EA analgesia. The CCK-B antagonist, L-365260 which normally potentiates EA analgesia in Wistar rats was no more effective in P77PMC rats [17], suggesting that in these rats there is very little CCK-8 functioning at the receptor level.

If a low CCK-8 content in CNS is the common denominator underlying the high susceptibility of audiogenic seizure and the high effectiveness of EA analgesia, one should expect a concomitant lowering of the two variables when the congenital defect of a low expression of CCK-8 in CNS is corrected by genetic engineering techniques. We constructed a pSV2-CCK vector which carries an insert of CCK cDNA in the pSV2 plasmid and encapsulated it with lipofectin. The plasmid-lipofectin complex was injected intracerebroventricularly (i.c.v.) to the P77PMC rat. Preliminary experiments using the reporter gene LacZ instead of CCK cDNA have shown that the blue staining of X-gal appeared in a

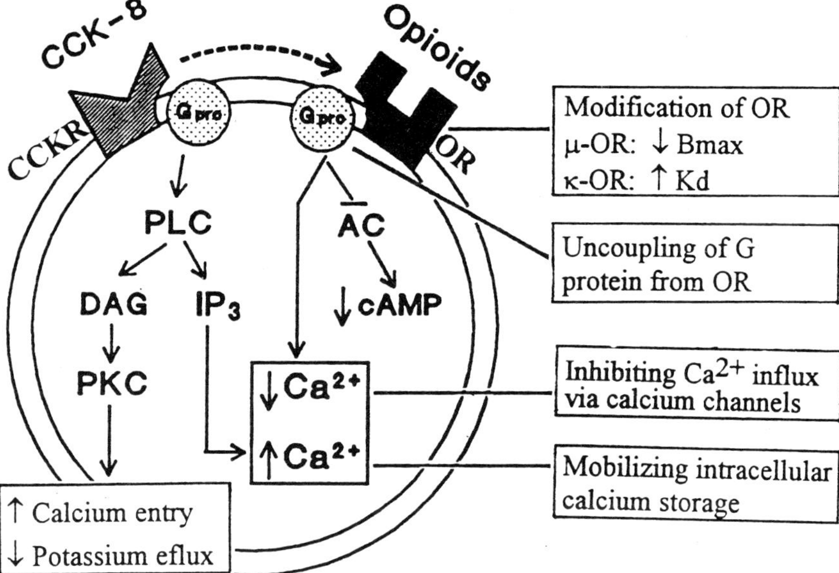

Fig. 1. Diagram showing the possible mechanisms of the anti-opioid effect of CCK-8. See text for details. AC: adenyl cyclase; CCKR: CCK receptor; DAG: diacyl glycerol; G pro: G protein; IP$_3$: inositol triphosphate; PKC: protein kinase C; PLC: phospholipase C; ↑: increase; ↓: decrease.

large amount in the ependymal cells and was distributed sporadically in the brain tissue. The staining was most prominent in day 2 through day 4 after its i.c.v. injection, and disappeared after 2–3 weeks, suggesting a temporal expression of the foreign gene in the CNS. Behavioral studies revealed a concomitant lowering of both the seizure susceptibility [18] and the effectiveness of EA analgesia (Zhang LX, Han JS, unpublished observations) in rats receiving pSV2-CCK vector as compared to the control rats receiving the empty pSV2 vector. The suppressive effects were most prominent in day 2 through day 4, and returned to normal level by days 7 to 12.

The message obtained from this gene transfer experiment is that it is feasible to modify the response of an animal to certain environmental changes by tilting the balance between neuropeptides controlling contradictory activities.

Figure 1 summarizes the molecular mechanisms underlying the anti-opioid effect of CCK-8.

Conclusions

(1) CCK-8 exhibits a potent antagonistic effect on opioid analgesia, especially on the analgesia induced by μ- and κ-opioid agonists. (2)

CCK-8 decreases the Bmax of μ-receptor and lowers the affinity of κ-opioid receptor, without affecting the δ-receptor. (3) The suppressive effect of CCK-8 on opioid binding may occur via interaction between opioid receptor and CCK receptor located on one and the same neuron. (4) The μ- and κ-opioid-mediated suppression of calcium current can be reversed by CCK-8 via the CCK-B receptor, as shown by the patch-clamp studies. (5) CCK-8 may uncouple the opioid receptor from its relevant G protein, resulting in a decrease of the binding capability of the receptor and a blockade of the transmembrane signal transduction. (6) CCK-8 stimulates the phosphatidylinositide signal system, which raises intracellular level of IP_3, thereby releasing free calcium from intracellular calcium storage and increases $[Ca^{2+}]_i$ to counteract the effect of opioids which decreases the $[Ca^{2+}]_i$. (7) Along with an increase in IP_3, CCK-8 may also cause an increase in PKC which increases the calcium conductance, causing a further increase of $[Ca^{2+}]_i$. (8) cAMP signal system seems not to be involved in the anti-opioid effect of CCK-8. (9) A balance between the functional activities of opioids and CCK in the CNS may serve as a cardinal determinant for certain behaviors including the responses to noxious stimulation.

Acknowledgements

This study was supported by the National Natural Science Foundation of China, and a grant from the National Institute of Drug Abuse, USA (DA 03983). The author wishes to thank Dr. MC Beinfeld of St. Louis University and Dr. JS Hong of the National Institute of Environmental Health Sciences, Research Triangle, NC, USA for the generous gifts of CCK antiserum, Squibb and Sons Inc. for the the donation of CCK-8, and Dr. RF Freidinger and Dr. SD Iversen of the Merck Sharp and Dohme for the supply of devezapide and L-365260 used in this study.

References

1. Itoh S, Katssura G, Maeda Y. Caerulein and CCK suppress β-endorphin-induced analgesia in the rat. Eur. J. Pharmacol. 1982; 80: 270–289.
2. Faris S, Komisaruk BR, Watkins LR, Mayer DJ. Evidence for the neuropeptide cholecystokinin as an antagonist of opioid analgesia. Science 1983; 219: 310–312.
3. Han JS. The role of CCK in electroacupuncture analgesia and electroacupuncture tolerance. In: Dourish CT, Cooper SJ, Iversen SD, Iversen LL, editors. Multiple cholecystokinin receptors in CNS. New York: Oxford University Press, 1992; 480–502.
4. Han JS, Ding XZ, Fan SG. Is CCK-8 a candidate for endogenous anti-opioid substrates? Neuropeptides 1985; 5: 399–402.
5. Han JS, Ding XZ, Fan SG. CCK-8: Antagonism to electroacupuncture analgesia and a possible role in electroacupuncture tolerance. Pain 1986; 27: 101–115.
6. Wang XJ, Wang XM, Han JS. CCK-8 antagonizes opioid analgesia mediated by μ- and κ- but not δ-receptors in the spinal cord of the rat. Brain Res. 1990; 523: 5–10.
7. Wang XJ, Fan SG, Ren MF, Han JS. CCK-8 suppressed 3H-etorphine binding to rat brain opiate receptors. Life Sci. 1989; 45: 117–123.
8. Wan XJ, Han JS. Modification by CCK-8 of the binding of μ-, δ- and κ-opioid receptors. J. Neurochem. 1990; 55: 1379–1382.

9. Zhang LX, Wang XJ, Han JS. Modification of opioid receptors and uncoupling of receptors from G protein as possible mechanisms underlying suppression of opioid binding by CCK-8. Chinese Med. Sci. J. 1993; 8: 1–4.
10. Cox B. Opioid receptor-G protein interactions: Acute and chronic effects of opioids. In: Herz A, editor. Opioids I. Berlin: Springer-Verlag, 1993; 145–188.
11. Sun SJ, Zhang LJ, Han JS. The anti-opioid effect of CCK-8 may not be mediated by cAMP in rat spinal cord. J. Beijing Med. Univ. 1993; 25: 108.
12. Zhang LJ, Lu XY, Han JS. Influence of CCK-8 on phosphoinositide turnover in neonatal-rat brain cells. Biochem. J. 1992; 285: 847–850.
13. Wang JF, Ren MF, Han JS. Mobilization of calcium from intracellular store as a possible mechanism underlying the anti-opioid effect of CCK-8. Peptides 1992; 13: 947–951.
14. Baraban JM, Snyder SH, Alger BE. Protein kinase C regulates ionic conductance in hippocampal pyramidal neurons. Electrophysiological effects of phorbol ester. Proc. Natl. Acad. Sci. USA 1985; 85: 2538–2542.
15. Xu T, Liu NJ, Li CQ, Shangguan Y, Yu YX, Kan HG, Han JS. CCK reverses the κ-opioid-receptor-mediated inhibition of calcium current in rat dorsal root ganglion neurons. Neuroscience. In press.
16. Zhang LX, Zhou Y, Du Y, Han JS. Effect of CCK-8 on audiogenic epileptic seizure in P77PMC rats. Neuropeptides 1993; 25: 73–76.
17. Chen XH, Han JS, Huang LT. CCK receptor antagonist L-365260 potentiated electroacupuncture analgesia in Wistar rats but not in audiogenic epileptic rats. Chinese Med. J. 1994; 107: 113–118.
18. Zhang LX, Wu M, Han JS. Suppression of audiogenic seizure by intracerebral injection of a CCK gene vector. NeuroReport 1992; 3: 700–703.

Pharmacology of Excitatory Amino Acid Receptors

Philip M. Beart[1], Peter D. Suzdak[2], Joel G. Bockaert[3],
Stephen F. Heinemann[4] and David Lodge[5]

[1]*Department of Pharmacology, Monash University, Clayton, Victoria 3168, Australia;* [2]*Novo Nordisk, 2760 Malov, Denmark;* [3]*CNRS UPR 9023, 34094 Montpellier Cedex 5, France;* [4]*Molecular Neurobiology, Salk Institute, PO Box 85800, San Diego, CA 92186-5800, USA;* [5]*Lilly Research Centre, Erl Wood Manor, Windlesham, GU20 6PH, UK*

Summary. Excitatory amino acids such as L-glutamate are the major excitatory neurotransmitters within the mammalian central nervous system. With advances in drug development and molecular neurobiology, the pharmacology of the individual glutamate receptors and their subunits is now being more fully delineated. Described here are the characteristics of heteromeric N-methyl-D-aspartate receptors, including the actions of various selective drugs. Agents such as cyclothiazide, GYKI 52466 and novel competitive antagonists have resulted in a better understanding of the pharmacology of the α-amino-3-hydroxy-5-methylisoxazole-4-propionate subtype of glutamate receptor. Metabotropic glutamate receptors are directly coupled to second messenger systems and their pharmacological characteristics and an analysis of their link to phospholipase C are reported. Pharmacotherapy using drugs directed at glutamatergic transmission is now a very real possibility.

Introduction

Simple acidic amino acids such as L-glutamate (Glu) and L-aspartate are the major carriers of excitatory information within the mammalian central nervous system. Receptors for Glu are widely distributed throughout the neuroaxis and there are now recognized to be two distinct families of Glu receptors; the first and most widely investigated are the ionotropic Glu receptors, which are ligand-gated ion channels, and which are classified into three subtypes by their selective agonists, N-methyl-D-aspartate (NMDA), α-amino-3-hydroxy-5-methylisoxazole-4-propionate (AMPA) and kainate (KA) [1]. Recently a second family of receptors for Glu, the metabotropic Glu receptors (mGluRs), that are directly coupled via G-proteins to second messenger systems have attracted appreciable attention [2, 3]. In the last 5 years molecular neurobiology has had a tremendous impact on the pharmacology of Glu receptors, not only through the molecular cloning studies, but also through the increasing usage of molecular biology techniques in conjunction with classical pharmacological methods. The discovery of mGluRs had led to the realization that Glu receptors are involved in the more subtle modulation of synaptic activity. Thus targets for pharma-

Correspondence to: Dr. Philip M. Beart, at the above address.

cotherapy would include neurological conditions, but also aspects of mood and behaviour considered under psychiatric illnesses [4].

NMDA Receptors

Several subunits of the NMDA receptor have been cloned and there are two subfamilies; NMDAR1 (NR1) and NMDAR2 (NR2) [1, 5]. NMDA receptors appear to be highly heterogeneous, being apparently composed of the essential NR1 subunit in a heteromeric combination with the subunits of the NR2 family (NR2A-2D), although the exact nature of the heteromeric assemblies has yet to be determined. Whilst the NR1 subunit is ubiquitously distributed through brain, the NR2 subunits have very different localizations [6]. The molecular heterogeneity and disparate localizations offer new challenges to the pharmacologist – exciting possibilities exist for targeting drugs not only to a particular NMDA heteromeric assembly, but also towards a discrete brain area, perhaps allowing the minimisation of side effects.

The NMDA receptor complex is a multi-domained receptor-ionophore complex consisting of a primary receptor site for agonists and antagonists, an ionophore, a redox site, plus binding sites for Mg^{2+}, Zn^{2+}, H^+, glycine and polyamines, which modulate receptor-ionophore coupling [5]. In the face of the incredible complexity of multiple heteromeric NMDA receptors, the minimalist approach of analysing NMDA ligands in terms of their action at the relevant receptor domain still remains a useful working model. Historically, investigations with the dissociative anaesthetics ketamine and phencyclidine, which were shown to be NMDA channel ligands, provided the first evidence for discrete domains within the NMDA receptor complex. There has been a re-awakening of interest in drugs like memantine, which unlike phencyclidine and dizocilipine (MK-801), have rapid actions in the ion channel. Memantine is clinically tolerated by humans and may prove a therapeutically useful NMDA receptor-directed drug [4]. Studies using NMDA receptor subunits in oocytes and transfected cell lines show there is considerable heterogeneity of the receptor channel [1, 5, 7].

Recombinant NMDA receptors also display different affinities for competitive NMDA antagonists, with the NR1-NR2A heteromer exhibiting the highest affinity for a number of competitive antagonists [1, 5, 7]. Chemically unique competitive NMDA antagonists may be selective probes for individual heteromers; LY233536 (Fig. 1) has high affinity for a subpopulation of NMDA receptors containing the NR2D subunit [7], whilst [^{125}I]CGP 55802A, (Fig. 1) binds preferentially to NR1-NR2A heteromer [8]. Other sterically demanding NMDA antagonists such as SDZ EAB 515 and MER-3273 may prove useful; these bi-

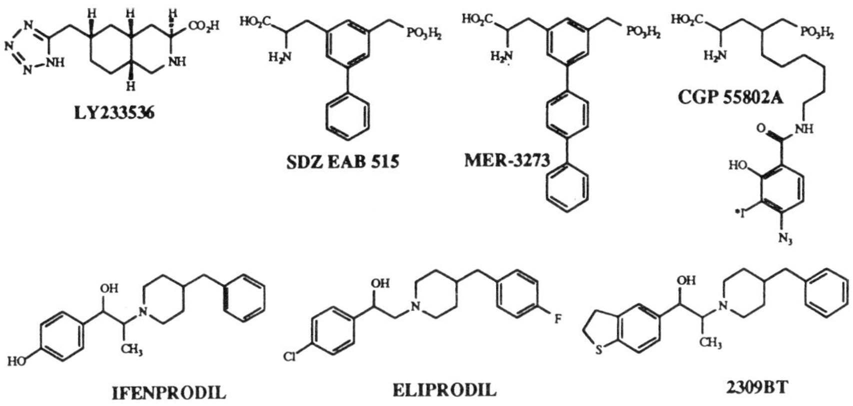

Fig. 1. Structures of NMDA receptor antagonists.

and triphenyl-aminophosphonocarboxylates (Fig. 1), possess high affinity for the NMDA receptor (K_is 180 & 40 nM, respectively) and require a deep lipophilic "pocket" to bind to the receptor.

Non-competitive NMDA antagonists appear more attractive than competitive antagonists and open-channel blockers for pharmacotherapy [4]. With a greater understanding of the glycine co-agonist site of the NMDA receptor there have been further advances in the development of glycine ligands [9]. The heterogeneity of glycine sites has been confirmed by various studies employing molecular biology techniques which indicate that the NR2C subunit has the highest affinity for glycine [1, 5, 7]. Antagonists acting at the polyamine domain of the NMDA receptor do not produce untoward behavioural stimulation, and ifenprodil and its more bioavailable derivative, eliprodil (Fig. 1), are neuroprotective in animal models of focal ischaemia and head trauma [4, 10]. Recently described heterocyclic aminoalcohols, which are ifenprodil analogues (e.g. 2309BT, Fig. 1), also have potential as cytoprotective agents [11]. Polyamine-sensitive actions of ifenprodil seem to be associated with NR1-NR2B heteromer [5, 12].

Stimulation of NMDA receptors leads to the subsequent activation of a number of second messenger systems including that involving nitric oxide (NO) [1, 5]. The involvement of NO at postsynaptic NMDA receptors has been well defined, but NO as a highly diffusible molecule plays many other roles as an inter- and intraneuronal messenger subsequent to the activation of NMDA receptors, including at both pre- and extra-synaptic NMDA release-regulating receptors [13]. The ability of extracellular NO to influence the redox state of the NMDA receptor-channel complex appears dependent upon the resting state of individual NMDA receptors [14]. Some NMDA heteromers have redox sites that seem to poorly potentiate channel opening and thus are relatively insensitive to exogenous NO [15]. Although the redox site seems associated with the NR2A subunit, its physiological significance remains to be determined [5, 15]. However, the redox site offers yet another target for pharmacotherapy and nitroglycerin, which can react with its sulphydryl groups, may be useful in controlling the size of cerebral infarcts [4].

Non-NMDA Receptors

Initially defined by agonist sensitivity, non-NMDA receptors were divided into AMPA and KA subtypes by the antagonistic effects of some early glutamate analogues and of quinoxalinediones [16, 17]. Cloning and expression of non-NMDA glutamate receptors have confirmed this division into AMPA- (GluR1–4) and KA-preferring (GluR5–7, KA1–2) subtypes [1].

In agreement with observations of *in situ* AMPA receptors, GluR1–4 expression results in receptor-channel complexes with a rapidly desensitising response to glutamate or AMPA and a smaller sustained response to kainate. As with most *in situ* observations, expression of GluR2 either alone or in combination with GluR1, 3 & 4 results in currents with linear current-voltage plots and poor permeability to calcium. Expression of the latter without GluR2, however, leads to rectifying channels with appreciable calcium permeability. This difference is due to an arginine (R) residue in the pore-lining TM2 segment of GluR2. In the other subunits this site is occupied by glutamine (Q). The R/Q substitution in GluR2 results from post-transitional editing; the ge-

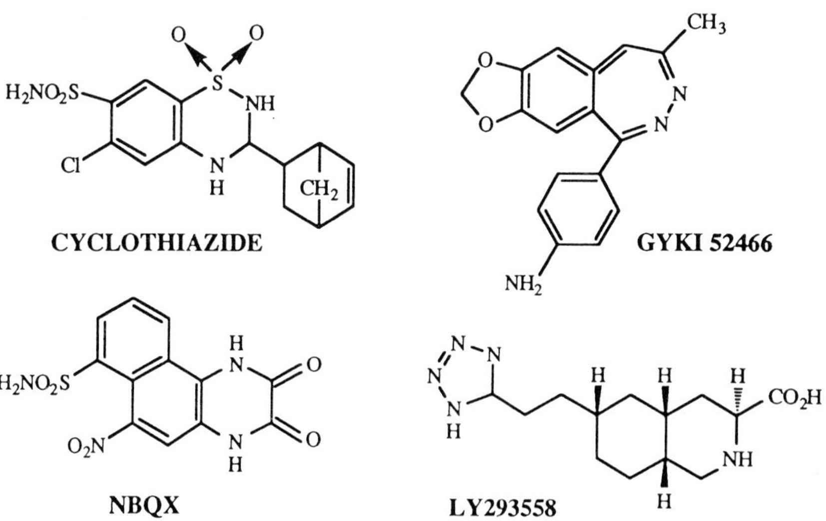

Fig. 2. Structures of drugs acting on the AMPA receptor.

nomic DNA is identical at this site in all 4 subunits. Why adenosine deaminase only edits GluR2 RNA is related to the intronic structure, inverted repeats and mRNA folding of this subunit.

For the KA-preferring subunits, expression of KA1, KA2, and GluR7 leads to high affinity KA binding sites but no functional channels whereas homomeric expression of GluR5 & 6 results in channels that desensitise more to KA than to AMPA. Co-expression of KA1 or 2 with GluR5 or 6 surprisingly results in increased sensitivity to AMPA relative to homomeric GluR5 & 6.

The pharmacological separation in both *in situ* and recombinant non-NMDA receptors is still at an early stage. The competitive antagonists, NBQX and LY293558 (Fig. 2), block responses to AMPA more potently than those to KA on cortical slices and this follows through into recombinant receptors [18]. LY293558 is particularly ineffective on GluR6 subunits. 2,3-Benzodiazepines, (e.g. GYKI 52466, Fig. 2), show similar preferences as AMPA antagonists both *in situ* and on cloned receptors. These compounds rapidly enter the CNS after an oral dose. No widely accepted KA antagonists are available although on cortical wedges both barbiturates and nickel were more effective versus AMPA- than KA-induced depolarisation [19].

A more striking difference between these two subclasses of non-NMDA receptor is in the pharmacology of the desensitisation process [20]. Concanavalin A, which enhances KA responses on dorsal root ganglion neurones, selectively reduces desensitisation of KA-preferring subunits whereas cyclothiazide (Fig. 2) has the same effect on AMPA-preferring subunits. Cyclothiazide enhances responses to AMPA (and KA) on neocortical, hippocampal, Purkinje and spinal neurones suggesting these neurones have a propensity of AMPA subunits. Non-NMDA channels can also be differentiated by their sensitivity to invertebrate polyamine toxins. Those subunits expressing unedited Q at the Q/R site are sensitive to argiotoxin and Joro spider toxin relative to the R-containing GluR2 subunit. Since AMPA responses of spinal neurones are easily blocked by the above toxins [20], such receptors would appear to be calcium permeable.

As the pharmacology of non-NMDA receptors unfolds the role of particular subunits combinations in physiological and pathological processes will become clearer and help the development of therapeutic agents.

Metabotropic Glutamate Receptors

mGluRs have been shown to affect multiple aspects on neuronal function, including changes in the activities of phospholipase C, phospholipase D, adenylyl cyclase, calcium channels, and potassium channels as well as excitatory postsynaptic potentials, long-term potentiation (LTP), and long-term depression (LTD) [2, 3]. In addition, mGluR ligands have been shown *in vitro* and *in vivo* either to enhance, or inhibit glutamatergic neurotransmission. Our understanding of the roles of the mGluR family has dramatically increased with the cloning and expression of the mGluR family. Seven mGluR subtypes have now been characterized [1, 2, 3, 21]. The mGluR family has been divided into 3 distinct groups based on amino acid sequence homology, signal transduction pathways and agonist selectivities [2]. These groups consist of (1) mGluR1/mGluR5 which are coupled to the phosphoinositide/Ca^{2+}-cascade and are highly sensitive to quisqualate; (2) mGluR2/mGluR3 which are negatively coupled to adenylate cyclase and are potently activated by Glu and (1S,3R)-ACPD, but not by L-AP4; and (3) mGluR4/mGluR8/mGluR7 which are negatively coupled to adenylate cyclase, but are highly sensitive to L-AP4. The lack of available mGluR subtype selective agonists and antagonists has greatly limited our understanding of the physiological roles of subtypes of the mGluR family. Recently a novel class of conformationally restricted analogs of Glu which consist of derivatives of phenylglycine, have been shown to antagonise many actions of (1S,3R)-ACPD *in vitro* (Table 1) and *in vivo* [3]. However, in such preparations the antagonism of mGluR-mediated responses may be mediated via several mGluR subtypes, or interactions at other classes of Glu receptors.

One such analog, (R,S)-4-carboxy-3-hydroxyphenylglycine ((R,S)-4C3HPG) was shown to be a potent competitive antagonist of mGluR1a ($K_B = 29\ \mu$M) [22]. Subsequently, (S)-4-carboxyphenylglycine and

Table 1. A summary of the potencies of phenylglycine analogs for subtypes mGluR1α and mGluR2 of the mGluR family

| Compound | mGluR1α | | mGluR2 | |
	IC_{50} (μM)	EC_{50} (μM)	IC_{50} (μM)	EC_{50} (μM)
(S)-4-Carboxy-3-hydroxyphenylglycine	15 ± 3	—	—	21 ± 4
(R,S)-α-Methyl-4-carboxyphenylglycine	155 ± 38	—	340 ± 59	—
(R)-4-Carboxyphenylglycine	> 1000	> 1000	> 1000	> 1000
(S)-4-Carboxyphenylglycine	65 ± 5	—	577 ± 74	—
(R)-3-Hydroxyphenylglycine	> 1000	> 1000	> 1000	451 ± 93
(S)-3-Hydroxyphenylglycine	—	68 ± 7	> 1000	> 1000
(R)-3-Carboxt-4-hydroxyphenylglycine	> 1000	> 1000	> 1000	> 1000
(S)-3-Carboxy-4-hydroxyphenylglycine	290 ± 47	—	—	97 ± 12

The values (mean $\pm$ S.E.M.) represent half-maximal concentrations for inhibiting (IC_{50}) or for stimulating (EC_{50}) functional responses in BHK cells expressing mGluR1α or mGluR2.

(R,S)-α-methyl-4-carboxyphenylglycine were shown to be competitive antagonists at both mGluR1α and mGluR2, while (S)-4-carboxy-3-hydroxyphenylglycine ((S)-4C3HPG) was a potent agonist at mGluR2 and a potent antagonist at mGluR1α (Table 1) [23]. None of the phenyglycine analogues tested had agonist or antagonist actions at mGluR4 [23]. In addition, the S-stereoisomers of phenyglycine analogues preferentially act at the mGluR family, showing no significant displacement of [^{3}H]AMPA, [^{3}H]KA or [^{3}H]CPP binding [23]. The competitive nature of these antagonists and their relative high potencies at the mGluRs indicate that these structures may prove useful for delineating the functions of this receptor family. However, their differential functional activities at subtypes mGluR1α and mGluR2, and possibly at additional subtypes of the mGluR family, should be considered when such experiments are evaluated.

The potential anticonvulsant properties of (S)-4C3HPG was examined using a rodent model of generalised epileptic seizures [24]. When audiogenic clonic and tonic convulsions were induced 15 min after intracerebroventricular infusion of (S)-4C3HPG, a dose-dependent protection against clonic and tonic convulsions was observed, showing an ED$_{50}$ value of 76 and 110 nmol per mouse, respectively. At the doses of (S)-4C3HPG that produced maximal protection against audiogenic-induced convulsions there was no significant impairment of motor activity as measured in the rotarod test. In a rat cingulate cortex-corpus callosum slice preparation, perfused with Mg^{2+}-free medium, (S)-4C3HPG partially decreased the frequency of spontaneous epileptic events. This is the first demonstration of an efficacious anticonvulsant action produced by a selective mGluR ligand. The anticonvulsant action of (S)-4C3HPG may be related to a depression of synaptic excitation, as was seen by a decrease in spontaneous epileptic spikes in the cingulate cortex-corpus callosum slice preparation, resulting from inhibition of excitatory amino acid release, probably via mGluR2 [25] and/or mGluR1a antagonism, resulting in an inhibition of Glu-mediated postsynaptic activity. These data suggest the potential importance of mGluR subtype selective ligands as therapeutic targets for the development of a new generation of antiepileptic agents.

Multiplicity of group 1 (mGluR1 & 5) is increased by the existence of splice variants [2, 3, 26]. In particular, there are three splice variants of mGluR1: mGluR1a, mGluR1b, mGluR1c. They have a common 46 aa tail following transmembrane domain (TMD) VII but differ by the total length of the tail (359,66 and 56 aa, respectively) (Fig. 3). Alignment of the intracellular loop sequences of the seven cloned mGluRs reveals that both i1 and i2 contain several amino acid residues which are conserved in PLC-coupled mGluRs (mGluR1 & 5) and are different from those conserved in AC-coupled mGluRs (mGluR2–4,6,7). The sequence of the intracellular domain located downstream of TMD VII is more

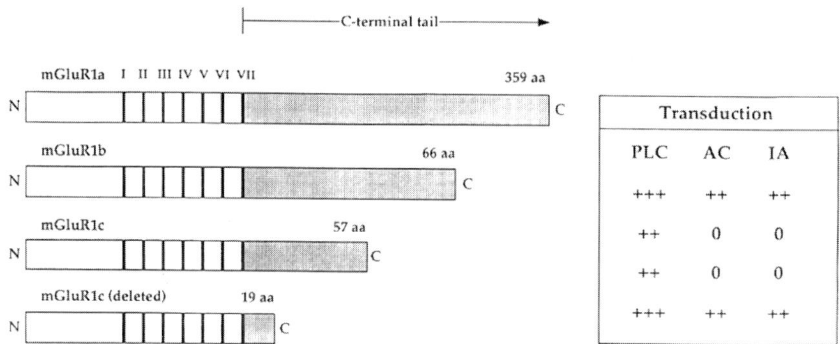

Fig. 3. Metabotropic glutamate receptors and signal transduction.

variable in AC-coupled mGluRs but is highly conserved in PLC-coupled mGluRs. Regions involved in the specific recognition of G proteins in most G-protein coupled receptors are likely to be amphiphilic α-helices. In Group I, both the N- and the C-terminal domains of i2 and the segment located downstream of TMD VII have a high α-helix amphipathicity value. We constructed chimeric receptors between mGluR3 and mGluR1c, and tested the ability of the resulting chimeric receptors to activate PLC in Xenopus ooctyes [27].

As previously reported mGluR1c clearly activates PLC, whereas mGluR3 is inactive. Replacing mGluR3's i2 by mGluR1's i2 does not enable the chimeric receptor R3/1-(i2) to activate PLC [27]. Similarly, the chimeric receptor R2/1C-C2 with the C-terminal intracellular domain of mGluR1c, does not activate PLC [27]. However, when both i2 and the C-terminal domain are exchanged between mGluR3 and mGluR2c, the resulting chimeric receptor activates PLC [27]. We also replaced mGluR1a's i2 by mGluR3's i2, and obtained a chimeric receptor which was inactive. The pharmacology of the active R3/1 chimeras resembles that of group II. As well as stimulating PLC, mGluR1a (359 aa in the C-terminal tail) activates AC and has an intrinsic activity (IA) (stimulation of PLC in the absence of Glu, and the presence of the antagonist 4-carboxyphenylglycine). In contrast, mGluR1b and 1c are unable to stimulate AC and have no IA (Fig. 3). Surprisingly, when the C-terminal tail of mGluR1 is further reduced (19 aa, Fig. 3), the receptor is able (as mGluR1a) to stimulate PLC, AC and has IA (Fig. 3). This indicates that it is the C-terminal tail of intermediary length (in mGluR1b & 1c) that inhibits the coupling to AC and the IA of the receptor and not the long C-terminal tail which confers specific coupling to Gs and the IA. The C-terminal tail suppresses the inhibitory effects of the C-terminal tails of intermediary length.

Perspective

Clearly there have been tremendous advances in our understanding of the pharmacology of receptors for excitatory amino acids. We can shortly expect new findings on how the various subunits of ionotropic Glu receptors are assembled into the physiologically active receptors. Together with mGluRs, these offer new targets for drug dvelopment. Drugs for Glu receptors should be a veritable panacea for the management of neurological and psychiatric disorders; we should look forward with some optimism to their successful use as pharmacotherapeutic agents.

References

1. Hollman M, Heinemann S. Cloned glutamate receptors. Ann. Rev. Neurosci. 1994; 17: 31–108.
2. Schoepp DD. Novel functions for subtypes of metabotropic glutamate receptors. Neurochem. Int. 1994; 24: 439–449.
3. Watkins JC, Collingridge GL. Phenylglycine derivatives as antagonists of metabotropic glutamate receptors. Trends Pharmacol. Sci. 1994; 15: 333–342.
4. Lipton SA, Rosenberg PA. Excitatory amino acids as a final common pathway for neurologic disorders. N. Eng. J. Med. 1994; 330: 613–622.
5. McBain CJ, Mayer ML. N-Methyl-D-aspartic acid receptor and function. Physiol. Rev. 1994; 74: 723–760.
6. Watanabe M, Inone Y, Sakimura K, Mishina M. Distinct distributions of five N-methyl-D-aspartate receptor channel subunit mRNAs in the forebrain. J. Comp. Neurol. 1993; 338: 377–390.
7. Laurie DJ, Seeburg PH. Ligand affinities at recombinant N-methyl-D-aspartate receptors depend on subunit composition. Eur. J. Pharmacol. 1994; 268: 335–345.
8. Marti T, Benke D, Mertens S, Heckendorn R, Pozza M, Allgeier H et al. Molecular distinction of three N-methyl-D-aspartate receptor subtypes in $situ$ and developmental receptor maturation demonstrated with the photoaffinity ligand ^{125}I-labelled CGP 55802A. Proc. Natl. Acad. Sci. USA 1993; 90: 8434–8438.
9. Kemp JA, Leeson PD. The glycine site of the NMDA receptor – five years on. Trends Pharmacol. Sci. 1993; 14; 20–25.
10. Carter CJ, Benavides J, Dana C, Schoemaker H, Perrault G, Sanger D et al. Non-competitive NMDA receptor antagonists acting on the polyamine site. In: Meldrum BS, editor. Excitatory amino acid antagonists. Oxford: Blackwell, 1991: 130–163.
11. Beart PM, Schousboe A, Fransden Aa. Blockade by polyamine NMDA antagonists related to ifenprodil of NMDA-induced synthesis of cyclic GMP, increases in calcium and cytotoxicity in cultured neurones. Br. J. Pharmacol. 1995; 114: 1359–1364.
12. Williams K. Ifenprodil discriminates subtypes of the N-methyl-D-aspartate receptor: selectivity and mechanisms at recombinant heteromeric receptors. J. Pharmacol. Exp. Ther. 1993; 44: 851–859.
13. Jones NM, Loiàcono R, Møller M, Beart PM. Diverse roles for nitric oxide in synaptic signalling after activation of NMDA release-regulating receptors. Neuropharmacology 1994; 33: 1351–1356.
14. Lipton SA, Choi Y, Pan ZH, Lei SZ, Chen HSV, Sucher NJ et al. A redox-based mechanism for the neuroprotective and neurodestructive effects of nitric oxide and related nitroso-compounds. Nature 1993; 364: 626–632.
15. Köhr G, Eckardt S, Lödens H, Monyer H, Seeburg PH. NMDA receptor channels: subunit-specific potentiation by reducing agents. Neuron 1994; 12: 1031–1040.
16. Watkins JC, Krogsgaard-Larsen P, Honore T. Structure-activity relationships in the development of excitatory amino acid receptor agonists and competitive antagonists. Trends Pharmacol. Sci. 1990; 11: 25–33.

17. Lodge D, Jones MG, Palmer AJ. Excitatory amino acids: new tools for old stories. Can. J. Physiol. Pharmacol. 1991; 69: 1123–1128.
18. Ornstein PL, Arnold MB, Augenstein NK, Lodge D, Leander JD, Schoepp DD. (3SR, 4aRS, 6RS, 8aRS)-6-[2(1H-Tetrazol-5-yl)ethyl]decahydroisoquinoline-3-carboxylic acid: a structurally novel, systemically active, competitive AMPA receptor antagonist. J. Med. Chem. 1993; 36: 2046–2048.
19. Lodge D, Palmer AJ, Zeman S. Methohexitone and nickel are selective antagonists of kainate on rat cortical slices. Mol. Neuropharmacol. 1992; 2: 43–45.
20. Partin KM, Patneau DK, Winters CA, Mayer ML, Buonanno A. Selective modulation of desensitization at AMPA versus kainate receptors by cyclothiazide and concanavalin A. Neuron. 1994; 11: 1069–1082.
21. Suzdak PD, Thomsen C, Mulvihill E, Kristensen P. Molecular cloning, expression and characterization of metabotropic glutamate receptors. In: Conn PJ, Patel J, editors. The metabotropic glutamate receptors. Totowa: Humana, 1994; 1–30.
22. Thomsen C, Suzdak PD. 4-Carboxy-3-hydroxphenylglycine, an antagonist at type 1 metabotropic glutamate receptors. Eur. J. Pharmacol. 1993; 245: 299–301.
23. Thomsen C, Boel E, Suzdak PD. Actions of phenylglycine analogs at subtypes of the metabotropic glutamate receptor family. Eur. J. Pharmacol. 1994; 267: 77–84.
24. Thomsen C, Klitgaard H, Sheardown M, Jackson HC, Eskesen K, Jacobsen P et al. (S)-4-Carboxy-3-hydroxyphenylglycine, an antagonist of metabotropic glutamate receptor (mGluR) 1a and an agonist of mGluR2, protects against audiogenic seizures in DBA/2 mice. J. Neurochem. 1994; 62: 2492–2495.
25. Lombardi G, Alesiana M, Leonardi P, Cherici G, Pellicciari R, Moroni F. Pharmacological characterization of the metabotropic glutamate receptor inhibiting D-[^{3}H]-aspartate output in rat striatum. Br. J. Pharmacol. 1993; 110: 1407–1412.
26. Pin J-P, Duvoisin R. The metabotropic glutamate receptors: structure and functions. Neuropharmacology 1995; 34: 1–26.
27. Pin J-P, Joly C, Heinemann SF, Bockaert J. Domains involved in the specificity of G protein activation in phospholipase C-coupled metabotropic glutamate receptors. EMBO J. 1994; 13: 342–348.

Pharmacological Sciences: Perspectives for
Research and Therapy in the Late 1990s
ed. by A.C. Cuello and B. Collier

GABA Receptors: Recent Advances

N.G. Bowery[1], K. Kuriyama[2], J. Lambert[3], R.W. Olsen[4] and
T.G. Smart[1]

[1]*Department of Pharmacology, The School of Pharmacy, University of London, 29/39
Brunswick Square, London WC1N 1AX, U.K.;* [2]*Kyoto Prefectural University of Medicine,
Kamikyo-ku, Kyoto 602, Japan;* [3]*Department of Pharmacology & Clinical Pharmacology,
Ninewells Hospital & Medical School, Dundee DD1 9SY, U.K.;* [4]*Department of Pharmacology,
University of California School of Medicine, Los Angeles, CA 90024, U.S.A.*

Summary. It is now nearly three decades since the full significance of GABA as an inhibitory
neurotransmitter was first established [1]. However, it is only for about the last 10 years that
we have known that the actions of GABA are mediated via multiple receptor subtypes, rather
than a single one [2]. Broadly the receptor-mediated effects of GABA can be classified as fast
or slow with the former relating to the $GABA_A$ receptor class and the latter to the $GABA_B$
class [3]. Further functional receptor types e.g. $GABA_C$ may well become established in future
as Johnston recently suggested [4] but for the purpose of this review, the material has been
confined to $GABA_A$ and $GABA_B$.

$GABA_A$ Receptor

The $GABA_A$ receptor gates neuronal Cl^- channels and is undoubtedly
the most abundant GABA receptor in the brain. Receptor activation
mediates a fast increase in membrane conductance to reduce neuronal
excitation. The pharmacological characteristics of this receptor have
been well documented over the years in relation to agonists and antagonists but more recently, with the advent of molecular biology, attention has been focused on the ultra-structure of the receptor and how this
relates to its overall function. Evidence indicates that the receptor
comprises five subunit protein sequences each possessing four membrane spanning domains (see [5]). A variety of subunits have been
isolated as indicated in Table 1 and in theory any combination of five
could form a $GABA_A$ receptor. However, in reality only a limited
number of combinations, probably 10–20, exist in the native form
although the stoichiometry of these subunits is still unknown. Moreover, the individual function of each type of subunit has still to be fully
defined but possible roles have been attributed as summarized in Table
1.

Numerous modulators are able to alter the function of $GABA_A$ sites
and these include the benzodiazepines, picrotoxin, barbiturates and
neurosteroids, all of which act at distinct sites. This distinction may be
defined by individual receptor subunits. For example, at least two types
of benzodiazepine sites have been described pharmacologically, BZ1

Table 1. GABA$_A$ receptor subunit classification

Subunit		Amino acid residues	Suggested functional role
alpha	α_1	428	Defines benzodiazepine
	α_2	423	pharmacology
	α_3	465	Type I or Type II
	α_4	521	Responsible for GABA
	α_5	433	affinity/efficacy
	α_6	434	
beta	β_1	449	Determines GABA response
	β_2S	450	amplitude?
	β_2L	467	Influences desensitization
	β_3	448	
	β_4	459	
	β_4'	463	
gamma	γ_1	430	Responsible for benzodiazepine
	γ_2S	428	functional action
	γ_2L	436	Sites for modulation –
	γ_3	450	protein Kinase C
	γ_4	436	and tyrosine kinase
delta	δ	433	
rho	$\rho 1$	458	Novel retinal receptors –
	$\rho 2$	465	bicuculline insensitive

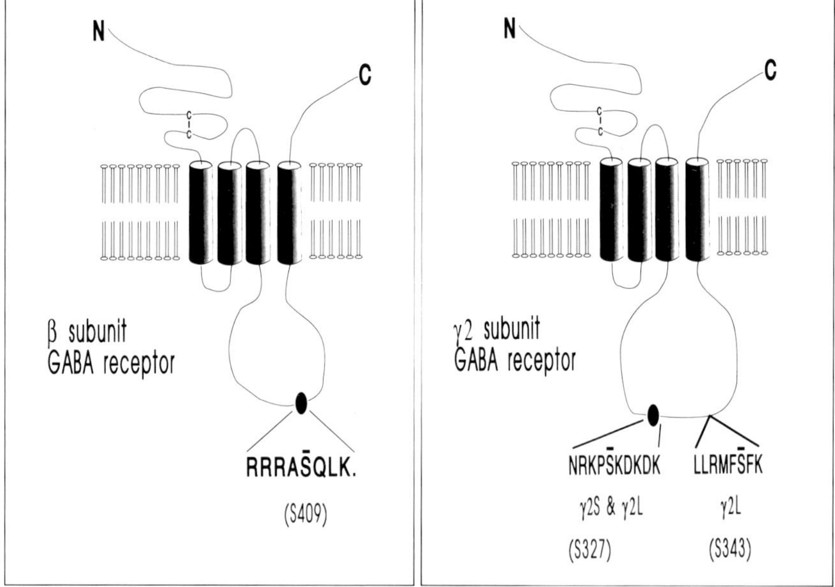

Fig. 1. Diagram indicating the location of the phosphorylation consensus sequences, including the serine residues [5], on the intracellular loops of the β and γ_2 subunits of the GABA$_A$ receptor.

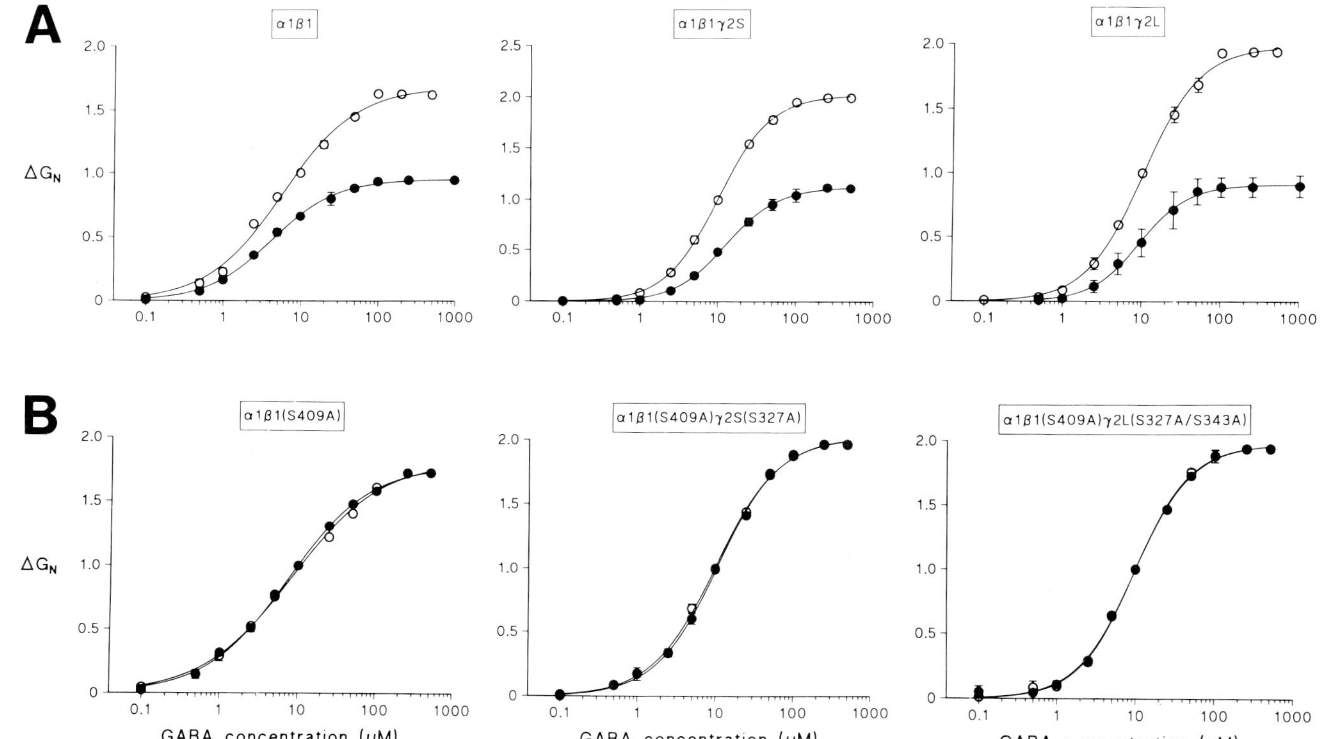

Fig. 2. Equilibrium concentration-response curves for GABA-induced membrane conductance recorded from oocytes expressing wild-type (A) or mutant (B) receptors. All curves (control, open symbols; +250 nM phorbol 12-myristate 13-acetate (PMA), closed symbols) were obtained from three to seven different oocytes for each receptor construct and have been normalized to the conductance induced by 10 μM GABA in control Ringer's solution. The points represent means $\pm$ SEM. The concentration-response curves were measured after the effect of PMA had attained a steady state, usually 30 min following application. Response curves for the mutant receptors were also measured after a 30 min exposure to PMA (taken from ref. [6]).

and BZ2, and this appears to correlate with the presence of the α_1 and α_2 or α_3 subunits respectively.

Modulatory sites additional to those attributed to the action of these drugs appear to exist on the intracellular loops of the receptor (see Fig. 1). These novel sites are points where the receptor protein may be phosphorylated by protein kinases [6]. Evidence thus far indicates that the consensus sequences concerned with the effect are located on the intracellular loops of the β and γ_2 subunits between the membrane-spanning domains 3 and 4 (Fig. 1) An important target would appear to be the serine residue (Fig. 1) and site-directed mutation with alanine substituting for serine removes the ability of phorbol esters to induce phosphorylation of the receptor. This is illustrated in Fig. 2 in which native or mutant GABA$_A$ receptors comprising $\alpha_1\beta_1$, $\alpha_1\beta_1\gamma_{2S}$ or $\alpha_1\beta_1\gamma_{2L}$ with or without serine substitution have been expressed in *Xenopus* oocytes. Measuring the steady-state GABA-induced Cl$^-$ conductance in these oocytes either in the absence or presence of phorbol 12-myristate 13-acetate indicated that the response in oocytes with mutant receptors was unaffected by the phorbol ester whilst the maximal conductance increase, induced in oocytes expressing native receptors, was decreased by approximately 50%. This latter effect is comparable with the response obtained in cells e.g. superior cervical ganglion, which possess native GABA$_A$ receptors. Phosphorylation of the β-subunits produces less reduction in the GABA current amplitude than occurs when the γ_{2L} subunit is phosphorylated.

The significance of these observations remains to be established but covalent modification may be important in determining receptor efficacy and may be relevant to any long-term alteration in receptor function. In addition, it may be important in the possible interaction between G-protein linked receptors and those coupled to fast channels. Of course, one of the major questions concerning our present detailed knowledge of the GABA$_A$ receptor (or other receptor) sequence is which of the amino acid residues, in any particular subunit, provides the agonist binding site. Does, for example, each subunit afford a distinct affinity profile for GABA or the benzodiazepines? Although the β-subunit has been thought to provide the major binding site for GABA$_A$ receptor agonists, it is still unclear whether each subunit alone contains an agonist binding site or whether adjacent identical subunits associate to create a binding site [7].

In an attempt to define the sequence(s) comprising the agonist binding site, Smith and Olsen [7] have used ^{3}H-muscimol to photoaffinity label the GABA$_A$ receptor protein followed by microsequencing of fragments produced by digestion with trypsin and chymotrypsin. Results indicate that ^{3}H-muscimol binds to a very large residue. However, by examining the differential level of binding in fragments of sizes 51–58 kDaltons, they were able to support the conclusion of Sigel et al. [8], obtained from point mutation studies, that a sequence containing a

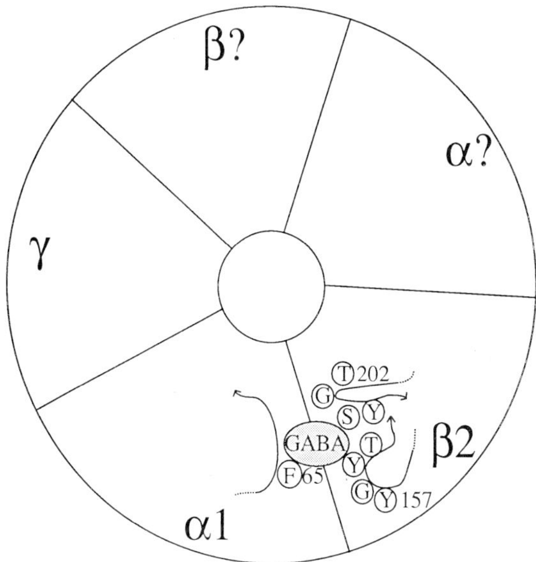

Fig. 3. Possible arrangement of coupling between GABA$_A$ receptor subunits in relation to the agonist binding site where phenylalanine 65 (F 65) is particularly significant.

phenylalanine residue appears to be a major site. In bovine brain, phenylalanine 65 within the α_1 subunit (Fig. 3) forms an important part of the GABA binding site and, interestingly, is conserved among all GABA$_A$ receptor α, γ_2 and δ subunits. In due course no doubt, the residues involved in the binding of modulators of the GABA$_A$ receptor will be determined as well.

These modulatory sites are quite diverse and more extensive than on any other fast channel receptor complex. For example, ligands for the neurosteroid site on the GABA$_A$ receptor complex exhibit unique selectivity for this channel. GABA responses can be potentiated at sub-micromolar concentrations whilst 1000-fold higher concentrations are required to alter glycine, glutamate or nicotinic receptor activation. But why are there so many modulatory sites on the GABA$_A$ receptor complex and what mechanisms underly their effects? Are they of physiological or pathological significance?

Numerous studies are in progress to answer these questions using, in particular, recombinant receptors expressed in a variety of cell types. The *Xenopus* oocyte has proved to be a popular expression system for intracellular recording of channel events. Using this recording system, it can be shown that activation of the neurosteroid site, which is quite distinct from the barbiturate site, not only enhances the action of GABA but also strongly enhances the direct GABA-like effect of pentobarbitone and propofol. Whilst the steroids enhance the maximum

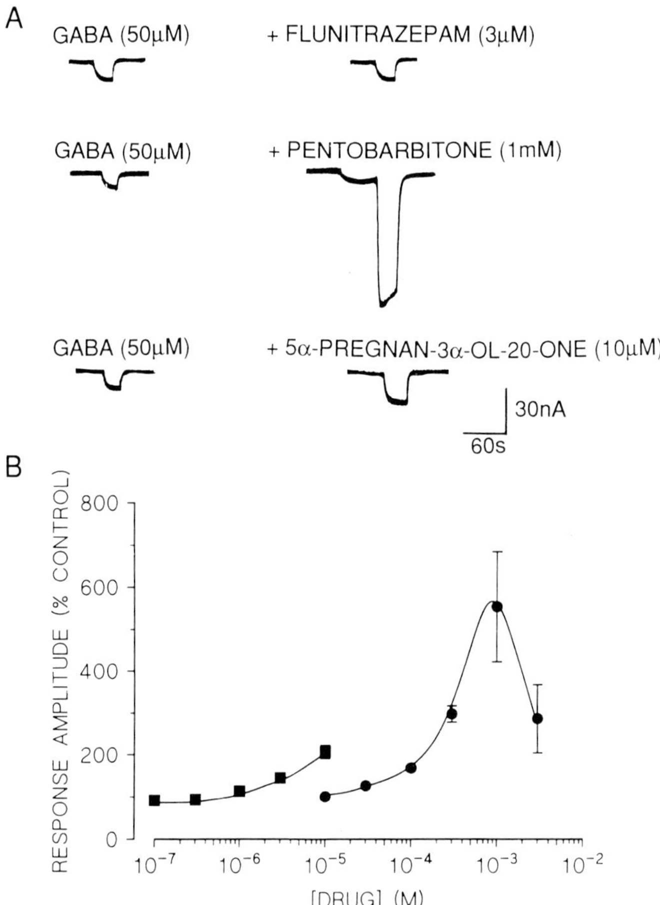

Fig. 4. Action of putative positive allosteric modulators on GABA-evoked currents recorded from *X. laevis* oocytes preinjected with cRNA for the adult *Drosophila* subunit. (A) Inward currents to bath-applied GABA (50 μM) were unaffected by flunitrazepam (3 μM) but were greatly enhanced by pentobarbitone (1 mM). The neurosteroid 5α-pregnan-3α-ol-20-one (10 μM) produced only a modest enhancement of the GABA-evoked current. (B) Concentration-dependent enhancement of GABA-evoked currents by 5α-pregnan-3α-ol-20-one (■) and pentobarbitone (●). Responses are expressed as percentages of the current produced by 50 μM GABA alone. Each point is the mean ± SEM determined from at least four oocytes (taken from ref. [9]).

current produced by GABA, this occurs to a lesser degree than that produced by pentobarbitone or propofol on GABA. Also, the steroids produce little or no direct effects on the GABA channel. All of these effects appear to be mediated via sites on the outer surface of the neuronal membrane since intracellular application produces no effect. In an attempt to examine the significance of individual subunits to the actions of modulators, Lambert and colleagues [9] have examined the functionality of GABA receptors cloned from *Drosphila melanogaster*. These receptors are homo-oligomeric and therefore may be of value in mutagenesis studies for defining modulatory, as well as agonist, binding sites. The data obtained from expression studies indicated the absence of benzodiazepine sites and the presence of only a weak, although stereospecific, action by the neurosteroid 5α-pregnan-3α-ol-20-one when compared to its effect on vertebrate recombinant receptors. By contrast, the response to pentobarbitone was very pronounced (Fig. 4) showing that this receptor subunit clearly discriminates between the action of the barbiturate and the neurosteroid. This supports the view that these drugs bind to separate sites [10].

The significance of the neurosteroid site in vertebrates has yet to be established but a variety of interesting possibilities arise including the contribution to stress and anxiety states. For example, are endogenous neurosteroid levels low in anxiety and can this be exploited for therapeutic use? Water soluble neurosteroids have recently been introduced and these appear to be excellent general anaesthetics in laboratory animals and potent enhancers of the actions of GABA. Presumably this could provide the basis for future therapeutic agents.

GABA$_B$ Receptors

GABA$_B$ receptors are distinct from GABA$_A$ receptors in every respect but with the exception that they are both activated by GABA (see [2]). Whilst we do not know the structural sequence of GABA$_B$ receptors it is clear that they are G-protein coupled and may therefore contain seven membrane spanning domains in line with other receptors of this class.

Perhaps, not surprisingly, the distribution of GABA$_B$ and GABA$_A$ receptors in mammalian brain are similar since they are activated by the same neurotransmitter. However, GABA$_B$ receptors are located in certain brain areas where only a low density of GABA$_A$ receptors exist e.g. the interpeduncular nucleus, where GABA$_B$ sites are present on cholinergic or glutamatergic nerve terminals. The GABA input to such heteroreceptors may derive from amino acid release from adjacent synapses [11] rather than from axo-axonic contacts for which anatomical evidence is lacking within higher centres. GABA$_B$ receptors also appear to have a primary role as autoreceptors on GABAergic terminals [12].

One brain region where presynaptic GABA receptors appear to be of special importance is the spinal cord where presynaptic innervation has been demonstrated and GABA fulfils the role of a presynaptic transmitter (see [13]). GABA$_B$ receptor activation of nerve terminals in the dorsal horn of rat spinal cord suppresses the evoked release of substance P [14] which may, in part, explain the acute antinociceptive activity of baclofen, the GABA$_B$ agonist. However, in chronic pain in man,

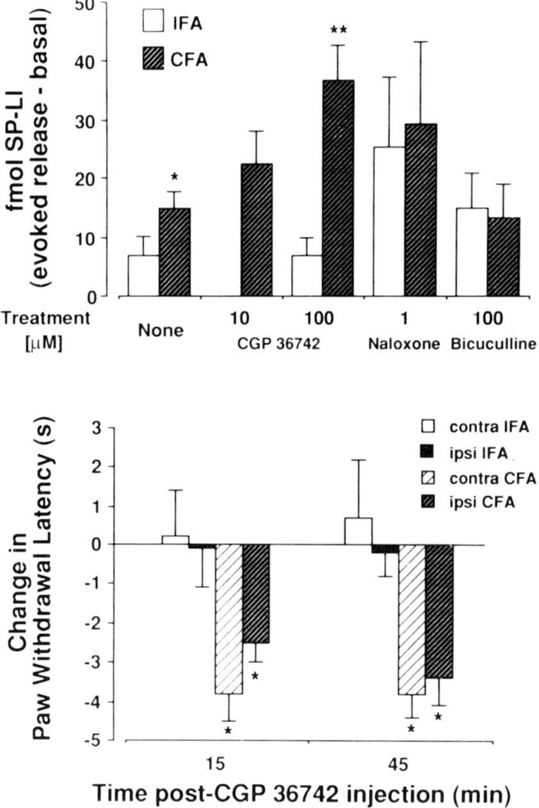

Fig. 5. Upper panel: Effect of the GABA$_B$ antagonist CGP 36742, naloxone and bicuculline on electrically-evoked substance P-like immunoreactivity (SP-L1) release from the spinal cord of incomplete (IFA) and complete Freund's adjuvant (CFA) rats. Basal outflow of SP-L1 from IFA (open bars) and CFA (hatched bars) rat spinal cord slices was 6.5 ± 1.5 fmol. 8 ml^{-1} per fraction (n = 26). Drugs were perfused 5 min prior to and during stimulation (20 V, 0.5 ms, 1 Hz for 8 min). Values are mean $\pm$ SEM of three to four slices for each group. Asterisks indicate significant difference between CFA and IFA. Lower panel: Effect of CGP 36742 on monoarthritis-induced hyperalgesia. CGP 36742 (100 mgkg^{-1} i.p.) was injected into all rats immediately after determination of pre-injection paw withdrawal latency. This was measured again at 15 and 45 min after injection. Pre-injection values were subtracted from the 15 and 45 min values. Data are mean $\pm$ SEM of six rats per group. Contralateral paw IFA - open bars; ipsilateral paw IFA – solid bars; contralateral paw CFA - widely hatched bars and ipsilateral paw CFA – closely hatched bars. Asterisks indicate significant difference between CFA and IFA. Student's t-test (taken from ref. [16]).

baclofen is ineffective as an analgesic and this may be due to an excessive endogenous activation of GABA$_B$ receptors in response to chronic pain. GABA immunoreactivity in the dorsal horn of spinal cord from monoarthritic rats is increased by 25% 21 days after injection of Freund's complete adjuvant into the footpad [15]. This may provide an increase in the source of transmitter GABA which is released to produce

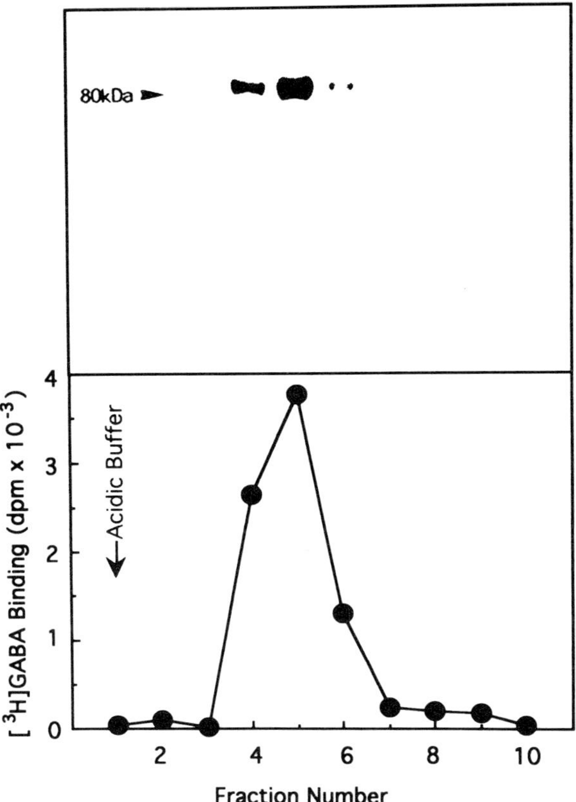

Fig. 6. Immunoaffinity purification of the 80-kDa GABA-binding protein. Solubilized synaptic membrane was incubated with monoclonal antibody-conjugated immunoaffinity beads. Elution was performed by the addition of an acidic buffer (50 mM citrate, pH 2.5) retaining 5 mM CHAPS and the protease inhibitors. Eluted fractions were analyzed by SDS-PAGE (upper panel) and by the [³H]-GABA binding assay (lower panel). The SDS-PAGE analysis shows that the major protein band is 80 kDa. Occasionally, one or two faintly stained bands were visible of 65 or 61 kDa (an example of the 65 kDa band is 2.5 nM) seen in fraction 5), for the ³H-GABA binding assay, ³H-GABA and purified 80-kDa GABA-binding protein were incubated for 1 h in ice-cold 50 mM Tris-HCl buffer, pH 7.4, containing 2 mM Mg_2Cl_2 and 2 mM $CaCl_2$ in the presence of a soybean phospholipid (final concentration 4 mM). The soybean phospholipid was added to minimize an inhibitory effect of CHAPS (which was contained in the immunoaffinity eluate, final concentration 0.5 mM) on [³H]-GABA binding.

maximal activation of $GABA_B$ receptors. The administration of baclofen can therefore produce no further response and injection of a $GABA_B$ antagonist to an arthritic animal produces hyperalgesia as well as an increase in substance P release from spinal cord slices [16] (Fig. 5). In normal animals, $GABA_B$ antagonists produce no nociceptive response or have any effect on the evoked release of substance P in the absence of baclofen. Thus, $GABA_B$ receptors may have a pathological rather than a physiological role in controlling nociception.

$GABA_B$ receptor mechanisms also appear to play a major role in the production and maintenance of absence epilepsy. $GABA_B$ receptor antagonists are particularly effective in reversing the symptoms of the absence syndrome in a variety of animal models [17]. The synaptic response to $GABA_B$ receptor activation in central neurones manifests as a reduction in presynaptic transmitter release coupled with the production of a late postsynaptic hyperpolarization. Within the thalamus this may provide the mechanism underlying seizure production by de-inactivating Ca^{++} "T" currents to facilitate the generation of Ca^{++} spikes producing the characteristic spike and wave discharges [18].

Recent evidence from pharmacological studies indicate that $GABA_B$ receptors are not homogeneous. However, no structural information is available to support or refute this possibility. Nevertheless, Kuriyama and colleagues [19] are actively pursuing this and have obtained a purified $GABA_B$ binding protein from bovine cerebral cortex (Fig. 6). The protein has a molecular weight of 80 kDaltons and exhibits characteristics consistent with a $GABA_B$ receptor. Production of a monoclonal antibody has been achieved using this purified receptor material. The antibody inhibits $GABA_B$ binding to neuronal membranes but, surprisingly, requires a long incubation of 10–15 h to achieve significant inhibition. The reason(s) for this is not clear. However, the antibody appears to act as an antagonist at $GABA_B$ receptors on intact neurones and exerts its effect on the outer surface of the membrane in line with a receptor locus of action. Despite this, the receptor structure remains elusive but why this should be is unclear. Perhaps the difficulty of expressing functional receptors contributes to the problem but close scrutiny of clones from cDNA libraries compiled from studies on the somatostatin and adenosine A1 receptors may provide a useful approach in view of the close pharmacological characteristics of these receptor classes and the $GABA_B$ receptor.

References

1. Krnjevic K, Schwartz S. The action of γ-aminobutyric acid on cortical neurones. Exp. Brain Res. 1967; 3: 320–326.
2. Bowery NG. $GABA_B$ receptor pharmacology. Annual Rev. Pharmacol. Toxicol. 1993; 33: 109–147.
3. Hill DR, Bowery NG. ^{3}H-baclofen and ^{3}H-GABA bind to bicuculline-insensitive $GABA_B$ sites in rat brain. Nature 1981; 290: 149–152.
4. Johnston GAR. GABA receptor pharmacology, pp. 11–16. This volume.
5. Olsen RW, Tobin AJ. Molecular biology of $GABA_A$ receptors. FASEB J. 1990; 4: 1469–1480.
6. Krishek BJ, Xie X, Blackstone C, Huganir RL, Moss SJ, Smart TG. Regulation of $GABA_A$ receptor function by protein kinase C phosphorylation. Neuron 1994; 12: 1081–1095.
7. Smith GR, Olsen RW. Identification of a [^{3}H]Muscimol photoaffinity substrate in the bovine γ-aminobutyric acid$_A$ receptor α subunit. J. Biol. Chem. 1994; 269: 20380–20387.

8. Sigel E, Baur R, Kellenberger S, Malherbe P. Point mutations affecting antagonist affinity and agonist dependent gating of GABA$_A$ receptor channels. EMBO J. 1992; 11: 2017–2023.

9. Chen R, Belelli D, Lambert JJ, Peters JA, Reyes A, Lan NC. Cloning and functional expression of a *Drosophila* γ-aminobutyric acid receptor. Proc. Natl. Acad. Sci. USA 1994; 91: 6069–6073.

10. Peters JA, Kirkness EF, Callachan H, Lambert JJ, Turner AJ. Modulation of the GABA$_A$ receptor by depressant barbiturates and pregnane steroids. Br. J. Pharmacol. 1988; 94: 1257–1269.

11. Isaacson JS, Solis JM, Nicoll RA. Local and diffuse synaptic actions of GABA in the hippocampus. Neuron 1993; 10: 165–175.

12. Bonnano G, Raiteri M. Multiple GABA$_B$ receptors. Trends in pharmacol. Sci. 1993; 14: 259–261.

13. Barber RP, Vaughn JE, Saito K, McLaughlin BJ, Roberts E. GABAergic terminals are presynaptic to primary afferent terminals in the substantia gelatinosa of the rat spinal cord. Brain Res. 1978; 141: 35–55.

14. Malcangio M, Bowery NG. γ-aminobutyric acid$_B$, but not γ-aminobutyric acid$_A$ receptor activation, inhibits electrically evoked substance P-like immunoreactivity release from the rat spinal cord in vitro. J. Pharm. Exp. Therap. 1993; 266: 1490–1496.

15. Castro-Lopez J-M, Tavares I, Tolle TR, Coito A. Increase in GABAergic cells and inflammation of the hindlimb of the rat. Eur. J. Neurosci. 1992; 4: 296–301.

16. Malcangio M, Bowery NG. Spinal cord SP release and hyperalgesia in monoarthritic rats: involvement of the GABA$_B$ receptor system. Br. J. Pharmacol. 1994; 113: 1561–1566.

17. Marescaux C, Vergnes M, Bernasconi R. GABA$_B$ receptor antagonists: potential new anti-absence drugs. J. Neural Transm. 1992; (Suppl.) 35: 179–88.

18. Crunelli V, Leresche N. A role for GABA$_B$ receptors in excitation and inhibition of thalamocortical cells. Trends Neurosci. 1991; 14: 16–21.

19. Nakayasu H, Nishikawa H, Mizutani H, Kimura K, Kuriyama K. Immunoaffinity purifaction and characterization of GABA$_B$ receptor from bovine cerebral cortex. J. Biol. Chem. 1993; 268: 8658–8664.

Pharmacological Sciences: Perspectives for
Research and Therapy in the Late 1990s
ed. by A.C. Cuello and B. Collier

Therapeutic Possibilities with Serotonergic Drugs

Pramod R. Saxena[1], David E. Clarke[2], Anthony P.D.W. Ford[2],
Daniel Hoyer[3], Ewan J. Mylecharane[4], Michael B. Tyers[5],
Jullie C. Barnes[5] and Frank D. Yocca[6]

[1]*Department of Pharmacology, Faculty of Medicine and Health Sciences, Erasmus University
Rotterdam, Post Box 1738, 3000 DR Rotterdam, The Netherlands;* [2]*Institute of Pharmacology,
Syntex Research, Palo Alto, CA 94303, U.S.A.;* [3]*Preclinical Research Division, Sandoz Pharma
Ltd., CH 4002 Basel, Switzerland;* [4]*Department of Pharmacology, University of Sydney,
Sydney, NSW, Australia;* [5]*Research Division, Glaxo Research and Development Ltd.,
Ware, U.K.;* [6]*CNS Drug Discovery, Bristol-Myers Squibb Pharmaceutical Research Institute,
Wallingford, CT 06492, U.S.A.*

Summary. The majority of serotonergic drugs act either via blocking uptake of serotonin
(5-hydroxytryptamine; 5-HT) into serotonergic neurons or by an action on a growing number
of serotonin receptors. Based on operational (agonist and antagonist rank order), transductional (second messenger coupling) and structural (gene and amino acid sequence) criteria,
four main types of serotonin receptors have been discerned. The responses mediated via many
serotonin receptors are now well understood and this, in turn, has resulted in the development
and use of serotonergic drugs in the therapy of several diseases, including anxiety and
migraine, and in the prevention of vomiting caused by anticancer agents. Many other
therapeutic avenues are being explored.

Introduction

Although serotonin (5-hydroxytryptamine; 5-HT) is known to act as a
neurotransmitter as well as a hormone, it is only relatively recently that
substantial efforts have been made to design and develop drugs that
selectively modify the activity of serotonergic systems. The majority of
serotonergic drugs act either via blocking uptake of serotonin into
serotonergic neurons or by an agonist or antagonist action at a growing
number of serotonin receptors.

Modern Serotonin Receptor Classification

In 1986, Bradley et al. [1] subdivided serotonin receptors into three
groups, 5-HT$_1$-like, 5-HT$_2$ and 5-HT$_3$. However, the wealth of biochemical and functional data, combined with the cloning of many serotonin
receptors, has made a revision of serotonin receptor classification necessary. The proposed scheme (Table 1) takes into account *structural* (gene

Correspondence to: Pramod R. Saxena, at the above address.

Table 1. Criteria (operational, transductional and structural) for classifying serotonin receptors, and the responses mediated by these receptors[a]

| Receptor subtype | Operational | | Transductional | Structural[b] | | | Functional responses | Comments |
	Selective agonist	Selective antagonist	Effector pathway	Gene code	nAA	nTM		
5-HT$_1$								
5-HT$_{1A}$	8-OH-DPAT	WAY 100135	cAMP ↓ K$^+$ channel ↑	X13556	421	7	Neuronal hyperpolarization, behavioral central hypotensive and endocrine effects.	Well characterized but potent silent, and selective antagonists awaited.
5-HT$_{1B}$	CP93,129		cAMP ↓	M89954	386	7	Autoreceptor in the rat brain.	Rodent equivalent of 5-HT$_{1D\beta}$ receptor.
5-HT$_{1D\alpha}$	Sumatriptan	GR127935	cAMP ↓	M81589	377	7	Not clearly established.	May be present on rabbit blood vessels.
5-HT$_{1D\beta}$	Sumatriptan	GR127935	cAMP ↓	M81590	390	7	Not clearly established.	May be related to 5-HT$_1$-like subtype.
5-HT$_1$-like	Sumatriptan	GR127935	cAMP ↓	Not yet available			Mainly cranial blood vessel contraction.	Previous name: 5-HT$_{1X}$.
5-ht$_{1E}$			cAMP ↓	M91467	365	7	Not known.	5-CT is a weak agonist.
5-ht$_{1F}$			cAMP ↓	L04962	366	7	Not known.	5-CT is a weak agonist.
5-HT$_2$								
5-HT$_{2A}$	α-Methyl-5-HT	Ketanserin LY53857	IP$_3$/DAG	X57830	471	7	Vascular and other smooth muscle contraction, platelet aggregation.	Previous names: 'D', 5-HT$_2$.
5-HT$_{2B}$	α-Methyl-5-HT	LY53857	IP$_3$/DAG	X66842	479	7	Rat stomach fundus contraction.	Previous name: 5-HT$_{2F}$.
5-HT$_{2C}$	α-Methyl-5-HT	Mesulergine LY53857	IP$_3$/DAG	M81778	458	7		Previous name: 5-HT$_{1C}$.
5-HT$_3$	2-Methyl-5-HT	Tropisetron Ondansetron Granisetron	Cation channel	M74425 Ion channel unit	487		Membrane depolarization, dermal pain and flare response, anticancer drug-induced vomiting.	Previous name: 'M'.

5-HT$_4$	Renzapride 5-Methoxy-tryptamine	GR113808 SB204070	cAMP ↑		407[c] 387[c]	7 7	Gastrokinesis, tachycardia in the pig and human.	Seems to be absent on the heart ventricle.
Recombinant/"Orphan" receptors								
5-ht$_{5A}$		Methiothepin		L11072	357	7	Not known.	Previous name: 5-HT$_{5\alpha}$; sumatriptan-insensitive, ergotamine-sensitive.
5-ht$_{5B}$		Methiothepin		L11073	370	7	Not known.	Previous name: 5-HT$_{5\beta}$; sumatriptan-insensitive, ergotamine-sensitive.
5-ht$_6$		Methiothepin	cAMP ↑	L03202	437	7	Not known.	
5-ht$_7$		Methiothepin	cAMP ↑	L21195	445	7	Not known.	Mediates vascular smooth muscle relaxation?
?	5-CT	Methiothepin	cAMP ↑	Not yet known			Vascular smooth muscle relaxation.	Previous names: 5-HT$_1$-like, 5-HT$_{1Y}$.
?	723C86	Cyproheptadine		Not yet known			Vascular relaxation.	Present on rabbit jugular vein endothelium.

[a]For references, see Hoyer et al. [2]; [b]only one species homologue is mentioned; [c]long (5-HT$_{4-L}$) and short (5-HT$_{4-S}$) splice variants of 5-HT$_4$ receptors [3]. Abbreviations: DAG, Diacylglycerol; IP$_3$, inositol (1,4,5) triphosphate; nAA, number of amino acids; nTM, number of transmembrane domains; 5-CT, 5-carboxamidotryptamine.

Compounds: CP 93,129, 5-hydroxy-3(4-1,2,5,6-tetrahydropyridyl)-4-azaindole; GR 113808, [1-[2-(methylsulphonyl)aminoethyl]-4-piperidinyl]methyl 1-methyl-1-H-indole-3-carboxylate; GR 127935, N-[methoxy-3-(4-methyl-1-piperazinyl)phenyl]-2′-methyl-4′-(5-methyl-1,2,4-oxadiazol-3-yl) [1,1,-biphenyl]-4-carboxamide; LY 53857, 4-isopropyl-7-methyl-9-(2-hydroxy-1-methylpropoxy carbonyl)-4,6,6A,7,8,9,11,11A-octahydroinodolo[4,3FG]quinolone; 8-OH-DPAT, 8-hydroxy-2-(di)-n-propylamino-tetralin; SB204070, (1-butyl-4-piperidinylmethyl)-8-amino-7-chloro-1,4-benzodioxan-5-carboxylate; WAY 100135, N-tert-butyl-3-(4-[2-methoxyphenyl] piperazin-1-yl)-2-phenylpropanamide dihydrochloride; 723C86, (±)1-[5-(2-thenyloxy-1H-indol-3-yl)propan]-2-amine hydrochloride.

Table 2. Therapeutic applications of drugs acting at serotonin receptors[a].

Drug	Disease	Status
5-HT$_{1A}$ receptor (partial) agonists		
Buspirone	Anxiety	Marketed
Ipsapirone	Anxiety	Clinical phase 3
Gepirone	Anxiety	Clinical phase 3
BMS-181101[b]	Anxiety	Clinical phase 3
Urapidil[c]	Hypertension	Marketed
5-HT$_{1A}$ receptor antagonists		
WAY 100135	Anxiety	Preclinical
WAY 100635	Anxiety	Preclinical
5-HT$_1$-like/5-HT$_{1D}$ receptor agonists		
Ergotamine[d]	Migraine	Marketed
Dihydroergotamine[d]	Migraine	Marketed
Methysergide[e]	Migraine	Marketed
Sumatriptan	Migraine	Marketed
Naratriptan	Migraine	Clinical phase 3
BW 311C90	Migraine	Clinical phase 3
MK 462	Migraine	Clinical phase 3
5-HT$_{1D}$ receptor antagonists		
GR127935	Depression	Preclinical
5-HT$_{2A}$ receptor antagonists		
Pizotifen[f]	Migraine	Marketed
Ketanserin[c]	Hypertension	Marketed
Risperidone[g]	Psychosis	Marketed
Ritanserin	Sleep disorders	Clinical phase 3
5-HT$_{2C}$ receptor antagonists		
SB 200646	Feeding disorders	Preclinical
SB 200646	Migraine	Preclinical
5-HT$_3$ receptor antagonists		
Ondansetron	Emesis	Marketed
Granisetron	Emesis	Marketed
Tropisetron	Emesis	Marketed
Ondansetron	Cognition disorders	Clinical phase 2/3
Alosetron	Irritable bowel syndrome	Clinical phase 2
5-HT$_4$ receptor agonists		
Cisapride[g]	Reduced gastric motility	Marketed
Renzapride	Reduced gastric motility	Clinical phase 3
SC 49518	Reduced gastric motility	Preclinical
5-HT$_4$ receptor antagonists		
GR 125487	Irritable bowel syndrome	Preclinical
SB 207711	Cardiac arrhythmias	Preclinical

[a]For references, see Saxena [4]; [b]also has serotonin uptake inhibiting and 5-HT$_{1D}$ receptor agonist properties; [c]α-adrenoceptor antagonist activity is apparently more important; [d]acts on several other receptors, including 5-HT$_2$ receptors and α-adrenoceptors; [e]partial agonist and also blocks 5-HT$_2$ receptors; [f]other properties are involved; [g]other properties are probably also involved.

Compounds: BMS-181101, 3-[3-[4-(5-methoxy-4-pyrimidinyl)-1-piperazinyl]propyl]-5-fluoro-1H-indole dihydrochloride; BW 311C90, N,N-dimethyl-2-[5-(2-oxo-1,3-oxazolidin-4-ylmethyl)-1H-indol-3-yl]-ethylamine GR 125487, [1-[2-[(methylsulphonyl)amino] ethyl]-4-piperidinyl]-methyl-5-fluoro-2-methoxy-1H-indole-3-carboxylate hydrochloride; GR 127935, N-[methoxy-3-(4-methyl-1-piperazinyl)phenyl]-2′-methyl-4′-(5-methyl-1,2,4-oxadiazol-3-yl)[1,1-biphenyl]-4-carboxamide; MK 462, N,N,-dimethyl-2-[5-(1,2,4-triazol-1-ylmethyl)-1H-indol-3-yl]ethyl-amine; SB 200646, N-(1-methyl-5-indolyl)-N-(3-pyridyl urea hydrochloride); SB 207711, [(1-butyl-4-piperidinylmethyl)-8-amino-7-iodo-1,4-benzodioxan-5-carboxylate]; SC 49518, N-[exo-(hexahydro-1H-pyrrolizine-1-yl)methyl]-2-methoxy-4-amino-5-chlorobenzamide hydrochloride; WAY 100135, N-tert-butyl-3-(4-[2-methoxyphenyl]piperazin-1-yl)-2-phenylpropan-amide dihydrochloride; 723C86, (±)1-[5-(2-thenyloxy-1H-indol-3-yl)-propan]-2-amine hydrochloride; WAY 100635, (N-[2-[4-(2-methoxyphenyl)-1-piperazinyl)ethyl-N-(2-pyridinyl)-cyclohexane carboxamide trihydrochloride.

sequence), *transductional* (second messenger system) and *operational* (pharmacological profile) information [2, 3].

Therapeutic Uses

There are currently more than 70 pharmaceutical companies with an interest in serotonergic drugs. Established as well as potential therapeutic applications of serotonin receptor agonists and antagonists are listed in Table 2 [4]. Some of these therapeutic applications are discussed below.

Central Nervous System Diseases

Affective Disorders
The role of serotonin-uptake inhibitors as well as serotonin receptor ligands in the treatment of affective disease states has been explored extensively [see 5, 6]. It appears that 5-HT_{1A} receptor partial agonists, such as buspirone, are clinically effective in generalized anxiety disorders (GAD) and in anxiety associated with depression, but not in panic disorders. Clinical results with the recently developed "silent" 5-HT_{1A} receptor antagonists (e.g., WAY 100135 and WAY 100635) are awaited with interest to help establish whether an agonist or antagonist activity is important in the anti-anxiety effect. Further, a novel agent which combines serotonin-uptake inhibition with 5-HT_{1A} and 5-HT_{1D} autoreceptor partial agonist activity (BMS-181101) is currently being evaluated in clinical trials as a rapidly acting antidepressant. Also, preliminary clinical trials suggest that 5-HT_2 and 5-HT_3 receptor antagonists may have beneficial effects in GAD-patients [6]. However, the evidence to date is far from compelling and the results of larger placebo-controlled clinical trials are awaited.

Cognitive Disorders
Preclinical evidence implicating a role for 5-HT in cognitive processing is now well established, but little progress has been made in elucidating the roles of many of the serotonin receptors in cognition or indeed in other physiological systems. However, the identification of highly selective antagonist ligands for the 5-HT_3 receptor, has allowed rapid advances in the knowledge of the localisation and functional significance of these receptors in the brain. Autoradiographic studies have highlighted the discrete distribution of 5-HT_3 receptors in many regions of the brain believed to control emotional and cognitive processing (see [7]). Recently, the cognitive-enhancing effects of a number of selective 5-HT_3 receptor antagonists, including ondansetron, zacopride and tropisetron

have been reported [8, 9]. In rodents, ondansetron reverses scopolamine- or lesion-induced cognitive deficits in habituation and T-maze tasks and ondansetron and DAU6215 have similar effects in the Morris water maze [8, 10]. Marked improvements in visual discrimination and reversal learning have also been shown in the marmoset [11]. These preclinical data await support from clinical studies, in particular investigations of the potential for the 5-HT$_3$ receptor antagonists to improve cognition in Alzheimer's disease and other neurodegenerative diseases. Early clinical data report encouraging results. Alosetron and zacopride partially reverse a cognitive deficit produced by scopolamine and ondansetron improves cognitive performance in patients showing age-associated memory impairment (see [7]).

Schizophrenia
Recent studies utilizing behavioral, electrophysiological and neurochemical techniques have shown that activation of 5-HT$_3$ receptors results in an increase in dopaminergic activity in the rat mesolimbic system [see 12]. Thus, 5-HT$_3$ receptor antagonists may have a selective inhibitory effect on mesolimbic dopaminergic function, and may prove to be effective in schizophrenia without producing the adverse extrapyramidal motor disturbances commonly associated with the dopamine receptor-blocking neuroleptic drugs. In uncontrolled studies, ondansetron has been found to be effective in schizophrenia without producing extrapyramidal effects [13]. However, in placebo-controlled trials results with ondansetron were inconclusive [14], and zacopride produced no significant improvement [15].

Migraine

A large number of studies has established conclusively that sumatriptan, an agonist at 5-HT$_1$-like (probably 5-HT$_{1D\beta}$) receptors, is effective in alleviating migraine [see 16]. Ergotamine and dihydroergotamine have also been shown to act via 5-HT$_1$-like/5-HT$_{1D}$ receptors [17–19]. This new awareness as well as the success of sumatriptan, both in a clinical and economic sense, has prompted a large number of pharmaceutical corporations to develop such drugs (e.g. naratriptan, BW 311C90, MK 462). In particular, the efforts are directed to make more lipid-soluble and selective compounds to improve oral bioavailability and avoid headache recurrence and coronary artery vasoconstriction.

  Some antimigraine drugs are potent antagonists at 5-HT$_{2A}$ receptors (methysergide, pizotifen, ergotamine, dihydroergotamine), but many other such agents (ketanserin, cyproheptadine, mianserin, sergolexole, ICI 169,369) have not been particularly helpful in migraine therapy, and those that have a proven efficacy in migraine (ergotamine, dihydroergotamine, methysergide) also have affinity for 5-HT_1-like/5-HT_{1D} receptors (see above). It has been advocated that 5-HT_{2C} receptor antagonism is important for antimigraine action, but drugs that block the 5-HT_{2C} receptor (e.g. mianserin, sergolexole) are not effective against migrainous headaches. Moreover, rather than being antagonists, the antimigraine drugs ergotamine and dihydroergotamine behave as potent (partial) agonists at the 5-HT_{2C} receptor (for references, see [4]).

In uncontrolled trials, 5-HT_3 receptor antagonists (MDL 72222 and granisetron) have been reported to be effective in migraine. However, carefully designed and extensive investigations with tropisetron, both as acute and prophylactic migraine therapy, were largely negative [20].

Emesis Following Anticancer Therapy

Chemotherapeutic agents (e.g. cisplatin) and radiation treatment evoke prolonged episodes of nausea and vomiting (over days) in cancer patients. From the patients' standpoint vomiting is the most feared effect of cancer treatment and, unless controlled, limits therapy. Evidence suggests that cisplatin and radiation therapy release 5-HT (and other endogenous substances) from intestinal enterochromaffin cells which then evokes afferent vagal discharge, via 5-HT_3 receptors, to areas of the hindbrain, such as the area postrema and nucleus tractus solitarius [21, 22]. Controversy exists over the role of blood-borne 5-HT, which theoretically could activate directly serotonin receptors in the area postrema, as this area lies outside the blood-brain barrier. Irrespective of locus of action, 5-HT_3 receptor antagonists (ondansetron, granisetron, tropisetron, dolasetron) have been shown to be an effective and safe group of agents for the prevention of acute emesis induced by cisplatin and several other chemotherapeutic agents. They are less effective, however, against delayed emesis, suggesting that this phenomenon may, to a greater extent, involve 5-HT independent mechanisms. Clinical efficacy of 5-HT_3 receptor antagonists is enhanced by simultaneous treatment with dexamethasone [21, 23]. More recently, these drugs have also been shown to be effective against post-operative vomiting.

Gastrointestinal Diseases

Functional dyspepsia
Functional dyspepsia refers to a chronic syndrome of upper gastrointestinal symptoms, which include epigastric discomfort (bloating), pain, nausea, post-prandial vomiting and, sometimes, heartburn and regurgitation (gastro-oesophageal reflux dyspepsia). Various placebo-controlled trials have proved the efficacy of metoclopramide and cisapride

in the treatment of functional dyspepsia [24]. The mechanism probably relates to 5-HT$_4$ receptor agonism and enhancement of gastric emptying. 5-HT$_3$ receptor blockade, which is present to a greater extent with metoclopramide, may not be an important factor as selective 5-HT$_3$ receptor antagonists have little effect on gastric emptying in man.

Gastroparesis

Gastroparesis (delayed gastric emptying) may be idiopathic or, more commonly, associated with diabetes mellitus (gastroparesis diabeticorum). Studies show that the delay in gastric emptying and associated symptomatology (nausea, vomiting, bloating, abdominal pain, and anorexia) can be reduced by treatment with cisapride [25] or renzapride [26].

Gastro-oesophageal reflux disease

Results of several clinical trials show that cisapride is as effective as metoclopramide or ranitidine (an histamine$_2$ receptor antagonist) measuring both global symptoms and endoscopic signs (healing of oesophagitis; see [27]). Restoration of oesophageal peristalsis, sphincter tone, and enhanced gastric emptying consistent with 5-HT$_4$ agonism, appear to provide the mechanistic base for the improvement [28].

Irritable bowel syndrome (IBS)

Although no consistent abnormality has been identified in IBS the condition appears to be due to disordered motility and secretion (constipation or diarrhoea), as well as altered visceral sensation, and afflicts the small intestine, rectum, and particularly, the colon. Mounting evidence shows hyperalgesia to balloon distension of the colon, rectum, and small intestine in IBS patients, probably due to a neural defect at the level of the gastrointestinal tract, or elsewhere in the visceral pain pathways [29].

5-HT$_3$ and 5-HT$_4$ receptor antagonists are obvious candidates for diarrhoea-predominant IBS. As yet, 5-HT$_3$ receptor antagonists have met with mixed success in the clinic [30] and 5-HT$_4$ receptor antagonists [28] are at the experimental stage or undergoing early clinical studies, and no data are available. It is important to point out that no objective evidence exists to link 5-HT with the pathology of IBS and, therefore, modulation of the 5-HT$_4$ receptor-effector mechanisms may not treat the underlying cause. Clinical experience with selective 5-HT$_4$ receptor antagonists will help to address this point.

Carcinoid Syndrome

Carcinoid syndrome emanates from tumours of enterochromaffin cells usually in ileum, stomach or bronchus. The principal features are flushing and diarrhoea, and, if located in lung, bronchoconstriction. Although symptoms may be due primarily to 5-HT, other agents also play a role. Obviously, in the face of such a barrage of released 5-HT, all subtypes of serotonin receptors may well become activated. Thus, $5\text{-}HT_1/5\text{-}HT_2$ (methysergide), $5\text{-}HT_2$ (cyproheptadine, ketanserin), $5\text{-}HT_3$ (ondansetron, granisetron, tropisetron) and, when available, $5\text{-}HT_4$ receptor antagonists may be used. The latter are indicated following observations that this class of antagonist can protect against 5-hydroxytryptophan-induced defecation [31] and diarrhoea [32] in mice.

References

1. Bradley PB, Engel G, Feniuk W, Fozard JR, Humphrey PPA, Middlemiss DN, et al. Proposals for the classification and nomenclature of functional receptors for 5-hydroxytryptamine. Neuropharmacology 1986; 25: 563–576.
2. Hoyer D, Clarke DE, Fozard JR, Hartig PR, Martin GR, Mylecharane EJ, Saxena PR, Humphrey PPA. International union of Pharmacology Classification of receptors for 5-hydroxytryptamine (serotonin). Pharmacol. Rev. 1994; 46: 157–203.
3. Gerald C, Adham N, Kao HT, Schechter IE, Olsen MA, Bard JA, et al. The $5\text{-}HT_4$ receptor: molecular cloning of two splice variants. In: Abstracts Book, Third IUPHAR Satellite Symposium on Serotonin, Chicago, 1990: 82.
4. Saxena PR. Modern 5-HT receptor classification and 5-HT based drugs. Exp. Opin. Inv. Drugs 1994; 5: 513–523.
5. Charney DS, Krystal JH, Delgado PL, Heninger GR. Serotonin-specific drugs for anxiety and depressive disorders. Ann. Rev. Med. 1990; 41: 437–446.
6. Murphy DL, Broocks A, Aulakh C, Pigott TA. Anxiolytic effects of drugs acting on 5-HT receptor subtypes. In: Vanhoutte PM, Saxena PR, Paoletti R, Brunello N, Jackson AS, editors. Serotonin, cell biology to pharmacology and therapeutics. Dordrecht: Kluwer Academic Publishers, 1993: 223–230.
7. Kilpatrick GJ, Ireland SJ, Tyers MB. Ondansetron and related $5\text{-}HT_3$ antagonists: recent advances. Prog. Med. Chem. 1992; 29: 239–270.
8. Barnes JM, Costall B, Coughlan J, Domeney AM, Gerrard PA, Kelly ME, et al. The effects of ondansetron, a 5-HT3 receptor antagonist, on cognition in rodents and primates. Pharmacol. Biochem. Behav. 1990; 35: 955–962.
9. Chugh Y, Saha N, Sankaranarayanan A, Datta H. Enhancement of memory retrieval and attenuation of scopolamine-induced amnesia following administration of 5-HT3 antagonist ICS 205-930. Pharmacol. Toxicol. 1991; 69: 115–116.
10. Pitikas N, Brambilla A, Borsini F. Effect of DAU 6215, a novel $5\text{-}HT_3$ receptor antagonist, on scopolamine-induced amnesia in the rat in a spatial learning task. Pharmacol. Biochem. Behav. 1994; 47: 95–99.
11. Domeney AM, Costall B, Gerrard PA, Jones DN, Naylor RJ, Tyers MB. The effect of ondansetron on cognitive performance in the marmoset. Pharmacol. Biochem. Behav. 1991; 38: 169–175.
12. Hagan RM, Kilpatrick GJ, Tyers MB. Interactions between $5\text{-}HT_3$ receptors and cerebral dopamine function: implications for the treatment of schizophrenia and psychoactive substance abuse. Psychopharmacology 1993; 112: S68–S75.
13. DeVeaugh-Geiss J, McBain S, Cooksey P, Bell JM. The effects of a novel $5\text{-}HT_3$ antagonist, ondansetron, in schi-ophrenia: results from uncontrolled trials. In: Meltzer HY, editor. Novel antipsychotic drugs. New York: Raven Press, 1992: 225–232.
14. McBain SL, Dineen M, Weller M, Taylor S, Cooksey P, Williams P, et al. Ondansetron: a double-blind placebo-controlled study in acute schizophrenia. Schizophren. Res. 1992; 6: 112.

15. Newcomer JW, Faustman WO, Zipursky RB, Csernansky JG. Zacopride in schizophrenia: a single-blind serotonin type 3 antagonist trial. Arch. Gen. Psychiat. 1992; 49: 751–752.
16. Ferrari MD, Saxena PR. Clinical and experimental effects of sumatriptan in humans. Trends Pharmacol. Sci. 1993; 14: 129–1233.
17. Hyer D, Schoeffter P. Interactions of dihydroergotamine (DHE), ergotamine and GR 43175 (sumatriptan) with 5-HT$_{1D}$ receptors. Naunyn-Schmiedebergs Arch. Pharmacol. 1991; 343: R111.
18. Bruinvels AT, Lery H, Nozulak J, Palacios JM, Hoyer D. 5-HT$_{1D}$ binding sites in various species: similar pharmacological profile in dog, monkey, calf, guinea-pig and human brain membranes. Naunyn-Schmiedebergs Arch. Pharmacol. 1992; 346: 243–248.
19. Müller-Schweinitzer E. Ergot alkaloids in migraine: is the effect via 5-HT receptors? In: Olesen J, Saxena PR, editors. 5-Hydroxytryptamine mechanisms in primary headache. New York: Raven Press, 1992: 297–304.
20. Ferrari MD. 5-HT$_3$ receptor antagonists and migraine therapy. J. Neurol. 1991; 238 (Suppl. 1): S53–S56.
21. Andrews PLR, Rapeport WG, Sanger GJ. Neuropharmacology of emesis induced by anti-cancer drugs. Trends Pharmacol. Sci. 1988; 9: 334–341.
22. Hesketh PJ, Gandara DR. Serotonin Antagonists: A new class of antiemetic agents. J. Natl. Cancer Inst. 1991; 83: 613–620.
23. Kois MG, Baltzer L, Pisters KRN, Tyson LB. Enhancing the effectiveness of the specific serotonin antagonists. Combination with dexamethasone. Cancer 1993; 72: 3436–3442.
24. Fumagalli I, Hamner B. Cisapride versus metaclopramide in the treatment of functional dyspepsia. A double-blind comparative trial. Scand. J. Gastroenterol. 1994; 29: 33–37.
25. Corinaldesi R, Stanghellini V, Tosetti C, Rea E, Corbelli C, Marengo M, et al. The effect of different dosage schedules of cisapride on gastric emptying in idiopathic gastroparesis. Eur. J. Clin. Pharmacol. 1993; 44: 429–432.
26. Mackie ADR, Ferrington C, Cowan S, Merrick MV, Baird JD, Palmer KR. The effects of renzapride, a novel prokinetic agent, in diabetic gastroparesis. Aliment. Pharmacol. Ther. 1991; 5: 135–142.
27. Arantikakis G, Nikopoulos A, Theohardis A, Giannoulis E, Vagios I, Anthopoulou H, et al. Cisapride and ranitidine in the treatment of gastro-oesophageal reflux disease – A comparative randomized double-blind trial. Aliment. Pharmacol. Ther. 1993; 7: 635–641.
28. Ford APDW, Clarke DE. The 5-HT$_4$ receptor. Med. Res. Revs. 1993; 13: 633–662.
29. Mayer EA, Gebhart GF. Functional bowel disorders and the visceral hyperalgesia hypothesis. In: Mayer EA, Raybould HE, editors. Basic and clinical aspects of chronic abdominal pain. New York: Elsevier Science Publishers, 1993: 3–28.
30. Steadman CJ, Talley JJ, Phillips SF, Zinsmeister AR. Selective 5-hydroxytryptamine type 3 receptor antagonism with ondansetron as treatment for diarrhoea-predominant irritable bowel syndrome: A pilot study. Mayo Clinic Proc. 1992; 67: 732–738.
31. Banner SE, Smith MI, Sanger GJ. 5-HT Receptors and 5-hydroxytryptophan-evoked defecation in mice. Br. J. Pharmacol. 1993; 111: 135P.
32. Hegde SS, Moy TM, Perry MR, Eglen RM. Involvement of 5-HT$_4$ receptors in 5-hydroxytryptophan-induced diarrhoea in mice. FASEB 1994; 8: A92.

Pharmacological Sciences: Perspectives for
Research and Therapy in the Late 1990s
ed. by A.C. Cuello and B. Collier

The Pharmacology of Neurotrophic Factors

A. Claudio Cuello[1] and Hans Thoenen[2]

[1]*Department of Pharmacology & Therapeutics, McGill University,
Montreal, Quebec, Canada, H3G 1Y6;* [2]*Department of Neurochemistry, Max Planck
Institute for Psychiatry, D-82152 Martinsried, Germany*

Neurotrophic Factors and their Potential Future Therapeutic Use

Since the early discoveries of Rita Levi-Montalcini and coworkers [1] the
field of neurotrophic factors has expanded dramatically, and has become
a pharmacological subject. We are facing a great number of neurotrophic
factors (NTFs) with defined biological actions and conceptual therapeutic potential. Of these, the neurotrophins (the name coined for the
members of the NGF gene family) and their receptors represent proto-
typical neurotrophic molecules (for review see [2–4]). It is our belief that
future clinical neurology will exploit the knowledge evolving from
current basic research for therapeutic use. This report summarizes the
main achievements of basic research and relevant clinical aspects of this
field, as covered by the presentations of Ted Ebendal (Uppsala, Sweden),
Fred Gage (San Diego, USA) and ourselves at this Congress.

The use of neurotrophic molecules for the treatment of degenerative
disorders of the nervous system is based on experimental observations
that neurotrophic molecules have the capability of protecting and also
restoring impaired functions of neurons which have been damaged by a
great variety of mechanical and chemical mechanisms and also by
genetic disorders of unknown molecular origin. In spite of great progress in the understanding of the biological actions of neurotrophic
molecules, including their interaction with the corresponding receptors
and the resulting signal-transduction cascade, the relationship between
these mechanisms and the protective and restorative effects of neurotrophic molecules remain largely elusive. In particular, an understanding of the relationship between the action of neurotrophic molecules
and apoptotic/anti-apoptotic mechanisms is just starting. It is important
to emphasize that the therapeutic benefits of neurotrophic molecules
may be obtained in a "nonspecific manner", i.e. independently of the
cause of the degenerative disorders. Thus, treatment with neurotrophic
molecules is a symptomatic rather than a causal one. As a case in point,
there is no evidence that Alzheimer's or Parkinson's Disease result from
a deficiency in trophic factors. The reported reduction of brain-derived

neurotrophic factor (BDNF) mRNA levels in the hippocampus of patients with Alzheimer's disease (AD) is likely to be a consequence rather than the cause of a degenerative disorder [5]. In order to deduce a causal relationship, it would be essential to demonstrate a reduction of BDNF mRNA at the very beginning or even before the functional manifestations of AD.

Although there are rational indications for the use of neurotrophic factors in the treatment of degenerative disorders of the central nervous system (CNS), the problem remains how to best administer these molecules. The blood vessels supplying the CNS display tight junctions which, in contrast to the periphery, do not allow the permeation of large molecules, such as NTFs, across the blood-brain barrier (BBB). Should they be administered into the cerebral ventricles via canulae connected to pumping systems? Should they be given in a large bolus of microencapsulated materials, including trophic factor producing cells? (see e.g. [6, 7]). Mechanisms to bring molecules across the BBB do exist. Indeed, there are specific transport receptors for molecules which are not synthesized in the brain, but play an essential biological role, as e.g. transferrin. These transfer mechanisms have been exploited in a conceptually elegant study in which anti-ferritin receptor antibodies have been covalently linked to nerve growth factor (NGF), resulting in a substantial tranfer of biologically active NGF across the BBB into the CNS [8, 9]. The effectiveness of this treatment has so far only been shown for NGF-responsive neuronal transplants into the anterior eye chamber. However, it would have been preferable – and the clinically minded scientific community is waiting for these experiments – if this approach had been used for the treatment of CNS lesions (see below), or on memory-deficient aged rats, for which a remarkable reversal of atrophic-degenerative changes in NGF-responsive cholinergic neurons along with a reduction in cognitive deficits has been demonstrated by the intraventricular injection of NGF [10]. Other possibilities might well become available, e.g. the use of activated T-lymphocytes as carriers which are capable of crossing the BBB [11]. Indeed, T-lymphocytes can be engineered to produce NTFs (as demonstrated for NGF) in comparable quantities to those produced by engineered fibroblasts. The exploitation of the transfer mechanisms across the BBB and the use of engineered T-cells are potentially noninvasive alternatives to the intraventricular infusion or the transplantation of engineered autologous fibroblasts or skeletal muscle cells to be grafted into the CNS (see below). Should we wait until small molecular weight trophic factor derivatives capable of crossing the blood-brain barrier are developed? In this regard it is worth noting that effective peptide analogues of NGF (antagonists) have been produced [12] and that non-peptide mimetics are on the horizon (e.g. [13]). For the NGF gene family, this might be a hopeless enterprise, because the intactness of the three-dimensional structure of the whole mature molecule might be necessary for the biological effects. Another potential complication of all these approaches is the fact that, regardless of the use of invasive or noninvasive procedures, the administered molecules "flood" the CNS indiscriminately, causing potential serious side-effects (discussed in [14]). Addressing these problems is an essential issue in the present preclinical and clinical evaluations of the therapeutic use of neurotrophic molecules.

Assuming that the trophic factor delivery problems can be solved satisfactorily, there are other questions which have to be addressed. Will these agents be supplied to the site of action in sufficient quantities to accomplish the envisaged therapeutic effect? The remarkable potency of currently known NTFs might anticipate a positive answer. On the other hand, administration of large quantities of neurotrophic molecules might elicit immunological responses. In the study carried out in Sweden (see below), this has not been the case so far. However, a number of other unwanted effects might result from the CNS application of NTFs. In the case of neurotrophins, an important issue to be addressed is the reported hyperalgesia induced by the administration of NGF [15, 16].

The invasive procedures are extremely valuable for establishing the principle therapeutic usefulness of a molecule. However, the implantation and the maintenance of pumps or the transplantation of engineered cells to hundreds of thousands or even millions of patients have their technical and also financial limitations. An alternative approach, better suited for the treatment of a larger number of patients, is the possibility of influencing the synthesis of endogenous NTFs. This also would have the advantage of producing neurotrophic molecules at the physiological site of action, consequently avoiding the "flooding" of the CNS by exogenous NTFs, be it by invasive or noninvasive methods. This would also preclude the disruption of normal synaptic connections which could have unpredictable functional consequences. Indeed, the intraventricular administration of NGF to young rats, in contrast to lesioned or aged memory-impaired rats, resulted in a deterioration rather than an improvement of the performance in the Morris water maze (Springer, personal communication). The possibility of regulating the synthesis of NTFs, at least for some members of the NGF gene family, is based on the observation that, in the CNS, the synthesis of NGF and BDNF is regulated by neuronal activity. Up-regulation occurs by glutamate via NMDA and non-NMDA receptors and by acetylcholine via muscarinic receptors, while GABA down-regulates NGF and BDNF mRNA synthesis via GABA-A$_1$ receptors [17]. These observations imply that, by these transmitter systems, a change in the synthesis and increased availability of these NTFs can be accomplished. The combination of glutamate and GABA receptor subunits with different functional properties, showing a regionally differential distribution in the CNS opens up the possibility of developing pharmacological compounds which only affect specific subpopulations of these receptors. Therefore, at least theoretically, specific regulatory effects on the synthesis of neurotrophins might be accomplished. Action via a gentle activation of all or part of the glutamatergic system, or a slight dampening of the GABAergic system might be suitable for enhancing the synthesis of NGF and BDNF. Certainly, this is a demanding and ambitious task, one nevertheless worthwhile in view of the potential clinical problems and limitations of other available methods mentioned above for the treatment of large numbers of patients suffering from degenerative disorders of the CNS.

Beyond the possibility of regulating BDNF and NGF in an activity-dependent manner, recent investigations have also demonstrated that neurotrophins, in particular NGF, can be released from neurons in an activity-dependent manner, and that the released NGF could initiate positive feedback effects reflected by an enhanced transmitted release from nerve terminals of neurons, exhibiting the corresponding trk receptors. Therefore, the local administration of NTFs may not only result in the re-arrangement of synapses, but may, in extreme cases, even result in the initiation of epileptic activity [18]. Therefore, both the local administration and transfer across the BBB of neurotrophic molecules will be a balancing act between attaining the desired neurotrophic effects while avoiding harmful side-effects.

Thus, an ideal situation would result from the stimulation of the synthesis of NGFs at their physiological site of action with a minimal perturbation of other sites. This is a highly demanding alternative, but worthwhile in view of the urgent clinical demands. In addition to these uses in the CNS, there is also a broad spectrum of potential use in the periphery for the treatment of various neuropathies, resulting from diabetes or from iatrogenic effects by cytostatic treatment by vincristin or taxol [4]. In this context, the use of engineered T-cells targeted to peripheral myelin constituents could open up very attractive possibilities for the treatment of neuropathies, including degenerative diseases of motoneurons (in particular amyotrophic lateral sclerosis), again, aiming at delivering the neurotrophic molecules at the site of the requested action to avoid side effects [14]. Another line of attack in seeking specificity would be to interfere with the mechanism of the trophic factor signal-transduction system. It is conceivable that future knowledge might allow the development of molecules which can interfere – stimulate or block – individual steps of the signal-transduction cascade. Thus, potential therapeutic use will not remain restricted to just the simple administration of these factors. A more profound knowledge of the molecular mechanism by which NTFs act would open up the possibility to bypass the ligand receptor interaction and to influence further "downhill' mechanisms in the resulting signal-transduction cascade. Therefore, the regulation of the synthesis release of NTFs and their modification with very specific steps in the signal-transduction cascades, should result in the development of therapies which are amenable to a more general application for a larger number of patients.

Trophic Factor Effects on Neuronal Phenotype and the Experimental Remodeling of Synapses in the Adult CNS

Among the many molecules capable of provoking trophic responses in neural tissues, NTFs have the best possibilities for becoming therapeutic agents as they have displayed great potency, possess specific receptors, and their biological actions are well defined in both *in vitro* and *in vivo* models. Besides the neurotrophins (NGF, BDNF, NT3 and NT4/5) there are potential therapeutical applications for Ciliary Neurotropic factor (CNTF) and the fibroblast growth factors – both basic and acidic (b-FGF and a-FGF) – among other putative NTFs. The field of NTFs emerged from developmental neurobiology, wherein each trophic factor is assumed to be responsible for the survival and differentiation of a specific sub-set of neurons in the CNS and PNS during different stages of development. It is thought that in the mature nervous system this activity is related to the maintenance of the phenotypic characteristic of the neurons possessing the respective NTFs receptors. However, in some extreme situations – wound lesions and other types of injuries – there might be an important up-regulation of expression of endogenous NTFs [19–23]. This production, however, is generally not sufficient to reach the threshold required to generate optimal trophic response. In other words, to obtain responses akin to the developmental stages, responses capable of profoundly changing the neuronal architecture, the application of exogenous trophic factor is required.

Much has been learnt from studies in forebrain cholinergic neurons regarding the potential application of NTFs in the CNS, (for reviews see [24, 25]). These neurons are ideally suited for these investigations because the forebrain cholinergic neurons of the medial septum and of the nucleus basalis (nbm) are particularly well endowed with the machinery to respond to NGF, the prototypic neurotrophin. They possess a low affinity receptor common to all neurotrophin sensitive neurons referred to as p75 LNGFR, and a high affinity receptor, the product of a proto-oncogen, a tyrosine kinase referred to as trkA. Other trks respond preferentially to BDNF (trkB) or to NT-3 (trkC). Following binding with the specific NTF receptor, autophosphorylation and a cascade of phosphorylating intracellular events results, leading to the generation of transcription factors for the expression of diverse proteins which ultimately represent the trophic response (e.g. structural proteins, biosynthetic enzymes, transporters, receptors, etc). It is not clear as yet whether the formation of homodimers of high affinity (trkA + trkA) are the main (exclusive?) modality of these responses or whether heteroreceptor dimers also occurs (trkA + p75) (for reviews see [26–28]). In the case of forebrain cholinergic neurons, the most dramatic events are the up-regulation of its biosynthetic enzyme *in vitro* and the salvaging of axotomized neurons of the medial septum *in vivo*, as illustrated by Hefti and collaborators (for review see [24]). In the nbm, a dramatic atrophy of cholinergic neurons occurs accompanied by a clear-cut depletion of choline acetyltransferase (ChAT), after 30 days of a partial unilateral infarction of the cerebral cortex [29, 30]. In this animal model of CNS cell degeneration, the neurons under study do not die but became markedly atrophic. This pathway (basalo-cortical) represents the bulk of the cholinergic innervation to the neocortex, and it is believed to participate in higher functions, such as attention, memory and learning [31].

This model has allowed the demonstration that the trophic responses of CNS neurons *in vivo* are indeed dose-dependent, and that doses as low as 1 μg/day for 7 days are capable of producing full biochemical protection, thus emphasizing the remarkable potency of NTFs [32]. It has been shown that both in the medial septum and the nbm there is some specificity of these responses: NGF being the most active compound while BDNF and NT3 have demonstrated less marked effects in the medial septum or have no effect (at equivalent doses) in the nbm [33–37]. Interestingly, other NTFs for which the basal forebrain cholinergic neurons do not display the expected receptor – such as CNTF and the FGFs – are capable of provoking the salvage-repair of these neurons both in the medial septum and nbm [38–41]. These findings would suggest the existence of yet undetected receptors (FGF receptor 1 or 2, gp130, LIFβ units) in these neurons or might indicate that indirect responses result from this type of trophic factor therapy.

The actions of NGF have been extensively analyzed in primate models, both in the medial septum-hippocampal and the basalo-coritical lesion models. Both mouse and recombinant NGF has been applied intraventricularly to *Macaca fascicularis* following the transection of the fimbria-fornix, a procedure which provokes the apparent disappearance of cholinergic neurons of the medial septum. Full protection of medial septum cholinergic cells has been observed after 2 and 4 weeks of continue administration [42–45]. In recent investigations on *Cercopithecus aethiops* utilizing the basalo-cortical lesion model, it was demonstrated that short-term application of human recombinant NGF resulted in long-term (up to 6 months) biochemical and morphological protection of the cholinergic neurons of the nbm [46].

One interesting observation concerning trophic factor therapy is that the application of NGF in the basalo-cortical lesion system provokes a robust up-regulation of the ChAT enzymatic activity in the remaining cerebral cortex, ipsilateral to the site of cortical infarcts [47]. This phenomenology is accompanied by several neurochemical indications of enhanced cholinergic prsynaptic function such as the increment in the Bmax of high affinity uptake sites [32] and improved release of endogenous acetylcholine [48]. Similarly, in partial deafferentations of the hippocampus, NGF is capable of facilitating the synthesis of ACh in the remaining terminals [49]. The question was raised whether these biochemical changes implied strutural modifications of the synaptology of the cerebral cortex. Thus, Garofalo and coworkers [50, 51] utilizing a combination of high resolution immunocytochemistry with image analysis and serial reconstructions, were able to demonstrate that, in the basalo-cortical lesion model, exogenously supplied NGF produces a hypertrophy of cholinergic presynaptic elements as well as the generation of new synapses: i.e. "synaptogenesis". These observations are of some relevance, in that some of the behavioral changes observed in aged or lesioned animals after the administration of NTFs, notably NGF, may be the result of synaptic remodelling rather than the mere preservation of the cell body morphology of NTF-sensitive neurons. This is an issue that requires attention since beneficial effects of NGF in the behavioural performance of aged rats ([10]; for review see [52]) as well as in cortically lesioned adult rats [53] have been well illustrated. In the first case, NGF can redress aged-related behavioural and cholinergic deficits in some cohorts of impaired rats [10] and, in the second case, rats retain fully the escape latency (about 10 s) of the Morris water maze while cortically lesioned rats with vehicle forget the previously acquired task (escape latency about 80 s) [53]. Future research efforts will have to focus on whether the NTF-induced synaptic reconnections are the correct ones or not, as well as how relevant these observations might be in a clinical scenario. In this regard it is interesting to point out that in the AD situation the pathology which best correlates with dementia is the loss of frontal cortex synapses (for reviews see [54, 55]).

Grafting Cells Genetically Modified to Secrete Trophic Factors to the Adult Mammalian Nervous System

One of the possible means of chronically supplying trophic factors, in a therapeutic context, would be genetic transfer, using a general strategy of providing the required gene in a region specific manner. At present, much of the research in CNS gene therapy focuses on the development of appropriate vectors to secure adequate levels of expression in a sustained fashion. Methods involving the injection of plasmids or vectors such as Herpes virus and other constructs are still in development (for review see [56–59]) but they have already shown that a moderate, transient expression can be assured in this manner. Another method which has been extensively used in the CNS is the *ex vivo* approach, by which cells are taken outside of the body, are genetically transformed in the laboratory to ensure the expresion of a given trophic factor, and are subsequently grafted within the CNS (for review see [60, 61]). Securing a good level of expression and ensuring that the transfected genes can be regulated by trophic factors are important elements in this approach. Most of the work done with NTFs has been with primary culture cells infected and selected for their level of gene expression. Classically this is achieved by utilizing a plasmid containing DNA of interest as well as a marker gene, which is transfected into "packaging lines". These packaging lines will create a virus, and the resulting lines will be selected by their ability to secrete a high titer virus. These viruses are single infective particles which can be harvested and utilized to infect primary culture cells with the adequate retrovirus. These cells are then selected by their ability to produce and secrete the desired NTF. Once these cells are characterized and grown as a cell line they can be utilized for transplantation purposes. Selection of the right type of cell is, however, important. The ideal long-term goal would be to transplant neurons capable of establishing synaptic connections, thus establishing reciprocal relations between the graft and the host. Presently, success of the grafting depends on the proximity of the implant with the target neurons and the presence of adequate receptors in them to secure specificity of the biological responses. Primary cells are preferred, as it has been seen that cell lines grow into tumors in the CNS, precluding their long-term utilization. When utilizing primary fibroblasts, the grafted cells can be easily detected by immunocytochemistry by their expression of fibronectin. When examined with electron microscopy months after grafting in the striatum [62], these fibroblasts display normal characteristics, secrete collagen and the implant is seen to be well vascularized. The collagen fibrils nest the grafted cells, while the newly formed endothelial cells migrate from the host CNS to the graft. These capillaries display tight junctions characteristic of endothelial cells participating in the BBB which have been shown to be fully functional [62].

Cells secreting NGF and BDNF have been already developed [63, 64]. A great deal of information is already available on the efficacy of this approach in recovering cholinergic neurons of the medial septum, following aspirative lesions of the fimbria-fornix in which resulting cavity the genetically engineered cells are grafted [65]. The model therefore allowed the analysis of the neurotrophic effects on the survival of medial septum neurons as well as the neurotropic effects on axonal elongation in the interrupted pathway. Grafted NTF-producing cells have been shown to elicit both neurotrophic and neurotropic effects. Thus, signficant rescue of the medial septum as well as axonal sprouting towards the grafted tissue occurs. These experiments also revealed a major role of CNS reactive astrocyte, which extended processes allowing elongation of NTF-induced axonal sprouting [65]. The graft might itself act as a bridge by using them as a plug, reconnecting the loose ends of the fimbria fornix cavity. Here the penetrating cholinergic axons can be monitored by their immunoreactivity to p75LNGFR and their movement towards the target site [65]. The deafferented hippocampus provokes sympathetic innervation, however when the bridging is over for the ingrowth of CNS cholinergic fibres, the sympathetic elements retract [66]. This observation would indicate that in circumstances when the regrowth is not impeded, the trophic factor should stimulate the correct type of reconnection.

From current studies it becomes clear that NTF therapeutic specificity might reside not just in the particular growth factor and its receptors, but in its regional or targeted administration (grafting, encapsulation, transfected and targeted T-cells and other). Also, the substrates or environment to secure appropriate reconnections, and not just the trophic effects, should be taken into account.

Neurotrophic Factors in Clinical Trials: Present and Future Models

The rationale for administering NTFs in human patients might be somewhat different from the experimental scenario. For example, NGF has so far been administered to a larger number of Parkinsonians than AD sufferers. In the former situation, the rationale is based on the support of the grafting in the CNS (*corpus striatum*) of the adrenomedullary cells from the same patient, a procedure pioneered at the Karolinska Institute. These cells are dependent on NGF support and in the presence of this NTF they become much more viable when grafted within the CNS. A very different situation is that of AD, where one could expect that NGF itself would stimulate cholinergic neurons bearing the corresponding receptors, and consequently, based on experimental studies, improve functions such as learning and memory. Finally one case of Huntington Chorea has been treated with NGF, as the large cholinergic neurons of the caudate and putamen are among the degenerating cells in this disease and they possess high affinity NGF receptors.

For these initial clinical attempts, the problems of circumventing the BBB were addressed as follows: For AD cases, NGF was pumped into the cerebroventricular space in the posterior horn of the lateral ventricle via an implanted canulae. The pump is a device similar to that used by diabetic patients and it is lodged subcutaneously. The patients received 70 μg of NGF per day for 3 months [67]. In the case of Parkinson disease (PD), NGF was administererd (120 μg/day) in the putamen in close approximation of the autografts, as experimental data in rodents indicate that grafted adremedullary cells develop a neuronal phenotype in the presence of NGF, improving their survival in the CNS and compensating for experimentally induced dopamine deficits [68–70]. In these cases the NGF was administered directly into the tissue stereotaxically, guided through holes in the skull. In all cases, pathogen-free, highly purified and well-characterized mouse NGF was administered. One of the aspects that was investigated from the start of these trials has been the monitoring of antibodies against the foreign NGF, as being indicative of an immunological response. In the Swedish experience, no significative reactions have been observed so far.

In PD cases, patients were monitored for some 2 years prior to grafting and NTF therapy. In patients with marked deterioration of their locomotor performance there was an immediate and dramatic improvement after grafting and NGF treatment [71], with some patients returning to states equivalent to those some years before treatment. The improvement observed in these PD-NTF treated patients in several motor performances was superior and longer lasting than that observed in patients which received grafts without NTF support. The motor readiness potential measured from the skull in NGF-treated, grafted patients, when asked to move their finger, improves significantly, particularly in the grafted side of the brain.

In AD cases, it was seen that the delivery of NGF into the CNS is very effective, as NGF material could be recovered from lumbar cerebrospinal fluid. The infusion must be continued, however, as the trophic factor is rapidly removed from the CNS after cessation of administration. PET evaluation showed that at the end of the 3-month period the nicotine binding increased significantly, diminishing after cessation of treatment. The nicotinic binding upregulation was taken as an indication of improvement of cholinergic function. More long lasting improvements were revealed in blood flow cortical areas as shown by ^{11}C-butanol signals. The blood flow approximately doubled in the cortical areas of this patient; also a shift in the electroencephalogram was noticed from slow to high frequencies. These features were considered indicative of an improvement of cortical function, however this was not matched by the patient's performance in the so-called MiniMental Test. The overall MiniMental score dropped during NGF perfusion and improved after cessation [67]. On the other hand the patient's performance in spatial memory improved moderately. Another complication observed in one AD patient and the Huntington patient has been a marked loss in body weight, an aspect predicted from animal experimentation [72].

The Huntington patient treated with NGF was followed for 1 year, and no improvement was observed in the rate of functional decline characteristic of this disease. No improvement was also observed in the functional marker of basal ganglia function as revealed by PET, but there was nevertheless an indication of enhanced cortical metabolism as indicated by deoxiglucose [73].

The lessons from these initial trials are that excellent distribution of the injected NGF can be attained in clinical settings, that the doses administered at present most probably exceed the therapeutic needs, and that they are most likely responsible for the unwanted effects such as the anorexia, herpes zoster and hyperalgesia observed in some patients. In some cases these unwanted effects might be severe, with an additional side-effect, insomnia, having been observed in NGF-treated patients. Improved therapeutic opportunities might be found with site-directed administration of lower doses, perhaps at earlier stages of the degenerative diseases. The latter possibility is at present precluded by the lack of adequate presymptomatic diagnoses (particularly in the case of AD) and ethical considerations. Further clinical and preclinical studies, the development of adequate animal models of the diseases (none available for AD, as yet) and progress in the biochemistry and cell biology of NTFs should better define the future therapeutic opportunities for these remarkably powerful agents.

References

1. Levi-Montalcini R. The nerve growth factor: thirty-five years later. EMBO Journal 1987: 6: 1145–1154.
2. Barde YA. The nerve growth factor family. Progr. Growth Factor Res. 1991; 2: 237–248.
3. Thoenen H. The changing scene of neurotrophic factors. Trends Neurosci. 1991; 14: 165–170.
4. Lindsay RM, Wiegannd SJ, Altar CA, DiStefano PS. Neurotropic factors: from molecule to man. Trends Neurosci. 1994; 17: 182–190.
5. Phillips HS, Hains JM, Armanini M, Laramee GR, Johnson SA, Winslow JW. BDNF mRNA is decreased in the hippocampus of individuals with Alzheimer's disease. Neuron 1991; 7: 695–702.
6. Aebischer P, Winn SR, Tresco PA, Jaeger CB, Green LA. Transplantation of polymer encapsulated neurotransmitter secreting cells: effects of the encapsulation technique. J. Biomech. Engr. 1994; 113: 178–183.
7. Maysinger D, Piccardo P, Filipovic-Grcic J, Cuelo AC. Microencapsulation of genetically engineered fibroblasts secreting nerve growth factor. Neurochem. Int. 1993; 23(2): 123–129.
8. Friden PW, Walus LR, Watson P, Doctrow SR, Kozarich JW, Bäckman C et al. Blood-brain barrier penetration and *in vivo* activity of an NGF conjugate. Science 1993; 259: 373–377.
9. Granholm AC, Backman C, Bloom F, Ebendal T, Gerhardt FA, Hoffer B et al. NGF and anti-transferrin receptor antibody conjugate: short and long-term effects on survival of cholinergic neurons in intraocular septal transplants. J. Pharm. Exp. Therapeu. 1994; 168(1): 448–459.
10. Fischer W, Wictorin K, Björklund A, Williams LR, Varon S, Gage FH. Amelioration of cholinergic neuron atrophy and spatial memory impairment in aged rats by nerve growth factor. Nature 1987; 329: 65–68.
11. Wekerle H, Linington C, Lassmann H, Meyermann R. Trends Neurosci. 1986; 9: 271–277.
12. LeSauteur L, Wei L, Gibbs B, Saragovi HU. Structural analogs of nerve growth factor bind specific receptors and mediate biological effects. Cdn J. of Phys. and Pharm. 1994; 72(1): 207.
13. Saragovi HU, Fitzpatrick D, Raktabuhr A, Nakanishi H, Kahn M, Greene MI. Design and synthesis of a mimetic of an antibody complementarity region. Science 1991; 253: 792–795.

14. Thoenen H, Castrén E, Berzaghi M, Blöchl A, Lindholm D. Neurotrophic factors: possibilities and limitations in the treatment of neurodegenerative disorders. Int. Acad. Biomed. Drug Res. 1994; 7: 197–203.
15. Lewin GR, Mendell LM. Trends Neurosci. 1993; 16: 353–359.
16. Petty BG, Cornblath DR, Adornato BT, Chaudry V, Flexner C, Wachsman M et al. The effects of systemically administered recombinant human nerve growth factor in healthy human subjects. Ann. Neurol. 1994; 36: 244–246.
17. Lindholm D, Castrén E, Berzaghi M, Blöchl A, Thoenen H. Activity dependent and hormonal regulation of neurotrophin mRNA levels in the brain – Implications for neuronal plasticity. J. Neurobio. 1994; 25(11): 1362–1372.
18. Blöchl A, Berninger B, Berzaghi M, Lindholm D, Thoenen H. Neurotrophins as mediators of neuronal plasticity. In: Ibàñez CF, Hökfelt T, Olsson L, Fuxe K, Jornvall H, Ottoson D (editors), Life and death in the nervous system – the role of neurotrophic factors and their receptors. London: Pergamon, 1995. In press.
19. Weskamp G, Gasser UE, David AR, Otten U. Fimbria-fornix lesion increases nerve growth factor content in adult rat septum and hippocampus. Neurosci. Letts 1986; 70(1): 121–126.
20. Bakhit C, Armanini M, Bennett GL, Wong WL, Hansen SE, Taylor R. Increase in glia-derived nerve growth factor following destruction of hippocampal neurons. Brain Res. 1991; 560(1–2): 76–83.
21. Oderfeld-Nowak B, Bacia A. Expression of astroglial nerve growth factor in damaged brain. Acta Neurobiologiae Experimentalis 1994; 54(2): 73–80.
22. Lapchak PA, Araujo DM, Hefti F. BDNF and trkB mRNA expression in the rat hippocampus following entorhinal cortex lesions. NeuroReport 1993; 4: 191–194.
23. Merlio JP, Ernfors P, Kokaia Z, Middlemas DS, Bengzon J, Kokaia M et al. Increased production of trkB protein tyrosine kinase redceptor after brain insults. Neuron 1993; 10: 151–164.
24. Hefti F, Hartikka J, Knusel B. Function of neurotrophic factors in the adult and aging brain and their possible role in the treatment of neurodegenerative disease. Neurobiol. Aging 1989; 10: 515–533.
25. Cuello AC. Trophic factor therapy in the adult CNS: remodelling of injured basalo-cortical neurons. In: Bloom F, (editor), Neuroscience: from the molecular to the cognitive. Progress in brain research, vol. 100. Amsterdam: Elsevier, 1994: 213–221.
26. Bothwell, M. Keeping track of neurotrophin receptors. Cell 1991; 65: 915–918.
27. Chao MV. Growth factor signaling: Where is the specificity. Cell 1992; 68: 995–997.
28. Meakin SO, Shooter EM. The nerve growth factor family of receptors. Trends Neurosci. 1992; 15: 323–331.
29. Sofroniew MV, Pearson RCA, Eckenstein F, Cuello AC, Powell TPS. Retrograde changes in cholinergic neurons in the basal forebrain of rat following cortical damage. Brain Res. 1983; 289: 370–374.
30. Stephens PH, Cuello AC, Sofroniew MV, Pearson RCA, Tagari P. The effects of unilateral decortication upon choline acetyltransferase and glutamate decarboxylase activities in the nucleus basalis and other areas of the rat brain. J. Neurochem. 1985; 45: 1021–1026.
31. Bartus RT, Dean RL, Beer B, Lippa S. The cholinergic hypothesis of geriatric memory dysfunction. Science 1982; 217: 408–417.
32. Garofalo L, Cuello AC. Pharmacological characterization of nerve growth factor and/or monosialoganglioside GM1 effects on cholinergic markers in the adult lesioned brain. J. Pharma. & Exp. Therap. 1995; 272: 527–545.
33. Skup M, Figueiredo BC, Cuello AC. Intraventricular application of BDNF and NT-3 failed to protect nucleus basalis magnocellularis cholinergic neurones. NeuroReport 1994; 5: 1105–1109.
34. Knüsel B, Beck KD, Winslow JW, Rosenthal A, Burton LE, Widmer HR et al. Brain-derived neurotrophic factor administration protects basal forebrain cholinergic but not nigral dopaminergic neurons from degenerative changes after axotomy in the adult rat brain. J. Neurosci. 1992; 21(11): 4391–4402.
35. Morse JK, Wiegand SJ, Anderson K, You Y, Cai N, Carnahan J et al. Brain-derived neurotrophic factor (BDNF) prevents the degeneration of medial septal cholinergic neurons following fimbria transection. J. Neurosci. 1993; 13: 4146–4156.

36. Widmer HR, Knüsel B, Hefti F. BDNF protection of basal forebrain cholinergic neurons after axotomy: complete protection of p75NGFR-positive cells. NeuroReport 1993; 4: 363–366.
37. Dekker AJ, Fagan AM, Gage FH, Thal LJ. Effects of brain-derived neurotrophic factor and nerve growth factor on remaining neurons in the lesioned nucleus basalis magnocellularis. Brain Res. 1994; 639(1): 149–155.
38. Hagg T, Quon D, Higaki J, Varon S. Ciliary neurotrophic factor prevents neuronal degeneration and promotes low affinity NGF receptor expression in the adult CNS. Neuron 1992; 8: 145–158.
39. Figueiredo BC, Piccardo P, Maysinger D, Clarke PBS, Cuello AC. Effects of acidic fibroblast growth factor on cholinergic neurons of nucleus basablis magnocellularis and in a spatial memory task following cortical devascularization. Neuroscience 1993; 56(4): 955–963.
40. Anderson KJ, Dam D, Lee S, Cotman CW. Basic fibroblast growth factor prevents death of lesioned cholinergic neurons in vivo. Nature 1988; 332: 360–362.
41. Otto D, Frotscher M, Unsicker K. Basic fibroblast growth factor and nerve growth factor administered in gelfoam rescue medial septal neurons after fimbria fornix transection. J. Neurosci. Res. 1989; 22: 83–91.
42. Koliatsos VE, Nauta HJW, Clatterbuck RE, Holtzman DM, Mobley WC, Price DL. Mouse nerve growth factor prevents degeneration of axotomized basal forebrain cholinergic neurons in the monkey. J. Neurosci. 1990; 10: 3801–3813.
43. Koliatsos VE, Clatterbuck RE, Nauta HJW, Knüsel B, Burton LE, Hefti FF et al. Human nerve growth factor prevents degeneration of basal forebrain cholinergic neurons in primates. Ann. Neurol. 1991; 30: 831–840.
44. Tuszynski MH, Sang UH, Amaral DG, Gage FH. Nerve growth factor infusion in the primate brain reduces lesion-induced neural degeneration. J. Neurosci. 1990; 10: 3604–3614.
45. Tuszynski MH, Sang UH, Yoshida K, Gage FH. Recombinant human growth factor infusions prevent cholinergic neural degeneration in the adult primate brain. Ann. Neurol. 1991; 30: 625–636.
46. Liberini P, Pioro EP, Maysinger D, Ervin FR, Cuello AC. Long-term protective effect of human recombinant growth factor and monosialoganglioside GM1 treatment on primate nucleus basalis of Meynert. Neuroscience 1993; 53: 635–637.
47. Cuello AC, Garofalo L, Kenisberg RL, Maysinger D. Ganglioside potentiate in vivo and in vitro effects of nerve growth factor on central cholinergic neurons. Proc. Natl. Acad. Sci. USA 1989; 86: 2056–2060.
48. Maysinger D, Herrera-Marshitz MN, Goiny M, Ungerstedt U, Cuello AC. Effects of nerve growth factor on cortical and striatal acetylcholine and dopamine release in rats with cortical devascularizing lesions. Brain Res. 1992; 577: 300–305.
49. Lapchak PA, Hefti F. Effect of recombinant human nerve growth factor on presynaptic cholinergic function in rat hippocampal slices following partial septohippocampal lesions: measures of [^{3}H]acetylcholine synthesis, [^{3}H]acetylcholine release, and choline acetyltransferase activity. Neuroscience, 1991; 42: 639–649.
50. Garofalo L, Ribeiro-da-Silva A, Cuello AC. Nerve growth factor-induced synaptogenesis and hypertrophy of cortical cholinergic terminals. Proc. Natl. Acad. Sci. USA 1992; 89: 2639–2643.
51. Garofalo L, Ribeiro-da-Silva A, Cuello AC. Potentiation of nerve growth factor-induced alterations in cholinergic fibre length and presynaptic terminal size in cortex of lesioned rats by the monosialoganglioside GM1. Neuroscience, 1993; 57(1): 21–40.
52. Williams LR, Rylett RJ, Ingram DK, Joseph JA, Moises HC, Tang AH et al. Nerve growth factor affects the cholinergic neurochemistry and behavior of aged rats. In: Cuello AC (editor), Cholinergic function and dysfunction. Progress In Brain Research, vol 98. Amsterdam: Elsevier, 1993: 251–263.
53. Garofalo L, Cuello AC. Nerve growth factor and the monosialoganglioside GM1: Analogous and different in vivo effects on biochemical, morphological and behavioral parameters of adult cortically lesioned rats. Exptl. Neurol. 1994; 125: 195–217.
54. Terry RD, Masliah E, Salomon DP, Butters N, DeTeresa R, Hill R et al. Physical basis of cognitive alterations in Alzheimer's disease: synapse loss is the major correlate of cognitive impairment. Ann. Neurol. 1991; 30: 572–580.

55. Cuello AC, Toward the repair of cortical synapses in Alzheimer's disease. In: Giacobini E, Becker R (editors), Alzheimer disease: therapeutic strategies. Boston: Birkhauser. 1994: 277–283.
56. Breakfield XO, Geller AI. Gene transfer into the nervous system. Molecular Neurobiol. 1987; 1: 339–371.
57. Geller AI, During MG, Neve RL. Molecular analysis of neuronal physiology by gene tranfer into neurons with herpes simplex virus vectors. Trends Neurosci. 1991; 14: 428–432.
58. Geller AI. Herpes viruses: Expression of genes in postmitotic brain cells. Curr. Opin. Genet. Dev. 1993; 3: 81–85.
59. Akli S, Cailland C, Vigne E, Stratford-Perricaudel LD, Poenaru L, Perricaudet M et al. Transfer of a foreign gene into the brain using adenovirus vectors. Nature Genet. 1993; 3(3): 224–228.
60. Gage FH, Kawaja D, Fisher LJ. Genetically modified cells: applications for intracerebral grafting. Trends Neurosci. 1991; 14: 328–333.
61. Suhr ST, Gage FH. Gene therapy for neurologic disease. Arch. Neurol. 1993; 50: 1252–1268.
62. Kawaja MD, Gage FH. Morphological and neurochemical features of cultured primary skin fibroblasts of Fischer 344 rats following striatal implantation. J. Comp. Neurol. 1992; 317(1): 102–116.
63. Lucidi-Phillipi CA, Gage FH, Shults CW, Jones KR, Reichardt LF, Kang UJ. Brain-derived neurotrophic factor-transduced fibroblasts: Production of BDNF and effects of grafting to the adult rat brain. J. Comp. Neurol. 1995; 351: 1–16.
64. Rosenberg MB, Friedmann T, Robertson RC, Tuszynsky M, Wolfe JA, Breakfield XO et al. Grafting genetically modified cells to the damaged brain: restorative effects of NGF expression. Science 1988; 242: 1575–1578.
65. Kawaja MD, Rosenberg MB, Yoshida K, Gage FH. Somatic gene tranfer of nerve growth factor promotes the survival of axotomized septal neurons and the regeneration of their axons in adult rats. J. Neurosci. 1992; 12(7): 2849–2864.
66. Eagle K, Chalmers G, Clary T, Gage FH. Morphological regeneration and limited functional recovery in the damaged rat septo hippocampal system. J. Comp. Neur. In press.
67. Olson L, Nordberg A, von Holst H, Bäckmann L, Ebendal T, Alafuzoff I et al. Nerve growth factor affects ^{11}C-nicotine binding, blood flow, EEG, and verbal episodic memory in an Alzheimer patient. J. Neural Transm. 1992; 4: 79–95.
68. Strömberg I, Herrera-Marschitz M, Ungerstedt U, Ebendal T, Olson L. Chronic implants of chromaffin tissue into the dopamine-denervated striatum: effects of NGF on graft survival, fiber growth, and rotational behavior. Exp. Brain Res. 1985; 60: 335–349.
69. Strömberg I, Hultgard-Nilsson A, Hedin U, Ebendal T. Fate of intraocular chromaffin cell suspensions: role of initial nerve growth factor support. Cell Tissue Res. 1988; 254: 487–497.
70. Strömberg I, Ebendal T. Aged adrenal medullary tissue survives intraocular grafting, forms nerve fibers, and responds to NGF. J. Neurosci. Res. 1989; 23: 162–171.
71. Olson L, Backlund EO, Ebendal T, Freedman R, Hamberger B, Hansson P et al. Intraputaminal infusion of nerve growth factor to support adrenal meduallary autografts in Parkinson's disease. Arch. Neurol. 1991; 48: 373–381.
72. Williiams LR. Hypophagia is induced by intracerebroventricular administration of nerve growth factor. Exp. Neurol. 1991; 113: 31–37.
73. Aquilonius SM, Carlson H, Ebendal T, Gustavsson L, Hartvig P, Lilja A et al. Intracerebroventricular NGF-infusion in Huntington's disease (HD). In: Abstracts of the Third International Congress of Movement Disorders; 1994 Nov 7–11; Orlando (FL), New York (NY): Raven Press, 1994.

Pharmacological Sciences: Perspectives for
Research and Therapy in the Late 1990s
ed. by A.C. Cuello and B. Collier
© 1995 Birkhäuser Verlag Basel/Switzerland

New Perspectives in the Pharmacology of Parenchimal Brain Anoxia-Ischemia

Domenico E. Pellegrini-Giampietro and Flavio Moroni

*Dipartimento di Farmacologia Preclinica e Clinica "Mario Aiazzi Mancini",
Università degli Studi di Firenze, 50134 Firenze, Italy*

Summary. A series of events is thought to be involved in the process linking brain ischemia to neuronal degeneration, and has been shown to be susceptible to modulation by pharmacological agents. These pathogenic events include glutamate release and glutamate receptor activation, loss of Ca^{2+} homeostasis, free radical formation, and proton accumulation (acidosis): some of them may interact with each other in a sort of vicious cycle terminating in neuronal death. In animal models, various types of molecules with different mechanisms of action have been shown to reduce the volume of an infarct following middle cerebral artery occlusion or prevent the death of vulnerable cells following transient global ischemia. This implies that of all the neurodegenerative mechanisms that are active in ischemic areas, many can be effectively restricted by drug intervention. In this chapter we discuss the experimental use of glutamate receptor antagonists and that of compounds, like 7-Cl-thio-kynurenic acid, possessing a dual mechanism of action (glutamate receptor antagonist and free radical scavenger). Molecules interfering with different pathogenetic events in ischemic tissues may be regarded as interesting prototypes of new and effective anti-ischemic drugs.

Introduction

Neurons have a high rate of oxygen metabolism but a relatively small reserve of high energy phosphates, which results in a rapid depletion of cellular energy when the delivery of oxygen and glucose to the CNS is reduced. The lack of energy phosphates, occurring in a matter of minutes after the interruption of blood supply, has been repeatedly proposed as the main cause of neuronal death [1, 2]. Such a concept may probably explain why, until a few years ago, no therapeutic approaches, other than a rapid restoration of blood flow, had been proposed to reduce ischemic or hypoxic brain damage [3]. However, in the last 10 years it has been clearly shown that a reduction in high energy store levels may not be sufficient to cause neuronal death and that neurons appear to be surprisingly resistant to oxygen and glucose deprivation, provided that their synaptic activity is depressed [4, 5]. This idea suggests that it might be possible to rescue neurons with pharmacological agents following deprivation of blood supply to the whole brain or part of it.

Correspondence to: Prof. Flavio Moroni at the above address.

In order to test this possibility, a number of animal models of cerebral ischemia have been developed, usually subdivided into two categories: models of focal and models of global ischemia (for review see 6). A typical model of focal ischemia is the surgical occlusion (with or without reperfusion) of the middle cerebral artery in rats or mice. Agents capable of reducing the volume of the resulting cortico-striatal infarction have been proposed as possible drugs for stroke. On the other hand, typical models of global ischemia include the transient occlusion of both common carotids in the gerbil (a rodent species lacking a complete circle of Willis) or the occlusion of both vertebral and both carotid arteries (four vessel occlusion) in rats. Global ischemia models induce selective and delayed damage of specific populations of neurons and are widely used to study the pathophysiology and possible treatment of brain damage associated with cardiac arrest, CO poisoning or massive hemorrhages. Other models of ischemic damage to the nervous tissue have been developed by using the photochemical reaction triggered by the fluorescent dye rose bengal when exposed to intense illumination. The reaction results in formation of free radicals, which leads to endothelial damage, platelet aggregation, and, eventually, thrombotic occlusion of vessels in the illuminated area. This strategy has been utilized for developing models of stroke [7] and of retinal ischemia [8, 9].

Pathogenesis of Ischemia-Induced Neuronal Damage

A series of events, mostly studied in neuronal primary cultures, are thought to be involved in the process linking brain ischemia to neuronal degeneration; some of them have been shown to be susceptible to modulation by pharmacological agents. These pathogenic mechanisms include glutamate release and glutamate receptor activation, loss of Ca^{2+} homeostasis, free radical formation, and proton accumulation (acidosis). The relative importance of each of these events is probably different in various pathological conditions having in common a reduced supply of oxygen and glucose to the brain. For example, in stroke models the extent of glutamate release varies in the different regions of the infarct, being maximal in the core and decreasing towards the periphery of the penumbral area. Similarly, the formation of free radicals is known to be particularly prominent during the early reperfusion phase following transient global ischemia. We have proposed [10] that the above-mentioned pathogenic events may interact with each other in a sort of vicious cycle terminating in neuronal death (Fig. 1). The cycle may be helpful in understanding the common final action of many of the compounds that have been shown to be effective in reducing the magnitude of ischemic neuronal damage (Table 1).

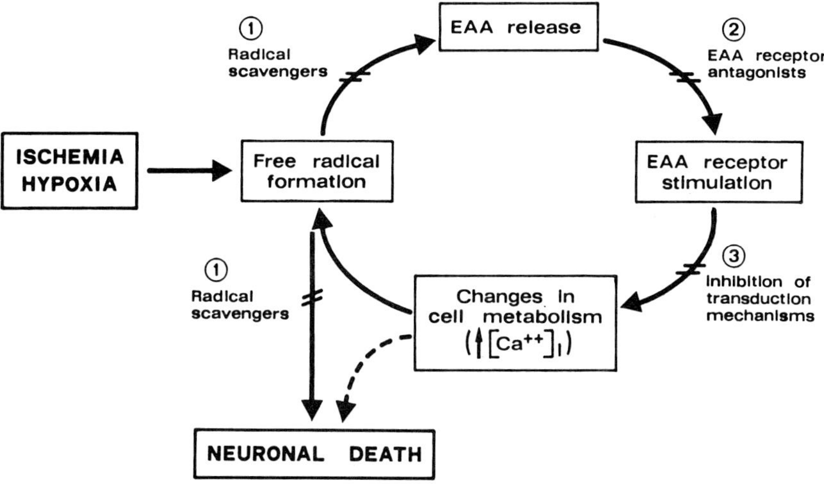

Fig. 1. Hypothetical vicious cycle illustrating the main biochemical events leading to hypoxic-ischemic damage. Free radicals, formed during or following brain ischemia, promote the release of excitatory amino acids, which in turn may exert their toxic effects by interacting with specific receptors. Stimulation of these receptors increases cytosolic free Ca^{2+}, resulting in both neuronal damage and further production of free radicals. Adapted from ref. [10].

Release of Glutamate

An increased release of glutamate has been observed when brain tissue is exposed to hypoxic/ischemic conditions both *in vitro* and *in vivo* [10, 11]; it cannot be detected in the absence of mature presynaptic terminals [12] and may occur via both Ca^{2+}-dependent and Ca^{2+}-independent mechanisms (for review see [13]). Increased glutamate release during and immediately after an ischemic episode is believed to be Ca^{2+}-independent and non-exocytotic, due to reversed operation of the cytoplasmic glutamate uptake carrier. The increase in $[Na^+]_i$ and $[K^+]_o$ that occurs during ischemia, together with anoxic membrane depolarization, tend to make the carrier run in reverse, in such a way that glutamate and Na^+ are pumped out of the cells into the extracellular space until a new equilibrium is reached at a neurotoxic level ([glutamate]$_o$ > 100 μM). This mechanism can be enhanced experimentally by blocking the Na^+/Ca^{2+} exchanger, the only other possible efflux pathway for intracellular Na^+ during ischemia [14]. Conversely, activation of the Na^+/Ca^{2+} exchanger in anoxic-glucopenic conditions, although leading to an increase in $[Ca^{2+}]_i$, may be neuroprotective because it reduces the outflow of Na^+ and glutamate through the carrier [15]. Other mechanisms, such as a reduced uptake of glutamate into presynaptic terminals or glial cells, may also contribute to the increase in extracellular glutamate during ischemia. Along this line, oxygen free

Table 1. Drugs that have proven to be neuroprotective in models of focal or global cerebral ischemia

Mechanism of action	Focal	Global
NMDA receptor		
channel blockers	MK-801	
	dextrorphan	
competitive antagonists	CGP 37849	
non competitive antagonists	ifenprodil	7-Cl-KYNA
	HA-966	HA-966
	7-Cl-thio-KYNA	7-Cl-thio-KYNA
oligonucleotides	anti-NMDAR1	
AMPA/kainate receptor		
competitive antagonists	NBQX	NBQX
non competitive antagonists	GYKI 52466	GYKI 52466
Free radical scavengers	SOD	oxypurinol
	tirilazad	tirilazad
	α-PBN	α-PBN
		dimethylthiourea
		LY231617
Ion channel drugs		
Ca^{2+} channel blockers	nimodipine	SNX-111
	SB 201823-A	SB 201823-A
Na^+ channel inhibitors	BW619C89	
	BW1003C87	
K^+ channel openers		cromakalim
Hypothermic drugs		U-80816E
Gangliosides	GM1	
	LIGA20	
Combination therapies	7-Cl-thio-KYNA	7-Cl-thio-KYNA
	MK-801 + NBQX	MK-801 + NBQX
		MK801 + nimodipine

radicals (known to be formed in ischemic areas, see below) have been shown to reduce the capacity of uptake systems to take up and remove glutamate from the extracellular space [16].

Loss of Calcium Homeostasis and Free Radical Formation

A delayed increase in cytosolic free Ca^{2+} following activation of glutamate receptors has been repeatedly demonstrated to be one of the most important mechanisms responsible for post-ischemic neuronal degeneration [17]. Glutamate may promote an increase in cytosolic free Ca^{2+} by at least three distinct mechanisms: *(i)* activation of Ca^{2+}-permeable NMDA receptors (but also non-NMDA receptors, see below); *(ii)* opening of voltage-dependent Ca^{2+} channels (indirectly, via membrane depolarization); or *(iii)* stimulation of metabotropic receptor-mediated events (resulting in release of Ca^{2+} from intracellular stores). Elevated cytosolic Ca^{2+} then activates several enzymes (including protein kinase C, calpain I, phospholipase A2, xanthine-oxidase, NO synthase) capable of either directly or indirectly destroying cellular components.

A number of Ca^{2+}-activated enzymes exert their toxic effects through the formation of reactive oxygen radical species, such as the superoxide ion or the hydroxyl radical. The direct measurement of free radicals in brain tissue following ischemia-reperfusion has been achieved by using magnetic resonance techniques and spin trap agents. Moreover, a number of studies have shown that free radical scavengers and/or lipid peroxidation inhibitors reduce brain damage following focal and global ischemia. Interestingly, one of these agents (the spin trap compound α-phenyl-tert-butyl-nitrone, α-PBN) appears to be neuroprotective in focal ischemia even when administered several hours (up to 12) after the occlusion [18], whereas its efficacy is limited to administration within 30 min following reperfusion in global ischemia [19]. In addition, the degree of cortical infarction induced by cerebral focal ischemia is reduced in transgenic mice overexpressing the CuZn-superoxide-dismutase-1 (CuZn-SOD-1) gene [20] or deficient in neuronal NO synthase [21].

Free radicals have been repeatedly shown to be produced upon glutamate receptor activation [22, 23]. Since oxygen radicals promote the release of glutamate [10] and inhibit its re-uptake [16], it appears that glutamate may also accumulate in the extracellular space as a consequence of free radical production, thus promoting a vicious cycle that could be responsible for the generation and propagation of neuronal death (Fig. 1). As mentioned, formation of free radicals in the vascular compartment triggered by rose bengal illumination has been proposed as a model of thrombotic ischemia in cortex and retina. It is interesting to note that glutamate receptor antagonists of both the NMDA and non-NMDA type significantly reduce photothrombotic neuronal damage in the retinal model [24].

Acidosis

Metabolic acidosis generated by anaerobic glycolysis in oxygen-deprived cerebral tissue has long been regarded as an important factor in causing ischemic brain damage [25]. Consistent with the acidosis hypothesis, pre-ischemic hyperglycemia and incomplete ischemia (presumably because of trickling glucose that fuels glycolysis) have both been shown to aggravate the post-ischemic pathological outcome. However, a link between severe acidosis and brain injury has been conclusively demonstrated only for cellular pannecrosis as seen in the core region of a focal infarct. The relationship between acidosis and the selective and delayed neuronal damage that occurs in the perifocal penumbra or following transient global ischemia is less clear.

Surprisingly, there is new evidence that mild acidosis, as opposed to severe acidosis, might play a neuroprotective role in ischemic brain damage. Several biochemical mechanisms, such as a reduced energy demand via depressed cell excitability, have been proposed to explain why mild acidosis stabilizes cell metabolism during ischemia (for review see [26]). The idea received decisive support when it was demonstrated that the probability of NMDA receptor-gated ion channel opening is decreased by protons [27]. Two *in vitro* studies then reported that exposure to acidified (pH 6.5–6.6) media protected against neuronal damage induced by ischemic-like conditions [28, 29]. More recently, Simon and co-workers have shown that brain acidosis induced by hypercarbic ventilation is beneficial in a model of focal ischemia *in vivo* and that maximal protection is observed at pH 6.8 [30]. These observations could have potential clinical applicability, especially in view of the fact that in moderate focal ischemia the minimal pH is 6.8 and is reached only after 60 min.

In contrast, when acidosis becomes severe it is no longer neuroprotective. In the same study by Simon et al. [30], it was observed that the protective effect of acidosis was lost at pH 6.5, presumably due to the effect of acidosis on glial glutamate uptake. In addition, pre-ischemic hypercapnia, which reduces brain extracellular pH to the low values normally encountered in hyperglycemia, dramatically enhances neuronal damage following global ischemia [31]. Among the mechanisms that have been proposed for the detrimental effects of severe acidosis on ischemic tissue are: facilitation of iron-catalyzed formation of free radicals, non-selective denaturation of proteins and nucleic acids, inhibition of mitochondrial energy metabolism, and promotion of Ca^{2+} release from intracellular stores.

Pharmacological Treatment of Cerebral Ischemia

Various types of molecules, with different mechanisms of action, have been shown to reduce the volume of an infarct following middle cerebral artery occlusion or prevent the death of vulnerable cells following transient global ischemia (Table 1). This implies that of all the neurodegenerative mechanisms that are active in ischemic areas, many can be effectively restricted by drug intervention. Parenchimal approaches should include simultaneous therapy with agents aimed at restoring cerebral blood flow post-ischemia (such as tissue plasminogen activator or urokinase): improved drug access may result, increasing the volume of ischemic tissue that responds to pharmacological treatment.

Glutamate receptor antagonists under selected conditions protect against cerebral ischemia *in vivo*. In models of focal ischemia, both NMDA receptor blockers, like MK-801 [32], and the AMPA/kainate receptor antagonist NBQX [33] have proven to be protective against neocortical damage. Whereas NBQX is also effective in preventing delayed CA1 cell death induced by transient global ischemia [34], the issue of protection afforded by NMDA receptor antagonists in global ischemia has been quite more controversial in the past few years. However, it is now quite clear that drugs like MK-801 may still be protective if the ischemic insult is incomplete or moderate [35], but they are not effective, or their effect is mediated by hypothermia, in models of severe forebrain ischemia [36, 37].

There are a number of data that could explain why AMPA/kainate receptors predominate in the injury process when the reduction in blood flow becomes severe, such as in rat global ischemia models. Firstly, it has been proposed that, 24 h following global ischemia in rats, the AMPA/kainate receptor channel might become permeable to Ca^{2+} ions because of a switch in subunit composition [38], and thus might mediate excitotoxic cell death. In addition, low pH inhibits currents flowing through NMDA receptors (see above): in severe global ischemia, a pH of 6.5 is reached within 1 min, probably precluding NMDA receptor activation. However, it must be noted that NMDA antagonists can still afford neuroprotection at pH 6.4 in an *in vitro* ischemic model [39]. Novel approaches, still dealing with the NMDA receptor, may include use of anti-sense oligonucleotides directed against the mRNA coding for the key subunit NMDAR1 [40] or of drugs acting at the glycine modulatory site. Among these, kynurenate (KYNA) has been shown to be neuroprotective in gerbils [35], while 7-Cl-KYNA is beneficial even in severe models [41].

An important concept emerging from the proposed cycle (Fig. 1) is that it may be possible to prevent cell death following ischemic injury to the brain by interfering with one or more of the following steps: *(i)* free radical formation and reactivity, *(ii)* EAA release, *(iii)* glutamate receptor stimulation, *(iv)* activation of transduction pathways leading to a rise in cytosolic free Ca^{2+}. For example, Oh and Betz [42] have shown that pretreatment with the radical scavenger dimethylthiourea and the NMDA receptor blocker MK-801 (alone or in combination) can reduce brain edema during the early stages of cerebral ischemia in rats. Some investigators have suggested that dual therapy of MK-801 plus a Ca^{2+} channel blocker may be helpful in cases where efficacy of MK-801 alone is limited [43].

We have tested the protective effect of a compound, 7-Cl-thio-kynurenic acid (7-Cl-thioKYNA), which is a potent antagonist at the glycine site of the NMDA receptor and, in addition, a free radical scavenger [44]. As such, 7-Cl-thioKYNA proved to be a more effective inhibitor of glutamate toxicity *in vitro* than 7-Cl-KYNA, which is equally potent as a glycine antagonist but fails to inhibit lipid peroxidation [44]. In a permanent middle cerebral artery occlusion stroke model in the rat,

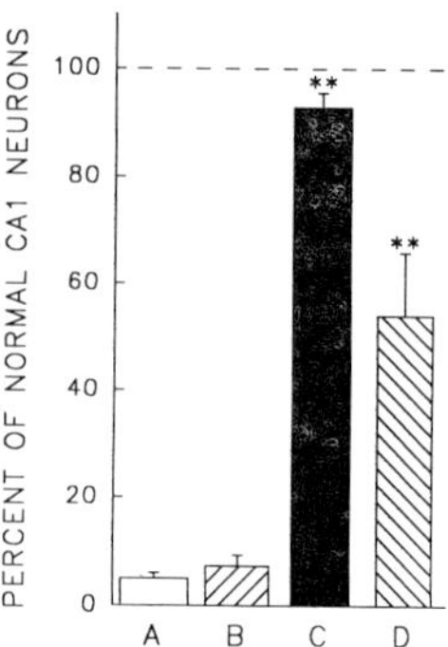

Fig. 2. CA1 neuronal damage in ischemic gerbils. **A:** Saline-treated ischemic gerbils. **B:** Saline-treated ischemic gerbils, in which temperature was experimentally decreased starting at 60 min after reperfusion. **C:** Ischemic gerbils treated with 7-Cl-thioKYNA (100 mg/kg × 5, i.p.). **D:** 7-Cl-ThioKYNA ischemic gerbils, in which temperature was maintained at 37°C until 360 min post-ischemia. $^{**}p < 0.01$ vs. A, ANOVA followed by Dunnett's test.

7-Cl-thioKYNA was effective and 5-Cl-thioKYNA (a good lipid peroxidation inhibitor but a poor glycine antagonist) was inactive in attenuating infarct volume [45]. Figure 2 shows that in gerbils subjected to 5 min of global ischemia, 7-Cl-thio-KYNA dramatically attenuated ischemia-induced CA1 cell loss; the protection was associated with a delayed and marked reduction in the animals' temperature. However, when the gerbils were maintained normothermic for at least 360 min, 7-Cl-thio-KYNA still provided partial but significant protection, indicating that the latter could not be ascribed to hypothermia alone. Moreover, no protection was observed when a reduction in temperature with a time-course similar to that caused by 7-Cl-thio-KYNA was experimentally induced in saline-treated ischemic animals (Fig. 2). Thus, it appears that molecules with dual mechanism of action, interfering with different pathogenetic events in ischemic tissues, may be regarded as interesting prototypes of new and effective anti-ischemic drugs.

Acknowledgements

This work was supported in part by the EEC Biomed 1 project number BMH1-CT93-1033. The technical assistance of Dr. A. Cozzi is gratefully acknowledged.

References

1. Vogt C, Vogt O. Sitz und Wesen der Krankheiten im Lichte der topischen Hirnforschung und der Variierens der Tiere. J. Psychol. Neurol. 1937; 47: 237–257.
2. Siesjö BK. Brain energy metabolism. London: John Wiley & Sons, 1978.

3. Grotta JC. Current medical and surgical therapy for cerebrovascular disease. New Engl. J. Med. 1987; 317: 1505–1516.
4. Rothman SM, Olncy JW. Glutamate and the patho-physiology of hypoxic-ischemic brain damage. Ann. Neurol. 1986; 19: 105–111.
5. Meldrum B, Garthwaite J. Excitatory amino acid neurotoxicity and neurodegenerative disease. Trends Pharmacol. Sci. 1990; 11: 379–387.
6. Ginsberg MD, Busto R. Rodent models of cerebral ischemia. Stroke 1989; 20: 1627–1642.
7. Watson BD, Dietrich WD, Busto R, Wachtel MS, Ginsberg MD. Induction of reproducible brain infarction by photochemically initiated thrombosis. Ann. Neurol. 1985; 17: 497–504.
8. Mosinger SL, Olney JW. Photothrombosis-induced ischemic neuronal degeneration in the rat retina. Exp. Neurol. 1989; 105: 110–113.
9. Moroni F, Lombardi G, Pellegrini-Faussone S, Moroni F. Photochemically-induced lesion of the rat retina: a quantitative model for the evaluation of ischemia-induced retinal damage. Vision Res. 1993; 33: 1887–1891.
10. Pellegrini-Giampietro DE, Cherici G, Alesiani M, Carlà V, Moroni F. Excitatory amino acid release and free radical formation may cooperate in the genesis of ischemia-induced neuronal damage. J. Neurosci. 1990; 10: 1035–1041.
11. Benveniste H, Drejer J, Schousboe A, Diemer N. Elevation of extracellular concentrations of glutamate and aspartate in rat hippocampus during transient cerebral ischemia monitored by intracerebral microdialysis. J. Neurochem. 1984; 43: 1369–1374.
12. Cherici G, Alesiani M, Pellegrini-Giampietro DE, Moroni F. Ischemia does not induce the release of excitotoxic amino acids from the hippocampus of newborn rats. Dev. Brain Res. 1991; 60: 235–240.
13. Szatkowski M, Attwell D. Triggering and execution of neuronal death in brain ischemia: two phases of glutamate release by different mechanisms. Trends neurosci. 1994; 17: 359–365.
14. Amoroso S, Sensi S, Di Renzo G, Annunziato L. Inhibition of the Na^+-Ca^{++} exchanger enhances anoxia and glucopenia-induced [^{3}H]aspartate release in hippocampal slices. J. Pharmacol. Exp. Ther. 1993; 264: 515–520.
15. Amoroso S, Iannotti E, Caso N, Russo G, Bassi A, Di Renzo GF, et al. Role of the Na^+-Ca^{2+} exchanger during anoxic and glucopenic conditions in C6 glioma cells. Soc. Neurosci. Abstr. 1994; 20: 1063.
16. Volterra A, Trotti D, Tromba C, Floridi S, Racagni G. Glutamate uptake inhibition by oxygen free radicals in rat cortical astrocytes. J. Neurosci. 1994; 14: 2924–2932.
17. Siesjö BK, Bengtsson F. Calcium fluxes, calcium antagonists and calcium-mediated pathology in brain ischemia, hypoglycemia and spreading depression. A unifying hypothesis. J. Cereb. Blood Flow Metab. 1989; 9: 127–140.
18. Cao X, Phillis JW. α-Phenyl-tert-butyl-nitrone reduces cortical infarct and edema in rats subjected to focal ischemia. Brain Res. 1994; 644: 267–272.
19. Phillis JW, Clough-Helfman C. Protection from cerebral ischemic injury in gerbils with the spin trap agent N-tert-butyl-α-phenylnitrone (PBN). Neurosci. Lett. 1990; 116: 315–319.
20. Kinouchi H, Epstein CJ, Mizui T, Carlson E, Chen SF, Chan PH. Attenuation of focal cerebral ischemic injury in transgenic mice overexpressing CuZn superoxide dismutase. Proc. Natl. Acad. Sci. USA 1991; 88: 11158–11162.
21. Huang Z, Huang PL, Panahian N, Dalkara T, Fishman MC, Moskowitz MA. Effects of cerebral ischemia in mice deficient in neuronal nitric oxid synthase. Science 1994; 265: 1883–1885.
22. Dykens JA, Stern A, Trenkner E. Mechanism of kainate toxicity to cerebellar neurons in vitro is analogous to reperfusion tissue injury. J. Neurochem. 1987; 49: 1222–1228.
23. Lafon-Cazal M, Pietri S, Culcasi M, Bockaert J. NMDA-dependent superoxide production and neurotoxicity. Nature 1993; 364: 535–537.
24. Lombardi G, Moroni F, Moroni F. Glutamate receptor antagonists protect against ischemia-induced retinal damage. Europ. J. Pharmacol. 1994; 271: 489–495.
25. Siesjö BK. Acidosis and ischemic brain damage. Neurochem. Pathol. 1988; 9: 31–88.
26. Tombaugh GC, Sapolsky RM. Evolving concepts about the role of acidosis in ischemic neuropathology. J. Neurochem. 1993; 61: 793–803.
27. Traynelis S, Cull-Candy S. Proton inhibition of N-methyl-D-aspartate receptors in cerebellar neurons. Nature 1990; 345: 347–350.

28. Giffard RG, Monyer H, Christine CW, Choi DW. Acidosis reduces NMDA receptor activation, glutamate neurotoxicity, and oxygen-glucose deprivation neuronal injury in cortical cultures. Brain Res. 1990; 506: 339–342.
29. Tombaugh GC, Sapolsky RM. Mild acidosis protects hippocampal neurons from injury induced by oxygen and glucose deprivation. Brain Res. 1990; 506: 343–345.
30. Simon RP, Niiro M, Gwinn R. Brain acidosis induced by hypercarbic ventilation attenuates focal ischemic injury. J. Pharmacol. Exp. Ther. 1993; 267: 1428–1431.
31. Katsura K, Kristian T, Smith ML, Siesjö BK. Acidosis induced by hypercapnia exaggerates ischemic brain damage. J. Cereb. Blood Flow Metab. 1994; 14: 243–250.
32. Park CK, Nehls DG, Graham DI, Teasdale GM, McCulloch J. The glutamate antagonist MK-801 reduces focal ischemic brain damage in the rat. Ann. Neurol. 1988; 24: 543–551.
33. Buchan AM, Xue D, Huang ZG, Smith KH, Lesiuk H. Delayed AMPA receptor blockade reduces cerebral infarction induced by focal ischemia. Neuroreport 1991; 2: 473–476.
34. Sheardown MJ, Nielsen EO, Hansen AJ, Jacobsen P, Honoré T. 2,3-Dihydroxy-6-nitro-7-sulfamoyl-benzo(F)quinoxaline: a neuroprotectant for cerebral ischemia. Science 1990; 247: 571–574.
35. Gill R, Woodruff GN. The neuroprotective actions of kynurenic acid and MK-801 in gerbils are synergistic and not related to hypothermia. Europ. J. Pharmacol. 1990; 176: 143–149.
36. Buchan A, Pulsinelli WA. Hypothermia but not the N-methyl-D-aspartete antagonist, MK-801, attenuates neuronal damage in gerbils subjected to transient global ischemia. J. Neurosci. 1990; 10: 311–316.
37. Nellgard B, Wieloch T. Post ischemic blockade of AMPA but not NMDA receptors mitigates neuronal damage in the rat brain following transient severe forebrain ischemia. J. Cereb. Blood Flow Metab. 1992; 12: 1–11.
38. Pellegrini-Giampietro DE, Zukin RS, Bennett MVL, Cho S, Pulsinelli WA. Switch in glutamate receptor subunit gene expression in CA1 subfield of hippocampus following global ischemia in rats. Proc. Natl. Acad. Sci. USA 1992; 89: 10499–10503.
39. Kaku DA, Giffard RG, Choi DW. Neuroprotective effects of glutamate antagonists and extracellular acidity. Science 1993; 260: 1516–1518.
40. Wahlestedt C, Golanov E, Yamamoto S, Yee F, Ericson H, Yoo H, et al. Antisense oligodeoxynucleotides to NMDA-R1 receptor channel protect cortical neurons from excitotoxicity and reduce focal ischemic infarctions. Nature 1993; 363: 260–263.
41. Wood ER, Bussey TJ, Phillips AG. A glycine antagonist reduces ischemia-induced CA1 cell loss in vivo. Neurosci. Lett. 1992; 145: 10–14.
42. Oh SM, Betz AL. Interaction between free radicals and excitatory amino acids in the formation of ischemic brain edema in rats. Stroke 1991; 22: 915–921.
43. Hewitt K, Corbett D. Combined treatment with MK-801 and nicardipine reduces global ischemic damage in the gerbil. Stroke 1992; 23: 82–86.
44. Moroni F, Alesiani M, Facci L, Fadda E, Skaper SD, Galli A, et al. Thiokynurenates prevent excitotoxic neuronal death in vitro and in vivo by acting as glycine antagonists and as inhibitors of lipid peroxidation. Europ. J. Pharmacol. 1992; 218: 145–151.
45. Chen J, Graham S, Moroni F, Simon R. A study of the dose dependency of a glycine receptor antagonist in focal ischemia. J. Pharmacol. Exp. Ther. 1993; 267: 937–941.

Pathophysiology and Future Pharmacotherapy of Chronic Pain

Andy Dray

Sandoz Institute for Medical Research, London, WC1E 6BN, UK

Pain and its Treatment Today

Pain produced by a mild, transient stimulus and associated with negligible tissue damage serves as a physiological warning. The pain associated with the pathophysiological processes of inflammation is more persistent, differs in quality, but can also be considered as a normal protective response to tissue injury which can resolve rapidly once the injury has healed, e.g. post-operative pain, toothache, cystitis. At the other extreme, chronic pain associated with inflammation, tissue damage or nerve lesions is long lasting, and may involve a chronic pathological lesion or degenerative process, but sometimes there may be no discernible pathology. These types of pain syndrome (e.g. rheumatoid arthritis, osteoarthritis, low back pain, pelvic and abdominal pain, cancer and neuropathic pain, migraine) are not well understood and are more difficult to treat. Indeed chronic pain suffers account for 10–20% of the adult population where approximately 5% experience pain which is poorly treated and is debilitating (loss of work, family crisis, depression, suicide) [1, 2]. The economic and medical costs to the community are proportional and obviously enormous.

Presently, treatments for resolvable pain have successfully relied on the use of NSAIDs (which combine anti-inflammatory and analgesic activity) and opioids. These drugs are less reliable in chronic pain treatment and their role in chronic inflammation is less secure as for example immune suppressants may neutralise the underlying mediation of pain. However both classes of drugs have serious drawbacks. NSAIDs produce GI disturbances, ulceration, renal damage and hypersensitivity reactions, while opioids induce nausea, constipation, confusion, respiratory depression and possible dependence. These problems have not been overcome despite considerable effort from the pharmaceutical industry, which has also failed to introduce any new class of analgesic. It is a pity that the failings of conventional drugs and the poor treatment of chronic pain sufferers has not sufficiently stimulated the medical profession or government bodies to demand and resource the delivery of more effective and safer pain therapies.

Over the last decade there has been a significant conceptual shift in our ideas of the pathophysiology of pain [2]. First, the belief that pain signals are aways transmitted along fixed communication lines has been replaced by the realization that pain signalling is a series of dynamic, ever changing events. Second, the idea that chronic pain is a natural, endurable symptom of an underlying pathology is being modified by the realization that unrelieved pain may induce its own organic or personality damage. The debate about these issues has followed the improved understanding of chronic pain mechanisms. This has stimulated a number of novel therapeutic approaches which are now being actively persued.

Mechanistic Studies: Peripheral and Central Signalling

Under normal physiological conditions, nociceptive signals are generated by intense thermal or mechanical stimuli and by irritant chemical which activate specialized C and $A\delta$ nerve fibre nociceptors. Action potentials, conducted to the spinal cord, are integrated and transmitted to the thalamus and cerebral cortex where further processing occurs, resulting in "pain awareness". In chronic pain states the superimposition of other processes changes the normal relationship between stimulus and response so that the biological usefulness of pain, as a warning signal, may be lost.

Disease, inflammation and injury to peripheral nerves and soft tissues induces changes in nociceptive pathways ranging from heightened excitability of sensory nerves to alterations in the cellular phenotype with the expression of new molecules including neurotransmitters, enzymes and receptors [3]. In addition, alterations of CNS neurochemistry produces an enhancement and prolongation of low levels of afferent input and allows normally innocuous stimuli to be perceived as painful (Fig. 1). Such modifications explain primary hyperalgesia, a heightened responsiveness at the site of injury, while secondary hyperalgesia, perceived as tenderness in the undamaged area around an injury is due to changes in the central processing of sensory signals [4, 5]. Structural changes also occur in central nociceptive pathways after peripheral nerve injury, including loss of spinal interneurones, inappropriate rearrangements of sensory nerve processes in the spinal cord and proliferation of sympathetic fibres into sensory ganglia which are not normally innervated to any significant degree [6]. It is also known that chronic pain results from affective disorders, infections or damage to the CNS. These processes are as yet poorly understood. Finally, chronic pain perception and behaviour has the overlay of complex human reactions which can be modified by mood, the environmental or sociological setting.

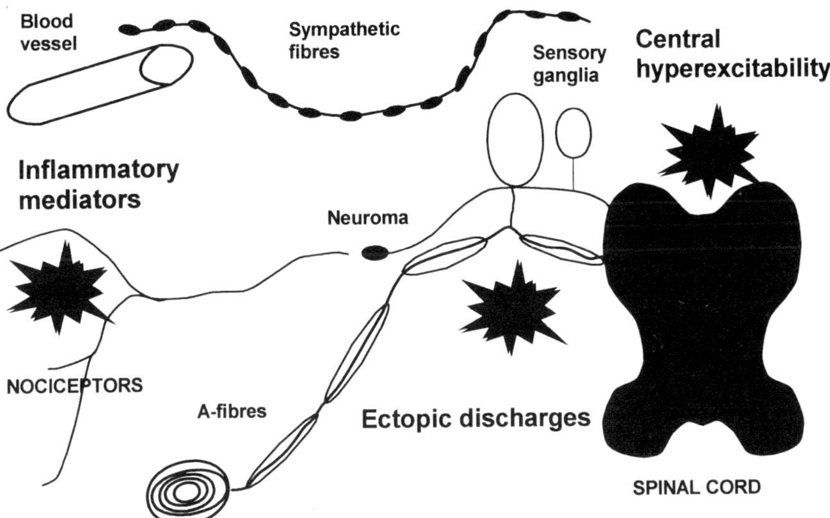

Fig. 1. Major targets for new analgesics are shown. In the periphery, strategies include inhibition of nociceptor activation by inflammatory mediators (released from damaged tissues, immune cells, blood vessels and sympathetic neurones) and inactivation of ectopic discharges in neuroma tissue and sensory ganglia (large and small fibres) after peripheral nerve damage. In the spinal cord, central hyperexcitability can be reduced by blocking sensory transmitters (with antagonists, enzyme inhibitors) and by enhancing central inhibitory mechanisms.

Targets of New Analgesic

Diverse chemicals produced during tissue damage and inflammation (Fig. 1) alter the excitability of nociceptors [7]. Nociceptor activation also induces an axon reflex and the release of substance P, neurokinin A (NKA) and CGRP. These sensory neuropeptides indirectly influence the excitability of sensory nerves and nearby postganglionic sympathetic fibres as they alter local blood flow, induce the release of other active substances by plasma extravasation or stimulate immune cells (Fig. 1). In addition, the secretion of a number of growth regulators, e.g. NGF, normally released by target tissues, may be altered [7]. Most mediators act in an organised fashion at receptors, concentrated at periphery and central sensory nerve terminals. Receptors may be coupled to membrane ion channels, cellular messengers and enzymes to regulate ion permeability and to change cell phenotype. These actions produce transient or long lasting alterations in cell excitability, biochemistry and structure [3, 6, 7].

Peripheral nerve injuries induce the formation of neuroma tissue at the lesion site which generates spontaneous and abnormal electrical activity [8]. Ectopic activity is also induced in undamaged sensory ganglion neurones, remote from the injured site (Fig. 1). Such activity contributes to the spontaneous pain and abnormal sensitivity to innocuous stimuli that occurs in painful neuropathies. The mechanisms responsible for these changes are unknown but it is believed that abnormal activation of sodium ion channels in large diameter peripheral nerves may underly these effects [8]. Abnormal calcium channel activity may also be important as blockers such a nimodipine improve experimental diabetic and cisplatin-induced neuropathy pain [9].

New Analgesic at Peripheral Targets

Of the numerous substances produced during injury and inflammation, kinins (bradykinin and kallidin) are amongst the first to be seen, and are of prime importance as mediators of pain. They also initiate a cascade of proinflammatory events such as plasma extravasation, the release of prostanoids and cytokines, degranulation of mast cells and activation of immune cells [10]. Bradykinin and kallidin act via the B2 receptor, for which several specific peptide (e.g. Hoe 140) and non-peptide (e.g. WIN 64338) antagonists have been made. There is abundant evidence that block of B2 receptors produces analgesia [11]. The B1 kinin receptor is expressed less frequently and is preferentially activated by the kinin metabolites desArg^{9}bradykinin and desArg^{10} kallidin and antagonised by Leu^{8}des-Arg^{9}bradykinin. It is upregulated or synthesised *de novo* during inflammation or infections and contributes significantly to hyperalgesia [10].

The cytokines IL1, IL6, IL8 and TNF-α are also important inflammatory agents which stimulate the production of prostanoids and induce B1 receptors expression [10]. Cytokine-induced hyperalgesia is reversed by indomethacin and by a novel IL1 antagonist Lys-D-Pro-Thr. In addition, IL8 and TNF-α produce hyperalgesia via sympathetic neurones while another cytokine, leukaemia inhibitory factor (LIF) induces phenotypic changes in sympathetic neurones resulting in the production of other proinflammatory mediators [7]. Thus block of abnormal sympathetic nerve activity may be of considerable therapeutic benefit. It is also well known that prostanoids (PGE_2, PGI_2, LTB_4) either excite nociceptors, or more usually, sensitize them to other stimuli, and that inhibition of the synthetic enzyme, cyclooxygenase (COX), is the basis for the analgesic and antiinflammatory actions of NSAIDs. The constitutive enzyme, COX-1 makes prostanoids for physiologically important functions, but COX-2, which is induced by inflammation, may be of greater pathophysiological significance. COX-2 is also susceptible to NSAIDs [12] and it is believed that a selective inhibitor of COX-2 should induce fewer side-effects.

NGF, produced by neural target tissues e.g. fibroblasts, Schwann cells, acts as a trophic factor and cellular regulator [13]. During inflammation NGF levels are increased and NGF induces hyperalgesia. Indeed antibodies to NGF, administered to animals exposed to NGF or inflammatory mediators, reduce response to painful and inflammatory stimuli and reverse the cellular neurochemical changes [14]. NGF acts on trk A, a specific receptor found on sensory neurones, and promotes gene regulation. The synthesis of several important proteins (the capsaicin receptor/ion channel, the TTX-resistant Na^+ channel) and neuropeptides (substance P and CGRP) are thus increased. NGF-induced axonal sprouting of sensory and sympathetic neurones may also contribute to the induction of hyperalgesia. Clearly an inhibitor of NGF is an attractive proposition for a novel analgesic and antiinflammatory agent.

Selective block of peripheral nociceptors can also be achieved with capsaicin and its analogues. Capsaicin, a pungent principle in *Capsicum* peppers, acts via highly specific receptors on sensory neurons to cause depolarization and an initial burning pain. This is followed by the selective inactivation of nociceptors inducing a reversible analgesic and antiinflammatory action [15]. Indeed topical capsaicin creams have been used clinically in a number of painful neuropathic disorders (diabetic neuropathy pain, post herpetic neuralgia pain) which responded poorly to other forms of therapy [16]. However the irritant properties of capsaicin have restricted its widespread acceptability and less pungent analogues have the capability of inducing analgesia without the concommitent irritation [16]. Finally, several substances (amitryptiline, carbamazepine, mexilitine, lidocaine and tocainide) have been used with limited success in the treatment of neuropathic pain and trigeminal neuralgia. This has stimulated interest in developing other sodium channel blocking drugs since these substances may block sodium channels to abolish abnormal peripheral nerve activity at concentrations which do not block nerve conduction [8].

New Analgesic at Central Targets

Although the search for better opioid analgesics continues, a variety of new approaches is being tried. Some of these include developing agonists and antagonists of sensory neuropeptides (CCK, galanin), adrenergic or serotonin receptors, as well as inhibitors for nitric oxide synthase and COX enzymes [9]. For some time however glutamatergic transmission has been shown to be important in the central hyperexcitablity associated with chronic pain and NMDA receptor antagonists are powerful inhibitors of this [6, 17]. NMDA receptor activation may further contribute to hyperexcitability by inducing a neurotoxic loss of spinal inhibitory neurones. Clinical studies have supported the efficacy of NMDA antagonists as analgesic [18], though presently available agents may cause motor and behavioural disturbances as side-effects.

Also of great importance are the neurokinins substance P (SP) and neurokinin A (NKA) which are released from sensory nerves both in the periphery and from central terminals to produce neurogenic inflammation and hyperalgesia and also to transmit pain signals in the spinal dorsal horn. Although neurokinins act on a number of receptors, the focus in hyperalgesia has been on the NK1 receptor which is upregulated by inflammation [17] and contributes, together with NMDA receptors, to the increased excitability of spinal dorsal horn neurones following repetitive stimulation of C-fibres. The NK1 receptor has been well characterized with a number of non-peptide antagonists (CP99994, RP67580, SR140333) which shown analgesic and antiinflammatory activity *in vivo* [19]. These advances have been encouraging in the development of therapeutic compounds.

In conclusion, enormous progress has been made in the last decade in characterizing the plastic changes that occur in nociceptive pathways during chronic pain conditions. This information has caused a fundamental shift in the way that pain therapy and analgesic drug development are being thought about. Mechanistic studies have highlighted new targets for new programs which are seeking inhibitors of inflammatory mediators (kinins, growth facors), newly induced proteins (B1 receptors, COS-2 enzymes) as well as blockers of afferent fibres (capsaicin analogues, ion channel blockers). In the CNS, a multiplicity of strategies is being pursued including the development of antagonists of NMDA and NK1 receptors which may eventually supplement or replace the traditional use of opioid analgesics.

References

1. von Korff M, Dworkin SF, Le Resche L. Graded chronic pain states: an epidemiological evaluation. Pain 1990; 40: 279–291.
2. Wall PD, Malzack R. Textbook of pain. Edinburgh, Churchill Livingstone: 1994.
3. Rang HP, Bevan SJ, Dray A. Nociceptive peripheral neurones: cellular properties. In: Wall PD, Melzack R. editors. Textbook of pain. Edinburgh, Churchill Livingstone, 1994: 57–78.
4. Handwerker HO. What peripheral mechanisms contribute to nociceptive transmission and hyperalgesia. In: Basbaum AI, Besson L-M. editors. Towards a new pharmacotherapy of pain. New York, Wiley: 1991: 5–20.
5. Treede R-D, Meyer RA, Raja SN, Campbell JN. Peripheral and central mechanisms of cutaneous hyperalgesia. Prog Neurobiol. 1992; 38: 397–421.
6. Woolf CJ, Doubell TP. The pathophysiology of chronic pain-increased sensitivity to low threshold Aβ-fibre inputs. Current Opinion in Neurobiology 1994; 4: 525–534.
7. Dray A. Tasting the inflammatory soup: the role of peripheral neurones. Pain Reviews 1994; 1: 153–171.
8. Devor M. The pathophysiology of damaged peripheral nerves. In: Wall PD, Melzack R, editors. Textbook of pain. Edinburgh, Churchill Livingstone: 1994: 79–100.
9. Dray A, Urban L, Dickenson A. Pharmacology of chronic pain. TIPS 1994: 15, 190–197.
10. Dray A, Perkins MN. Bradykinin and inflammatory pain. Trends Neurosci. 1993; 16: 99–104.

11. Steranka RR, Manning D, DeHass CJ, Ferkany JW, Borosky SA, Connor JR et al. Bradykinin as a pain mediator: receptors are localized to sensory neurons, and antagonists have analgesic actions. Proc. Natl. Acad. Sci. USA 1988; 85: 3245–3249.
12. Mitchell JA, Akarasereenont P, Thiemermann C, Flower RJ, Vane JR. Selectivity of nonsteroidal antiinflammatory drugs as inhibitors of constitutive and inducible cyclooxygenase. Proc. Natl. Acad. Sci. USA 1993; 90: 11693–11697.
13. Lewin GR, Mendall LM. Nerve growth factor and nociception. Trends in Neurosciences 1993; 16: 353–358.
14. Woolf CJ, Safieh-Garabedian B, Ma Q-P, Crilly P. Winter J. Nerve growth factor contributes to the generation of inflammatory sensory hyperalgesia. Neuroscience 1994; 62: 327–331.
15. Dray, A. Neuropharmacological mechanisms of capsaicin and related substances. Biochem Pharmacol. 1992; 44: 611–615.
16. Campbell E, Bevan SJ, Dray A. Clinical applications of capsaicin analogues. In: Wood J, editor. Capsaicin in the study of pain. New York, Academic Press: 1993: 255–272.
17. Urban L, Thompson SWN, Dray A. Modulation of spinal excitability: co-operation between neurokinin and excitatory amino acid neurotransmitters. TINS 1994; 17: 432–438.
18. Kristensen JD, Svensson U, Gordh T. The NMDA-receptor antagonist CPP abolishes neurogenic 'wind-up' pain after intrathecal administration in humans. Pain 1992; 51: 249–253.
19. Birch PJ, Harrison SM, Hayes AG, Rogers H, Tyers MB. The non-peptide NK1 receptor antagonist, (±)-CP-96,345, produces antiociception and anti-oedema effects in the rat. Br. J. Pharmacol. 1992; 115: 508–510.

Pharmacological Sciences: Perspectives for
Research and Therapy in the Late 1990s
ed. by A.C. Cuello and B. Collier
© 1995 Birkhäuser Verlag Basel/Switzerland

Neurohormonal Systems Underlying Drug Addiction: Relevance for Treatment Strategies

Jan M. van Ree[1], George F. Koob[2], Guy A. Higgins[3],
Claudio A. Naranjo[4] and Edwin E. Zvartau[5]

[1]*Department of Medical Pharmacology, Rudolf Magnus Institute for Neurosciences,
Utrecht University, Universiteitsweg 100, 3584 CG Utrecht, The Netherlands;*
[2]*Department of Neuropharmacology, The Scripps Research Institute, 10666 North Torrey Pines
Road, La Jolla, Ca 92037, USA;* [3]*Glaxo Research and Development Ltd., Park Road, Ware,
Herts SG12 0DP, United Kingdom;* [4]*Psychopharmacology Research Program, Sunnybrook
Health Science Centre, 2075 Bayview Avenue, Toronto ON M4N 3M5, Canada;*
[5]*Pavlov Medical Institute, St. Petersburg 197089, Russia*

Introduction

Substance abuse and dependence have many serious adverse economical, social and medical consequences. However, effective pharmacotherapies for drug and alcohol dependence are still lacking. Clinical research has been stimulated in recent years, particularly by the progress in the understanding of the neurobiological mechanisms in substance dependence. Drug dependence may be described as a compulsive desire for a psychoactive substance for no apparent therapeutic benefit, usually associated with tolerance and withdrawl symptoms when the drug is removed. It may be sustained by the positive reinforcing properties of the drug, the influence of secondary cues associated with drug taking, or by an avoidance of a negative state associated with the cessation of drug taking [1]. These and other aspects pertinent to human drug dependence have been reliably studied in experimental animal research. Drug self-adminstration procedures have been an effective tool for the study of drug-seeking behavior and reinforcement, and studies involving intravenous self-adminstration of psychomotor stimulants, opiates and alcohol have identified specific neural substrates that mediate the reinforcing actions of these drugs, and suggest that these neural systems constitute part of the central nervous systems that evolved for mediating motivated behavior and reinforcement in general [2]. Other relevant procedures employed are place preference conditioning and drug discrimination. Place preference is characterized by a positional bias for an environment associated with the drug of interest; it has been proposed to represent a conditioned response to the approach eliciting properties of abused drugs. Drug discrimination provides an opportunity to examine the subjective properties of drugs of interest. The involvement of distinct neurochemical systems in substance dependence will be briefly reviewed together with the available data from clinical research.

Dopamine

Studies on intravenous self-administration have strongly implicated the mesocorticolimbic dopamine system and its connections in the region of the nucleus accumbens in the reinforcing effects of cocaine and amphetamine [3]. Partial blockade of dopamine receptors by systemic adminstration of low doses of dopamine receptor antagonists reliably increase cocaine and amphetamine self-adminstration in rats, suggesting a partial blockade of the reinforcing actions of these drugs. Different dopamine receptor types are probably involved, since D1 and D2 antagonists appear to block the effect of coaine and the D3 agonist, 7-OHDPAT, has been shown to potentiate the effect of cocaine. Dopamine denervation of the nucleus accumbens by 6-hydroxydopamine (6-OHDA) produces extinction-like responding, and a long-lasting decrease in self-adminstration of cocaine and amphetamine. Other studies suggest that dopamine denervation of the nucleus accumbens blocks the motivation to respond for cocaine but not food reward. The brain dopamine systems in the nucleus accumbens may also be involved in the reinforcing properties of low doses of ethanol. Thus, dopamine antagonists injected into the nucleus accumbens decrease oral ethanol self-adminstration in non-water deprived rats. Moreover, extracellular dopamine levels increase in rats orally self-administering low doses of ethanol and rats will self-administer ethanol directly into the ventral tegmental cell body regions of the mesocorticolimbic dopamine system. However, dopamine denervation of the nucleus accumbens failed to alter voluntary response for alcohol and dopamine antagonists injected into the region of the nucleus accumbens decrease responding for alcohol, but also for saccharin. Thus the role of dopamine in alcohol reinforcement may reflect more general incentive-motivational functions. Similar evidence exists suggesting a role for dopamine in opiate reward. Opiates are self-administered into the ventral tegmental area and produce an increase in dopamine release in the nucleus accumbens. However, blockade of dopamine receptors in the nucleus accumbens or destruction of dopamine terminals in this area have little effect on heroin self-administration. Human studies have shown that dopamine agonists, i.c. bromocriptine and amantidine, can reduce craving for cocaine and symptoms of withdrawal and the dopamine receptor blocker flupenthixol can decrease cocaine craving. Tricyclic antidepressants decrease cocaine use and increase abstinence in some, but not all studies. Alcohol consumption may be decreased by treatment with the dopamine agonist bromocriptine.

Endorphins

Soon after the discovery of endogenous opioid peptides in the brain, endorphins were implicated in addiction and reward [4]. Accordingly, β-endorphin and related opioid peptides are self-administered in animals and opioid antagonists decreased, under certain conditions, brain stimulation reward. Initiation of intravenous cocaine self-administration in rats is attenuated by opioid antagonists, but only when a threshold unit dose of cocaine is administered. This effect of naltrexone is probably exerted by blocking opioid receptors in the ventral tegmental area, which may link the endorphin and dopamine systems in this respect. Biochemical studies have shown that β-endorphin levels in the anterior part of the limbic system are decreased during initiation of cocaine or heroin self-administration in rats at the moment the "craving" for the drug was thought to be high. There is evidence for an interaction between alcohol and endogenous opioid activity in animals as well as in humans. This has stimulated research to explore the significance of endogenous opioids for alcohol addiction. Naltrexone reduces free choice alcohol drinking in monkeys in a dose-dependent manner [5]. After a period of 2 days of imposed abstinence, which resulted in an increased alcohol consumption after renewed access, the monkeys were more sensitive to naltrexone with respect to its decreasing effect on alcohol consumption. Recent clinical observations have shown that chronic oral treatment of alcoholics with naltrexone decreased the craving for alcohol and resulted in a diminished relapse rate [6]. Consistent with the monkey studies is the finding that in human studies drinking alcohol under naltrexone treatment led less frequently to relapse than under placebo treatment. Naltrexone can also prevent relapse in opiate addicts. An opioid-free interval is necessary before beginning a naltrexone regime in order to avoid precipitated withdrawal. Although naltrexone is accepted by certain subgroups of patients, it does not always control the craving for the drug experienced by many addicts.

Serotonin

Serotonergic drugs such as fluoxetine and dexfenfluramine decrease feeding behavior and amphetamine, cocaine, heroin and alcohol intake in animals [7]. However, dose response studies could not reveal any specific action in this respect, suggesting an effect on positively reinforced behavior in general. It has been proposed that serotonergic systems may serve to oppose forebrain dopamine systems that are integral for the initiation and maintenance of reward related behaviors. Accordingly, central serotonin depletion has been reported to enhance amphetamine and alcohol self-adminstration and the alcohol-preferring rat has a deficiency in central serotonergic activity. Experimental studies using the place conditioning paradigm suggest that selective 5-HT3 receptor antagonists may diminish the rewarding effects of various drugs of abuse [8]. However, subsequent studies using models of drug self-administration and discrimination have, with the exception of alcohol, generally failed to advance this view (e.g. [7]). Reduction in alcohol self-administration following 5-HT-3 antagonist pretreatment has been described. The profile of 5-HT-3 antagonists seems quite different from indirect 5-HT agonists, since the effect of 5-HT-3 antagonists seems to be limited to alcohol intake and not generalized to food, opioid or psychomotor stimulant intake. Also, other specific 5-HT receptor antagonists, such as ritanserin (5-HT-2 antagonist) have been shown to attenuate alcohol intake in animals under certain conditions. Clinical studies have also revealed some relation between serotonin and alcohol dependence. Serotonin uptake inhibitors, i.e., zimeldine, citalopram, viqualine and fluoxetine, decrease somewhat – but significantly – short-term alcohol intake [9]. This effect may be due to a decrease in desire or urge to drink. The 5-HT-1A agonist buspirone has been reported to reduce alcohol intake in some, but not all studies, and to decrease anxiety and alcohol craving in recently detoxified anxious alcoholics. The 5-HT-2 antagonist reduced the desire to drink but did not decrease alcohol intake, while a low dose of the 5-HT-3 antagonist ondansetron seems to reduce alcohol consumption [10].

γ-Amino Butyric Acid (GABA)

This neurotransmitter has been implicated in drug dependence through at least two mechanisms [3]. First, GABA has long been hypothesized to have a role in the intoxicating effects of alcohol based on the effectiveness of GABA-ergic antagonists to reverse the behavioral effects of alcohol and the effectiveness of GABA-mimetic drugs to increase alcohol's actions. Of particular interest is the finding that the partial inverse benzodiazepine agonist Ro 15-4513 produces a dose-dependent reduction in alcohol self-administration. GABA also seems to play an important functional role in the neuropharmacology of the physical signs of alcohol withdrawl. Second, GABA in the region of the ventral pallidum have been implicated in the reinforcing effects of drugs. Lesions in this area decrease baseline self-administration of cocaine and heroin. Of particular importance in this respect may be the efferent GABA containing connections from the nucleus accumbens to the region of the ventral pallidum.

Calcium Antagonists

Calcium has been implicated in the central effects of different abused drugs. Recent studies have shown that self-administration of cocaine, morphine, amphetamine, phencyclidine, nicotine and ethanol is inhibited by the treatment with calcium channel blockers [11]. Of interest is that the calcium channel activator Bay K 8644 potentiates cocaine and morphine reward [12]. Among the different types of calcium blockers, only the dihydropyridines (e.g., nimodipine, isradipine, nifedipine) are active in blocking drug-induced rewards. The specific action of the dihydropyridine has been evaluated using different animal model i.c. drug self-administration, drug-induced conditioned place preference and drug-induced activation of brain stimulation reward. Thus, it seems that there is a dihydropyride sensitive mechanism involved in the rewarding action of all kinds of drugs and alcohol. Whether this mechanism is related to dopamine or endorphins or to other neurotransmitters remains unknown.

Corticotropin-Releasing Hormone (CRH)

Stress is a significant component of drug dependence in humans and is a particularly important part of drug abstinence during the course of drug dependence. The neurological basis for stress and anxiety associated with drug dependence is largely unknown. Logical candidates would be the neurotransmitter linked to each drug class, e.g. GABA for ethanol and benzodiazepines, dopamine for cocaine, endorphins for opiates, acetylcholine for nicotine. However, a common element may also be involved. A candidate in this respect is CRH, which is thought to have a role in mediating behavioral responses to stress. Recently, some evidence has been presented showing that CRH in the central nervous system and perhaps particularly in the amygdala may have a role in the more motivational effects of alcohol withdrawl [13]. Whether this hormone has a similar role in the case of other drugs remains to be elucidated.

Replacement Therapies

To alleviate the acute withdrawal syndrome pure and partial μ-opioid agonists are used. In general, methadone, a pure opioid agonist is the treatment of choice. Buprenorphine, a partial μ-opioid agonist, seems also to be effective as pharmacotherapy of opioid withdrawal, and some, but not all withdrawal symptoms can be suppressed by the α-2 noradrenergic agonist clonidine. Opioid dependence is further characterized by the persistence of some withdrawal symptoms after a long period of abstinence (6–12 months). As a maintenance treatment of opioid dependence methadone is prescribed [14]. Because patients' retention rate is high, methadone is also prescribed in order to reduce illicit opioid use, reduce needle use, and to achieve medical and psychosocial functioning. Buprenorphine could become an alternative to methadon maintenance and levo-α-acetylmethadol (LAAM), a long-acting congener of methaon is currently under investigation in maintenance pharmacotherapy.

Replacement therapies are also employed in the field of nicotine dependence. Nicotine can be administered in chewing gum or a skin patch. The most successful replacement so far is the transdermal nicotine patch, but the success rate is not high.

In recently detoxified alcoholics, the administration of calcium acetylhomotaurinate (acamprosate) reduces relapse as compared to placebo treatment. Acamprosate may act through modulation of the glutamate receptor complex.

Concluding Remarks

New and effective pharmacotherapies are needed for drug-dependent patients [10]. To develop these therapies reliable and validated animal models are available. These models have much contributed to our present knowledge of mechanisms in the brain implicated in drug reward and addiction. Recent progress in basis and clinical research has increased the number of potential neuropharmacological treatment for alcohol, nicotine, cocaine and opioid dependence. However, the promising results in animal studies are not always replicated in humans. Most researchers have focused on one neurotransmitter or biological mechanism, but it is unlikely that a behaviour as complex as human drug-seeking is singularly controlled. Furthermore, many people are dependent on two or more drugs, or have a comorbid psychiatric illness. The role of pharmacotherapy in drug dependence, perhaps in conjunction with psychosocial or cognitive behavioral treatments, need further investigation.

References

1. Koob GF, Bloom FE. Cellular and molecular mechanisms of drug dependence. Science 1988; 242: 715–723.
2. Van Ree JM. Reinforcing stimulus properties of drugs. Neuropharmacol. 1979; 18: 963–969.
3. Koob GF. Drugs of abuse: anatomy, pharmacology, and function of reward pathways. Trends Pharmacol. Sci 1992; 13: 177–184.

4. Van Ree JM. Reward and abuse: opiates and neuropeptides. In: Engel J, Oreland L, editors. Brain reward systems and abuse. New York: Raven Press, 1987: 75–88.

5. Kornet M, Goosen C, Van Ree JM. The effect of naltrexone on alcohol consumption during chronic alcohol drinking and after a period of imposed abstinence in free-choice drinking rhesus monkeys. Psychopharmacol. 1991; 104: 367–376.

6. Volpicelli JR, Alterman AI, Hayashida M, O'Brien CP. Naltrexone in the treatment of alcohol dependence. Arch. Gen. Psychiatry 1992; 49: 876–880.

7. Higgins GA, Wang Y, Corrigall WA, Sellers EM. Influence of 5-HT3 receptor antagonists and the indirect 5-HT agonist, dexfenfluramine, on heroin self-adminstration in rats. Psychopharmacol. 1994; 114: 611–619.

8. Carboni E, Acquas E, Leone P, DiChiara G. 5-HT3 receptor antagonists block morphine – and nicotine – but not amphetamine-induced reward. Psychopharmacol. 1989; 97: 175–178.

9. Naranjo CA, Bremner KE. Evaluation of the effects of serotonin uptake inhibitors in alcoholics: a review. In: Naramjo CA, Sellers EM, editors. Novel pharmacological interventions for alcoholism. New York: Spring Verlag, 1992: 105–117.

10. Naranjo CA, Bremner KE. Pharmacotherapy of substance use disorders. Can. J. Clin. Pharmacol. 1994; 1: 55–71.

11. Zvartau EE, Kuzmin A, Patkina N. Calcium entry blockers and drug addiction. Eur. Neuropsychopharmacol. 1993; 3: 220–221.

12. Kuzmin AV, Patkina NA, Zvartau EE. Analgesic and reinforcing effects of morphine in mice. Influence of Bay K8644 and nimodipine. Brain Res. 1994; 652: 1–8.

13. Rassnick S, Heinrichs SC, Britton KT, Koob GF. Microinjection of a corticotropin-releasing factor antagonist into the central nucleus of the amygdala reverses anxiogenic-like effects of ethanol withdrawal. Brain Res. 1993; 605: 25–32.

14. Zweben JE, Payte JT. Methadone maintenance in the treatment of opioid dependence. A current perspective. West J. Med. 1990; 152: 588–599.

Pharmacological Sciences: Perspectives for
Research and Therapy in the Late 1990s
ed. by A.C. Cuello and B. Collier

Molecular and Cellular Mechanisms in Neurosecretion

J.-M. Trifaró[1] and A.G. García[2]

[1]*Department of Pharmacology, University of Ottawa, Ottawa, Canada K1H 8M5;*
[2]*Department of Pharmacology, Universidad Autónoma de Madrid, Madrid 28029, Spain*

Summary. Recent published evidence has increased our understanding on the cellular and molecular mechanisms involved in secretion. It has been demonstrated that the ratios among soluble components of secretory vesicles can be changed according to specific demands and that vesicle contents are released in response to an increase in $[Ca^{2+}]_i$, which is the result of $[Ca^{2+}]_0$ entry through some but not all Ca^{2+} channels present in neurosecretory cells. It is also clear that target cells modulate the handling of $[Ca^{2+}]_i$ by neurosecretory cells. One site of action for Ca^{2+} is the cytoskeleton. This organelle controls in neurons as well as in other secretory cells, the delivery of secretory vesicles to exocytotic sites. A fine regulation of this process is provided by second messengers and actin associated proteins such as scinderin and synapsin I.

Introduction

The process of exocytosis is a fascinating interplay between cellular components and the secretory vesicle. It has become clear in recent years that secretion by cells can take two forms: constitutive and regulated. Constitutive secretion is unregulated and closely follows the rates of synthesis of secretory products. This form of secretion occurs in almost all cell types. The other form of secretion is highly regulated and characteristic of endocrine and exocrine cells as well as neurons. These cells store their secretory products in membrane-bound secretory vesicles [1]. Regulated secretion is triggered by an increase in intracellular Ca^{2+}. However, despite the fact that the role of Ca^{2+} in secretion was observed many years ago, the exact mechanism by which Ca^{2+} triggers secretion is still poorly understood. One attractive hypothesis is that the Ca^{2+} effect in the secretion of neurons and other secretory cells is mediated through its control of the cytoskeleton network [2]. In both neurons and secretory cells, Ca^{2+} plays a pivotal role in the control of cytoskeleton dynamics during secretion [3, 4]. Ca^{2+} enters neurosecretory cells through at least four different channel types [5]. However, only some types are responsible for the delivery of Ca^{2+} required for exocytosis [5]. The regulation of intracellular Ca^{2+} levels by neurosecretory cells is also modulated by trophic substances released by target cells

Correspondence to: Dr. J.-M. Trifaró, at the above address.

[6]. In the following pages new contributions to the understanding of cellular and molecular events involved in neurosecretion are discussed.

Modification of Secretory Content of Large Dense Core Vesicles by Regulation of Biosynthesis

The large dense core vesicles (LDV) of chromaffin cells (CC) and neurons have a complex composition [1]. Their secretory content con-

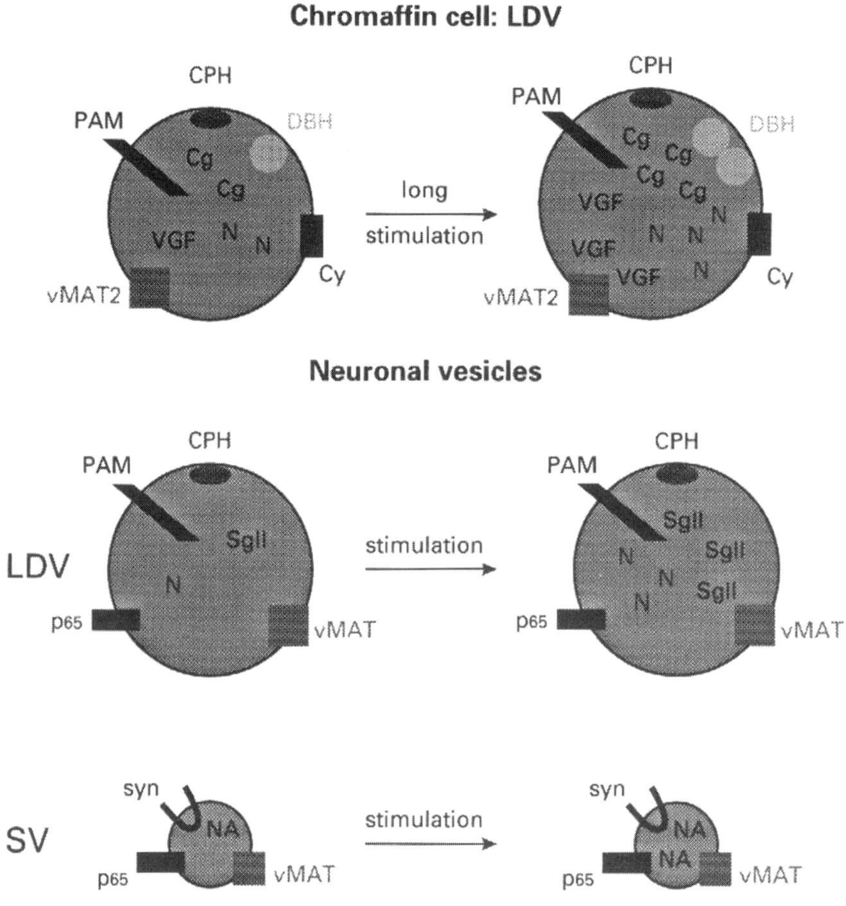

Fig. 1. Control of secretory vesicle content. LDV: large dense core vesicle, SV: synaptic vesicle; Cg: Chromogranin; Cy: cytochrome b_{651}; SgII: secretogranin II; NA: noradrenaline; $VMAT_2$: vacuolar membrane ATPase 2; DβH: dopamine β-hydroxylase; Syn: synaptophysin; N: neuropeptide; p65: synaptotagmin; PAM: peptidyl α-amidating monooxygenase; CPH: carboxypeptidase H. Taken from [1].

sists of small molecules (ATP, catecholamines, ions) and various peptides including chromogranins A and B, secretogranin II and neuropeptide precursors [1]. LDV membranes contain proteins involved in neurotransmitter and nucleotide uptake, in neuropeptide processing (PC_1, PC_2, PAM) and proteins of unknown function (glycoprotein II and III, SV_2, p65) [7]. LDV biosynthesis has been studied in several models: CC stimulated by hypoglycemia [8] and by reserpine [9], brain neurons stimulated by salt loading [10], etc. From the results obtained in these studies a model on differential changes in the biosynthesis of SV components in response to stimualation has been proposed (Fig. 1). This model suggests that the synthesis of secretory proteins such as chromogranins and other neuropeptides is up-regulated, whereas the synthesis of membrane-bound or intrinsic membrane proteins, with the exception of dopamine β-hydroxylase, is practically unchanged. Thus, after prolonged stimulation either the number of vesicles or their size does not change significantly whereas the number of molecules in the vesicle content, such as chromogranins and neuropeptides, is increased (Fig. 1) [1]. Similar changes occur in neuronal LDV during stimulation. Moreover, during stimulation, up-regulation of tyrosine hydroxylase increases the synthesis of amines with the fast refilling of the vesicles with more molecules of neurotransmitter per vesicle (Fig. 1). In conclusion, the available evidence indicates that prolonged stimulation of CC and neurons leads to increased secretory quanta. Thus, further stimulation can release a higher quantum per each exocytotic event. This is an adaptive modulatory mechanism which operates when the demand for secretory products is high.

Control of the Neurosecretory Process by Cytosolic Calcium

Exocytosis of neurotransmitters is triggered by an abrupt increase in cytoplasmic calcium ($[Ca^{2+}]_i$) concentrations. However, changes in the rate of secretion do not strictly parallel changes in $[Ca^{2+}]_i$ [5]. When CC are superfused with Ca^{2+}-free solutions containing 100 mM K^+ and extracellular Ca^{2+}($[Ca^{2+}]_0$) is offered as single steps or as continuous increasing ramps, a clear separation between increases in $[Ca^{2+}]_i$ and amine release is observed [5]. On the other hand, secretion induced by Ba^{2+} steps or ramps starts more slowly but does not desensitize [11]. Dialysis of Ba^{2+} into a single voltage-clamped bovine CC produced a concentration-dependent increase in cell membrane capacitance, reflecting an enhanced rate of exocytotic events. Moreover, external Ba^{2+} application to a single CC led to a long-lasting secretory response, as measured with a carbon fiber electrode and amperometry. Although Ba^{2+} is 100 times less potent than Ca^{2+}, the overall secretory response is much greater with Ba^{2+}, suggesting that Ca^{2+} but not Ba^{2+} inactivates some of the steps leading to secretion [12]. The whole-cell current through bovine Ca^{2+} channels has at least four components, which are sensitive, respectively, to ω-conotoxin GVIA (CTx-VIA; a marker for N-type Ca^{2+} channels), ω-agatoxin IVA (Aga-IVA; a marker for P-type Ca^{2+} channels), ω-conotoxin MVIIC (CTx-MVIIC; a marker for Q-type Ca^{2+} channels), and to the 1,4-dihydropyridine derivative furnidipine (a marker for L-type Ca^{2+} channels) [13]. The study of the contribution of these channels to the secretory response indicates a lack of correlation between the blockade of Ca^{2+} currents and the blockade of secretion, as shown by experiments in which 10 s pulses of 70 mM K^+ in the presence of 2 mM $[Ca^{2+}]_0$ were applied as 5 min intervals to superfused bovine CC. Furnidipine (3 μM) produces a 50% decrease in the secretory response and CTx-MVIIC (3 μM) added on top of furnipidine fully inhibits catecholamine release. When CTx-GVIA (1 μM) and Aga-IVA (0.2 μM) are combined, the secretory response remains unchanged. In other species, such as the cat, the control of secretion is dominated by L-type Ca^{2+} channels [14]. Although present, N channels contribute little to secretion [5].

The evidence so far accumulated clearly indicates that of the various Ca^{2+} channel subtypes present in CC, only some are responsible for the immediate delivery of the $[Ca^{2+}]_0$ required to trigger catecholamine release (Fig. 2). Taking into account that $[Ca^{2+}]_i$ must quickly rise to several tenths micromolar at specialized exocytotic sites, it is not surprising that Ca^{2+} channels are highly localized nearby those sites. In cat

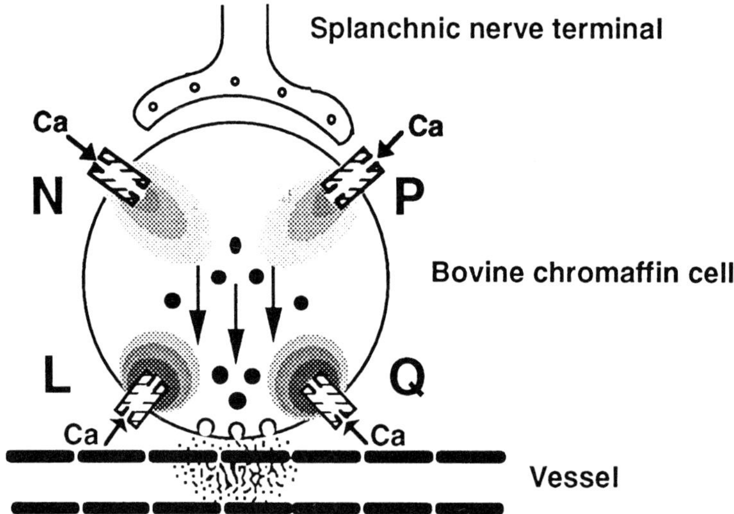

Fig. 2. Ca^{2+} channel distribution in chromaffin cell. N: channel sensitive to ω-conotoxin GVIA; P: channel sensitive to ω-agatoxin IVA; L: channel sensitive to dihydropyridines and Q: channel sensitive to ω-conotoxin MVIIC. Taken from [5].

CC, L-type Ca^{2+} channels are likely to be localized at exocytotic sites, while in bovine CC both Q- and L-type Ca^{2+} channels should be preferentially localized at those sites (Fig. 2). Other channel types (N and P) are probably responsible for delivering $[Ca^{2+}]_0$ not required for the fast and immediate secretion of catecholamines; this $[Ca^{2+}]_0$ entering the cell would serve other Ca_i^{2+}-dependent functions such as vesicle transport or catecholamine synthesis.

Scinderin and Cytoskeleton Dynamics during Exocytosis

The actin microfilament network localized underneath the plasma membrane of CC [15, 16] acts as a barrier to the movement of secretory vesicles (SV), blocking their access to exocytotic sites [16]. Nicotinic stimulation of CC induces cortical filamentous actin (F-actin) disassembly and allows the free movement of SV and their subsequent interaction with the plasma membrane [4, 17]. The existence of actin binding proteins such as scinderin (Sc), which regulate G-actin/F-actin equilibrium [18] suggest a role for these proteins in the re-organization of cortical F-actin networks during secretion. Sc is present in adrenal CC [18] as well as in other tissues with high secretory activity [19]. Immunofluorescence microscopy studies revealed that under resting conditions Sc shows a diffuse cytoplasmic staining and continuous cortical fluorescent ring, suggesting some interaction of the protein with plasma membrane elements. Nicotinic stimulation causes a Ca^{2+}-dependent redistribution of Sc along with F-actin disassembly, processes which precede exocytosis [17]. Sites of exocytosis are preferentially localized to cortical areas of F-actin disassembly [17]. Thus, stimulation-evoked $[Ca^{2+}]_0$ influx induces the association of Sc with actin filaments and promotes its severing activity. Under resting conditions, cortical Sc is not associated with F-actin but interacts with other membrane components such as phospholipids [20] and even though Ca^{2+} has a key role in this process, other intracellular messengers modulate Sc distribution and activity [4]. Sc is a Ca^{2+}-dependent F-actin severing protein with two Ca^{2+}-binding sites [18]. Molecular cloning of Sc cDNA reveals 67% and 51% sequence homology with gelsolin and villin respectively, other Ca^{2+}-dependent F-actin severing proteins [21, 22]. The N-terminal half of the molecule is the functional domain of Sc and two sequences corresponding to actin-binding domains are present here together with two other additional PIP_2 binding domains (Fig. 3). Like gelsolin, Sc has internal repeats of shorter motifs which occur six times at approximately equal intervals (Fig. 3).

Several lines of evidence suggest the participation of protein kinase C (PKC) in catecholamine secretion. One possibility is that PKC is involved in the re-organization of the cortical cytoskeleton preceding

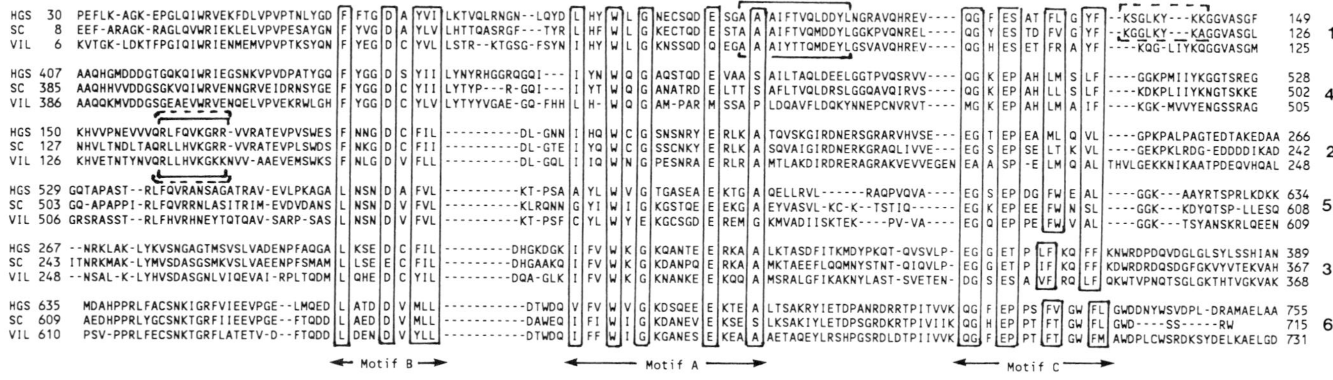

Fig. 3. Scinderin sequence and its segmental alignment with gelsolin and villin. Alignments are based on pairwise analysis of all combinations of sequences as performed by Way and Weeds [33] for gelsolin and villin. Highly conserved motifs (B, A and C) are shown in boxes. Proteins are abbreviated as follows: HGS, human plasma gelsolin; SC, bovine scinderin; VIL, chicken villin. The numbers of either end indicate the positions of amino acid residues. The large numbers at the right side indicate the domain number (1–6) for HGS, SC and VIL. These domains together with motifs B, A and C are represented in the diagram at the bottom of the figure. Each motif is found once in a domain of each protein (HGS, SC, VIL) and it is repeated five additional times along the protein. (▭): Actin binding sites; (▭---): PIP_2 binding sites. Taken from [21].

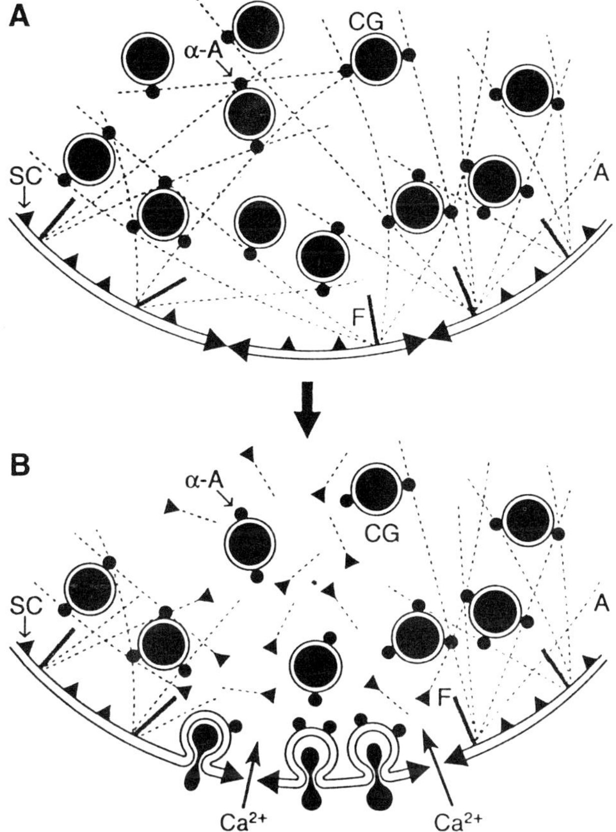

Fig. 4. The possible involvement of the cortical actin filament network in secretory granule exocytosis from chromaffin cells. (A) Resting and (B) stimulated cells. A: actin filaments; F: fodrin; Cg: chromaffin granules; α-A: α-actinin; Sc: scinderin. Taken from [2].

exocytosis [4]. PKC stimulation with the phorbol ester PMA produces Sc redistribution, F-actin disassembly and removes the barrier to SV mobility with an increase in the number of SV at release sites (release-ready vesicle pool) [4]. This effect is responsible for the increase in the initial rate of amine release to nicotine in cells preincubated with PMA [23]. When membrane capacitance recording technique is used to measure exocytosis, a three fold increase in the number of SV fusing with the plasma membrane is also observed in PMA-pretreated cells [4]. In conclusion, the results suggest that CC stimulation and $[Ca^{2+}]_0$ entry activates surface Sc with the consequent severing of F-actin, leaving subplasmalemmal areas devoid of F-actin and with decreased cytoplasmic viscosity (Fig. 4). These are the areas of high SV mobility in which exocytotic sites are preferentially localized [2, 4].

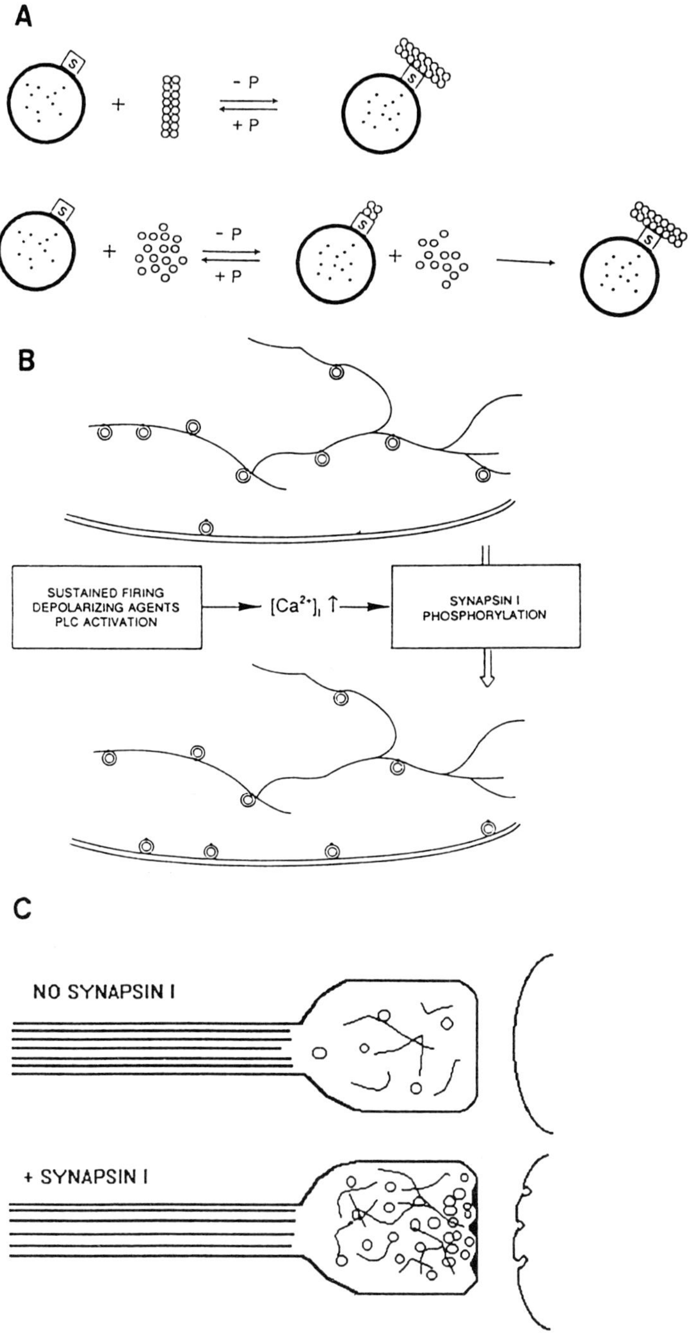

Fig. 5. Role of synapsin I in neurotransmitter A) storage, B) release and C) synapse formation. S: synapsin I. Taken from [3].

Regulation of Neurotransmitter Release and Synapse Formation by Synapsins

The synapsins (Sy) are a family of four nerve ending-specific phospho-proteins which are implicated in the regulation of neurotransmitter release and synapse formation [3]. Sy are involved in the interaction of SV membranes with various cytoskeletal components. These interactions are regulated by phosphorylation and suggest a role for these proteins in the targeting of SV to exocytotic sites [24]. Moreover, Sy have been recently implicated in synapse development, contributing to the maturation of developing nerve endings [3]. SyI binds to the SV membrane and its binding is inhibited after phosphorylation by CaM kinase II [25]. SyI also binds to actin monomers and filaments in a phosphorylation-dependent fashion (Fig. 5A) [25]. SyI is involved in the anchorage of SV to actin and this interaction is disrupted when SyI is phosphorylated by CaM kinase II. Thus, SyI inhibits neurotransmitter release, an effect which is abolished by its phosphorylation [26]. There-fore, SyI controls the number of SV available for exocytosis in response to contant stimulus (Fig. 5B) [27]. The appearance of Sy in neurons correlates with the onset of synaptogenesis [3]. Studies on *Xenopus* embryos loaded with SyI by the early-blastomere injection method, suggest the involvement of the protein in functional maturation of neuromuscular synapsis [28]. A rearrangement in the ultrastructure of the nerve endings with a large number of SV clustered close to the plasma membrane and an earlier development of active zones is ob-served in nerve endings of SyI-loaded neurons (Fig. 5C). Therefore, SyI by interacting with SV and the actin cytoskeleton might induce and control the assembly of molecular networks necessary for the matura-tion of nerve endings. In conclusion, SyI may have a dual role: firstly in the early developing of nerve endings and secondly in the modulation of neurotransmitter release in mature neurons.

Target Cells Control Catecholamine Release by Sympathetic Neurons

It is known that increases in the frequency of stimulation of mature sympathetic neurons (SyN) enhances the amount of neurotransmitter released [6]. Moreover, a tetraethylammonium (TEA) block of K^+ channels facilitates neurotransmitter release from SyN [29]. On the other hand, SyN in culture do not have a positive frequency-release response [6]. These abnormal release properties are dramatically altered when SyN are co-cultured with cardiac cells (a physiologic sympathetic target). Under these conditions, the release of neurotransmitter at 10 Hz stimulation is greater than 1 Hz. Because changes in intracellular $Ca^{2+}([Ca^{2+}]_i)$ are responsible for stimulating release, it is possible that

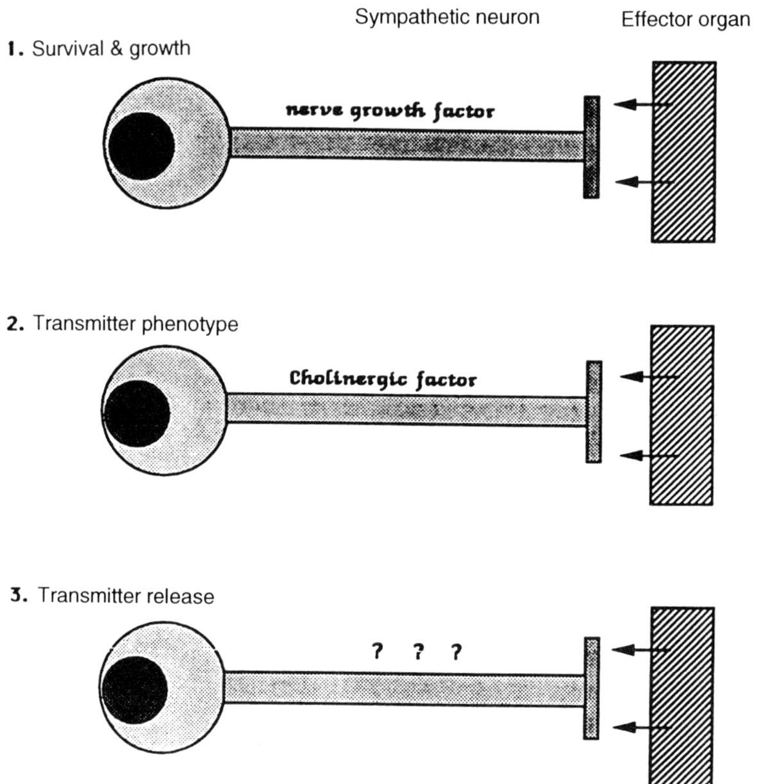

Fig. 6. Trophic influences on sympathetic neuron. Trophic influences from target cells control survival and growth, neurotransmitter phenotype and release. Taken from [6].

the effects of cardiac cells are due to changes in $[Ca^{2+}]_i$ handling. When indo-1 loading SyN are studied, a single stimulus produces a three fold increase in $[Ca^{2+}]_i$, whereas in the presence of cardiac cells in the culture the increase in $[Ca^{2+}]_i$ is hardly detectable [6]. Voltage-clamped Ca^{2+} currents evoked by a step depolarization to 0 mV from a holding potential of -70 mV in SyN grown alone or with cardiac cells have been determined [30]. Whole cell Ca^{2+} current is decreased as much as four fold when SyN are co-cultured with cardiac cells [6]. However, evoked action potentials in SyN under the two culture conditions are essentially the same [6]. These observations suggest that cardiac cells (target tissue) cause a decrease in SyN Ca^{2+} currents without changes in action potential duration. Therefore, frequency and TEA responses of SyN grown in the presence of cardiac cells are characteristic of those responses seen *in vivo*. Figure 6 summarizes recent findings about environmental influences on SyN. Trophic substances such as nerve growth factor released from effector organs are known to maintain survival and growth of SyN [31]. It is also known that factors released from effector organs determine the phenotype of SyN in a way that they become cholinergic by the influence of their target tissue [32]. Furthermore, the evidence discussed above indicates that target cells also modulate Ca^{2+} homeostasis and neurotransmitter release from SyN. The exact nature of this new trophic influence remains to be elucidated.

Abbreviations

SV: secretory vesicle; LDV: large dense core vesicle; Sc: scinderin; PKC: protein kinase C; F-actin: filamentous actin; CC: chromaffin cell; Sy: synapsin; SyN: sympathetic neuron.

References

1. Winkler H. Modifications of the secretory content of large dense core vesicles by regulation of biosynthesis. Can. J. Physiol. Pharmacol. 1994; 72(1): 56.
2. Trifaró J-M, Vitale ML. Cytoskeleton dynamics during neurotransmitter release. TINS 1993; 16: 466–472.
3. Benfenati F, Valtorta F, Greengard P. The synapsins, a family of synaptic vesicle proteins regulating neurotransmitter release and synapse formation. Can. J. Physiol. Pharmacol. 1994; 72(1): 56.
4. Trifaró J-M, Marcu M, Vitale ML, Rodriguez Del Castillo A. Scinderin and cytoskeleton dynamics during exocytosis. Can. J. Physiol. Pharmacol. 1994; 72(1): 56.
5. Garcia AG, Michelena P, Artalejo AR, Gandfa L, Albillos A, von Riiden L et al. Control of the neurosecretory process by cytosolic calcium. Can. J. Physiol. Pharmacol. 1994; 72(1): 56.
6. Wakade AR, Przywara DA, Wakade TD. Cardiac cells control norepinephrine release in cultured sympathetic neurons. Can. J. Physiol. Pharmacol. 1994; 72(1): 56.
7. Winkler H, Fischer-Colbrie R. Common membrane proteins of chromaffin granules, endocrine and synaptic vesicle: properties, tissue distribution, membrane topography and regulation of synthesis. Neurochem. Int. 1990; 17: 245–262.
8. Laslop A, Wohlfarter T, Fischer-Colbrie R, Steiner HJ, Humpel Ch, Saria A et al. Insulin hypoglycemia increases the levels of neuropeptide Y and calcitonin gene-related peptide, but not of chromogranins A and B, in rat chromaffin granules. Reg. Peptides 1989; 26: 191–202.
9. Sietzen M, Schober M, Fischer-Colbrie R, Scherman D, Sperk G, Winkler H. Rat adrenal medulla: levels of chromogranins, enkephalins, dopamine 0-hydroxylase and of the amine transporter are changed by nervous activity and hypophysectomy. Neuroscience 1987; 22: 131–139.
10. Mahata SK, Mahata M, Steiner H-J, Fischer-Colbrie R, Winkler H. In situ hybridization: mRNA levels of secretogranin II, neuropeptides and carboxypeptidase H in brains of salt loaded and Brattleboro rats. Neuroscience 1992; 48: 669–680.
11. Michelena P, Garcia-Perez L-E, Artalejo AR, Garcia AG. Separation between cytosolic calcium and secretion in chromaffin cells superfused with calcium ramps. Proc. Natl. Acad. Sci. USA 1993; 90: 3284–3288.
12. Lopez MG, Albillos A, de la Fuente MT, Borges R, Gandía L, Carbone E et al. Localized L-type calcium channels control exocytosis in cat chromaffin cells. Pfliigers Arch. Europ. J. Physiol. 1994; 427: 348–354.
13. Albillos A, Garcia AG, Gandía L. ω-Agatoxin-IVA-sensitive calcium channels in bovine chromaffin cells. FEBS Lett. 1993; 336: 259–262.
14. Garcia AG, Sala F, Reig JA, Viniegra S, Frias J, Fonteriz RI, Gandía L. Dihydropyridine Bay-K-8644 activates chromaffin cell calcium channels. Nature 1984; 308: 69–71.

15. Lee RWH, Trifaró J-M. Characterization of anti-actin antibodies and their use in immunocytochemical studies on the localization of actin in adrenal chromaffin cells. Neuroscience 1981; 6: 2087–2108.

16. Trifaró J-M, Lee RWH, Kenigsberg RL, Côté A. Contractile proteins and chromaffin cell function. Adv. Biosc. 1982; V: 151–158.

17. Vitale ML, Rodriguez Del Castillo A, Tchakarov L, Trifaró J-M. Cortical filamentous actin disassembly and scinderin redistribution during chromaffin cell stimulation precede exocytosis: a phenomenon not exhibited by gelsolin. J. Cell Biol. 1991; 113: 1057–1067.

18. Rodriguez Del Castillo A, Lemaire S, Tchakarov L, Jeyapragasan M, Doucet JP, Vitate ML, Trifaró J-M. Chromaffin cell scinderin: a novel calcium-dependent actin-filament severing protein. EMBO J. 1990; 9: 43–52.

19. Tchakarov L, Vitale ML, Jeyapragasan M, Rodriguez Del Castillo A, Trifaró J-M. Expression of scinderin, an actin filament severing protein in different tissues. FEBS Lett. 1990; 268: 209–212.

20. Roderiguez Del Castillo A, Vitale ML, Trifaró J-M. Ca^{2+} and pH determine the interaction of chromaffin cell scinderin with phosphatidylserine and phosphatidylinositol 4,5 bisphosphate and its cellular distribution during nicotine-receptor stimulation and protein kinase C activation. J. Cell Biol. 1992; 119: 797–810.

21. Marcu MG, Rodriguez Del Castillo A, Trifaró J-M. Molecular cloning of bovine chromaffin cell scinderin (Sc) cDNA reveals actin polyphosphoinositide (PPI) binding domains. Can. J. Physiol. Pharmacol. 1994; 72(1): 264.

22. Marcu MG, Rodriguez Del Castillo A, Vitale ML, Trifaró J-M. Molecular cloning and functional expression of chromaffin cell scinderin indicates that it belongs to the family of Ca^{2+}-dependent F-actin severing proteins. Mol. Cell. Biochem. 1994; 141: 153–165.

23. Vitale ML, Rodriguez Del Castillo A, Trifaró J-M. Protein kinase C activation by phorbol esters induces chromaffin cell cortical filamentous actin disassembly and increases the initial rate of exocytosis in response to nicotinic receptor stimulation. Neuroscience 1992; 51: 463–474.

24. Greengard P, Valtorta F, Czernik AJ, Benfenati F. Synaptic vesicle phosphoproteins and regulation of synaptic function. Science 1993; 259: 780–785.

25. Bedfenati F, Valtorta F, Chieregatti E, Greengard P. Interaction of free and synaptic vesicle-bound synapsin I with F-actin. Neuron 1992; 8: 377–386.

26. Llinas R, Gruner JA, Sugimori M, McGuinness TL, Greengard P. Regulation by synapsin I and Ca^{2+}/calmodulin-dependent kinase II of transmitter release in squid giant synapse. J. Physiol. 1991; 436: 257–282.

27. Benfenati F, Valtorta F, Greengard P. Computer modelling of synapsin I binding to synaptic vesicles and F-actin: implications for regulation of neurotransmitter release. Proc. Natl. Acad. Sci. USA 1991; 88: 575–579.

28. Lu B, Greengard P, Poo M-m. Exogenous synapsin I promotes functional maturation of developing neuromuscular synapses. Neuron 1992; 8: 521–529.

29. Wakade AR, Wakade TD. Comparison of transmitter release properties of embryonic sympathetic neurons growing *in vivo* and *in vitro*. Neuroscience 1988; 27: 1007–1019.

30. Przywara DA, Bhave SV, Bhave A, Wakade TD, Wakade AR. Dissociation between intracellular Ca^{2+} and modulation of ^{3}H-noradrenaline release in chick sympathetic neurons. J. Physiol. 1991; 437: 201–220.

31. Levi-Montalcini R, Hamburger. Selective growth-stimulating effects of mouse sarcoma on sensory and sympathetic nervous system of chick embryo. J. Exp. Zool. 1951; 116: 321–361.

32. Patterson PH. Environmental determination of autonomic neurotransmitter functions. Ann. Rev. Neurosci, 1978; 1: 1–17.

33. Way M, Weeds A. Nucleotide sequence of pig plasma gelsolin. Comparison of protein sequence with human gelsolin and other actin-severing proteins shows strong homologies and evidence for large internal repeats. J. Mol. Biol. 1988; 203: 1127–1133.

Purines

Pharmacological Sciences: Perspectives for
Research and Therapy in the Late 1990s
ed. by A.C. Cuello and B. Collier
© 1995 Birkhäuser Verlag Basel/Switzerland

Role of Purines in the Central Nervous System

Thomas D. White[1], J.A. Ribeiro[2], Frances A. Edwards[3],
Bertil B. Fredholm[4] and John W. Phillis[5]

[1]*Department of Pharmacology, Dalhousie University, Halifax, Nova Scotia, Canada B3H 4H7;*
[2]*Laboratory of Pharmacology, Gulbenkian Institute of Science, 2781 Oeiras, Portugal;*
[3]*Department of Pharmacology, University of Sydney, Sydney, NSW 2006, Australia;*
[4]*Department of Pharmacology, Karolinska Institute, PO Box 60400, S-104 01 Stockholm,
Sweden;* [5]*Department of Physiology, Wayne State University, Detroit, MI 48201, U.S.A.*

Summary. There is increasing evidence that adenosine and ATP perform important functions
in the central nervous system. ATP, acting at P2X receptors, appears to be an excitatory
neurotransmitter in the brain. Extracellular adenosine can be formed from released ATP or
released as such during excitatory amino acid (EAA) transmission, hypoxia and ischemia.
Adenosine, acting at inhibitory A_1 receptors, diminishes the release of neurotransmitters
including EAAs and may diminish excitotoxicity. Acting at A_2 receptors, adenosine appears
to have excitatory effects in certain brain regions. Finally, purines such as hypoxanthine may
form free radicals which could cause some of the damage associated with cerebral ischemia.

Sources of Extracellular Purines in the CNS and the Therapeutic Potential of Altering their Levels

In conditions such as excitotoxicity, seizure disorders, ischemia, hypoxia
and hypoglycemia, some of the pathology is due to the release of
excitatory amino acids (EAAs), which then produce excessive EAA
transmission. Adenosine is released when EAA receptors are activated
[1] and also during cerebral ischemia/hypoxia [2]. It can then act
presynaptically to diminish the release of glutamate and postsynapti-
cally to diminish the excitatory actions of glutamate, thus providing
inhibitory influences and protecting against excessive EAA transmission
and excitotoxicity. Although there is evidence that activation of A_2
receptors can cause excitation in certain brain regions, there is no
evidence that endogenous adenosine plays any role in excitotoxicity.
Acting at A_1 receptors, endogenous adenosine provides only partial
protection against seizures, excitotoxicity and ischemic damage since the
administration of additional adenosine or other agonists affords further
protection. However, augmenting the accumulation of extracellular
adenosine that forms when EAA receptors are activated might provide
additional protection against excitotoxicity.

Correspondence to: Dr. Thomas D. White at the above address.

The major routes of removal of extracellular adenosine include uptake on nucleoside transporter(s), conversion to inosine by adenosine deaminase and conversion to AMP by adenosine kinase. Block of adenosine transport with dipyridamole greatly augments cortical adenosine release evoked by NMDA (derived from a released nucleotide), and surprisingly also release evoked by AMPA and kainate (released as adenosine *per se*) [1]. Apparently, activation of non-NMDA receptors triggers the efflux of adenosine via a dipyridamole-insensitive process whereas adenosine's removal from the extracellular space takes place on a dipyridamole-sensitive transporter. Perhaps adenosine release occurs from one cell type and removal of extracellular adenosine into another cell type. Inhibiting adenosine kinase with either 5'-amino-5'-deoxyadenosine (NH_2dADO) or iodotubercidin slightly increases the basal rate of adenosine release but greatly augments adenosine release evoked by NMDA, AMPA or kainate. Inhibition of adenosine deaminase with either EHNA or deoxycoformycin modestly augments NMDA-evoked adenosine release. Finally, AICA riboside (AICAr) has no effect on basal adenosine release from cortical slices, nor does it affect NMDA or AMPA-evoked accumulation of extracellular adenosine. However, AICAr potentiates kainate-evoked adenosine release. The above findings are strikingly similar to the anticonvulsant actions of these drugs in a bicuculline model of epilepsy [3], where block of adenosine uptake on the nucleoside transporter or its metabolism by adenosine kinase completely prevented seizures, whereas block of adenosine deaminase was less efficacious and AICAr was without effect.

Excitatory Effects of Adenosine on The Central Nervous System

There has been a general consensus that adenosine is mainly an inhibitory substance in the nervous system. However, we showed that adenosine increases carotid body chemoreceptor activity, an action mediated by A_2 adenosine receptors [4]. Other studies have confirmed that adenosine increases acetylcholine release at the neuromuscular junction [5].

There is also evidence that adenosine has excitatory actions in the CNS. Spignoli et al. [6] first reported excitatory actions of adenosine on acetylcholine release in the CNS and Okada et al. [7] reported excitatory actions of adenosine in the hippocampus. CGS 21680, an A_{2a} adenosine receptor agonist, increases the amplitude of population spikes or the slope of field excitatory post-synaptic potentials recorded from the CA1 area after stimulation of the Schaeffer collaterals. This effect is antagonized by the A_2 adenosine receptor antagonist, dimethyl propargylxanthine (DMPX) [8]. We studied ^{3}H-ACh release from three different areas of the hippocampus: CA1, CA3 and dentate gyrus (DG). These sub-slices of the hippocampus were electrically stimulated and it was found that both inhibitory and excitatory effects of adenosine on ^{3}H-ACh release were different according to the area, as follows: (1) there are inhibitory receptors (A_1) in the CA1, CA3 and DG areas; (2) there are excitatory receptors (A_{2a}-CGS 21680 sensitive) in CA3 and DG areas; (3) endogenous adenosine tonically inhibits the CA1 area, both inhibits and excites the CA3 area, and despite both A_1 and A_{2a}-adenosine receptors being present in the DG area, has little effect in this region.

In conclusion, the excitatory effects of adenosine in the nervous system could be important in the design of molecules that, by themselves or in combination with A_1 adenosine receptor antagonists, could be therapeutically useful in the following conditions:

(1) To improve neuromuscular transmission in situations of neuromuscular depression (e.g. myasthenia gravis and myasthenic syndromes).
(2) To stimulate respiration in conditions of respiratory depression (e.g. drug respiratory depression of the CNS, respiratory depression originated peripherally, neonate apnea).
(3) By increasing ACh release to be used as cognitive enhancers to help in learning and memory and therefore , to be useful in Alzheimer's disease.

It is also possible that A_2-adenosine receptor activation could be deleterious in increasing synaptic transmission in the hippocampus which is glutamate-dependent and, therefore, possibly contributing to cell death caused by glutamate during stroke. In conclusion, the excitatory effects of adenosine certainly deserve as much attention as the inhibitory effects have had during the past 20 years.

ATP As a Fast Transmitter in the Rat Central Nervous System

In slices of rat medial habenula nucleus, various evoked and spontaneous currents can be detected using whole-cell recording. Many of these currents are blocked by glutamate and GABA antagonists showing that, like in most neurones, strong glutamate and GABA inputs are present. With these currents blocked however, a residual population of small synaptic currents remain which are seldom more than 50 pA in amplitude but feature fast rise times (<1 ms) and decay times of around 15 ms, intermediate between the glutamatergic (5 ms) and GABAergic (>30 ms) currents. Thus medial habenula neurones have synaptic currents which are clearly mediated by ligand-gated ion channels, not gated by glutamate or GABA. The only other ligand-gated channels known to exist in the brain are glycine, $5HT_3$, ATP or nicotinic ACh receptor/channels. We thus tested which of these channels were present on medial habenula cells and of these four receptor/channels only ATP and acetylcholine receptor/channels were found. We attempted and failed to block the synaptic currents with a variety of nicotinic antagonists (hexamethonium, mecamylamine and trimethaphan) but reversibly inhibited both evoked and spontaneous transmission with the competitive ATP-receptor antagonist, suramin, and also with the desensitising ATP-receptor agonist, α,β-methylene ATP. It thus seemed clear that these novel synaptic currents were mediated by ATP-gated receptor/channels [9].

Three main features stand out which make ATP-mediated transmission different from the main fast excitatory transmission in the brain, mediated by glutamate: (1) In some preparations ATP receptor/channels have been shown to have significant Ca^{2+} permeability, even at hyperpolarised potentials. (2) ATP is broken down to adenosine in the synaptic cleft. Thus if adenosine receptors are present this could provide a novel form of autoinhibition. (3) In the peripheral nervous system, ATP has always been shown to be a cotransmitter with ACh or noradrenaline. It may be that ATP will also turn out to be a cotransmitter (for example with glutamate) in the CNS.

Our preliminary results indicate that in some (but not all) cells the permeability of the channel to Ca^{2+} is in the order of 10 times the permeability to Na^+. This is a substantial permeability, comparable to the nicotinic ACh receptor/channel and the NMDA receptor/channel. No nicotinic synaptic currents have yet been found in the brain (though the receptors are commonly present) and the NMDA channel has an unusual current voltage relation due to being effectively blocked by Mg^{2+} at resting membrane potentials. In contrast the current/voltage relationship for central ATP-mediated synaptic currents is linear and thus represents the only means known for Ca^{2+} to enter central neurones in a non voltage sensitive manner. The only likely exception to this, unless nicotinic transmission is found somewhere in the brain, is a particular subunit combination of AMPA receptor (those lacking GluR-B) which also shows a degree of calcium permeability ($P_{Ca}/P_{Na} \sim 1.4$) and may be found to mediate synaptic transmission in some cell types. So far Jonas has reported that the pyramidal and granule cells in the hippocampus do not show substantial Ca^{2+} permeability [10].

Finally it is interesting to note that the development of fast ATP-mediated synaptic currents seems to occur quite late and very rapidly, at about 3 weeks of age. This is reminiscent of the development of the excitatory junction potential in mesenteric artery. This synaptic potential, which is now thought to be mediated by ATP-receptors, was shown many years ago to develop very rapidly at about 2 weeks of age [11].

The developmental change seen in the two systems is rather similar with small, barely detectable slow rising responses converting to a fast rising robust response over a period of a few days. Thus central and peripheral synapses which use ATP may share similar developmental features.

Purinoceptors in the Nervous System

The effects of both adenosine and ATP are exerted via cell surface receptors, collectively known as purinoceptors [see (12] for review). The purinoceptors are subdivided into two major classes: adenosine receptors (P_1 purinoceptors) and P_2 purinoceptors (also called nucleotide receptors). There is a fundamental problem in the P_2 nomenclature in that it does not provide for the possibility that there are receptors interacting with pyrimidines under physiological conditions. Moreover, the current classification scheme for nucleotides uses designations that do not conform with the IUPHAR rules. Therefore an interim usage is proposed as structural information becomes available [12]. The IUPHAR subcommittee proposal is that when *adequate* data to enable classification exist, the receptors may be grouped into P2X and P2Y families. The P2X receptor should designate ligand gated cation channels whereas P2Y receptors are G-protein coupled receptors.

All subforms of adenosine receptors belong to the family of rhodopsin-like G-protein coupled receptors. The A_1 receptor couples to members of the pertussis-toxin sensitive G-protein family (G_{i1-3}, G_o). The G-proteins can cause the inhibition of adenylyl cyclase (via both the α- and the β,γ-subunits), the activation of several types of K^+-channels, the inactivation of at least some types of voltage dependent Ca^{2+}-channels, and the activation of phospholipase C (via the β,γ-subunits of the G-protein), with subsequent activation of protein kinase C and increases in intracellular Ca^{2+} [12]. In addition, there are effects on phospholipase D, on Cl^--channels and on gene expression which may, at least in part, be secondary to the above-mentioned actions. A_1 receptors are widely distributed in the CNS and appear to physiologically regulate transmitter release and neuronal firing rates.

Particularly in dopamine-rich regions of the CNS, there is expression of A_{2a} mRNA. In the striatum, A_{2a} receptors are expressed in the same cells as those that express dopamine D_2 receptors – the so-called medium sized spiny neurones [13]. These neurons use GABA as their primary transmitter and also express met-enkephalin. There is evidence that the A_{2a} and D_2 receptors interact functionally. Thus, activation of A_{2a} receptors decreases the affinity of D_2 agonists for their receptors [14]. Furthermore, caffeine and theophylline, which act as non-specific adenosine receptor antagonists, potentiate D_2-receptor mediated responses and produce dopamine-like effects on their own as a consequence of their block of A_{2a} receptors. At least part of caffeine's stimulatory effects on motor behaviours could be due to the selective enhancement of a component of dopaminergic transmission. The net effect of caffeine is complicated by the fact that not only A_{2a} but also A_1 receptors, located presynaptically on both the excitatory glutamatergic inputs and the dopamine terminals, are affected.

Activation of A_{2b} receptors results in an increase in cAMP. These receptors, abundant in astrocytes [15], may be responsible for most of the adenosine induced increase in cAMP in brain slices. The A_{2b} receptor usually requires rather high concentrations of adenosine to be activated. This suggests that glial adenosine receptors are predominantly activated in pathological conditions that involve an increased adenosine formation.

A_3 receptors, resistant to antagonism by methylxanthines especially in rodents, are located in the CNS [12]. Activation of these receptors results in decreased cAMP formation and apparently decreases locomotor activity in mice. There is also some evidence that there may be other types of adenosine receptors. We found that the purportedly A_{2a} receptor selective agonist, CGS 21680, also binds, albeit with somewhat lower affinity, to cortical brain areas. These binding sites show atypical affinities not only for agonists, but also for antagonists (high affinity for the A_1 receptor antagonist DPCPX but also a high affinity for the A_{2a} agonist CGS 21680) [16]. The functional significance of this putative receptor is not known.

Beneficial Role of Adenosine in Cerebral Ischemia

Decreases in cerebral blood flow (CBF) below a critical threshold result in cerebral ischemia and associated damage. Two major hypotheses have been developed to account for the phenomenon of ischemia/reperfusion induced neuronal death. The EAA neurotransmitter hypothesis is largely aimed at events during the actual period of ischemia, whereas the free radical hypothesis is directed largely at events during reperfusion. There is convincing evidence both for glutamate and aspartate release during cerebral ischemia and that excessive activation of NMDA and AMPA receptors by these EAAs promotes depolarization and the accumulation of intracellular Ca^{++} leading to neuronal injury and death.

The brain appears to be particularly susceptible to oxidative damage in that it contains high levels of readily peroxidizable fatty acids. Furthermore, the brain is relatively poorly endowed with protective antioxidant enzymes or antioxidant compounds. Free radicals are able to initiate damage to many cellular elements such as lipids, proteins and nucleic acids. The damage to membrane lipids can result in changes in fluidity and permeability as well as in the function of membrane bound receptors, ion channels, and enzymes. A potential source of free radicals is from purine metabolism. During ischemia, ATP is utilized but cannot be resynthesized due to the cessation of mitochondrial oxidative phosphorylation. Its metabolites, adenosine, inosine, hypoxanthine and xanthine accumulate in the extracellular space. In the normoxic brain, hypoxanthine is metabolized by xanthine dehydrogenase to xanthine and ultimately to uric acid. With the onset of ischemia, the rise in intracellular Ca^{2+} results in a conversion of xanthine dehydrogenase to xanthine oxidase (XO), which uses molecular oxygen instead of the nucleotide radical of NAD as its electron acceptor, catalyzing the formation of O_2 during reperfusion.

There are potentially several mechanisms by which adenosine, either released from cells within the brain or administered from external sources, might exert cerebroprotective activity. Adenosine, acting at A_2 receptors dilates cerebral arterial resistance vessels, increasing cerebral blood flow. This action could be particularly important in neutralizing the "no-reflow" phenomenon which occurs in reperfused ischemic brains, as a result of the formation and release of vasoconstrictor agents such as prostaglandins. Adenosine, acting at A_1 receptor, inhibits the ischemia-evoked release of the EAAs glutamate and aspartate and should diminish excitotoxicity. A_1 receptor agonists have been demonstrated to reduce the damage caused by kainic acid injections, in part as a result of a reduction in kainate-elicited glutamate release. Adenosine, again acting at A_1 receptors, reduces membrane Ca^{2+} permeability and increases K^+ permeability, thereby depressing neuronal excitability. Neutrophil aggregation and migration into the ischemic/reperfused brain is inhibited by activation of A_2 receptors, reducing the potentially injurious consequences of the invasion of these cells. Finally, adenosine itself, after re-uptake into cerebral neurons and glia, is the substrate for resynthesis of ATP.

Potentially damaging consequences of elevated extracellular adenosine concentrations have recently become apparent. Activation of adenosine A_2 receptors results in an enhancement of glutamate and aspartate release from the ischemic brain [17] and we have recently demonstrated that the adenosine A_2 receptor antagonist, CGS 15943, attenuates ischemia-evoked damage to the CA1 hippocampal pyramidal cell layer of the Mongolian gerbil [18]. The conversion of the adenosine metabolite hypoxanthine to xanthine and uric acid by xanthine oxidase is associated with the generation of injurious free radicals, which likely exacerbate ischemia-evoked neuronal injury and death. Adenosine induces apoptosis of astroglial cells in rat brain primary cultures by a receptor-independent process [19] and it is possible that, if a similar event occurs in the intact brain, this could be a contributing factor to the phenomenon of delayed neuronal death following ischemia/reperfusion.

References

1. Craig CG, White TD. N-Methyl-D-aspartate- and non-N-methyl-D-aspartate-evoked adenosine release from rat coritcal slices: distinct purinergic sources and mechanisms of action. J. Neurochem. 193; 60: 1073–1080.
2. Rudolphi KA, Schubert P, Parkinson FE, Fredholm BB. Adenosine and brain ischemia. Cerebrovasc. Brain Metab. Rev. 1992; 4: 346–369.
3. Zhang G, Franklin PH, Murray TF. Manipulation of endogenous adenosine in the rat prepiriform cortex modulates seizure susceptibility. J. Pharmacol. Exp. Ther. 1993; 264: 1415–1424.
4. McQueen DS, Ribeiro JA. Pharmacological characterization of the receptor involved in the chemoexcitation induced by adenosine. Br. J. Pharmacol. 1986; 88: 615–620.
5. Correia-de-sá P, Sebastião AM, Ribeiro JA. Inhibitory and excitatory effects of adenosine receptor agonists on evoked transmitter release from phrenic nerve endings of the rat. Br. J. Pharmacol. 1991; 103: 1614–1620.
6. Spignoli G, Pedata F, Pepeu G. A_1 and A_2 adenosine receptors modulate acetylcholine release from brain slices. Dur. J. Pharmacol. 1984; 97: 341–342.
7. Okada Y, Nishimura S, Miyamoto T. Excitatory effect of adenosine on neurotransmission in the slices of superior colliculus and hippocampus of guinea pig. Neurosci. Lett. 1990; 120: 205–208.
8. Cuhna RA, Milusheva E, Vizi ES, Ribeiro JA, Sebastião AM. Excitatory and inhibitory effects of A_1 and A_{2a} adenosine receptor activation on the electrically evoked $[^3H]$ acetylcholine release from different areas of the rat hippocampus. J. Neurochem. 1994; 63: 207–214.
9. Edwards FA, Gibb AJ, Colquhoun D. ATP receptor-mediated synaptic currrents in the central nervous system. Nature 1992; 359: 144–147.
10. Jonas P. Glutamate receptors in the central nervous system. Ann. NY Acad. Sci. 1993; 707: 126–135.
11. Hill CE, Hirst GDS, Van Helden DF. Development of sympathetic innervation to proximal and distal arteries of the rat mesentery. J. Physiol. 1983; 338: 129–147.
12. Fredholm BB, Abbracchio MP, Burnstock G, Daly JW, Harden TK, Jacobson KA, Leff P, Williams M. Nomenclature and classification of purinoceptors: a report from the IUPHAR subcommittee. Pharmacol. Rev. 1994; 46: 143–156.
13. Fink JS, Weaver DR, Rivkes SA, Peterfreund R, Pollak R, Adler E, Reppert SM. Molecular cloning of a rat A_2 adenosine receptor: Selective co-expression with D_2 dopamine receptors in rat striatum. Mol. Brain Res. 1992; 14: 186–195.
14. Ferré S, von Euler G, Johansson B, Fredholm BB, Fuxe K. Stimulation of adenosine A_2 receptors decreases the affinity of dopamine D_2 receptors in rat striatal membranes. Proc. Natl. Acad. Sci. USA 1991; 88: 7238–7241.
15. Altiok N, Balmforth AJ, Fredholm BB. Adenosine receptor induced cAMP changes in D384 astrocytoma cells and the effect of bradykinin thereon. Acta. Physiol. Scand. 1992; 144: 55–63.
16. Cunha RA, Johansson, Sebastião AM, Ribeiro JA, Fredholm BB. The binding sites of the A_{2a} adenosine receptor agonist $[^3H]$CGS 21680 are different in the hippocampus and cerebral cortex and in the striatum of the rat. Drug. Develop. Res. 1994b; 31: 260.
17. Simpson RE, O'Regan MH, Perkins LM, Phillis JW. Excitatory transmitter amino acid release from the ischemic rat cerebral cortex: effects of adenosine receptor agonists and antiagonists. J. Neurochem. 1992; 58: 1683–1690.
18. O'Regan MH, Simpson RE, Perkins LM, Phillis JW. The selective A_2 adenosine receptor agonist CGS 21680 enhances excitatory transmitter amino acid release from the ischemic rat cerebral cortex. Neurosci. Lett. 1992; 138: 169–172.
19. Gao Y, Phillis JW. CGS 15943, an adenosine A_2 receptor antagonist, reduces cerebral ischemic injury in the Mongolian gerbil. Life Sciences 1994; 55: PL61–PL65.

Pharmacological Sciences: Perspectives for
Research and Therapy in the Late 1990s
ed. by A.C. Cuello and B. Collier

Adenosine: Some Therapeutic Applications and Prospects

Trevor W. Stone[1], Michael G. Collis[2], Michael Williams[3],
Leonard P. Miller[4], Akira Karasawa[5], and D. Hillaire-Buys[6]

[1]*School of Pharmacology, Institute of Biomedical and Life Sciences, University of Glasgow,
Scotland;* [2]*Cardiovascular Biology, Pfizer Central Research, Sandwich, Kent, UK;*
[3]*Neuroscience Discovery, Abbott Laboratories, Abbott Park, IL., USA;* [4]*Gensia Inc.,
San Diego, CA., USA;* [5]*Pharmaceutical Research Laboratories, Kyowa Hakka Kogyo Ltd.,
Shimotagari, Japan;* [6]*Laboratoire de Pharmacologie, Institute de Biologie, F-34060 Montpellier,
France*

Summary. The presence of all four types of adenosine receptor, A_1, A_{2a}, A_{2b} and A_3, in the
cardiovascular system raises the possibility of developing adenosine analogues for use in a
variety of clinical disorders. The production of anginal pain and tachyphylaxis have limited
applications in angina or hypertension, but adenosine has found use in the diagnosis and
treatment of supraventricular tachycardias. Selective adenosine receptor agonists may be used
to mimic cardiac preconditioning protection and protect against the consequences of myocardial ischaemia. A_1 receptors in the kidney can mediate renal afferent arteriole constriction with
a reduction of glomerular filtration rate, and antagonists show great promise in the treatment
of acute renal failure. In the CNS, purine agonists show activity as analgesics, antipsychotics,
cerebral neuroprotectants, anticonvulsants and antidepressants, as well as other actions. The
development of adenosine regulatory agents, such as acadesine, to augment local concentrations or activity of endogenous adenosine, also shows great promise in tissue protection
against ischaemia. In the pancreas, receptors for adenosine and adenine nucleotides play an
important role in the regulation of hormone secretion, and agonists selective for the P_{2y}
population of ATP receptors may be valuable new additions to the class of oral antidiabetic
drugs.

Introduction

Since the discovery by Sattin and Rall [1] that adenosine could have
actions on central neurones which were quite independent of its conversion to cyclic AMP and which could be blocked by theophylline, the
study of adenosine receptors and their physiological roles has developed
apace. In parallel, academic and pharmaceutical laboratories have been
developing new chemical entities with which to probe receptor function
and which might be of value in medical treatment. Therapeutic areas in
which adenosine receptor agonists, antagonists and enhancers might
soon appear to be of clinical benefit include cardiovascular, central
nervous system, kidney and pancreatic disorders. This chapter will
briefly review aspects of purine potential use in these areas.

Correspondence to: T.W. Stone, School of Pharmacology, Institute of Biomedical and Life
Sciences, West Medical Building, University of Glasgow, Glasgow, Scotland G12 8QQ.

Cardiovascular System

The vasodilator, cardio-inhibitory and renal actions of adenosine have been known for many years. The release of the purine from metabolically stressed tissues has prompted the concept that it is a "retaliatory metabolite" which redresses metabolic imbalance by modulating both blood supply and metabolic demand. Adenosine exerts its effects via receptors which have been sub-divided into four types (A_1, A_{2a}, A_{2b}, A_3) based on pharmacological and molecular characteristics [2]. All four occur in the cardiovascular system but A_1 and A_{2a} are predominantly responsible for the cardio-inhibitory and vasodilator actions of the purine. The role of A_{2b} and A_3 receptors is less clear: the former may mediate relaxation of large conduit arteries and cutaneous veins whereas the latter may have cardiac effects in some species.

A large number of therapeutic applications for drugs which mimic, block or potentiate the effects of the purine have been recognised (Table 1). Candidate drugs based on these mechanisms have undergone or are undergoing clinical evaluation. However, only adenosine and the non-selective adenosine receptor antagonist theophylline are currently available for medical use.

Adenosine has a short plasma half-life [measurable in seconds] and is used acutely as a coronary vasodilator during thallium scintigraphy. Attempts to exploit the vasodilator actions of stable adenosine analogues in the treatment of hypertension or angina have not been successful due to tachyphylaxis in the former case and to a paradoxical provocation of anginal pain in the latter. Adenosine also inhibits cardiac S-A nodal rate and A-V nodal conduction and has thus found use in the acute treatment of supraventricular arrhythmias. Conversely, excessive concentrations of endogenous adenosine may account for some cardiac rhythm disturbances caused by myocardial ischaemia. The non-selective adenosine antagonist theophylline has been used to treat such dysrhythmias.

Table 1. Cardiovascular applications of adenosine receptor activation

Effect	Application	Receptor	Ligand	Status
Coronary vasodilator	Scintigraphy	A_{2a}	Adenosine	Market
Coronary vasodilator	Angina	A_{2a}	NECA	—
Systemic vasodilator	Hypertension	A_{2a}	CGS21680	— (Receptor down-regulation)
Exercise hyperaemia	Ischaemic disease	A_{2a}	ZD 4398	Early clinical
Renal vasoconstriction	Renal failure	A_1	KW-3902	Early clinical
Negative dromotropic	PSVT	A_1	Adenosine	Market
Negative dromotropic	A-V block	A_1	Theophylline	Market
Cardioprotective	Myocardial ischaemia	A_1/A_2	Acadesine	Phase III
Pre-conditioning	Myocardial ischaemia	A_1/A_3	—	—
Venodilator	Angina?	A_{2b}?	—	—

Theophylline is also beneficial in the treatment of angina and intermittent claudication. It is unclear whether this beneficial effect is due to the prevention of adenosine mediated vascular steal phenomena or to blockade of the nociceptive effects of the purine. Current clinical evaluation of vascular selective adenosine receptor antagonists should resolve this question.

A number of the actions of adenosine such as coronary vasodilator, anti-platelet, anti-neutrophil and anti-adrenergic effects, could be beneficial during acute myocardial infarction. Its use in this clinical situation is prevented by systemic vasodilatation, which can provoke vascular steal and hypotension. Acadesine is a nucleoside analogue that enhances local adenosine release from ischaemic tissues and therefore avoids systemic effects. A recent phase III U.S. clinical trial with acadesine suggests that infusion during cardiac CABG surgery may be useful in reducing the incidence of myocardial infarction.

Recently, it has been shown that adenosine mediates cardiac preconditioning in many species including man. Preconditioning is the phenomenon by which a short period of ischaemia, preceding a longer period, protects the heart from the damaging consequences of the latter. The adenosine receptor mediating this effect is either A_1 or A_3 depending on species. Further studies are needed to assess the feasibility of mimicking preconditioning, using adenosine receptor agonists.

Thus, the wide variety of cardiovascular effects of adenosine offers significant potential for therapeutic application. Several compounds which interact with adenosine receptors are currently undergoing clinical evaluation, some of which may realise this potential. The increasing number of adenosine receptor sub-types that are being recognised should enhance the opportunities for discovery of new therapeutic agents with improved selectivity of action.

Renal Effects

Adenosine plays a major role in the pathogenesis of some forms of acute renal failure (ARF) by its action to constrict the afferent renal arteriole and mesangial cells causing a reduction in glomerular filtration rate (GFR). These effects are mediated via A_1 receptors and selective antagonists are in clinical development for treatment of ARF. The novel adenosine A_1 receptor antagonist KW-3902 induces diuresis and natriuresis in various animal species without affecting glomerular filtration rate. Following the chronic administration of KW-3902, serological abnormalities are less common as compared with furosemide and trichlormethiazide. Treatment with KW-3902 prevented the NaCl-induced hypertension in Dahl-NaCl sensitive rats, and ameliorated the ascites in rats caused by puromycin aminonucleoside. In animal models of acute renal failure induced by glycerol, cisplatin, gentamycin, cephaloridine or contrast media, KW-3902, but not furosemide or trichlormethiazide, significantly prevented the development of ARF. Moreover, KW-3902, in contrast to the other agents, induced consistent natriuretic effects even in the ARF state. This adenosine A_1 receptor antagonist could be a novel type of therapeutic agent for the treatment of hypertension, edema and ARF.

Central Nervous System (CNS)

Adenosine functions as a homeostatic agent in the brain regulating neurotransmitter release and the level of alertness of the organism. This is clearly shown by the sedative actions of adenosine and the worldwide consumption of the adenosine antagonist, caffeine, as both a CNS stimulant and cognition enhancer [3, 4].

From a pathophysiological viewpoint, very little is known regarding the involvement of adenosine in disease processes although there are a number of psychopharmacological and neurological states in which adenosine has been implicated. These include: anxiety, depression, schizophrenia, epilepsy, analgesia, sleep disorders, stress cognition, stroke, attention deficit disorder, Alzheimer's and Parkinson's diseases.

Many of the experimental paradigms used to define the role(s) of adenosine have been limited in nature. They have, due to a historical lack of tools to delineate actions at A_1, A_{2a} and A_{2b} receptors, focused exclusively on A_1 receptor-associated processes such that very little is known regarding the role of A_{2a} and A_{2b} receptors or the newly discovered A_3 receptor [5]. To date, potent, selective and bioavailable antagonists for non-A_1 receptors have not been available, which has limited the progression from the molecular to the whole animal, functional level.

Nonetheless, the role of adenosine in CNS responses to insult is well defined. Cerebral ischemia or hypoxia lead to significant increases in exogenous adenosine levels that conceptually can act homeostatically to limit excitotoxic cell death [6]. Similarly, the epileptic episodes are associated with increased adenosine levels [7]. Therapeutically, adenosine A_1 receptor selective agonists as well as non-selective agents like NECA are effective in inhibiting ischemia/hypoxia – evoked release of glutamate and will protect cultured cells exposed to anoxic conditions which normally result in cell death. However, extension of these investigations to whole animal studies has not led to universally convincing observations. This probably relates to the model of ischemia examined, e.g. global versus focal, and the approach utilized for activating adenosine receptors. These approaches include: (a) administration of A_1 receptor agonists, or (b) utilization of adenosine regulating agents

(ARAs) to potentiate endogenous adenosine levels. Side-effects, such as hypotension and bradycardia, associated with the administration of agonists, have hindered the progression of this approach beyond preclinical investigations. On the other hand, ARAs applied in recent preclinical experiments have not shown cardiovascular side-effects and have resulted in significant reduction in infarct volume in both focal and photothrombotic rat stroke models. Presently there is a small number of clinically available drugs which, to varying degrees, mediate their effects through activation of endogenous adenosine receptors. However, only acadesine has been examined in a clinical trial where stroke as defined by signs or symptoms of significant neurologic deficit that persisted for 24 h was an outcome measurement. In an American phase III trial, use of acadesine was associated with a dose-related reduction in the incidence of stroke following cardiac bypass graft surgery.

Another ARA, GP-1-668, has been examined in a preclinical model of focal stroke with reperfusion in rats. Ischemia was effected by occlusion of blood flow at the origin of the middle cerebral artery. Perfusion was restored after 105 min and animals were killed 24 h later for the examination of infarct volume using triphenyltetrazolium chloride stain. Infusion of GP-1-668 was initiated 30 min before occlusion and continued for 24 h. Infarct volume was reduced by 56% ($p = 0.002$, $n = 16$), supporting the potential utility of ARAs in ischemic brain injury.

In addition to dilation of cerebral vessels and inhibition of platelet aggregation, adenosine may also modulate cell adhesion processes and may have an additional role in attenuating the consequences of the inflammatory response such as blocking free radical formation from neutrophils recruited as part of the host defence process [8]. A_1 and A_{2a} receptor agonists are also effective anticonvulsant agents. Adenosine A_1 antagonists are, like caffeine, effective cognition enhancers.

Findings made some 20 years ago in classical animal models related to the antiparkinsonian actions of methylxanthines have been recently complemented by molecular biological findings. These have shown a close cellular and pharmacological association between adenosine A_{2a} receptors and dopamine D_2 receptors [9].

Stroke, epilepsy and Parkinson's disease thus reflect robust targets for adenosine therapeutics. Since agonist effects are subject to attenuation, adenosine antagonists and agents that potentiate adenosine actions represent the most promising treatment modalities. Currently in clinical trials are the selective A_1 receptor antagonists KFM19 and MDL 102,234 as cognition enhancers, and KW-6002 (an A_{2a} receptor antagonist) for the treatment of Parkinson's disease and GP 1-515 (an adenosine kinase inhibitor) for use in septic shock [10].

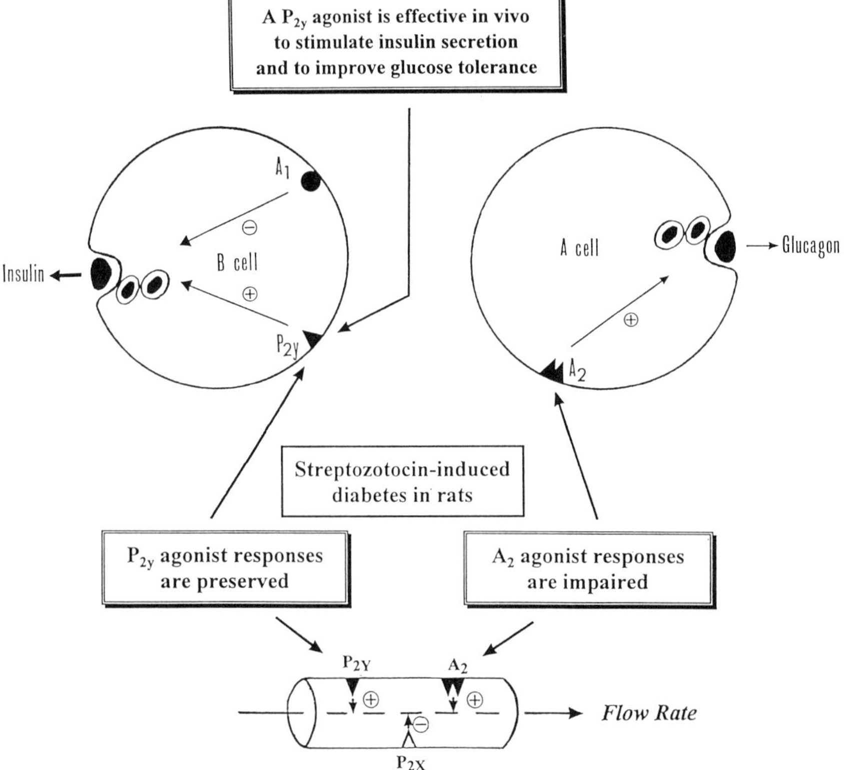

Fig. 1. Sites of action of adenine nucleotides on pancreatic cells and vessels.

Pancreatic Disorders

The insulin-secreting B cell is provided with two types of purinoceptors: P_2 purinoceptors (for ATP and/or ADP), of the P_{2y} subtype, activation of which stimulates insulin secretion, and P_1 purinoceptors of the A_1 subtype, activation of which decreases insulin secretion. The glucagon-secreting A cell possesses P_1 purinoceptors (A_2 subtype), the activation of which increases glucagon secretion. In pancreatic vessels, resistance is either decreased by the activation of both an A_2 and a P_{2y} purinoceptor or increased by the activation of a P_{2x} purinoceptor (Fig. 1).

In diabetic states, complications are frequently observed such as an impairment of the A-cell response and vascular disturbances. In the isolated perfused pancreas from normal rats and 5-week diabetic rats (streptozotocin-induced) adenosine-induced glucagon secretion and vasodilation are abolished and strongly reduced respectively. A 2-week chronic insulin treatment restores both A-cell and vessel responses whereas an acute insulin treatment (a 30 min *in vitro* infusion) is without effect. Thus, an impairment in P_2 purinoceptors may play a role in the pathophysiology of some diabetic complications.

The stable analogue of ADP, adenosine-5′-O-(2-thiodiphosphate) (ADPßS), is about 100 times more potent than ATP in stimulating insulin secretion from isolated perfused rat pancreas. In the diabetic model, insulin secretion from remnant B cells and vasodilation elicited by ADPßS are preserved. P_{2y} agonists may therefore be useful in the treatment of non-insulin dependent diabetes mellitus for their ability to stimulate insulin release. In normal animals (anesthetized rats and conscious dogs), in basal conditions and during glucose tolerance tests ADPßS stimulates insulin secretion and improves glucose tolerance after intravenous injection or oral administration. Thus the P_{2y} purinoceptors of B-cells may be considered a target for a new class of oral antidiabetic drugs in the treatment of non-insulin-dependent diabetes.

References

1. Sattin A, Rall TW. The effect of adenosine and adenine nucleotides on the cyclic AMP content of guinea-pig cerebral cortex slices. Molec. Pharmacol. 1970; 6: 13–20.
2. Fredholm BB, Abbrachio MP, Burnstock G, Daly JW, Harden TK, Jacobson KA et al. Nomenclature and classification of purinoceptors: a report from the IUPHAR subcommittee. Pharmacol. Rev. 1994; 46: 143–156.
3. Stone TW, editor. Adenosine in the nervous system. London: Academic Press, 1991.
4. Williams M. Purinoceptors in CNS function: targets for new therapeutic agents. In: Bloom FE, Kupfer DJ (editors). Psychopharmacology: Fourth generation of progress. New York: Raven Press. In press.
5. Ji X-D, von Lubitz D, Olah ME, Stiles GL, Jacobson KA. Species differences in ligand affinity at central A_3 receptors. Drug Develop. Res. In press.
6. Miller LP, Hsu C. Therapeutic potential for adenosine receptor activation in ischemic brain injury. J. Neurotrauma 1992; 9: S563–S577.
7. During MJ, Spencer DD. Adenosine: a potential mediator of seizure arrest and postictal refractoriness. Ann. Neurol. 1992; 32: 618–624.
8. Cronstein BN. Adenosine, an endogenous anti-inflammatory agent. J. Appl. Physiol. 1994; 76: 5–13.
9. Ferre S, Fuxe K, von Euler B, Johansson B, Fredholm BB. Adenosine-dopamine interactions in the brain. Neuroscience 1992; 51: 501–512.
10. Firestein GS, Boyle D, Bullough DA, Gruber HE, Sajjadi FG, Montag A, Sambol G, Mullane KM. Protective effect of an adenosine kinase inhibitor in septic shock. J. Immunol. 1994; 152: 5853–5859.

Cardiovascular Pharmacology

Pharmacological Sciences: Perspectives for
Research and Therapy in the Late 1990s
ed. by A.C. Cuello and B. Collier
© 1995 Birkhäuser Verlag Basel/Switzerland

Endothelins: Recent Progress

Timothy D. Warner[1], Michael J. Dunn[2], Sadao K. Kimura[3],
Duncan J. Stewart[4], Masashi Yanagisawa[5], Tomoh Masaki[6] and
John R. Vane[1]

[1]*The William Harvey Research Institute, The Medical College of St. Bartholomew's Hospital,
London EC1M 6BQ, U.K.;* [2]*Department of Physiology and Biophysics, Case Western Reserve
University, Cleveland, Ohio, U.S.A.;* [3]*Center for Biomedical Science, Chiba University School of
Medicine, Chiba, Japan;* [4]*McGill Vascular Biology Group, McGill University, Montréal,
Canada;* [5]*Department of Molecular Genetics, Howard Hughes Medical Institute, The University
of Texas, Southwestern Medical Center at Dallas, Dallas, Texas, U.S.A.;* [6]*Department of
Pharmacology, Faculty of Medicine, Kyoto University, Kyoto 606, Japan;* [7]*The William Harvey
Research Institute, The Medical College of St. Bartholomew's Hospital, London EC1M 6BQ,
U.K.*

Summary. Some recent progress in endothelin (ET) research is reviewed. Numerous antagonists for ET_A/ET_B receptors and endothelin (ET) analogues have been developed recently and
used to elucidate the roles of ET-receptor subtypes in vascular and non-vascular systems and
in the physiological and pathophysiological effects of ET. Concurrently, the intracellular
signalling pathways stimulated by ET have become better characterised. For instance,
activation of MAP kinase is involved in the mitotic action of ET and ET-1 induces the
expression of COX-2. Most interestingly, the cDNA for endothelin-converting enzyme (ECE)
has been cloned recently by several laboratories and this will greatly assist in establishing the
true structure and function of ECE. Initial gene knock-out experiments have also shown that
ET-1 and ET_A receptors are important for normal pharyngeal development in mice. Finally,
ET-1 has been implicated in numerous pathologies. For instance, ET-1 and ET_A receptors are
abundantly produced in pulmonary tissues and ET-1 may play an important role in the
development of pulmonary hypertension.

Introduction

Endothelin (ET-1) is a potent vasoconstrictor peptide, originally isolated from conditioned media of cultured porcine aortic endothelial cells
[1]. Subsequently, two other isoforms of endothelin, ET-2 and ET-3
were described [2], although it appears that vascular endothelial cells
produce exclusively ET-1. In addition to the endothelin peptides at least
four distinct, but structurally very similar, peptides named sarafotoxin
(SX) 6a, SX6b, SX6c, and SX6d have been identified in a rare snake
venom. The ET/SX peptides have a common structure of 21 amino acid
residues linked by two intrachain disulfide bridges. Using these agonists
as tools much effort has been put into discerning which receptors [3, 4]
mediate the effects of the endothelins in different tissues. This work has
been accompanied by great progress in the identification of endothelin
processing pathways, endothelin intracellular signalling pathways [5],
and physiological and pathophysiological roles for the endothelins.

Endothelin Receptor Subtypes and Pharmacological Responses

The effect of the ET/SX peptides, and of synthetic analogues such as IRL1620 and BQ3020 are mediated by specific endothelin receptors, of which two have been cloned and expressed in mammalian species. These are the ET_A receptor, which is selective for ET-1 or ET-2 over ET-3, and on which SX6c has little or no effect [3], and the ET_B receptor on which the ET/SX peptides act equally [4]. Initial experiments had to rely on the differences in agonist potencies to characterise endothelin receptors. However, more recently ET_A receptor-selective antagonists such as BQ-123 and FR139317, and $ET_{A/B}$ receptor non-selective antagonists such as PD142893, PD145065, Ro46-2005 and Ro47-0203 have greatly assisted in this study. From early experiments it appeared that within the vasculature ET_A receptors mediated the vasoconstrictor effects of the ET/SX peptides and ET_B receptors the endothelium-dependent vasodilator effects. However, it has now become clear that ET_A and/or ET_B receptors may be present in different amounts on vascular smooth muscle (see ref. [6]). For instance, in the rat kidney vasoconstrictions induced by ET-1, *in vivo* or *in vitro*, are only partially blocked by BQ-123, whereas they are completely blocked by PD145065, showing mediation by a mixed population of ET_A and ET_B receptors. However, vasodilatations appear to be exclusively mediated by ET_B receptors. For instance, in the rat mesentery, *in vitro*, SX6c is equiactive with ET-1 at inducing relaxations, an effect which is antagonised by PD142983, but not by BQ-123. Interestingly, further studies characterising these ET_B dilator receptors and comparing them to ET_B constrictor receptors suggest that they are differentially sensitive to blockade by e.g. PD142893. These different receptors have been tentatively named ET_{B1}, present on endothelium, and ET_{B2}, present on smooth muscle [7]. Possibly the ET_{B1}, receptors correlate with super high affinity sites reported in cerebral membranes (see ref. [7]). In addition, some functional studies have suggested that endothelial cells may express ET_C receptors, selective for ET-3 over ET-1. However, molecular biological techniques have so far only identified such receptors in non-mammalian species.

ET receptors are also present throughout non-vascular smooth muscles, where, as in the vasculature, ET_A and/or ET_B receptors can mediate contraction. For instance, in the guinea-pig constriction of airways induced by the ET/SX peptides is mediated predominately by ET_B receptors, while ET_A receptors play a minor role [8]. Interestingly, *in vivo*, in this same species, there is also an indirect bronchoconstrictor response to ET-1 that is mediated by the release of thromboxane A_2, following the activation of ET_B receptors [9]. Thus, as in the vasculature, the endothelins produce effects upon non-vascular smooth muscle directly, and indirectly via stimulation of the release of other mediators.

Most recently the characterization of ET receptors, and more importanly the elucidation of the physiological and pathophysiological roles of the endothelins, has been greatly accelerated by the development of orally active ET receptors antagonists such as BMS-182874, CGS27830, SB209670 and Ro47-0203 (bosentan). Experiments using these compounds have given increasing support to the idea that ET is associated with hypertension, renal ischemia and reperfusion, and subarachnoid haemorrhage [7].

Endothelins Activate Postreceptor Signalling Through the Mitogen-Activated Protein Kinase Cascade

Endothelins stimulate cell growth, protein synthesis and matrix development, probably via stimulation of mitogen-activated protein kinase (MAPK). MAPKs are a family of serine/threonine kinases with multiple isoforms of which p42 and p44 MAPK are activated by growth factors and vasoconstrictor peptides. The upstream kinases linking receptor occupancy to MAPK activation include MAPKK, MAPKKK (MAPK kinase, MAPK kinase kinase) and rab kinase activated by ras. MAPK phosphorylates, for instance, tyrosine and threonine with MAPKK. To investigate this further, CHO cells have been selectively transfected with either ET_A or ET_B receptors after which p42 and p44 MAPK were found to be activated by ET-1. Interestingly, p42 was only poorly activated and phosphorylated by ET-3 suggesting this is an ET_A receptor-mediated event. In cultured rat glomerular mesangial cells ET-1 is mitogenic and stimulates p42 biphasically [10] such that there is a peak in activity at 5 min, and then a sustained increase for up to 4 h. In these same cells p44 MAPK is also acutely stimulated over 5–30 min, but there is no sustained activation. In CHO cells transfected with ET_A receptors, ET-1 activates p42 MAPK with a similar time-course following activation of immediate upstream kinase p45 MAPKK. Interestingly, stimulation of mesangial cells by ET-1 leads to the translocation of MAPK to the nucleus, whereas MAPKK remains cytosolic in location.

ET-1 increases the activity of cystolic phospholipase A_2 leading to an increase in prostaglandin E_2 and I_2 production in CHO cells stably expressing ET_A or ET_B receptors [11]. This ET-1 induced activation in phospholipase A_2 is independent of protein kinase C and not dependent upon an elevation in intracellular free calcium level. However, 1–2 h treatment of the cells with ET-1 causes an increased expression of COX-2, which is inhibited by heparin, dexamethasone, and herbimycin, suggesting that a tyrosine kinase pathway may be involved in this mechanism.

Endothelin Converting Enzyme

Processing of the 212 amino acid human proendothelin-1 yields a 38 amino acid residue intermediate called big ET-1. Conversion of big ET-1 to mature ET-1 is then catalyzed by an enzyme named endothelin converting enzyme (ECE). Since the vasoconstrictor activity of ET-1 is several hundred-fold higher than that of big ET-1, this enzyme is a key enzyme in the production of the potent vasoconstrictor peptide. ECE appears most likely to be a membrane-bound, phosphoramidon-sensitive, glycosylated metalloproteinase with a relative molecular weight of 130 kDa, which is active at neutral pH, and inhibited by EDTA. However, other peptidases have also been suggested as candidates for the physiologically relevant form of ECE.

Recently, several laboratories have reported the cloning of cDNA for ECE from vascular endothelial cells and deduced the amino acid sequence of the mature protein [12–14]. Thus, the deduced amino acid sequences of rat and bovine ECE consist of 754 and 758 amino acid residues, respectively. The two proteins have a 91% sequence identity, and show that ECE is structurally related to neutral endopeptidase and human Kell blood group protein. It has a transmembrane domain at its N-terminal side and the C-terminal, putative, extracellular domain, has ten N-linked glycosylation sites. ECE also has a consensus sequence for a zinc binding motif. Interestingly, it was also demonstrated that glutamic acid at position 752 is important in ECE activity, for when this was replaced by arginine, the activity dramatically decreased.

In COS or CHO cells the activity of transfected and expressed ECE was inhibited by phosphoramidon but not by thiorphan as seen in native endothelial cells. In further studies, when CHO K1 cells were transfected with prepro ET-1 and ECE cDNAs the cells were found to secrete mature ET-1, a process which was inhibited by phosphoramidon. Thus, this ECE can progress big ET-1 within intact cells. The cloned ECE converts big ET-1 well, but big ET-2 or big ET-3 only poorly. Interestingly, FR90153 was found to inhibit ECE in broken cells, but not in intact cells. This suggests that big ET-1 is processed within the cells, as FR901533 most probably does not penetrate cell membranes.

As may be expected, Northern blot analysis of ECE cDNA showed it to be widely distributed in the ovaries, testes, adrenal gland, liver, cerebrum, lungs and heart, and, of course, endothelial cells.

Physiological and Pathophysiological Meanings of ETs

To elucidate the physiological and pathophysiological roles of endothelins, some investigators have disrupted the ET or ET-receptor genes in embryonic stem cells and then observed changes produced in the animals. These studies have shown that ET-1, or ET_A receptor-deficient mice die of respiratory failure at birth and have morphological abnormalities of the mandible, ear, tongue and palate [15]. Interestingly, ET-1 gene heterozygotic mice, which produce lower levels of ET-1 than wild-type mice, actually show an elevated blood pressure and increased heart rate relative to controls, suggesting that ET-1 is implicated in the regulation of blood pressure by mechanism other than vasoconstriction.

The plasma ET concentration is elevated in many cardiovascular diseases states, possibly as a result of vascular injury. For instance, patients with pulmonary hypertension have substantial and complex alterations in the levels of plasma immunoreactive ET-1, which may reflect changes in the net release or clearance of ET-1 by the lung [16]. Thus, in patients with primary pulmonary hypertension, there are high levels of arterial compared with venous ET-1, suggesting a net pulmonary production of ET-1 which may elevate pulmonary vascular resistance. Such an increase in ET-1 production may also cause the abnormalities in parenchymal lung function seen in patients with pulmonary fibrosis. Indeed, further studies using immunocytochemical and *in situ* hybridization techniques showed that there is abundant ET-1 within the endothelium of markedly hypertrophied muscular pulmonary arteries and plexogenic lesions, the levels of which correlated with the hemodynamic severity of disease in pulmonary hypertensives [17]. In addition, a high expression of ET-1 was observed within hypertrophied, type-II, alveolar, epithelial cells in patients suffering from idiophatic pulmonary fibrosis, which was related neither to pulmonary hypertension nor to the vascular expression of ET-1. These results suggest a role for increased expression of ET-1 in the pathophysiology of diverse pulmonary disorders characterized by increased vascular tone and/or abnormal cellular proliferation and fibrosis.

Perspectives

In the 6 years following the discovery of the endothelins there has been an explosion of research in this field. However, there are still major questions to which answers are needed. For instance: (1) which receptors mediate the effects of the endothelins in pathological conditions; (2) what is the true nature of the physiological ECE; (3) which are the important signal transduction mechanisms in the mediation of the vasoconstrictor and mitotic actions of the endothelins; and (4) what is the true extent of the interactions of the endothelins with other vasoactive mediators? Although these questions cannot be answered at present, it is clear that the current speed of endothelin research coupled with newly available research tools will lead to the answers and will reveal the true role of the endothelins in both vascular and non-vascular systems.

References

1. Yanagisawa M, Kurihara H, Kimura S, Tomobe Y, Kobayashi M, Mitsui Y et al. A novel potent vasoconstrictor peptide produced by vascular endothelial cells. Nature 1988; 332: 411–415.
2. Inoue A, Yanagisawa M, Kimura S, Kasuya Y, Miyauchi T, Goto K. et al. The human endothelin family: three structurally and pharmacologically distinct isopeptides predicted by three separate genes. Proc. Natl. Acad. Sci. USA 1989; 86: 2863–2867.
3. Arai H, Hori S, Aramori I, Ohkubo H, Nakanishi S. Cloning and expression of a cDNA encoding an endothelin receptor. Nature 1990; 348: 730–732.
4. Sakurai T, Yanagisawa M, Takuwa Y, Miyazaki H, Kimura S, Goto K et al. Cloning of a cDNA encoding a non-isopeptide-selective subtype of the endothelin receptor. Nature 1990; 348: 732–735.
5. Masaki T. Endothelins: Homeostatic and compensatory actions in the circulatory and endocrine systems. Endocrine Review 1993; 14: 256–268.
6. Warner TD. Characterization of endothelin synthetic pathways and receptor subtypes: physiological and pathophysiological implications. Eur. Heart J. 1993; 14(Suppl I): 42–47.
7. Warner TD, Battistini B, Doherty A, Corder R. Endothelin receptor antagonists: actions and rationale for their development. Biochem. Pharmacol. 1994; 48: 625–635.
8. Battistini B, Warner TD, Fornier A, Vane JR. Comparison of PD145065 and Ro46-2005 as antagonists of contractions of guinea-pig airways induced by endothelin-1 or IRL1620. Eur. J. Pharmacol. 1994; 252(3): 341–345.
9. Warner TD, Sorrentino R, Allcock GH, Battistini B, Vane JR. The non-selective endothelin receptor antagonist, PD145065, inhibits bronchoconstrictions induced by endothelin-1 or sarafotoxin 6c in the anaesthetised guinea-pig. Pharmacol. Commun. 1995; 5: 163–169.
10. Wang Y, Pouyssegur J, Dunn MJ. Endothelin stimulates mitogen activated protein kinase p42 actively through the phosphorylation of the kinase in rat mesangial cells. J. Cardio-vasc. Pharmacol. 1993; 22(Suppl. 8): S164–167.
11. Schramek H, Wang Y, Konieczkowski M, Rose PM, Sedor JR, Dunn MJ. Endothelin-1 stimulates cystolic phospholipase A_2 in chinese hamster ovary cells stably expressing the human ET_A or ET_B receptor subtype. Biochem. Biophys. Res. Commun. 1994; 199: 992–997.
12. Shimada K, Takahashi M, Tanzawa K. Cloning and functional expression of endothelin-converting enzyme from rat endothelial cells. J. Biol. Chem. 1994; 269: 18275–18278.
13. Xu D, Emoto N, Giaid A, Slaughter C, Kaw S, deWit D, Yanagisawa M. ECE-1: A membrane-bound metalloprotease that catalyzes the proteolytic activation of big endothe-lin-1. Cell 1994; 78(1): 1–20.
14. Ikura T, Sawamura T, Shiraki T, Hosokawa H, Kido T, Hoshikawa H, Shimada K, Tanzawa K, Kobayashi S, Miwa S, Masaki T. cDNA cloning and expression of bovine endothelin converting enzyme. Biochem. Biophys. Res. Commun. 1994; 203: 1417–1422.
15. Kurihara Y, Kurihara H, Suzuki H, Kodama T, Maemura K, Nagai R et al. Elevated blood pressure and craniofacial abnormalities in mice deficient in endothelin-1. Nature 1994; 368: 703–710.
16. Stewart DJ, Levy RD, Cernacerk P, Langleben D. Increased plasma endothelin-1 in pulmonary hypertension: Marker or mediator of disease? Ann. Intern. Med. 1991; 114: 464–469.
17. Giaid A, Yangisawa M, Langleben D, Michel RP, Levy R, Shennib H. Expression of endothelin-1 in the lungs of patients with pulmonary hypertension. N. Engl. J. Med. 1993; 328(24): 1732–1739.

Drugs Directly Affecting The Arterial Wall

Rodolfo Paoletti and Maurizio R. Soma

Institute of Pharmacological Sciences, University of Milan, Via Balzaretti 9, 20122 Milan, Italy

Introduction

Great progress has been made over the last three decades in identifying the most important risk factors for cardiovascular disease (CVD), and we have learned to apply measures that successfully lower the incidence of this disease. The interest focused on smoking in the late 1960s and on hypertension in the 1970s, while the 1980s saw paramount attention on the diagnosis and management of hypercholesterolemia. The declining rates of CVD mortality and morbidity throughout the industrialized world during the same period clearly reflect the identification and proper management of these three major risk factors [1]. The recent advances in epidemiological, clinical, and basic research on primary and secondary dyslipidemias have allowed both investigators and physicians to gain new insights into the pathogenesis and clinical management of CVD [2]. Atheroscelerosis can be viewed as the result of a multiplicity of complex interactions that involve repeated injury with healing or reparative responses. Atherosclerosis plays a crucial role in many cardiovascular diseases, including myocardial infarction and cerebro-vascular insufficiency [3], and has been involved, together with other mechanisms, in the restenosis after successful coronary angioplasty [4]. This latter emerges as a critical clinical problem for the future, in view of the explosive growth of coronary angioplasty [5], the still high rate of restenosis, and the poor efficacy of current treatments [6].

Therapeutic strategies for the prevention of "spontaneous" or induced atherosclerosis in man are essentially based on the correction of major risk factors, either elevated plasma lipids or arterial blood pressure. Such interventions have been successful in reducing cardiovascular risk following long-term treatment and can induce a regression of existing lesions. However, it appears that the prevention of cardiovascular disease in the future will involve also the direct pharmacological control of the processes occurring in the arterial wall [7]. For these reasons, more attention should be paid to the development and clinical use of drugs acting directly on the arterial walls. Calcium antagonists and HMG-COA reductase inhibitors (statins) will be discussed herein.

Calcium Antagonists

The calcium antagonists show several important anti-atherosclerotic activities *in vitro* and *in vivo*. In cultured cells these drugs protect arterial smooth muscle cells (SMC) against cholesterol deposition and control cellular proliferation. In addition, the matrix synthesis by SMC is reduced. In *in vivo* models calcium antagonists protect against lesions induced by cholesterol feeding, endothelial injury and experimental calcinosis [8].

The antiatherosclerotic effect of calcium antagonists does not involve a reduction of plasma cholesterol or blood pressure thus indicating a direct effect on the arterial wall [8]. In a series of animal experiments the influence of calcium antagonists on calcium deposition within the vessel wall and on the development of artifically-induced atherosclerosis proved to be favourable [9–11]. Two randomised, placebo-controlled studies [12, 13] have reported on the effects of calcium antagonists on atheroma in the coronary arteries. The first study is the International Nifedipine Trial on Antiatherosclerotic Therapy (INTACT). Four hundred and twenty-five patients were controlled and randomised to treatment with nifedipine. The study lasted for 3 years. No regression, or at least lack of progression, was reported, but a 28% reduction in the number of new lesions could be recorded. In the second study, by Waters et al. [13], 383 patients were randomised to treatment with nicardipine. The study lasted for 2 years and the findings were essentially the same as those reported in the INTACT study: no effect on the regression or progression of established coronary lesions, but a reduction in the progression of small or new lesions. In both studies the inhibitory effect on progression of new lesions was accomplished without a correlated effect on blood lipid levels. The results of these studies indicate that some as yet unidentified biological processes in the early development of atheromatous lesions are sensitive to calcium antagonists. While the efficacy of calcium antagonists as coronary and peripheral vascular dilators can be accounted for in terms of their inhibitory effects on calcium influx throughout the voltage-sensitive calcium channels, their ability to act as antiatherosclerotic agents is less understood.

Table 1. Effect of lacidipine (3 mg/kg/day) on proliferative lesions induced by perivascular manipulation of hypercholesterolemic normotensive rabbit carotid arteries

Treatment	I:M	SD	% of control	p value
Control	0.56	0.11	—	—
Sham	0.03	0.02	5	—
Lacidipine	0.32	0.10	57	0.01

Ten rabbits each in the control and lacidipine treatment groups; 20 rabbits in the sham group. I:M, intimal:medial tissue ratio (mean ± SD); SD, standard deviation; p value, lacidipine vs, control treatment.

Several calcium-dependent processes contribute to the antherogenesis, including lipid infiltration and oxidation, endothelial injury, action of chemotactic and growth factors, smooth muscle cell migration and proliferation. Proliferation of vascular smooth muscle cells is an early and key event in the formation of an antherosclerotic lesion. Smooth muscle cells accumulate in the intima in atherogenic conditions. In response to experimental arterial injury, SMC migrate from the media into the intima, where they proliferate and secrete large amounts of extracellular matrix. This reaction results in rapid neointimal hyperplasia, a convenient means for studying proliferative atherosclerotic lesions and stenotic processes frequently observed in patients undergoing percutaneous transluminal coronary angioplasty (PTCA) [14].

An *in vivo* model for the study of SMC in the carotid walls of rabbits has been used in our laboratory [14].

To characterize the antiatherogenic activity of the new calcium antagonist lacidipine, we studied the effect of this drug on the atherosclerotic response of the hypercholesterolemic rabbit carotid artery 14 days after perivascular manipulation of the vessel. This model, described by Booth et al. [15], allows direct following, *in vivo*, and independently of other factors, of the effect of a drug on arterial myocyte proliferation in normotensive hypercholesterolemic rabbits.

The use of lacidipine, a lipophylic dihydropyridine, clearly indicates the possibility to use this, and similar, calcium antagonists as protective agents against lesions to the arterial walls (Table 1).

The availability of quantitative procedures to assess in man the thickening of the carotid arterial walls in pathological condition, also opens the way for clinical studies leading to the protection of the carotid artery in presence of one or more major risk factors and not only in hypertensive patients.

HMGCoA Reductase Inhibitors (Statins)

In recent years the development of drugs that inhibit cholesterol synthesis, thus lowering plasma cholesterol levels, has, to a large degree, centred on controlling the HMG-CoA reductase activity. However, since mevalonic acid, the product of the enzyme reaction, is the precursor of numerous metabolites, inhibition of HMG-CoA reductase has the potential to result in pleiotropic effects. These possibilities are supported by several *in vivo* and *in vitro* observations which have shown how some HMG-CoA reductase inhibitors can in fact inhibit cellular proliferation and modulate changes in the cell cycle and cell morphology [4, 16–18]. Based on these findings this class of drugs have received increasing attention as pharmacological tools for controlling abnormal pathological cell growth, such as in tumours and in proliferative processes contributing to atherosclerosis.

Table 2. Effect of different vastatins on intimal thickening induced by perivascular manipulation of rabbit carotid arteries after 2 weeks

Treatment	I/M		% of control	P value
	(mean)	± SD		
Positive control	0.38	0.04	—	—
Sham	0.03	0.02	3	—
Pravastatin	0.34	0.04	89	NS
Lovastatin	0.23	0.03	61	0.001
Simvastatin	0.19	0.03	50	0.001
Fluvastatin	0.18	0.03	47	0.001

I/M, intimal/medial layer ratio; SD, standard deviation; NS, not significant. n = 10. The sham value is the mean ± SD of all 50 rabbits.

Pharmacological approaches using vastatins to inhibit smooth muscle cell proliferation, an early event in atherogenesis, have already been tried. Simvastatin inhibits smooth muscle cell growth *in vitro* [19]. Aggressive lipid-lowering treatment with lovastatin prevented restenosis after balloon angioplasty in hypercholesterolemic rabbits [20] and humans [21]. However, no correlations between serum cholesterol and restenosis were found in percutaneous transluminal coronary angioplasty patients.

Several vastatins have been tested in our laboratory in order to establish if they directly affect neointimal formation in the carotid arteries of normocholesterolemic rabbits independent of the lowering of plasma cholesterol concentration. For this purpose, we have used a scheme for the drug-treatment regimen that did not modify the plasma cholesterol level in normolipemic rabbits. Intimal thickening was induced by inserting a flexible extra-arterial collar around the common carotid artery, as already described.

The effect of vastatins was evaluated 14 days after collar insertion. The sham-operated contralateral arteries showed no thickening of the intima either in positive control or drug-treated rabbits (Table 2). A marked increase in intimal thickness (mostly cellular) was evident in the

Table 3. Total plasma cholesterol levels through the study in control and vastatin-treated groups

Treatment	Total cholesterol before surgery	Total cholesterol at sacrifice
Positive control	28.4 ± 3.6	31.5 ± 4.2
Pravastatin	26.5 ± 4.8	24.6 ± 6.3
Lovastatin	29.6 ± 5.2	27.8 ± 6.4
Simvastatin	29.7 ± 12.5	31.1 ± 7.8
Fluvastatin	32.6 ± 7.7	28.9 ± 5.5

Values are in milligrams per decilitre and are mean ± SD.

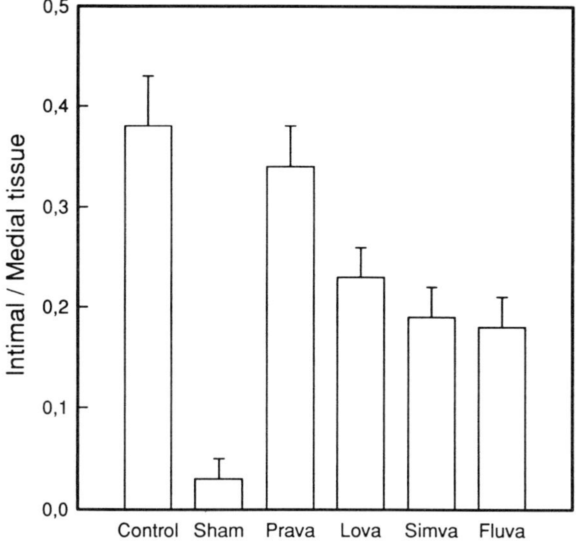

Fig. 1. Bar graph of vastatins' effect on neointimal formation induced by perivascular manipulation of normocholesterolemic rabbit carotid arteries. Vastatins were administered for 3
weeks mixed with food at a daily dose of 20 mg/kg body weight starting on the day of collar
placement.

carotid arteries with the collar of untreated animals (positive control
group). The mean value of intimal carotid thickening in the positive
control group, expressed as the I/M ratio, was 12 times greater than
that observed in the contralateral sham-operated carotid artery (Table
2).

The levels of total plasma cholesterol measured before surgery and at
the time the animals were killed (14 days later) did not change, and the
values were comparable in positive control and vastatin-treated animals
throughout the study (Table 3).

Figure 1 reports the effect of the different vastatins on thickening
induced in the carotid intimas of rabbits: all drugs at the dose of
20 mg/kg body wt per day limited the increase of the I/M ratio (when
compared with the positive-control-group mean value). Fluvastatin was
the most effective drug in this regard, followed by simvastatin, lovas-

Table 4. Effect of mevalonate on arterial SMC proliferation inhibited by Fluvastatin

	Control	Fluva	Fluva + MVA
I/M	0.22	0.12	0.20
LI	3.6	1.6	3.1

Fluva = Fluvastatin 5 mg/kg/day I.P./or 5 days.
MVA = Sodium Mevalonate 8 mg/kg/day (local delivery) for 5 days.
I/M = Intimal/medial tissue (ratio).
LI = Labelling index (INTIMA*MEDIA).

tatin, and pravastatin in decreasing order. Differences versus positive controls were statistically highly significant for all drugs with the exception of pravastatin (Table 2). In all rabbits treated with vastatins, the inhibition of hyperplasia consisted of fewer layers of intimal cells.

The inhibition of SMC proliferation induced by statins is totally abolished by local infusion of mevalonate (Table 4).

These *in vivo* data are more in accordance with an antiproliferative action of statins than a cholesterol-lowering effect.

Conclusions

The examples of two classes of drugs which protect arterial walls even in the presence of risk factors or after endothelial and SMC damage shows new ways to prevent atherosclerosis and its cardiovascular consequences.

New clinical approaches using quantitative and repeated observations of the arterial wall thickness and the progression and regression of plaques in man are needed to study those and newly developed drugs at a clinical level.

Combinations of two drugs (such as a calcium antagonist and a statin) may be tested in human subjects with a variety of risk factors (hypertension, dislipoproteinemias and diabetes). This new approach is needed to avoid multiplication of drugs which affect individually risk factors without a well demonstrated global impact on the atherosclerotic disease.

References

1. Paoletti R, Soma MR. Pharmacological approach of individual and combined risk factors. In: Van Zwieten PA, Mancia G, Brodde OE (editors). Hypertension atherosclerosis, and lipids. London, New York: Royal Soc. Med. Serv. 1992: 27–34.
2. Ross R. The pathogenesis of atherosclerosis: a perspective for the 1990s. Nature 1993; 362: 801–809.
3. Corsini A, Mazotti M, Raiteri M, Soma MR, Gabbiani G, Fumagalli R. et al. Relationship between mevalonate pathway and arterial myocyte proliferation: *in vitro* studies with inhibitors of HMG-CoA reductase. Atherosclerosis 1993; 101: 117–125.
4. Soma MR, Donetti E, Parolini C, Mazzini G, Ferrari C, Fumagalli R. et al. HMG-CoA reductase inhibitors: *in vivo* effects on carotid intimal thickening in normocholesterolemic rabbits. Arterioscl. Thromb. 1993; 13: 571–578.
5. Goldstein JL, Brown MS. Regulation of the mevalonate pathway. Nature 1990; 343: 425–430.
6. Maltese WA, Sheridan KM. Isoprenoid synthesis during the cell cycle. J. Biol. Chem. 1988; 263: 10104–10110.
7. Faust JR, Goldstein JL, Brown MS. Synthesis of ubiquinone and cholesterol in human fibroblast: regulation of branched pathway. Arch. Biochem. Biophys. 1979; 192: 86–99.
8. Soma MR, Donetti E, Parolini C, Barberi L, Paoletti R, Fumagalli R et al. Effect of lacidipine on the carotid intimal thickening induced by cuff injury. J. Cardiovasc. Pharmacol. 1994; 23 (suppl 5): S71–S74.

9. Schmitz G, Hankovitz J, Kovacs EM. Cellular processes in atherogenesis: potential targets of calcium channel blockers. Atherosclerosis 1991; 88: 109–132.

10. Fronek K. Calcium antagonists and experimental atherosclerosis. Cardiovasc Drug Rev 1990; 8: 229–237.

11. Scheneider W, Kober G, Roebruck P, Noack H, Alle M, Cieslinski G et al. Retardation of development and progression of coronary atherosclerosis: a new indication for calcium antagonists? Eur. J. Clin. Pharmacol. 1990; 39: S17–S23.

12. Lichtlen PR, Hugenholtz PG, Hecker H, Jost S, Deckers JW. Retardation of angiographic progression of coronary artery disease by nifedipine, Results of international nifedipine trail on atherosclerotic therapy (INTACT). Lancet 1990; 335: 1109–1113.

13. Waters D, Lespérance J, Francetich M, Causey D, Theroux P, Chiang Y-K et al. A controlled clinical trial to assess the effect of calcium channel blocker on the progression of coronary atherosclerosis. Circulation 1990; 82: 1940–1953.

14. Popma JJ, Califf RM, Topol EJ. Clinical trials of restenosis after coronary angioplasty. Circulation 1991; 84: 1426–1436.

15. Booth RGF, Martin JF, Honey AC, Hassall DG, Beesley JE, Moncada S. Rapid development of atherosclerotic lesions in the rabbit carotid artery induced by perivascular manipulation. Atherosclerosis 1989; 76: 257–268.

16. Soma MR, Baetta R, De Remzis R, Mazzini G, Davegnov C, Magrassi L, Butti G, Perrotia S, Paoletti R, Fumagalli R. In vivo enhanced antitumor activity of carmustine [N,N'-Bis(2-chloroethyl)-N-nitrosourea] by simvastatin. Cancer Res 1995; 55: 597–602.

17. Soma MR, Baetta R, Paoletti R, Fumagalli R, Ferrari C, de Renzis MR et al. Effect on cell cycle of hypocholesterolemic drugs. Eur. J. Histochem. 1993; 37 (suppl.): 88. [abstract].

18. Soma MR, Pagliarini P, Butti G, Paoletti R, Paoletti P, Fumagalli R. Simvastatin, an inhibitor of cholesterol biosynthesis, shows synergistic effect with N,N'-Bis(2-chloroethyl)-N-nitrosourea and beta interferon on human glioma cells. Cancer Res. 1992; 52: 1–9.

19. Corsini A, Raiteri M, Soma MR, Fumagalli R, Paoletti R. Simvastatin but no pravastatin inhibits the proliferation of rat aorta myocytes. Pharm. Res. 1991; 23: 173–180.

20. Gellman J, Ezekowitz MD, Sarembock IJ, Azrin MA, Nochomowitz LE, Lerner E et al. Effect of lovastatin on intimal hyperplasia after balloon angioplasty. A study in an atherosclerotic hypercholesterolemic rabbit. J. Am. Coll. Cardio. 1991; 17: 251–259.

21. Sahni R, Maniet AR, Voci G, Banka VS. Prevention of restenosis by lovastatin after successful coronary angioplasty. Am. Heart. J. 1991; 121: 1600–1608.

Pharmacological Sciences: Perspectives for
Research and Therapy in the Late 1990s
ed. by A.C. Cuello and B. Collier

Chronic Drug Treatment of Essential Hypertension

P.A. van Zwieten[1] and J. de Champlain[2]

[1]*Departments of Pharmacotherapy and Cardiology, University of Amsterdam, Academic
Medical Centre, Meibergdreef 15, 1105 AZ Amsterdam, The Netherlands;* [2]*Faculté de
Médecine, CP 6128, Département de Physiologie, Université de Montréal, Montréal H3C 3J7,
Canada*

Summary. Changes in life style and drug treatment are the cornerstones in the long-term
management of essential hypertension (EHT). In the present symposium attention was paid to
the following issues:

(1) Presynaptic mechanisms in the antihypertensive effect of ß-blockers.
(2) The long-term effect of calcium antagonists in antihypertensive treatment.
(3) A survey of endothelin, endothelin receptors and endothelin receptor antagonists as
     potential antihypertensives.
(4) Antihypertensive drugs interacting with the renin-angiotensin-aldosterone system: ACE-
     inhibitors, renin inhibitors and angiotensin II-receptor (AT) antagonists.
(5) Therapeutic trials in the drug treatment of EHT.

Introduction

Essential hypertension (EHT) represents at least 90% of the diagnosed
cases of high blood pressure. Its treatment is not only aiming at
lowering or even normalizing the elevated blood pressure but also at
preventing and repairing the damage to the cardiovascular system, such
as cardiac and vascular hypertrophy. However, the final goal of EHT-
treatment is protection against stroke, coronary heart disease (in particular myocardial infarction, MI), heart failure and renal disease.

Treatment of EHT has to be performed for several years, and in most
patients for life.

Treatment is currently based on two principles:

(1) changes in life style, such as: moderation of the consumption of
     sodium and alcohol; cessation of smoking; correction of overweight;
     regular physical exercise;
(2) drug treatment. Currently used drugs are β-blockers, diuretics,
     α_1-adrenoceptor antagonists, calcium antagonists and ACE-in-
     hibitors. Monotherapy with one of these agents can be achieved in
     globally half of the patients. Concomitantly, the other half of the
     patients requires treatment with two or sometimes even three of the
     aforementioned categories of drugs.

The present symposium in the frame of the XIIth IUPHAR Congress
at Montréal dealt with both well-known, currently used drugs and with
experimental agents and new approaches.

β-Blockers continue to occupy an important position in the long-term drug treatment of EHT. In spite of this, many details of their mode of antihypertensive action remain unknown, notwithstanding the vast research effort invested over the years.

Evidence in favour of a mechanism involving presynaptic β-adrenoceptors was discussed during the symposium.

Calcium antagonists are not only effective vasodilators and antihypertensives, but also interfere in a beneficial manner with the hypertension-induced changes in the cardiovascular system, and may hence lead to a certain degree of repair of the damage associated with long-existing EHT.

Endothelin, an important and extremely potent endogenous vasoconstrictor agent may be associated directly or indirectly with several diseases, possibly including EHT. Endothelin receptors have been identified in the vascular system.

Endothelin receptor antagonists are now indeed available and they deserve our attention as experimental antihypertensive drugs.

In order to suppress the activity of the renin-angiotensin-aldosterone-system (RAAS) not only *ACE-inhibitors* but also *renin-inhibitors* and *angiotensin II-receptor antagonists* may be used and hence considered as therapeutics in EHT or congestive heart failure (CHF).

Finally, attention was paid to the outcome of the major large-scale intervention trials in EHT, including those concerning elderly patients. So far the available data on an epidemiological scale are limited to β-blockers and/or diuretics, although trials concerning calcium antagonists and ACE-inhibitors are ongoing.

Participation of Presynaptic Mechanisms in the Antihypertensive Effect of β-Blockers

The precise mode of action of β-blockers as antihypertensives so far remains unknown. One of the mechanisms discussed to explain the antihypertensive action of these compounds involves an important role of peripheral presynaptic β-adrenoceptors, which are predominantly of the β_2-type.

Presynaptic β-adrenoceptors when stimulated by adrenaline or noradrenaline will enhance the release of endogenous noradrenaline. Conversely, β-adrenoceptor antagonists will block presynaptic β-receptors and hence suppress the release of endogenous noradrenaline from the sympathetic nerve endings.

The antihypertensive effect of β-blockers is most pronounced in animals or human EHT patients in a hypersympathetic state. With respect to animal models this enhanced response to β-blocker treatment is found in particular in the DOCA-salt rat, which is characterized by an elevated sympathetic tone and reactivity, as well as by an enhanced sensitivity of the β-adrenergic presynaptic facilitatory mechanism.

In EHT-patients the response to β-blockers is most pronounced in those patients with an activated sympathetic system and high reactivity to catecholamines, as reflected by high levels of circulating noradrenaline and of heart rate at rest (pretreatment). Both in DOCA-salt rats and in hyperadrenergic EHT-patients the acute and chronic treatment with β-blockers causes an antihypertensive effect which runs parallel with a decrease in the level of circulating noradrenaline and with a fall in heart rate. These findings strongly favour an important role of presynaptic β-receptor blockade in the antihypertensive action of β-blockers [1, 2].

A problem in this connection are the selective β_1-blockers such as atenolol, bisoprolol or metoprolol, which as antihypertensives are as effective as the non-selective antagonists, which also block β_2-adrenoceptors; in this context it should be recalled that the presynaptic β-adrenoceptors predominantly belong to the β_2-subtype. It can be imagined that the selectivity of β_1-adrenoceptor blockers in therapeutic doses is limited, in particular at the receptor level.

Long-Term Effects of Calcium Antagonists in Hypertension

Calcium antagonists, in particular the dihydropyridines (DHP), are potent vasodilators and antihypertensives. Nifedipine is the prototype of the DHP-calcium antagonists. Its pharmacokinetic profile has been improved by means of sophisticated slow release formulations such as the GITS-preparation. Newer DHP-compounds like amlodipine and lacidipine are characterized by a slow onset and long duration of the antihypertensive action. Accordingly, such compounds are virtually devoid of reflex tachycardia, and when given once daily adequate control of blood pressure is achieved. Owing to a certain degree of vascular selectivity the newer DHP are virtually devoid of negative inotropic activity [3].

As found for several types (but not all) of antihypertensive agents, long-term treatment with DHP-calcium antagonists is associated with a regression of left ventricular hypertrophy (LVH). An unexpected and so far unexplained finding is the increased *right* ventricular wall thickness associated with long-term treatment with calcium antagonists.

Long-term treatment of EHT patients or hypertensive animals with calcium antagonists also influences the *vascular* changes characteristic for long-standing EHT [4]. Both the molecular and the structural changes in the vascular wall appear to be counteracted by antihypertensive treatment with calcium antagonists. An increase in vascular compliance has been described as a result of treatment with verapamil. A certain improvement of the microcirculation has been found as a result of long-term treatment with calcium antagonists. DHP-calcium antagonists appear to counteract vascular smooth muscle cell proliferation and the vascular hypertrophy which are characteristic for EHT.

At present there exists considerable interest for the potential antiatherogenic activity of the DHP-calcium antagonists. This activity can be readily demonstrated in animal and biochemical models of atherosclerosis. In patients with coronary heart disease some evidence has been obtained for the inhibitory effect of nifedipine or nicardipine on the formation of new lesions, although no regression of existing atherosclerotic plaques was observed. The large-scale MIDAS trial in EHT patients yielded inconclusive results with respect to the influence of isradipine on the development of atherosclerosis in the carotid arteries.

In conclusion, a clinically relevant antiatherogenic effect of calcium antagonists in CHD-patients and/or hypertensives remains to be demonstrated.

Endothelin and Endothelin Receptor Antagonists

The endogenous peptide *endothelin* causes transient vasodilatation, mediated by NO and/or prostacyclin when infused intravenously, but on long-term the compound is known to be a very potent vasoconstrictor. The vasoconstrictor effect of endothelin is known to be mediated by both ET_A and ET_B-receptors. Endothelin has been associated, directly or indirectly, with several diseases, such as migraine, primary pulmonary hypertension, coronary spasm, myocardial infarction, congestive heart failure, cardiogenic shock and renal failure. However, the role of endothelin in EHT remains controversial. In spite of this there exists considerable interest for endothelin receptor antagonists as potential antihypertensive drugs.

Both ET_A and ET_A/ET_B receptor antagonists have been developed. *Bosentan* is an example of an orally effective $ET_A + ET_B$-receptor antagonist. This drug has been shown to lower blood pressure in various animal models, such as SHR, hypertensive monkeys and rats with renal ischaemia.

Concomitantly, endothelin receptor antagonists deserve our attention as potentially interesting antihypertensives [5].

Clinical Experience with *ACE*-Inhibitors, Renin Inhibitors and Angiotensin II (*AT*)-Receptor Antagonists in Essential Hypertension

It remains a controversial issue whether the renin-angiotensin-aldosterone system (RAAS) plays a causative role in the pathogenesis of

EHT. In spite of this there exists ample evidence that the suppression of the activity of the RAAS by means of drug treatment is a useful measure in the management of EHT and congestive heart failure. ACE-inhibitors are the best known example of antihypertensive drugs interacting with the RAAS. However, it should not be forgotten that most β-blockers will reduce renin release, whereas spironolactone and canrenoate are antagonists of aldosterone at the receptor level. Conversely, newer compounds have been introduced which suppress RAAS-activity at the level of the enzyme renin (renin inhibitors) or the angiotensin II-receptor (angiotensin II-receptor antagonists).

It should be realized that during chronic treatment with an ACE-inhibitor both plasma-renin-activity (PRA) and angiotensin II levels are elevated as a result of a feedback mechanism. Accordingly, a certain degree of RAAS-activation and partial tolerance to drug treatment appears to exist during the long-term treatment with ACE-inhibitors. Apart from this issue hardly any new insights have emerged concerning the use and role of ACE-inhibitors in antihypertensive drug treatment.

However, the use of ACE-inhibitors in congestive heart failure and as a measure of secondary prevention following acute MI have considerably increased. The successful application of ACE-inhibitors is increasingly more substantiated by large scale clinical trials, such as the SAVE- and SOLVD-studies.

Renin inhibitors have been studied for well over a decade, but so far hardly any clinically attractive compounds have emerged. Most of the newly introduced compounds contain peptidergic bonds which makes them difficult to handle under clinical conditions, as a result of low bioavailability and short duration of action. Compound RO-42-5892 is an example of a non-peptidergic, orally active renin inhibitor. Its antihypertensive effect appears not to run parallel with the immediate suppression of RAAS-activity. An advantage may be that, as observed and expected for other ACE-inhibitors, PRA is not increased during treatment. However, like the peptidergic renin inhibitors its low bioavailability and short duration of action are unsatisfactory.

Angiotensin II receptor (AT)-antagonists, as a logical approach in antihypertensive therapy, are now being developed by several pharmaceutical companies. The antihypertensive activity of these compounds is mediated by vascular AT_1-receptors. As an example of a recently introduced selective AT_1-receptor blocker, we mention compound SR 47346, which unlike losartan, the prototype of AT_1-receptor antagonists, is not converted into an active metabolite. SR 47346 is an active antihypertensive agent with a long duration of action. During treatment with this compound there occurs no change in heart rate and postural hypotension was not observed. As found for other compounds of this type the antihypertensive effect was accompanied by a dose-dependent rise in PRA.

The question now arises of whether AT_1-receptor blocking agents have more to offer than the very successful ACE-inhibitors. On theoretical grounds major differences with the ACE-inhibitors can hardly be expected, but this matter can only be settled by means of appropriate, comparative clinical trials. A potential advantage of the AT_1-receptor blockers is the absence of cough as an adverse reaction.

A matter of concern is the high level of plasma angiotensin II associated with AT_1-blocker treatment, the pathophysiological implications of this factor being unknown. Since during the use of selective AT_1-receptor blockers the AT_2-receptor is freely accessible to circulating angiotensin II, there may be some interest for the development of non-selective $AT_1 + AT_2$-receptor blockers that will protect the AT_2-receptor.

Therapeutic Trials in the Treatment of Hypertension

Antihypertensive drug therapy aims at protection against stroke, coronary heart disease (in particular MI), heart failure and renal disease. The protection achieved by drug therapy and changes in lifestyle can only be judged by means of appropriately designed clinical trials on a sufficiently large scale. The classical trials on intervention have been performed in the 1970s and 1980s, with the MRC-trial as a well-known example. These trials, where diuretics and β-blockers were used, showed an important reduction in stroke and heart failure, whereas there was only marginal protection against coronary heart disease and its sequelae. In a later stage the protective effect, predominantly on the incidence of stroke, was also shown in elderly hypertensives (SHEP, STOP and MRC-Elderly trials). Again, diuretics and β-blockers were used.

No data on an epidemiological scale are available for the newer antihypertensive drugs such as calcium antagonists, ACE-inhibitors or selective α_1-adrenoceptor antagonists. Accordingly, we are at present experiencing a paradoxical situation, in the sense that the armamentarium of new antihypertensive drugs is ever increasing, whereas the evidence for protection (morbidity and mortality) is only available for diuretics and β-blockers.

A few ongoing trials concerning the effects of calcium antagonists and ACE-inhibitors are, for instance, the STOP II, SYSTEUR, HOT, MIDAS and ELSA studies. Their results and implications for drug therapy will be known within a few years' time. For reviews see [9–11].

References

1. Bouvier M, de Champlain J. Effects of acute and chronic administration of sotalol on the blood pressure and the sympathoadrenal activity of anesthetized deoxycorticosterone acetate-salt hypertensive rats. Can. J. Physiol. Pharmacol. 1986; 64: 1164–1169.

2. de Champlain J. Pre- and post-synaptic dysfunctions in hypertension. J. Hypertens. 1990; 8(suppl. 5): S77–S85.
3. Van Zwieten PA. Developments and trends in the field of calcium antagonists. High Blood Press. (Milano). 1994; 3: 356–366.
4. Giannattasio C, Cleroux J, Cuspidi C et al. Cardiac and vascular hypertrophy in essential hypertension. High Blood Press. 1993; 2: 299–306.
5. Masaki T, Kimura S, Yanagisiwa M, Katsutoshi G. Molecular and cellular mechanism of endothelin regulation. Circulation 1991; 841: 1457–1468.
6. Fisher NDL, Allan D, Kifor I et al. Responses to converting enzyme and renin inhibition. Hypertens. 1994; 23: 44–51.
7. Timmermans PBMWM, Wong PC, Chin AT et al. New perspectives in angiotensin system control. J. Human Hypertens. 1993; 7 (suppl. 2): S19–S32.
8. Lee RJ, Brunner HR. Clinical experience with angiotensin II receptor antagonists. J. Human Hypertens. 1993; 7 (suppl. 2): S33–S36.
9. Collins R, Peto R, Mac Mahon S et al. Blood pressure, stroke and coronary heart disease. Part II: Short term reduction in blood pressure: overview of randomised drug trials in their epidemiological context. Lancet 1990; 335: 827–838.
10. Sever P. 1985 – The year of the hypertension trials: interpreting the results. Trends in Pharmacol. Sci. 1986; 6: 134–139.
11. Doyle AE. Treating hypertension in older patients. Blood Press. 1992; 1: 72–74.

Role of Potassium Channel Blockers in the Treatment of Cardiac Arrhythmias

B.I. Sasyniuk[1] and E. Carmeliet[2]

[1]*Department of Pharmacology, McGill University, Montreal, Quebec, Canada, H4A 3N7;*
[2]*Laboratory of Physiology, University of Leuven, Gasthuisberg, 49, 3000 Leuven, Belgium*

Introduction

The results of the Cardiac Arrhythmia Suppression Trial (CAST) highlighted the inadequacy of therapy with class I antiarrhythmic drugs in patients with ischemic heart disease and served as an impetus for the search for new antiarrhythmic strategies [1]. One such strategy is action potential prolongation. It is far from clear which of the many channels active during cardiac repolarization are best targeted in the implementation of this strategy. Potassium channel blockade, in particular blockade of the i_{kr} channel, is one strategy which is being actively pursued at the present time.

Prolongation of action potential duration as an antiarrhythmic principle is not new. It was advanced over two decades ago with the initial characterization of the electrophysiologic actions of amiodarone and sotalol [2]. One of the first compounds to be recognized as a relatively pure action potential prolonging agent was NAPA, the major metabolite of procainamide. NAPA and sotalol were the prototypes for the synthesis of many of the new selective class III antiarrhythmic drugs [3]. Compounds were synthesized by many companies worldwide but their mechanism was not very clear until a rapidly activated delayed rectifier current (i_{kr}) was identified in guinea pig ventricle which was selectively blocked by E-4031 [4]. The sensitivity to the blocking action of E-4031 was actually used to define the rapid component of the delayed rectifier current.

The CAST trials also highlighted the important interaction among infarct, drug and arrhythmogenic triggers such as myocardial ischemia and led to a critical reassessment of risks versus benefits of antiarrhythmic drug therapy [5]. A major concern regarding further development of selective K channel blockers is the occasional, and apparently unpre-

Correspondence to: Dr. B.I. Sasyniuk, Department of Pharmacology, McGill University, 3655 Drummond Street, Montreal, Quebec, Canada H4A 3N7.

dictable, development of marked QT prolongation and torsade de pointes ventricular arrhythmias in some patients which could offset any advantage conferred by these drugs. Advances in our understanding of the mechanism of this important arrhythmia may help alleviate this problem. To what extent the newer selective K channel blockers will meet the requirements of an ideal antifibrillatory agent that reduces mortality in patients with ischemic heart disease must await ongoing clinical trials.

Potassium Channel Blockade as an Antiarrhythmic Strategy

Experimental and clinical evidence suggests that K channel blockers exert their antiarrhythmic effects mainly via slowing of repolarization. The theoretical basis for this action is that slowing of cardiac repolarization with no alteration of excitability and conduction velocity will result in an increased wavelength of the reentrant circuit and reduce the excitable gap, an effect that is beneficial in treating reentrant arrhythmias. The tachycardia cycle length will be prolonged and some arrhythmias will be prevented from deteriorating into fibrillation. Lengthening of repolarization in ventricular and atrial tissues by K channel blockers is always associated with a negative chronotropic effect in sinus nodal tissues. Since the delayed rectifier current also greatly contributes to repolarization in sinus nodal cells, drugs which block this current induce an increase in the spontaneously beating cycle length by prolonging action potential duration without altering diastolic depolarization [6].

Much attention is being paid to K channel blockade as an antiarrhythmic strategy [6]. One reason is the remarkable efficacy of these drugs in preclinical trials in decreasing the incidence of ischemic ventricular fibrillation [7–9]. In a canine model of previous myocardial infarction, flecainide was no more effective than placebo in preventing ischemia-related ventricular fibrillation, whereas several of the class III agents, including E-4031, sematilide and dofetilide, were all greater than 50–60% effective.

Unlike most other antiarrhythmic drugs, the new selective agents produce a moderate positive inotropic effect [6, 10]. This property is particularly important since severe ventricular arrhythmias are frequently associated with depressed ventricular function such as in congestive heart failure. No deterioration of left ventricular function occurred following infusion of d-sotalol, dofetilide and almokalant in dogs with acute ischemic heart failure at concentrations which increased the QT interval and the effective refractory period [6, 11].

In contrast to Na channel blockers, these agents have been shown to decrease the energy required to defibrillate the heart both in experimental animals and humans [12, 13] suggesting that they may be a useful adjunct in patients with implantable cardioverter defibrillators.

The ability of the K channel blockers to lengthen action potential duration may be significantly attenuated in hypoxic or ischemic myocardium [6, 14, 15]. The action of hypoxia to induce a large conductance of current carried via K_{ATP} channels most likely overwhelms the effect of K channel blockers to inhibit the delayed rectifier current. Whether these effects will significantly limit the antiarrhythmic properties in the presence of ischemia must await further studies. It may be that prolongation of refractoriness in normal tissue is the mechanism of their antiarrhythmic effect.

Most drugs with class III antiarrhythmic properties lose their action potential prolonging effect at high heart rates [16]. Amiodarone and dofetilide may be the exception. Furthermore, there is evidence to suggest that catecholamines may directly antagonize drug-induced action potential prolongation (via activation of i_{ks} channels) in addition to increasing the heart rate [17, 18]. The important interaction between the sympathetic nervous system and class III antiarrhythmic drug effect suggests that further studies are needed to sort out the mechanism of this interaction at different heart rates and different levels of sympathetic activation.

Molecular Mechanisms of i_{kr} Block in Cardiac Cells

The cardiac action potential is the time-dependent voltage signature of the behaviour of multiple individual ion channels, ion pumps and exchangers [19]. During the plateau a fine balance exists between inward and outward currents. The net membrane conductance at plateau potentials is extremely small such that minor changes in any of the conductances (e.g., the delayed rectifier current) can significantly alter action potential duration. The characteristics of the delayed rectifier current varies with species. In the guinea pig [4] this current is complex and consists of two components: a rapidly activating component termed, i_{kr}, and a second slower component, termed, i_{ks}. The faster component is the only one present in rabbit and cat ventricular myocytes [20]. The two components are affected differently by drugs. In general, i_{kr} is the most sensitive to the recently developed class III agents [4, 19–24].

The term "reverse" use dependence was coined to account for greater prolongation of cardiac action potentials by class III drugs at low compared with high frequencies of stimulation [16]. To avoid early afterdepolarizations and torsade de pointes arrhythmias, the prolongation at low frequencies should be minimal or absent. It was suggested that drugs which display open channel block should have a greater effect in prolonging action potential duration at fast heart rates and thus should be more "ideal" antiarrhythmic drugs during a rapid tachycardia. This concept is not borne out by recent data. Most of the newer selective class III drugs block the open state of the K channel and either exhibit use dependent or rate independent block [20–23]. Yet none seems to fulfill the requirements for the "ideal" drug as most of them exhibit reverse frequency dependent effects on action potential duration.

Why then do open state i_{kr} channel blockers still produce reverse rate-dependent lengthening of action potential duration? One suggestion is that the rate-dependent effects reflect the diminished contribution of I_{kr} and an enhanced contribution if i_{ks} to net repolarizing current during rapid pacing. Incomplete deactivation of i_{ks} during short diastolic intervals will result in a marked summation of current with frequency [22, 23]. Furthermore, a given block of i_{kr} current cannot be translated directly into a change of action potential duration [22]. A drug with positive use dependence may cause a greater prolongation of action potentials for two reasons. Firstly, a smaller net current maintains the action potential plateau at low frequencies. Even a small decrease in i_{kr} will effectively prolong the action potential. Secondly, the total action potential duration is longer at low frequencies increasing the amount of open channel block. The hypothesis has been put forth that certain characteristics of block may be needed to achieve the "ideal" drug, viz., a slow onset of block development and a rapid recovery from block [20–22]. Such a combination is not available in the current drugs. Recovery from block with dofetilide and almokalant is very slow.

The relative importance of I_{kr} to phase 3 repolarization in a specific tissue may be more important in determining the propensity toward exaggerated action potential prolongation at slow heart rates than the mechanism of block per se. Specialized cells within the ventricles (Purkinje fibers and M cells) have a greater sensitivity to the actions of these drugs than ventricular muscle cells [25–29]. The reason for this enhanced sensitivity is not known but is most probably related to differences in the ionic current mechanism defining the action potential plateau. Regional variations in the magntidue of i_{kr} current between Purkinje cells and endocardial and epicardial muscle may play a role (unpublished observations).

Potassium Channel Blockade as a Therapeutic Principle

The electrocardiographic manifestations of i_{kr} blockade on the surface ECG would be predicted to be an increase in QT interval without any lengthening of PR interval or QRS duration. Clinical electrophysiological testing would predict increases in atrial and ventricular effective refractory period, with an uncertain effect on the A-V node. Until recently, it has not been possible to assess clinically the effects of pure i_{kr}

blockade because the clinically available class III drugs such as amiodarone and d,1-sotalol have important additional electrophysiologic properties which may be responsible for their antiarrhythmic effects.

The new i_{kr} blockers such as E-4031, dofetilide, almokalant and d-sotalol have now come into clinical evaluation and have confirmed the predicted effects of i_{kr} blockade on cardiac electrophysiology [30–34]. On the basis of these effects, it may be predicted that pure i_{kr} blockers would have a role in prophylaxis or termination of atrial and ventricular arrhythmias. Their effect on arrhythmias involving the A-V node is likely to be lessened by virtue of the lack of specific effect on nodal tissue. These predictions have largely been confirmed in preliminary clinical observations.

The future for i_{kr} blockade as an important therapeutic modality remains uncertain. Class III antiarrhythmic action may be altered under pathological conditions, and this alteration may diminish the antiarrhythmic efficacy and may enhance the chance of proarrhythmia. In addition, the phenomenon of reverse use dependence may limit the benefits of class III effect. In this context, it is worth remembering that the class III drugs of proven therapeutic benefit, amiodarone and sotalol, have multiple modes of action, and it may be the combination of these rather than class III action alone which is responsible for their efficacy [35, 36]. For this reason, the available clinical data are insufficient to answer the fundamental question of whether class III antiarrhythmic activity due to i_{kr} blockade will prove of therapeutic benefit. Undoubtedly, sotalol and amiodarone are useful agents but the value of i_{kr} blockade as a therapeutic principle awaits the completion of the currently ongoing studies.

Potential Proarrhythmic Effects of i_{kr} Channel Blockers

It has been suggested that more selective i_{kr} blockers might lack the proarrhythmic effects of the nonselective K channel blockers, such as quinidine [37]. Torsade de pointes has been documented during clinical trials with most if not all of the newly synthesized selective i_{kr} blockers, indicating that such selectivity has not proven beneficial as far as potential proarrhythmic effects are concerned [3, 5, 38, 39].

The proarrhythmic effects of various class III agents have been studied in a variety of relevant experimental models [26, 28, 29, 40–44]. These models have led to a better understanding of the mechanism of the arrhythmogenic effect. In a conscious dog model of chronic A-V block and diuretic-induced hypokalemia only drugs with class III actions produced long QT-dependent arrhythmogenic effects and only at high doses [41]. In this model development of ventricular hypertrophy during chronic A-V block may be an important risk factor.

In the alpha-chloralose anesthetized rabbit model the combination of a class III agent and continuous alpha$_1$ stimulation with methoxamine produces a high incidence of torsade de pointes-like polymorphic ventricular tachycardia [28, 29, 42, 43]. Incidence of proarrhythmia is preceded by a significant lengthening of QT-interval and associated with increased QT-dispersion and high rates of drug infusion. Clinical incidence of torsade de pointes has also been associated with high infusion rates [38]. The rabbit exhibits an extreme sensitivity to the actions of repolarization prolonging agents. Although this model appears to lack specificity, ibutilide, which prolongs action potential duration by activating a slow Na$^+$ current produces a lower incidence of arrhythmias as compared with other class III agents at doses which produce maximum class III effects [44].

For many of the recently developed class III agents it has been convincingly shown that certain structures within the heart, in particular the Purkinje fiber network and the recently described M cells respond with a much more marked prolongation of action potential duration that the ventricular muscle, thus causing increased dispersion of repolarization between the two tissue types [25–29]. The marked lengthening of the plateau in Purkinje fibers eventually leads to EAD-induced triggered activity which is conducted to ventricular muscle. Under certain conditions the greater prolongation of Purkinje fiber action potentials can create a gradient of refractoriness between these cells and ventricular muscle, possibly facilitating a reentrant mechanism [26]. The conditions which predispose to arrhythmia clinically (bradycardia, hypokalemia) are similar to the conditions under which these drugs induce marked prolongation of action potentials in isolated Purkinje fibers. Evidence for increased dispersion of repolarization has also been documented in the whole heart using monophasic action potential recordings (48).

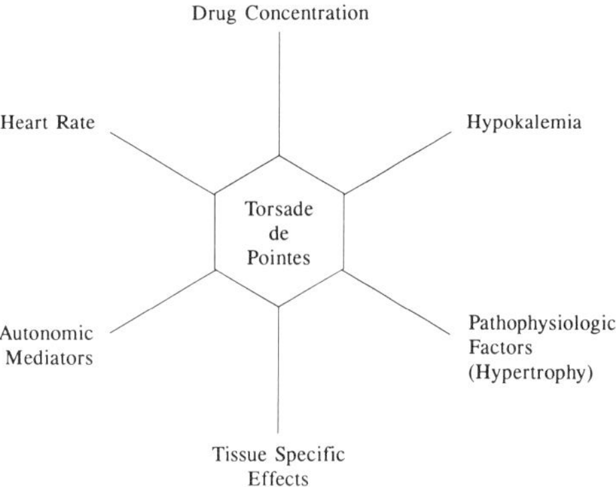

Fig. 1. Important variables modulating the predisposition to torsade de pointes by potassium channel blocking drugs

Action potential prolongation exhibits simple dose-dependence for i_{kr} channel blockers in contrast to class IA drugs which exhibit reverse dose dependence [45]. Dofetilide and ibutilide may be exceptions [46]. Whereas class IA drugs are proarrhythmic at plasma levels within the therapeutic range, pure class III agents appear to be proarrhythmic at concentrations well above the therapeutic range [28, 38]. The generation of EADs and triggered activity *in vitro* occurs only at high concentrations. A large concentration alone may be sufficient to induce triggered action potentials. Addition of precipitating factors (eg., hypokalemia) decreases the concentration and the stimulation frequency which is necessary to generate EADs.

The role of modulating factors for developing torsade de pointes are incompletely understood. Figure 1 lists some of the factors which may predispose the ventricles to the development of this lethal arrhythmia clinically. A combination of predisposing factors may be necessary to trigger the arrhythmia during therapy with class III agents.

Can torsades de pointes be minimized or eliminated? Class Ib drugs such as lidocaine [25, 43] have been shown to inhibit the initiation of rhythm abnormalities related to delayed repolarization. The mechanism of lidocaine's effect may be related to targeting the site of origin of the arrhythmia in the Purkinje system. ATP-sensitive K channel openers have also been shown to be effective and may provide a novel and useful clinical intervention [29]. Studies are underway to try and determine what markers may be useful in identifying patients which may be prone to development of drug-induced torsade de pointes. The presence of marked QT-dispersion has been suggested as a possibility. Targeting the modulatory factors may be a reasonable strategy. Risk of induction of this potentially lethal arrhythmia must be estimated at an early stage of drug development and, thus, justifies the search for a suitable model to test its predictability with a group of drugs which may have promise as effective antiarrhythmic agents.

Conclusion

The new potent selective i_{kr} blockers may have many important advantages over existing antiarrhythmic drugs. Prolongation of action potential duration secondary to block of the delayed rectifier current seems an effective way to silence reentry arrhythmias. Their efficacy against atrial and ventricular arrhythmias in which reentry is the likely mechanism must await the results of ongoing clinical trials. Proarrhythmia is a major concern. Understanding the basic mechanisms underlying torsade de pointe and the triggering factors that can sensitize the ventricles to class III drug action may help to minimize this risk. Advances in ion channel molecular biology hold the promise of improving our understanding of arrhythmogenesis in the absence and presence of antiarrhythmic agents and may identify completely novel targets for antiarrhythmic drug development [47].

References

1. Echt DS, Liebson PR, Mitchell LB, Peters RW, Obias-Manno D, Barker AH et al. Mortality and morbidity in patients receiving encainide, flecanide, or placebo: The Cardiac Arrhythmia Suppression Trial. N. Engl. J. Med. 1991; 324: 781–788.
2. Singh BN, Vaughan Williams EM. A third class of anti-arrhythmic action: effects on atrial and ventricular intracellular potentials, and other pharmacologic actions on cardiac muscle, of MJ 1999 and AH 3747. Br. J. Pharmacol. 1970; 39: 675–689.
3. Roden DM. Current status of class III antiarrhythmic drug therapy. Am. J. Cardiol. 1993; 72: 44B–49B.
4. Sanguinetti MC, Jurkiewicz NK. Two components of cardiac delayed rectifier K^+ current. Differential sensitivity to block by class III antiarrhythmic agents. J. Gen. Physiol. 1990; 96: 195–215.
5. Roden DM. Risks and benefits of antiarrhythmic therapy. New Eng. J. Med. 1994; 331: 785–791.
6. Mortensen E, Yang T, Refsum H. Potassium channel blockade as an antiarrhythmic principle. Cardiovasc. Drug Rev. 1993; 11: 370–384.
7. Lynch JJ Jr, Heaney LA, Wallace AA, Gehret JR, Selnick HG, Stein RB. Suppression of lethal ischemic ventricular arrhythmias by the class III agent E-4031 in a canine model of previous myocardial infarction. J. Cardiovasc. Pharmacol. 1990; 15: 764–775.
8. Chi L, Mu D, Driscoll EM, Lucchesi BR. Antiarrhythmic and electrophysiologic actions of CK-3579 and sematilide in a conscious canine model of sudden coronary death. J. Cardiovasc. Pharmacol. 1990; 16: 312–324.
9. Black SC, Lucchesi BR. UK-68,798, a class III antiarrhythmic drug with antifibrillatory properties. Cardiovasc. Drug Rev. 1992; 10: 170–181.
10. Wallace AA, Stupienski RF, Brookes LM, Selnick HG, Claremon DA, Lynch JJ Jr. Cardiac electrophysiologic and inotropic actions of new and potent methanesulfonanilide class III antiarrhythmic agents in anesthetized dogs. J. Cardiovasc. Pharmacol. 1991; 18: 687–695.
11. Mortensen E, Yang T, Refsum H. Class III antiarrhythmic action and inotropy: effects of dofetilide in acute ischemic failure in dogs. J. Cardiovasc. Pharmacol. 1992; 19: 216–221.
12. Echt DS, Black JN, Barbey JT, Coxe DR, Cato EL. Evaluation of antiarrhythmic drugs on defibrillation energy requirements in dogs: sodium channel block and action potential prolongation. Circulation 1989; 79: 1106–1117.
13. Dorion P, Newman D. Effect of sotalol on ventricular fibrillation and defibrillation in humans. Am. J. Cardiol. 1993: 72; 72A–79A.
14. Yang T, Tande PM, Refsum H. Electromechanical action of dofetilide and d-sotalol during simulated metabolic acidosis in isolated guinea pig ventricular muscle. J. Cardiovasc. Pharmacol. 1992; 20: 889–894.
15. Cobbe SM. Modification of class III antiarrhythmic activity in abnormal myocardium. Cardiovasc. Res. 1988; 22: 847–854.
16. Hondeghem LM, Snyders DJ. Class III antiarrhythmic agents have a lot of potential but a long way to go: Reduced effectiveness and dangers of reverse use dependence. Circulation 1990; 81: 686–690.
17. Sanguinetti MC, Jurkiewicz NK. Isoproterenol antagonizes prolongation of refractory period by the class III antiarrhythmic agent, E-4031, in guinea pig myocytes: mechanism of action. Circ. Res. 1991; 68: 77–84.
18. Newman D, Dorian P, Feder-Elituv R. Isoproterenol antagonizes drug-induced prolongation of action potential duration in humans. Can. J. Physiol. Pharmacol. 1993; 71: 755–760.
19. Carmeliet E. Mechanisms and control of repolarization. Eur. Heart. J. 1993b; 14: 3–13.
20. Carmeliet E. Use-dependent block of the delayed K current in rabbit ventricular myocytes. Cardiovasc. Drugs Ther. 1993c; 7: 599–604.

21. Carmeliet E. Voltage- and time-dependent block of the delayed K^+ current in cardiac myocytes by dofetilide. J. Pharmacol. Exp. Ther. 1993a; 262: 809–817.

22. Carmeliet E. Use-dependent block and use-dependent unblock of the delayed rectifier K^+ current by almokalant in rabbit ventricular myocytes. Circ. Res. 1993b; 73: 857–868.

23. Jurkiewicz NK, Sanguinetti MC. Rate-dependent prolongation of cardiac action potentials by a methanesulfonanilide class III antiarrhythmic agent: specific block of rapidly activating delayed rectifier K^+ current by dofetilide. Circ. Res. 1993; 72: 75–83.

24. Wettwer E, Grundke M, Ravens U. Differential effects of the new class III antiarrhythmic agents almokalant, E-4031 and d-sotalol, and of quinidine, on delayed rectifier currents in guinea pig ventricular myocytes. Cardiovasc. Res. 1992; 26: 1145–1152.

25. Sasyniuk BI, Valois M, Toy W. Recent advances in understanding the mechanisms of drug-induced torsades de pointe arrhythmias. Am. J. Cardiol. 1989; 64: 29–32.

26. Sasyniuk BI, Brunet S. Proarrhythmic effects of d-sotalol in rabbit ventricle associated with differential effects on endocardial cells at slow heart rates. Circulation 1994; 90: I-146.

27. Antzelevitch C, Sicouri S. Clinical relevance of cardiac arrhythmias generated by afterde-polarizations: role of M cells in the generation of U waves, triggered activity and torsade de pointes. Am. J. Cardiol. 1994; 23: 259–277.

28. Carlsson L, Abrahamson C, Anderson B, Duker G, Schiller-Linhardt G. Proarrhythmic effects of the class III agent almokalant: importance of infusion rate, QT dispersion, and early afterdepolarizations. Cardiovasc. Res. 1993; 27: 2186–2193.

29. Carlsson L, Abrahamson C, Drews L, Duker G. Antiarrhythmic effects of potassium channel openers in rhythm abnormalities related to delayed repolarization. Circulation 1992; 85: 1491–1500.

30. Fujiki A, Tani M, Mizumaki K, Shimono M, Inoue H. Electrophysiologic effects of intravenous E-4031, a novel class III antiarrhythmic agent, in patients with supraventricular arrhythmias. J. Cardiovasc. Pharmacol. 1994; 23: 374–378.

31. Wong W, Pavlou HN, Birgersdotter UM, Hilleman DE, Mohiuddin SM, Roden DM. Pharmacology of the class III antiarrhythmic agent sematilide in patients with arrhythmias. Am. J. Cardiol. 1992; 69: 206–212.

32. Suttorp MJ, Polak PE, Van'tHof A, Rasmussen HS, Dunselman PH, Kingma JH. Efficacy and safety of a new selective class III antiarrhythmic agent dofetilide in paroxysmal atrial fibrillation of atrial flutter. Am. J. Cardiol. 1992; 69: 417–419.

33. Sedgwick ML, Rasmussen HS, Cobbe SM. Effects of the class III antiarrhythmic drug dofetilide on ventricular monophasic action potential duration and QT interval dispersion in stable angina pectoris. Am. J. Cardiol. 1992; 70: 1432–1437.

34. Isomoto S, Shimizu A, Konoe A, Kaibara M, Centurion AO, Fukatani M, Yano K. Electrophysiologic effects of E-4031, a new class III antiarrhythmic agent, in patients with supraventricular tachyarrhythmias. Am. J. Cardiol. 1993; 71: 1464–1467.

35. Teo KK. Evaluation of therapeutic modalities in patients with life threatening arrhythmias. Can. J. Cardiol. 1994; 10: 333–341.

36. Mason JW. For ESVEM Investigators. A comparison of seven antiarrhythmic drugs in patients with ventricular tachyarrhythmias. N. Engl. J. Med. 1993; 329: 452–458.

37. Colatsky TJ, Follmer CH, Starmer CF. Channel specificity in antiarrhythmic drug action. Mechanism of potassium channel block and its role in suppressing and aggravating cardiac arrhythmias. Circulation 1990; 82: 2235–2242.

38. Wiesfeld ACP, Crijns HJGM, Bergstrand RH, Ollmgren O, Hillage HL, Lie KI. Torsades de pointes with Almokalant, a new class III antiarrhythmic drug. Am. Heart J. 1993; 126: 1008–1011.

39. Hohnloser SH, Arendts W, Quart B and Bristol Myers Squibb Research Institute. Incidence, type and dose-dependence of proarrhythmic events during sotalol therapy in patients treated for sustained VT/VF. PACE 1992; 15: 551.

40. Weissenburger J, Davy JM, Chezalviel F. Experimental models of torsades de pointes. Fundam. Clin. Pharmacol. 1993; 7: 29–38.

41. Weissenburger J, Davy J-M, Chézalviel F, Ertzbischoff O, Poirier JM, Engel F et al. Arrhythmogenic activities of antiarrhythmic drugs in conscious dogs with atrioventricular block: comparison between quinidine, lidocaine, flecainide, propanolol and sotalol. J. Pharmacol. Exp. Ther. 1991; 259: 871–883.

42. Carlsson L, Almgren O, Duker G. QTU-prolongation and torsades de pointes induced by putative class III antiarrhythmic agents in the rabbit: etiology and interventions. J. Cardiovasc. Pharmacol. 1990; 16: 276–285.

43. Carlsson L, Drews L, Duker G, Schiller-Linhardt G. Attenuation of proarrhythmias related to delayed repolarization by low-dose lidocaine in the anesthetized rabbit. J. Pharmacol. Exp. Ther. 1993; 267: 1076–1080.
44. Buchanan L, Kabell G, Brunden M, Gibson J. Comparative assessment of ibutilide, d-sotalol, clofilium, E-4031, and UK-68,798 in a rabbit model or proarrhythmia. J. Cardiovasc. Pharmacol. 1993; 22: 540–549.
45. Wyse KR, Ye V, Campbell TJ. Action potential prolongation exhibits simple dose-dependence for sotalol, but reverse dose-dependence for quinidine and disopyramide: implications for proarrhythmia due to triggered activity. J. Cardiovasc. Pharmacol. 1993; 21: 316–322.
46. Lee KS, Tsai TD, Lee EW. Membrane activity of class III antiarrhythmic compounds; a comparison between ibutilide, d-sotalol, E-4031, sematilide and dofetilide. Eur. J. Pharmacol. 1993; 234: 43–53.
47. Roden DM, Tamkun MM. Toward a molecular view of cardiac arrhythmogenesis. Trends in Cardiovasc. Med. 1994; 4: 278–285.
48. Sasyniuk BI, Brunet, S. Torsade de pointes induced by quinidine, d-sotalol and E-4031 in the isolated rabbit heart: importance of interval dependent dispersion of repolarization. PACE. 1995; 18: 11–904.

Nitric Oxide

Pharmacological Sciences: Perspectives for
Research and Therapy in the Late 1990s
ed. by A.C. Cuello and B. Collier
© 1995 Birkhäuser Verlag Basel/Switzerland

Nitric Oxide in the Cardiovascular System

S. Moncada and E.A. Higgs

The Wellcome Foundation Limited, Langley Court, Beckenham, Kent, BR3 3BS, UK

Introduction

Between 1987 and 1988, we found that vascular endothelial cells generate nitric oxide (NO) from the amino acid L-arginine. The formation of NO by this L-arginine: NO pathway was subsequently found to occur in many other cells and tissues, where NO acts as a transcellular signal. In recent years, interest in the physiological, pathological and therapeutic implications of this novel mediator has increased exponentially, revealing the diversity of biological roles played by this simple molecule both in the cardiovascular and in other systems (see [1–3]).

Physiological Roles of Nitric Oxide in the Cardiovascular System

The search for the identity of the labile vasodilator known as endothelium-derived relaxing factor (EDRF) led us to discover in the vasculature an enzyme, NO synthase, that generates NO from L-arginine. This endothelial isoform of the enzyme was found to be constitutive, Ca^{2+}/calmodulin-dependent and not to occur in vascular smooth muscle cells. Certain analogues of L-arginine, such as N^G-monomethyl-L-arginine (L-NMMA), were shown to compete with the substrate for the enzyme and to inhibit the synthesis of NO. Use of such inhibitors has provided much information on the relevance of NO in biological processes (see [1]).

L-NMMA, which itself has no intrinsic constrictor activity on vascular smooth muscle, constricts vascular beds and elevates blood pressure, indicating that there is a physiological, NO-dependent vasodilator tone that is essential for the regulation of blood flow and pressure (see [4]). This tone is maintained by the physical action of pulsatile flow and shear stress and by the chemical action of vasoactive mediators on endothelial cells. Nitric oxide released by non-adrenergic, non-cholinergic (NANC) nerves in the vasculature may also contribute to the regulation of blood flow and pressure [5].

Correspondence to: Dr. S. Moncada, at the above address.

These observations using L-NMMA revealed the existence of an endogenous nitrovasodilator, the actions of which are mimicked by drugs such as glyceryl trinitrate and sodium nitroprusside (see [4]). These compounds, which have long been in clinical use, act following their conversion into NO [6]. The reaction of NO with the ferrous iron in the haem prosthetic group of the soluble guanylate cyclase in the vascular smooth muscle cells increases their concentration of cyclic GMP, thus leading to vascular relaxation [7].

In addition to its vasodilator properties, NO has other actions on the cardiovascular system. It alters cardiac contractility by shortening contraction time and by its negative inotropic action, and it also inhibits platelet aggregation and adhesion. Platelets themselves generate NO and its release may act as a negative feedback mechanism to inhibit platelet activation. Nitric oxide may also protect against vascular damage since it inhibits leukocyte activation and inhibits vascular smooth muscle cell proliferation. These actions of NO are all mediated by diffusion of NO to neighbouring cells where it activates the soluble guanylate cyclase, thus elevating cyclic GMP levels within these target cells (see [8]).

Clinical Implications of the Role of Nitric Oxide in Cardiovascular Disease

Decreased synthesis or action of NO has been implicated in most diseases or conditions associated with increased vascular tone, vasospasm or enhanced adhesion of platelets and white cells to the vessel wall. Thus, reduced endothelium-dependent vasodilation has been demonstrated in patients with hypertension, hypercholesterolemia, diabetes or atherosclerosis (see [8]).

Since compounds that donate NO mimic the actions of the endogenous mediator, they can be used as replacement therapy to treat its impaired production in conditions such as hypertension and atherosclerosis. Nitrovasodilators have been in clinical use for over a century in the treatment of hypertension and are still widely used in a number of conditions including angina pectoris, congestive heart failure and pulmonary hypertension. The beneficial effects of these drugs in cardiovascular disease result from their dilator action on arterial and venous smooth muscle and their ability to dilate large coronary vessels and thus increase coronary flow. The resultant decrease in myocardial oxygen demand and preload produces symptomatic relief in angina pectoris and congestive heart failure [9].

Nitrovasodilators also inhibit platelet aggregation *in vivo*, although this can only be demonstrated *in vitro* using those compounds (e.g. sodium nitroprusside) that liberate NO spontaneously (see [6]) since the platelet itself appears not to be able to metabolize organic nitrates to produce NO [10]. The clinical use of NO donors as inhibitors of platelet aggregation in the treatment of thrombotic disorders such as atherosclerosis and angioplasty restenosis has been limited by their concomitant hypotensive effect. Recently, however, S-nitroso-glutathione, a stable nitrosothiol that is metabolized to NO within platelets, has been shown to inhibit platelet aggregation and adhesion in animals [11] and humans [12] at doses that cause only minimal vasodilatation. This selective targeting of the platelet without accompanying hypotension may prove to be beneficial in the clinic.

It has recently been shown that glyceryl trinitrate increases uterine blood flow in pregnant women with abnormal uterine blood flow [13]. This suggests the possibility that nitrovasodilators might be useful in the treatment of the hypertension, reduced placental blood flow and increased platelet aggregability of women with pre-eclampsia.

Infusion of L-arginine, the precursor of NO formation, improves endothelium-dependent vasodilatation in hypercholesterolemic patients [14]. Administration of L-arginine may inhibit intimal hyperplasia and consequent restenosis after balloon angioplasty, since such treatment in rabbits attenuated the development of intimal hyperplasia by $\sim 40\%$ [15]. In addition, L-arginine has been reported to cause a rapid reduction in systolic and diastolic pressures when infused into healthy humans and patients with various forms of hypertension [16].

Nitric oxide gas, inhaled at low concentrations (6–80 ppm), can reverse pulmonary hypertension induced by hypoxia or after surgery, persistent pulmonary hypertension of the newborn and chronic pulmonary hypertension [17–19]. These reductions in pulmonary artery pressure were not associated with the changes in mean arterial pressure or systemic vascular resistance that occur using nitrovasodilators or vasodilator prostaglandins. The beneficial effects of NO lasted throughout the inhalation period (up to several weeks) and in some cases persisted after termination of treatment.

Inhalation of NO by patients with severe adult respiratory distress syndrome (ARDS) alleviates the pulmonary hypertension and hypoxemia associated with this condition [20–22]. This is achieved by the NO being distributed selectively to the ventilated areas, thus increasing blood flow preferentially to the well-ventilated alveoli. Continuous inhalation of low concentrations of NO (2–130 ppb) for many days consistently lowered the pulmonary artery pressure and increased arterial oxygenation without producing systemic vasodilatation or inducing tachyphylaxis.

Nitric Oxide Synthase

As research has developed in the L-arginine: NO field it has become apparent that the NO synthase is not just one enzyme but a family of enzymes that have sequence similarity with cytochrome P-450 reductase (see [23, 24]). All the isoforms require a number of cofactors (NADPH, flavin mononucleotide (FMN), flavin adenine dinucleotide (FAD), tetrahydrobioterin) and the presence of calmodulin. The most notable difference between the isoenzymes is the fact that those in the vascular endothelium (eNOS) and nervous tissue (nNOS) are constitutive, whereas another isoform can be induced in many cells such as macrophages following activation by lipopolysaccharide (LPS) and/or certain cytokines [see 25]. Induction of the inducible NO synthase (iNOS) requires a process of *de novo* protein synthesis and can be inhibited by glucocorticoids [26]; this inhibition might contribute to the anti-inflammatory actions of these compounds. Nitric oxide synthases have now been purified from many sources including human tissue. The constitutive enzymes from the brain [27] and vascular endothelium [28] and the inducible enzyme from macrophages [29] have been cloned, sequenced, and found to have molecular weights ranging from 130 000 to 155 000.

The early classification of NO synthases into constitutive and inducible may need to be revised in the light of recent evidence showing that chronic exercise enhances eNOS gene expression in aortic endothelial cell extracts [30] and that both pregnancy and estradiol treatment elevate mRNAs for eNOS and nNOS, leading to an increased expression of NO synthase in the vasculature and other tissues [31]. Thus elevated production of NO may account for adaptive changes in the vasculature following chronic exercise, as well as the reduced vascular tone and contractility of pregnancy.

Nitric Oxide and Cardiovascular Pathophysiology

The inducible NO synthase proved to be expressed not only in cells of the reticuloendothelial system but also in many somatic cells, as was first demonstrated in hepatocytes [32]. Treatment with interferon-γ, tumour necrosis factor and LPS induces NO synthase in vascular smooth muscle cells as well as endothelial cells.

It is now clearly established that the generation of large quantities of NO by vascular tissue is responsible for the hypotension and hyporesponsiveness to vasoconstrictors that occurs in septic shock [33] or following cytokine therapy in cancer patients [34]. Inhibitors of NO synthase in animals can reverse or prevent the hypotension induced by LPS or cytokine administration or in animals with hemorrhagic or anaphylactic shock (see [35]). In patients with septic shock, L-NMMA, added to standard therapy, restores the blood pressure, although its effects on morbidity and mortality are not yet known [36]. The degree of inhibition of NO production may be crucial for the outcome of treatment, since high doses of L-NMMA in animals lead to severe vasoconstriction, end organ damage and death (see [35]). This presumably occurs because in these circumstances the eNOS is inhibited as well as the iNOS, thus the actions of the increased circulating concentrations of vasoconstrictors that occur in septic shock are not counteracted by NO. This problem might be overcome by inhibiting both eNOS and iNOS while at the same time administering a nitrovasodilator to reverse the hypertension and reduce platelet reactivity. Alternatively, the development of selective inhibitors of the inducible NO synthase would be clearly beneficial in conditions in which there is overproduction of NO.

The hyperdynamic state of cirrhosis is associated with vasodilatation, hyporesponsiveness to vasoconstrictors, and high circulating concentrations of endotoxin. This led to the suggestion that this condition may result from excessive generation of NO by the inducible enzyme [37]. Evidence is accumulating in favour of this hypothesis, since it has been shown that patients with cirrhosis have elevated serum concentrations of nitrite and nitrate (the stable oxidation products of NO metabolism) and these correlate with the degree of endotoxemia [38]. Venous tissues and the myocardium also express the inducible NO synthase, which may explain the increase in venous compliance [39] and the specific cardiac dysfunction [40] of septic shock. Expression of inducible NO synthase is also increased in the myocardium of patients with dilated cardiomyopathy [41].

Nitric oxide produced for long periods by the inducible NO synthase is cytostatic/cytotoxic for invading microorganisms as well as for those cells that produce it and for neighbouring cells (see [8, 25]). Part of the cytotoxic action of NO has been attributed to the interaction between NO and superoxide (O_2^-) to form peroxynitrite ($ONOO^-$). This, when protonated, forms OH^- and NO_2, thus leading to increased tissue damage [42]. Whether $ONOO^-$ is actually formed *in vivo* remains to be established. Moreover, if formed, $ONOO^-$ may be damaging only in certain situations, for it has recently been demonstrated that there are very efficient mechanisms for detoxification of this compound [43].

Conclusions

Nitric oxide is an important mediator of several homeostatic processes in the cardiovascular system. Its production may be impaired or excessively enhanced, both of which abnormalities result in pathological conditions. Manipulation of the L-arginine: nitric oxide pathway may be achieved in a number of ways, giving rise to the possibility of novel therapies for cardiovascular diseases in the future.

References

1. Moncada S, Palmer RMJ and Higgs EA. Nitric oxide: physiology, pathophysiology and pharmacology. Pharm. Revs. 1991; 43: 109–142.
2. Furchgott RF. Studies on endothelium-dependent vasodilation and the endothelium-derived relaxing factor. Acta. Physiol. Scand. 1990; 139: 257–270.
3. Ignarro LJ. Biosynthesis and metabolism of endothelium-derived nitric oxide. Annu. Rev. Pharmacol. Toxicol. 1990; 30: 535–560.
4. Moncada S, Palmer RMJ, Higgs EA. The discovery of nitric oxide as the endogenous nitrovasodilator. Hypertension 1988; 12: 365–372.
5. Toda N, Okamura T. Regulation by nitroxidergic nerve of arterial tone. NIPS 1992; 7: 148–152.
6. Feelisch M. Biotransformation to nitric oxide of organic nitrates in comparison to other nitrovasodilators. European Heart J. 1993; 14: 123–132.
7. Waldman SA, Murad F. Cyclic GMP synthesis and function. Pharmacol. Revs 1987; 39: 163–196.
8. Moncada S, Higgs A. The L-arginine-nitric oxide pathway. New. Engl. J. Med. 1993; 329: 2002–2012.
9. Fung H-L. Do nitrates differ? Br. J. Clin. Pharmacol. 1992; 34: 5S–9S.
10. Benjamin N, Dutton JAE, Ritter JM. Human vascular smooth muscle cells inhibit platelet aggregation when incubated with glyceryl trinitrate: evidence for generation of nitric oxide. Br. J. Pharmacol. 1991; 102: 847–850.
11. Radomski MW, Rees DD, Dutra A, Moncada S. S-nitroso-glutathione inhibits platelet activation *in vitro* and *in vivo*. Br. J. Pharmacol. 1992; 107: 745–749.
12. de Belder AJ, MacAllister R, Radomski MW, Moncada S, Vallance PJ. Effects of S-nitroso-glutathione in the human forearm circulation: Evidence for selective inhibition of platelet activation. Cardiovasc. Res. 1994; 28: 691–694.
13. Ramsey B, de Belder A, Campbell S, Moncada S, Martin JF. A nitric oxide donor improves uterine artery diastolic blood flow in normal early pregnancy and in women at high risk of pre-eclampsia. Eur. J. Clin. Invest. 1994; 24: 76–78.
14. Creager MA, Gallagher SJ, Girerd XJ, Coleman SM, Dzau VJ, Cooke JP. L-arginine improves endothelium-dependent vasodilation in hypercholesterolemic humans. J. Clin. Invest. 1992; 90: 1248–1253.
15. McNamara DB, Bedi B, Aurora H, Tena L, Ignarro LJ, Kadowitz et al. L-arginine inhibits balloon catheter-induced intimal hyperplasia. Biochem. Biophys. Res. Commun. 1993; 193: 291–296.
16. Hishikawa K, Nakaki T, Suzuki H, Kato R, Saruta T. L-arginine as an antihypertensive agent. J. Cardiovasc. Pharmacol. 1992; 20: S196–S197.
17. Pepke-Zaba J, Higenbottam TW, Dinh-Xuan AT, Stone D, Wallwork J. Inhaled nitric oxide as a cause of selective pulmonary vasodilatation in pulmonary hypertension. Lancet 1991; 338: 1173–1174.
18. Roberts JD, Polaner DM, Lang P, Zapol WM. Inhaled nitric oxide in persistent pulmonary hypertension of the newborn. Lancet 1992; 340: 818–819.
19. Frostell CG, Blomqvist H, Hedenstierna G, Lundberg J, Zapol WM. Inhaled nitric oxide selectively reverses human hypoxic pulmonary vasoconstriction without causing systemic vasodilation. Anesthesiology. 1993; 78: 427–435.
20. Rossaint R, Falke KJ, Lopez F, Slama K, Pison U, Zapol WM. Inhaled nitric oxide for the adult respiratory distress syndrome. N. Engl. J. Med. 1993; 328: 399–405.
21. Gerlach H, Pappert D, Lewandowski K, Rossaint R, Falke KJ. Long-term inhalation with evaluated low doses of nitric oxide for selective improvement of oxygenation in patients with adult respiratory distress syndrome. Intensive Care Med. 1993; 19: 443–449.
22. Gerlach H, Rossaint R, Pappert D, Falke KJ. Time-course and dose-response of nitric oxide inhalation for systemic oxygenation and pulmonary hypertension in patients with adult respiratory distress syndrome. Eur. J. Clin. Invest. 1993; 23: 499–502.
23. Sessa WC. The nitric oxide synthase family of proteins. J. Vasc. Res. 1994; 31: 131–143.
24. Knowles RG, Moncada S. Nitric oxide synthases in mammals. Biochem. J. 1994; 298: 249–258.
25. Nathan CF, Hibbs JB Jr. Role of nitric oxide synthesis in macrophage antimicrobial activity. Curr. Opin. Immunol. 1991; 3: 65–70.

26. Radomski MW, Palmer RMJ, Moncada S. Glucocorticoids inhibit the expression of an inducible, but not the constitutive, nitric oxide synthase in vascular endothelial cells. Proc. Natl. Acad. Sci. USA 1990; 87: 10043–10047.
27. Bredt DS, Hwang PM, Glatt CE, Lowenstein C, Reed RR, Snyder SH. Cloned and expressed nitric oxide synthase structurally resembles cytochrome P-450 reductase. Nature 1991; 351: 714–718.
28. Lamas S, Marsden PA, Li, GK, Tempst P, Michel T. Endothelial nitric oxide synthase: molecular cloning and characterization of a distinct constitutive enzyme isoform. Proc. Natl. Acad. Sci. USA 1992; 89: 6348–6352.
29. Xie Q-W, Cho JJ, Calaycay J. Cloning and characterization of inducible nitric oxide synthase from mouse macrophages. Science 1992; 256: 225–228.
30. Sessa WC, Pritchard K, Seyedi N, Wang J, Hintze TH. Chronic exercise in dogs increases coronary vascular nitric oxide production and endothelial cells nitric oxide synthase gene expression. Circ. Res. 1994; 74: 349–353.
31. Weiner CP, Lizasoain I, Baylis SA, Knowles RG, Charles IG, Moncada S. Induction of calcium-dependent nitric oxide by sex hormones. Proc. Natl. Acad. Sci. USA 1994; 91: 5212–5216.
32. Curran RD, Billiar TR, Stuehr DJ, Hofmann K, Simmons RL. Hepatocytes produce nitrogen oxides from L-arginine in response to inflammatory products from Kupffer cells. J. Exp. Med. 1989; 170: 1769–1774.
33. Parratt JR. Myocardial and circulatory effects of *E. coli* endotoxin: modification of response to catecholamines. Br. J. Pharmacol. 1973; 47: 12–25.
34. Hibbs JB Jr, Westenfelder C, Taintor R, Vavrin Z, Kablitz C, Baranowski RL et al. Evidence for cytokine-inducible nitric oxide synthesis from L-arginine in patients receiving interleukin-2 therapy. J. Clin. Invest. 1992; 89: 867–877.
35. Vallance P, Moncada S. Role of endogenous nitric oxide in septic shock. New Horizons 1993; 1: 77–86.
36. Petros A, Lamb G, Leone A, Moncada S, Bennett D, Vallance P. Effects of a nitric oxide synthase inhibitor in humans with septic shock. Cardiovasc. Res. 1994; 28: 34–39.
37. Vallance P, Moncada S. Hyperdynamic circulation in cirrhosis: a role for nitric oxide? Lancet 1991; 337: 776–778.
38. Guarner C, Soriano G, Thomas A, Bulbena O, Novella MT, Balanzo J et al. Increased serum nitrite and nitrate levels in patients with cirrhosis: relationship to endotoxemia. Hepatology 1993; 18: 1139–1143.
39. Vallance P, Palmer RMJ, Moncada S. The role of induction of nitric oxide synthesis in the altered responses of jugular veins from endotoxaemic rabbits. Br. J. Pharmacol. 1992; 106: 459–463.
40. Schulz R, Nava E, Moncada S. Induction and potential biological relevance of a Ca^{2+}-independent nitric oxide synthase in the myocardium. Br. J. Pharmacol. 1992; 105: 575–580.
41. de Belder AJ, Radomski MW, Why HJF, Richardson PJ, Bucknall CA, Salas E et al. Nitric oxide synthase activities in human myocardium. Lancet 1993; 341: 84–85.
42. Beckman JS, Beckman TW, Chen J, Marshall PA, Freeman BA. Apparent hydroxyl radical production by peroxynitrite: implications for endothelial injury from nitric oxide and superoxide. Proc. Natl. Acad. Sci. USA 1990; 87: 1620–1624.
43. Moro MA, Darley-Usmar VM, Goodwin DA, Read NG, Zamora-Pino R, Feelisch M et al. Paradoxical fate and biological action of peroxynitrite on human platelets. Proc. Natl. Acad. Sci. USA 1994; 91: 6702–6706.

Pharmacological Sciences: Perspectives for
Research and Therapy in the Late 1990s
ed. by A.C. Cuello and B. Collier

Role of the L-Arginine-NO-Cyclic GMP Pathway in NANC Neurotransmission

Michael J. Rand and Chun Guang Li

*Pharmacology Research Laboratory, Department of Medical Laboratory Science,
Royal Melbourne Institute of Technology, Melbourne 3001, Australia*

Summary. Some noradrenergic noncholinergic (NANC) autonomic and enteric nerves utilize
NO or an NO-like factor as a transmitter and have been termed nitrergic. They contain a
characteristic neuronal isozyme of NO synthase (NOS). Synthesis of NO from L-arginine
requires a number of cofactors, of which haem and terahydrobiopterin are the most recently
recognized. The entry of Ca^{2+} produced by stimulation of nitrergic nerves activates NOS and
results in release of the transmitter. It acts on the soluble guanylate cyclase of effector smooth
muscle cells, resulting in a rise in the cyclic GMP level and hence relaxation. Nitrergic
neuroeffector junctions in the gastrointestinal tract may also include interstitial cells. Cotransmitters, such as ATP and VIP, may be released with nitrergic transmitter. Further work is
required to determine its functional interactions with other transmitters, on the modulation of
NO synthesis and release, and the identification of exact nature of the nitrergic transmitter.

Introduction

The generation of nitric oxide (NO) from L-arginine by NO synthases
(NOS) constitutes an essential part of signalling systems utilized by
vascular endothelial cells, in which the messenger role of NO was first
recognized [1], and in central [2], autonomic and enteric neurons [3–5].
The NOS of endothelial cells differs slightly from that of neurons, but
both are constitutively expressed and the overall reaction and the
cofactors are the same.

Components of the System

Chemistry of NO in Relation to Biological Function

The biological activity of NO as a transcellular messenger is predominantly based on its chemistry. The free radical is water-soluble and
lipophilic. Reaction of NO with molecular oxygen is second order with
respect to NO, and is therefore only significant at high NO concentra-

Correspondence to: Professor M.J. Rand, at the above address.
Based on a Symposium at the Twelfth International Congress of Pharmacology, Montreal, 29
July, 1994. The speakers, in order of their presentations, were B. Mayer (Graz, Austria), K.
Sanders (Reno, Nevada, USA), N. Toda (Ohtsu, Japan), M.J. Rand (Melbourne, Australia),
and G. Burnstock (London, UK).

tions. However, NO reacts rapidly with superoxide, even at low NO concentrations, to form peroxynitrite, which decomposes at physiological pH to yield highly reactive, cytotoxic intermediates. The cytotoxicity of NO may also involve rapid reactions with haem-proteins leading to formation of nitrosyl-haem complexes, and with non-haem iron-containing proteins leading to depletion of protein-bound iron. These matters are only of theoretical interest in the present context since there is no evidence suggesting NO emanating from peripheral nitrergic nerves mediates cytotoxicity.

The Soluble Guanylate Cyclase Target for Nitric Oxide

Activation of the haem-containing enzyme soluble guanylate cyclase (sGC), resulting in the formation of the second messenger cGMP from GTP, is the most prominent physiological effect of messenger NO [1]. Molecular biology studies of sGC revealed that it is a 150 kDa heterdimeric protein consisting of an α- and a β-subunit [6]. Both subunits have similar amino acid sequences to the C-terminal part of ANF-sensitive particulate GC and to the adenyl cyclases, suggesting that these domains participate in catalytic activity. The basal activity of sGC is low, but it is increased several hundred-fold by nanomolar concentrations of NO. This activation is most likely due to binding of NO to the haem prosthetic group located in the N-terminal part of the β-subunit: a haem-free form in which the histidine was exchanged for phenylalanine at position 105 in the β-subunit by site-directed mutagenesis still had basal activity but was completely insensitive to NO [7].

Neuronal Biosynthesis of Nitric Oxide

The activation of neuronal NOS, which is a calmodulin-dependent enzyme, follows from the rise in intracellular Ca^{2+} resulting from the depolarization-induced opening of voltage-dependent Ca^{2+} channels. All NOS isozymes contain a catalytically active haem group that is involved in reductive activation of the molecular oxygen required for the oxidation of L-arginine to NO and citrulline [8]. The required reducing equivalents are derived from cosubstrate NADPH and are shuttled to the haem by the flavins FAD and FMN. Thus NOS isozymes can be considered as self-sufficient relatives of cytochrome P_{450} enzymes in which oxygenase and reductase domains are located within a single polypeptide chain.

The Function of the Tetrahydrobiopterin Cofactor of NOS

In contrast to typical cytochrome P_{450} enzymes, the cofactor tetrahydrobiopterin (THB) is required by NOS. This may serve a dual role as an allosteric modifier and as a reactant. An allosteric action is suggested by the dimerization it induces in macrophage NOS and the positive cooperativity between the L-arginine and THB binding sites of neuronal NOS [9]. The role of THB as a reactant is suggested by the finding that the oxidized derivative, dihydrobiopterin, binds with high affinity to neuronal NOS but it is inactive as a cofactor [10].

It has recently been shown (Mayer B, unpublished observations) that sub-micromolar concentrations of THB prevent the feed-back inhibition of NOS by NO. This could be due to the formation of superoxide by auto-oxidation of THB and subsequent oxidation of NO to peroxynitrite since the effect of THB was prevented by a high amount (> 1000 U/ml) of superoxide dismutase. Thus, the physiological role of THB as a cofactor for NOS isozymes may be to protect NOS from feed-back inhibition by NO. If this should be the case, the active product formed by NOS would be peroxynitrite rather than NO. Conceivably, as yet unrecognised enzymes or chemical reactions could be involved in the further conversion of peroxynitrite to a biologically active NO-like species, and this would be the transmitter ultimately generated by neuronal NOS.

Enteric Nitrergic Transmission

Nitrergic nerves contribute significantly to enteric inhibitory neurotransmission in all species tested as part of the mechanisms for regulation of gastrointestinal (GI) motility. Enteric inhibitory nerves contribute to important reflexes such as receptive relaxation and the descending inhibition following distension [3–5]. Morphological evidence for an enteric nitrergic innervation has been demonstrated by immunohistochemical and histochemical (using the NADPH diaphorase reaction) methods, which reveal the presence of NOS in Dogiel type I nerves in enteric ganglia and in varicose nerve fibres within the circular and lonitudinal muscle layers. Pharmacodynamic evidence is provided by the findings that enteric inhibitory neurotransmission can generally be reduced by arginine analogues that inhibit NOS, and their effects can be reversed by excess L-arginine; furthermore, enteric inhibitory neurotransmission can also be blocked by agents, such as oxyhaemoglobin, that sequester or inactivate NO. Exogenous NO and donors of NO mimic the effects of enteric inhibitory neurotransmission in producing relaxation, inhibition of tonic contractions or phasic contractions and hyperpolarization of smooth muscle cells.

Targets of the Nitrergic Transmitter in the GI Tract

A significant part of the effects of NO is mediated by activation of guanylate cyclase and hence the production of cGMP, as evidenced by the findings that addition of cGMP mimics the effects of NO, levels of cGMP in smooth muscle are enhanced by NO, and prolongation of the persistence of cGMP by blockade of the cGMP-specific Type V phosphodiesterase with M&B 22948 prolongs responses to NO.

Immunohistochemical localization of cGMP-like immunoreactivity indicates that there are at least three cellular sites at which NO released by nitrergic nerve stimulation could act: (i) smooth muscle cells; (ii) a sub-population of enteric neurones other than the neurones that synthesize NO; and (iii) interstitial cells of Cajal.

Transduction: Activation of Guanylate Cyclase

The contractile activity of smooth muscle is related to the intracellular Ca^{2+} concentration ($[Ca^{2+}]_i$). There is no unique relationship between $[Ca^{2+}]_i$ and force in smooth muscles and many agonists are known to alter the relationship. Both cGMP and cAMP reduce the sensitivity of the contractile apparatus to $[Ca^{2+}]_i$ such that an increase in cyclic nucleotide production reduces the amount of force generated at a given level of $[Ca^{2+}]_i$. Part of the reduction in $[Ca^{2+}]_i$ in smooth muscle cells by NO may be due to a cGMP-dependent increase in Ca^{2+} uptake into the sacroplasmic reticulum. However, hyperpolarization of these cells may also contribute to the reduction of $[Ca^{2+}]_i$ by closing voltage-dependent Ca^{2+} channels.

Hyperpolarization and Relaxation of Smooth Muscle

The hyperpolarization of GI smooth muscle cells produced by NO and enteric inhibitory nerve stimulation is mediated by the activation of various K^+ channels. The activation of large conductance Ca^{2+}-activated K^+ channels appears to involve a cGMP-dependent protein kinase C. Recent reports also suggest that Ca^{2+}-activated K channels can be directly activated by NO. Smaller conductance K^+ channels are also activated, suggesting that the effects of NO on the membrane potential are complex and a single K^+ channel blocker (except very non-specific blockers like quinine) may not be able to block totally NO-dependent hyperpolarization.

Role of Interstitial Cells of Cajal in Nitrergic Enteric Transmission

The neuroeffector junction for enteric nitrergic axon terminals may include interstitial cells of Cajal (ICs), which are found in close association with enteric nerve fibres and are often interposed between axon varicosities and smooth muscle cells. Recent studies have shown that ICs express a constitutive form of NOS which has been identified by immunohistochemistry using antibodies raised against endothelial NOS. Stimulation of ICs with a number of agonists including NO leads to an increase in $[Ca^{2+}]_i$. This signal might be expected to increase synthesis of NO, and when $[Ca^{2+}]_i$ is raised in ICs, a diffusible substance, identified as NO by bioassay, is released. These data suggest that ICs could contribute to enteric inhibitory neurotransmission by providing a post-junctional source of NO and by amplifying inhibitory neurotransmission.

NO and NANC Neurotransmission in Cerebral Arteries

The presence of a NANC vasodilator innervation of cerebral arteries has been apparent for some years, but the nature of transmitter has only been clearly elucidated in the past 4 years. Relaxations induced by nicotine or transmural electrical stimulation were not affected when arteries were rendered unresponsive to a number of possible mediators (substance P, VIP, CGRP); however, it had been already established in 1975 that these relaxations were abolished by oxyhaemoglobin and methylene blue, which in retrospect is presumptive evidence for the involvement of NO [11].

More recent studies [12] have demonstrated that NO synthase inhibitors abolished the relaxations of cerebral arteries caused by stimulation of NANC vasodilator nerves in a variety of mammals including monkeys and dogs, and the responses were restored by L- but not D-arginine, and were associated with an increase in the tissue content of cyclic GMP. The release of the NO-like transmitter occurred in endothelium-denuded cerebral arterial strips, indicating it was directly of neural origin. Neurally-induced relaxations were not influenced by dihydropyridine Ca^{2+}-entry blockers but were suppressed by cadmium and calmodulin inhibitors, in agreement with the role of Ca^{2+} and calmodulin in activating neuronal NOS.

Abundant nerve bundles and fibres containing NOS immunoreactivity have been observed in the adventitia and outer media of cerebral and temporal arteries of monkeys and originate unilaterally from the pterygopalatine ganglion [13].

Nature of the Nitrergic Transmitter

There are still some concerns about the extent to which NO fulfils the criteria for acceptance as the nitrergic transmitter since the transmitter that traverses the neuroeffector junction has distinctly different properties from NO in aqueous solution [4, 5]. For example, in the rat anococcygeus muscle, in which nitrergic transmission was first demonstrated, hydroxocobalamin (HC) and the spin trap agent carboxy PTIO block responses to NO but not those to nitrergic nerve stimulation. These agents, however, do block responses mediated by EDRF. In addition, the nitrergic transmitter seems to be more resistant to inactivation by superoxide than is free NO [14], therefore, the nitrergic transmitter does not totally behave like free NO. A further difference is that responses to NO are greatly enhanced by cysteine (1 mM), which presumably forms the relatively stable NO-donating adduct nitrosocysteine, but responses to nitrergic nerve stimulation are not enhanced: it may be relevant to note that the smooth muscle relaxant action of NOCys, like that of the nitrergic transmitter, is not blocked by HC.

Despite the apparent qualitative differences between NO in aqueous solution and the nitrergic transmitter, it has been suggested that the difference is purely quantitative. This is based on a kinetic analysis which indicates that the rate of spread of effective concentrations of NO from a source such as a neurone is extremely rapid on account of the small molecular size of NO, and it is hypothesised that reductions in the half-life of NO (for example, by reaction with HC, carboxy PTIO, or superoxide) are not sufficient to lower the concentration of NO in the neuroeffector junction below the effective level [15]. Whether there is such a quantitative difference or whether the nitrergic transmitter is qualitatively different from free NO requires further study.

Other NANC Transmitters

NANC Transmission

It is likely that several neurotransmitters are employed in mediating autonomic and enteric inhibitory neurotransmission, including not only NO or a NO-donating compound, but also vasoactive intestinal polypeptide (VIP), and adenosine triphosphate (ATP). Earlier studies of NANC transmission have played an important role in recognizing the multiplicity of NANC transmitters in the autonomic and enteric systems. The demonstrations of adenosine triphosphate (ATP) and vasoactive intestinal polypeptide (VIP) as possible neurotransmitters in NANC inhibitory nerves in the 1970s led to the concept of cotransmission [16], and to the "chemical coding" of autonomic and enteric nerves in terms of the combination of different transmitters in different nerve types, their projections and central connections [17]. The amount of influence exerted by each transmitter appears to vary depending upon the frequency of nerve stimulation, the nature of the effector tissue, the region of the GI tract, and the species. There is histochemical evidence for colocalisation of NOS with VIP and/or quinacrine (which is a marker for high levels of ATP) in sub-populations of neurones supplying various regions of the gut [18]. It also seems likely that NO acts as a cotransmitter with ATP and/or neuropeptides in sub-populations of intrinsic nerves in the heart which project to coronary microvessels, in the bladder which project to the urethra, and in the airways which project to the tracheal musculature and pulmonary blood vessels.

Plasticity of Autonomic Nerves

Autonomic nerves show a remarkable degree of plasticity, changes in density and expression of cotransmitters occurring under a variety of conditions [19], although changes in the contribution of NO as a transmitter under these conditions have not yet been described. It has recently been shown that NO and ATP co-released from NANC nerves in the myenteric plexus appear to act synergistically with growth factors to promote growth of nerve fibres in the brain following transplantation (Höpker, Tew, Anderson, Saffrey and Burnstock, unpublished observations); thus trophic actions of NANC cotransmitters should also be considered.

Conclusions

Since the first demonstration of nitrergic transmission about 5 years ago, considerable progress has been made in understanding the role of NO in neurotransmission. A large amount of evidence now indicates that NO is a principle mediator of an inhibitory NANC transmission in many autonomically innervated organs and tissues, including vascular smooth muscle in the cerebral circulation, and throughout the enteric nervous system. In some cases, NO may also act as a cotransmitter. The main target so far established for neuronally generated NO is the guanylate cyclase of innervated effector cells. However, more work is required on modulation of NO synthesis, interactions of the nitrergic transmitter with other transmitters, possible targets other than soluble guanylate cyclase, and the identification of the exact nature of the nitrergic transmitter.

Acknowledgements

We are grateful to the speakers in the Symposium for supplying brief accounts of their presentations, most of which were incorporated in the text of this article. The authors' work was supported by grants from the National Health & Medical Research Council and the Smoking & Health Research Foundation of Australia.

References

1. Moncada S, Palmer MJ, Higgs EA. Nitric oxide: physiology, pathophysiology and pharmacology. Pharmacol. Rev. 1991; 43: 109–142.
2. Garthwaite J. Glutamate, nitric oxide and cell–cell signalling in the nervous system. Trends Neurosci. 1991; 14: 60–67.
3. Sanders KM, Ward SM. Nitric oxide as a mediator of nonadrenergic noncholinergic neurotransmission. Am. J. Physiol. 1992; 262: G379–G392.
4. Rand MJ, Li CG. Nitric oxide in the autonomic and enteric nervous systems. In: SR Vincent (editor), Nitric oxide in the nervous system. London: Academic Press, 1995: 227–279.
5. Rand MJ, Li CG. Nitric oxide as a neurotransmitter in peripheral nerves: nature of transmitter and mechanism of transmission. Ann. Rev. Physiol. 1995; 57: 659–682.
6. Koesling D, Böhme E, Schultz, G. Guanylyl cyclase, a growing family of signal transducing enzymes. FASEB J 1991; 5: 2785–2791.
7. Wedel B, Humbert P, Harteneck C, Foerster J, Maldewitz J, Böhme E, Schulz G, Koesling D. Mutation of His-105 of the β1-subunit yields a nitric oxide-insensitive form of soluble guanylyl cyclase. Proc. Natl. Acad. Sci. USA 1994; 91: 2592–2596.
8. Mayer B. Molecular characteristics and enzymology of nitric oxide synthase and soluble guanylyl cyclase in the CNS. Semin. Neurosci. 1993; 5: 197–205.
9. Klatt P, Schmid M, Leopold E, Schmidt K, Werner ER, Mayer B. The pteridine binding site of brain nitric oxide synthase – tetrahydrobiopterin binding kinetics, specificity, and allosteric interaction with the substrate domain. J. Biol. Chem. 1994; 269: 13861–13866.
10. Mayer B, Schmidt E, Leopold E, Klatt P, Schmidt K, Werner ER. Allosteric interactions of the tetrahydrobiopterin and substrate binding sites of neuronal nitric oxide synthase. Can. J. Physiol. Pharmacol. 1994; 72 (Suppl. 1): 480.
11. Toda N. Nicotine-induced relaxation in isolated canine cerebral arteries. J. Pharmacol. Exp. Ther. 1975; 193: 376–384.
12. Toda N. Nitric oxide and the regulation of cerebral blood flow. In: Vincent SR editor, London: Academic Press, 1995: 207–225.
13. Yoshida K, Okamura T, Toda N. Histological and functional studies on the nitroxidergic nerve innervating monkey cerebral, mesenteric and temporal arteries. Jpn. J. Pharmacol. 1994; 65: 351–359.
14. Gibson A, Brave, Tucker JF. Differential effect of xanthine:xanthine oxidase on NANC- and NO-induced relaxations of the mouse anocyccygeus. Can. J. Physiol. Pharmaol. 1994; 72 (Suppl. 1): 475.
15. Wood J, Garthwaite J. Models of the diffusional spread of nitric oxide (NO): implications for neural NO signalling and its pharmacological properties. Neuropharmacology 1994; 33: 1235–1244.
16. Burnstock G. Do some nerve cells release more than one transmitter? Neuroscience 1976: 1: 239–248.
17. Burnstock G. Co-transmission. The Fifth Heymans Lecture. Arch. Int. Pharmacodyn. Ther 1990; 304: 7–33.
18. Belai A, Burnstock G. Evidence for coexistence of ATP and nitric oxide in non-adrenergic, non-cholinergic (NANC) inhibitory neurones in the rat ileum, colon and anococcygeus muscle. Cell Tissue Res. 1994; 278: 197–200.
19. Burnstock G. Changes in expression of autonomic nerves in aging and disease. J. Auton. Nerv. Syst. 1990; 30: 525–534.

Endocrine Pharmacology and Immunopharmacology

Pharmacological Sciences: Perspectives for
Research and Therapy in the Late 1990s
ed. by A.C. Cuello and B. Collier

Neuroendocrine Hormones and the Immune System

Paul A. Kelly[1], J. Edwin Blalock[2], George P. Chrousos[3], Li Yu-Lee[4],
and Vincent Geenen[5]

[1]*INSERM Unité 344, Faculté de Médecine Necker, 75730 Paris Cedex 15, France;* [2]*Department
of Physiology & Biophysics, University of Alabama, Birmingham, AL 35295, U.S.A.;* [3]*NICHD,
Pediatric Endocrinology, National Institutes of Health, Bethesda, MD 20820-892, U.S.A.;*
[4]*Department of Medicine, Section Rheumatology, Baylor College of Medicine, Houston,
TX 77030, U.S.A.;* [5]*Centre Hospitalier Université de Liege, Service d'Endocrinologie,
Domaine Universitarie du Sart Tilman, 35 B-4000 Liege 1, Belgium*

Identification of Neuroendocrine-immune Interactions

From ancient times up to the present, there have been anecdotal
observations of an association between "personality," health and disease. Today, we are confronted with the apparent reality of the association as evidenced by classical Pavlovian conditioning of immune
responses as well as immunomodulatory effects of stress and central
nervous system lesioning (for review, see [1]). However, a molecular
basis for these phenomena has only recently emerged.

To explain this intricate communication between the nervous, endocrine and immune systems, Blalock [2] has suggested that the immune
and neuroendocrine systems represent a totally integrated circuit by
virtue of sharing a common set of hormones and neurotransmitters,
such as corticotropin (ACTH), thyrotropin (TSH), chorionic go-
nadotropin (CG), growth hormone (GH), substance P, and endorphins,
as well as similar receptors for these peptides. The immune system-
derived hormones seem very similar, if not identical, to their neuroendocrine counterparts. With regard to ACTH, the sequences are, in fact,
identical. While immunocyte production of these peptides was originally
described in response to immunostimulants, it now appears that their
synthesis is also controlled by hypothalamic releasing factors (HRFs).
For instance, corticotropin releasing hormone (CRH) induced and
dexamethasone blocked the *de novo* synthesis of immunoreactive ACTH
and β-endorphin. Although regulation of ACTH production in im-
munocytes seems quite similar to that seen in the pituitary, it differs by
having interleukin (IL)-1 as an intermediate in the axis. Thus, CRF
actually induces IL-1 and the IL-1 causes β lymphocytes to make
ACTH and β-endorphin. In addition to responding to HRFs, cells of
the immune system also produce HRFs, such as CRH and growth

hormone releasing hormone (GHRH). Moreover, lymphokines and monokines are produced by and can elicit hormone and HRF production by pituitary cells and hypothalamic neurons. Other studies have revealed that the immune and neuroendocrine systems also seem to share a similar group of structurally-related receptors for cytokines as well as neuroendocrine peptides. Such shared ligands and receptors may well represent the molecular mechanism of communication between these systems [2–4]. There is also important interplay between the shared molecules. For instance, CRH has been shown to elevate corticotroph IL-1 receptors and to sensitize the pituitary gland to the direct action of IL-1. Such sensitization represents a novel interactive point between the nervous, endocrine, and immune systems whereby very mild psychological or physical stress could profoundly impact an inflammatory response by increasing pituitary sensitivity to immunologic mediators such as IL-1 [5].

An important conceptual advance from these studies is the idea that the immune system serves as a sensory organ for stimuli not recognized by our central and peripheral nervous systems [1]. Recognition of such "non cognitive stimuli" leads to immunologic production of the aforementioned biochemical signals which in turn changes one's physiology. Thus in some instances, biological responses such as *stress* may ultimately originate in the immune system rather than the brain. In contrast, the immune system is made aware of nervous system recognition of cognitive stimuli by detecting hormones and neurotransmitters.

Direct Modulation of Inflammation by CRH and Other Neuropeptides

Corticotropin-releasing hormone is a major regulator of the hypothalamic-pituitary-adrenal (HPA) axis and the principal coordinator of the stress response. Increased HPA activity normally would lead to increased susceptibility of an individual to a host of infectious agents and tumors but resistance to autoimmune/inflammatory diseases. Two highly inbred rat strains (Fischer and Lewis) have been selected out of Sprague-Dawley rats for their resistance and for their susceptibility to inflammatory diseases, respectively [6]. Further characterization of the defect in the Lewis rat localized it in the hypothalamic CRH neuron, which was globally defective in its response to all neurotransmitters [7]. Recently, CRH receptors were described in peripheral sites of the immune system and CRH was shown to promote several immune functions in vitro. Interestingly, hypothalamic expression of AVP and peripheral AVP concentrations are elevated in Lewis rats, and neutralizing antisera to either CRP or AVP could ameliorate carrageenin-induced inflammation [8]. Rabbit anti-CRH caused suppression of both inflammatory exudate volume and cell concentration by 50–60%. Large amounts of immunoreactive CRH were detected by immunocytochemistry and/or radioimmunoassay in the inflammatory area, but not in the concurrent plasma samples from the systemic circulation. Glucocorticoid and somatostatin inhibited the production of CRH at the inflammatory site.

There are human homologs of the Fischer-Lewis rat paradigm. A subgroup of patients with rheumatoid arthritis have low or normal circadian concentrations of ACTH and cortisol, in spite of elevated concentrations of IL-1β and IL-6, poor responses to stress of joint replacement, elevated levels of circulatory AVP and markedly elevated concentrations of CRH in their inflamed joints [9, 10]. In addition, in Hashimoto's thyroiditis, inflamed tissues show elevated concentrations of CRH. The development of agents that specifically regulate the local production or action of these neuropeptides will have important therapeutic benefits.

In summary, at the periphery, CRH can act directly on cells of the immune system to stimulate the inflammatory reaction, in contrast to its central, indirect immunosuppressive/anti-inflammatory actions.

Prolactin-Growth Hormone and Immune Function

Prolactin (PRL) and growth hormone (GH) have been shown to affect immune functions in experimental animals and humans. In hypophysectomized rats or mice, the production of antibodies is reduced, and this effect can be counteracted by the administration of PRL or GH. Such effects are mediated by hormones normally produced in the pituitary gland, acting via an endocrine pathway. Alternatively, GH and PRL have been shown to be produced directly by lymphocytes, and thus act in a paracrine or autocrine fashion. Using reverse transcriptase PCR analysis of RNA prepared from various lymphoid organs, and actual measurement of prolactin secretion, Pellegrini et al. [11] demonstrated that only T-cells produce human prolactin, but that all immune cells contain prolactin receptors. More recently, Touraine et al. [12] have shown, using reverse transcriptase PCR analysis of RNA prepared from various lymphoid organs, that both the long and short forms of PRL receptors are expressed in rats and mice in varying amounts. The different forms of receptors may have specific functions in immune cells and be involved in specific immune responses of the animal. Recently, FACS analysis using an anti-receptor monoclonal antibody, confirmed the expression of PRL receptors at the cell surface of all immune cells thus far tested in rodents and man [13, 14].

The mechanism of action of PRL and GH has been shown to involve the activation of an associated tyrosine kinase, Jak2. In the Nb2 lymphoma cell line, which is dependent upon PRL for growth, Jak2 and the receptor itself are rapidly tyrosine phosphorylated following ligand binding. A juxtamembrane region conserved among many members of the cytokine/GH/PRL receptor family, known as Box 1, is important for activation of Jak2 activity and for stimulation of milk protein gene transcription [15]. Although the interaction of Jak2 and the PRL receptor probably occurs via Box 1, the interaction may be indirect and involve an adaptor protein.

It will be important to investigate domains of the receptor involved in specific prolactin actions such as cell proliferation and gene transcription, and identify effector proteins that are involved in the signal transduction pathway.

Hormone Response Element of IRF-1 in T-cell Activation

Prolactin (PRL) is a novel cytokine which is co-mitogenic for T-cell proliferation. The PRL receptor is found on hematopoietic cells and is a member of the hematopoietin/cytokine receptor superfamily. PRL acts as both a competence factor needed for quiescent T-cells to enter the cell cycle, as well as progression factor needed for activated T-cells to enter S-phase and commit to DNA synthesis [16]. Stimulation of a quiescent T-cell line, Nb2, by PRL leads to the rapid transcriptional induction of T-cell activation genes, including the transcription factor interferon regulatory factor-1 (IRF-1) [17]. IRF-1 is induced twice by PRL in a single-cell cycle, first during G1 at 30 to 60 min and again over early S-phase at 10 to 12 h [18]. Thus, IRF-1 is an excellent early response marker for PRL signaling leading to cell proliferation. The present discussion focuses on the PRL-inducible expression of the IRF-1 gene, the identification of cis-regulatory sequences in the IRF-1 flanking/promoter DNA that mediate the PRL-inducible response, and the identification of trans-acting factors which may constitute PRL signaling molecules to the IRF-1 promoter 19. These include cytokine signaling molecules called Stat and cell cycle-related proteins.

To understand how PRL regulates the expression of the early response gene IRF-1, we measured the transcription rate of the IRF-1 gene in response to PRL stimulation [20]. The transcription rate of the IRF-1 gene is rapidly induced by PRL over 20-fold at 30 to 60 min. This induction is independent of new protein synthesis, suggesting that pre-existing factors are involved in the transcriptional induction of the IRF-1 gene by PRL. This induction is also transient, as it returns to basal levels by 4 h. Between 8 to 10 h during early S-phase, IRF-1 transcription rate is again increased. Thus, both peaks of IRF-1 expression are due to transcriptional activation of the IRF-1 gene and further suggest the presence of PRL responsive elements (PRL-REs) in the IRF-1 gene. To localize PRL-REs, 5′ flanking IRF-1 DNA was shown to contain elements that mediate both G1 and S-phase expression. The

−200 bp IRF-1 promoter DNA contains elements that respond primarily to G1 PRL stimulation.

Further promoter deletion and transfection analyses delineated a minimal PRL response region between −113 and −205 bp. Two distinct sequences within this region exhibit binding with nuclear proteins in a PRL-inducible, cell cycle-dependent manner. A GAS site consisting of two inverted GAAA motifs (−123/−113), previously shown to be a functional IFNγ response element which binds the cytokine signaling molecule Stat91, showed interactions with Stat-related proteins in response to PRL stimulation. Curiously, a Stat-like protein appears to interact with IRF-1 GAS during G1 activation while the more widely-used Stat91 is found at the IRF-1 promoter mainly during S-phase induction [20]. The biphasic binding of different Stat molecules to the IRF-1 promoter corresponds very well with the biphasic expression of the IRF-1 gene in response to PRL stimulation. A second sequence (−205/−144), a 72 bp DNA, lies upstream of the GAS site, and showed interactions with cell cycle-regulated proteins including retinoblastoma (Rb), cyclin A, cyclin D1 and cdc2 kinase [20]. Rb is known to associate with cyclins and kinases in a cell cycle-dependent manner, and their combined presence at the IRF-1 promoter suggests a novel role for cell cycle-related proteins in the regulation of gene transcription. The biphasic interaction of Rb-containing proteins with the IRF-1 promoter again correlated very well with the cell cycle-dependent expression of the IRF-1 gene. Further, nuclear proteins binding to the G1 PRL responsive region are likely to be involved in protein/protein interactions with proteins binding at promoter distal regions, as both promoter-proximal and distal elements have been demonstrated to be required for mediating IRF-1 gene transcription during S phase of the cell cycle.

These studies indicate that both Stat proteins and cell cycle-regulated proteins participate in PRL receptor signal transduction to regulate the biphasic expression of the IRF-1 gene in PRL-stimulated T-cells. They further show that cytokine signals are integrated with cell cycle controls to regulate early growth-response genes in activated T-cells.

Cryptocrine Cell-to-Cell Signaling in the Thymus

The thymus shapes the whole T-cell repertoire in two distinct ways. The most specific role of the thymus is in the induction of central T-cell tolerance, which follows the negative selection of self-reactive T-cells that emerge during the random recombination of gene segments for the T-cell antigen receptor (TCR). The thymus is also responsible for the developmental program and positive selection of peripheral T-cell repertoire. This dual physiological role constitutes one major paradox in contemporary immunology.

Thymic epithelial cells (TEC) from different species synthesize polypeptide precursors belonging to the neurohypophysial (NHP), insulin and tachykinin families [21, 22]. A subpopulation of TEC, thymic nurse cells (TNC), probably derive from the embryonic neural crest, since they express markers of the diffuse neuroendocrine system, and synthesize NHP as well as insulin-related (IGF-11) peptides [23]. The classical models of neuroendocrine signaling and secretion, however, cannot be applied to TEC/TNC. The term cryptocrine has been introduced to describe a novel type of cell-to-cell signaling within specialized micro-environments (such as TNC) where mobile cells, in this case T-cell precursors, migrate and differentiate in close contact with large epithelial "nurse" cells. The expression of functional NHP peptide receptors by pre T-cells as well as the mitogenic properties of NHP-related peptides on pre-T-cells strongly supports a cryptocrine signaling mediated by NHP-related peptides and receptors between TEC/TNC and pre-T-cells [24]. This novel type of signaling probably serves *in vivo* as an accessory pathway in the process positive selection.

Intrathymic cell-to-cell signaling is closely associated with the presentation of molecular structure of "self" to the developing T-cell system. Specific sequences can be identified to be major self antigens of protein families that could be presented by the major histocompatibility complex (MHC)-derived thymic proteins. Moreover, preliminary data strongly support that the presentation of neuroendocrine selfantigens in the thymus is performed through a pathway different from the peripheral presentation of allo- or auto-antigens [25]. This specific pathway for the intrathymic presentation of neuroendocrine selfantigens offers the selective advantage of bypassing MHC allelic antigen restriction. The available data also suggest the existence of tissue-specific genetic mechanisms underlying the expression of thymic oxytocin self-peptide and its subsequent presentation to pre-T-cells.

The cryptocrine cell-to-cell communication demonstrated in the thymus may be a primitive, intermediary step between cell adhesion and paracrine types of signaling. The immune system has primarily evolved to protect the integrity of the self against aggression from nonself infectious invaders. The primary lymphoid organs thus seem to exert a radical anti-hazard constraint by purging the immune system of self-reactive components which could represent a serious threat for survival. In the same global perspective, pathological autoimmunity may be considered as the tribute paid by the human species for the very potent efficiency of its immune defences.

References

1. Blalock JE editor. Chemical Immunology, vol. 52, Neuroimmunoendocrinology. Basel: Karger, 1992.

2. Blalock JE. The immune system as a sensory organ. J. Immunol. 1984; 132: 1067–1070.
3. Blalock JE. A molecular basis for bidirectional communication between the immune and neuroendocrine systems. Physiol. Rev. 1989; 69: 1–32.
4. Blalock JE. The immune system: our sixth sense. The Immunologist 1994; 2: 8–15.
5. Payne LC, Weigent DA, Blalock JE. Induction of pituitary sensitivity to interleukin-1: a new function for corticotropin releasing hormone. Biochem. Biophys. Res. Comm. 1994; 198: 480–484.
6. Sternberg E, Young WS Jr, Bernardini R, Calogero A, Chrousos GP, Gold PW et al. A central nervous defect in the stress response is associated with susceptibility to streptococcal cell wall arthritis in Lewis rats. Proc. Natl. Acad. Sci. USA 1989; 86: 4771–4775.
7. Calogero A, Sternberg E, Bagdy G, Smith G, Bernardini R, Aksentijevich S et al. Neurotransmitter-induced hypothalamic-pituitary-adrenal axis responsiveness is defective in inflammatory disease-susceptible Lewis rats: *in vivo* and *in vitro* studies suggesting globally defective hypothalamic CRH secretion. Neuroendocrinology 1992; 55: 600–608.
8. Patchev V, Kalogeras K, Zelazowski P, Wilder RL, Chrousos GP. Increased plasma concentrations and hypothalamic content and *in vitro* release of arginine-vasopressin in inflammatory disease-prone, hypothalamic CRH-deficient Lewis rats. Endocrinology 1992; 131: 1453–1457.
9. Chikanza IC, Petrou P, Chrousos GP, Kingsley G, Panayi GS. Defective hypothalamic response to immune/inflammatory stimuli in patients with rheumatoid arthritis. Arthritis Rheum. 1992; 35: 1281–1288.
10. Crofford LJ, Sano H, Karalis K, Freidman TC, Epps HR, Remmers EF et al. Corticotropin-releasing hormone in synovial fluids and tissues of patients with rheumatoid arthritis. J. Immunol. 1993; 151: 1–10.
11. Pellegrini I, Lebrun JJ, Ali S, Kelly PA. Expression of prolactin and its receptor in human lymphoid cells. Mol. Endocrinol. 1992; 6: 1023–1031.
12. Touraine P, Leite de Moraes MC, Dardenne M, Kelly PA. Expression of the short and long forms of prolactin receptor in murine lymphoid tissues. Mol. Cell Endocrinol. In press.
13. Gagnerault MC, Touraine P, Savino W, Kelly PA, Dardenne M. Expression of prolactin receptors in murine lymphoid cells in normal and autoimmune situations. J. Immunol. 1993; 150: 5673–5681.
14. Dardenne M, Leite de Morales, MdC, Kelly PA, Gagnerault MC. Prolactin receptor expression in human hematopoietic tissues analyzed by flow cytofluorometry. Endocrinology 1994; 134: 2108–2114.
15. Lebrun JJ, Ali S, Sofer L, Ullrich A, Kelly PA. Prolactin-induced proliferation of Nb2 cells involves tyrosine phosphorylation of the prolactin receptor and its associated tyrosine kinase. J. Biol. Chem. 1994; 269: 14021–14026.
16. Stevens AM, Yu-Lee L-y. Multiple prolactin-responsive elements mediate G1 and S phase expression of the interferon regulatory factor-1 gene. Mol. Endocrinol. 1994; 8: 345–355.
17. Yu-Lee L-y, Hrachovy JA, Stevens AM, Schwarz LA. Interferon-regulatory factor-1 is an immediate-early gene under transcriptional regulation by prolactin in Nb2 T cells. Mol. Cell. Biol. 1990; 10: 3087–3094.
18. Stevens AM, Yu-Lee L-y. The transcription factor interferon regulatory factor-1 is expressed during both early G1 and G1/S transition in the prolactin-induce lymphocyte cell cycle. Mol. Endocrinol. 1992; 6: 2236–2243.
19. Horseman N, Yu-Lee L-y. Transcriptional regulation by the helix bundle peptide hormones: GH, PRL, and hematopoietic cytokines. Endocrine Review 1994; 15: 627–649.
20. Stevens AM, Wang Y-f, Sieger KA, Lu H, Yu-Lee L-y. Biphasic transcriptional regulation of the interferon regulatory factor-1 gene by prolactin: Involvement of STAT- and cell cycle-related proteins. Mol. Endocrinol. 1994; 8: 345–355.
21. Geenen V, Legros JJ, Franchimont P, Baudrihaye M, Defresne MP, Boniver J. The neuroendocrine thymus: coexistence of oxytocin and neurophysin in the human thymus. Science 1986; 232: 508–511.
22. Geenen V, Cormann-Goffin V, Martens H, Vandersmissen E, Robert F, Benhida A et al. Thymic neurohypophysial-related peptides and T-cell selection. Regul. Pept. 1993; 45: 273–278.
23. Geenen V, Defresne MP, Robert F, Legros JJ, Franchimont P, Boniver J. The neurohormonal thymic microenvironment: immunocytochemical evidence that thymic nurse cells are neuroendocrine cells. Neuroendocrinology 1988; 47: 365–368.

24. Martens H, Robert F, Legros JJ, Geenen V, Franchimont P. Expression of functional neurohypophysial peptide receptors by murine immature and cytotoxic T-cell lines. Prog. NeuroEndocrinImmunol 1992; 5: 31–39.
25. Geenen V, Vandersmissen E, Cormann-Goffin N, Martens H, Legros JJ, Degiovanni G et al. Membrane translocation and relationship with MHC class I of a human thymic neurophysin-like protein. Thymus 1993; 22: 55–65.

Pharmacological Sciences: Perspectives for
Research and Therapy in the Late 1990s
ed. by A.C. Cuello and B. Collier
© 1995 Birkhäuser Verlag Basel/Switzerland

Selected Aspects of the Immunopharmacology of Cytokines

Alberto Mantovani[1], Roland Mertelsman[2], Hans Schreiber[3],
Marc Feldmann[4] and Enrico Mihich[5]

[1]Istituto Di Ricerche Farmacologiche, Mario Negri, Milano, Italy; [2]University of Freiburg,
Medical Center, Dept. Hematology/Oncology, Freiburg, Germany; [3]University of Chicago,
Chicago, IL, USA; [4]Kennedy Institute of Rheumatology, Stanley Bldg, Hammersmith,
London, UK; [5]Roswell Park Cancer Institute, Grace Cancer Drug Center, Buffalo NY
14263-0001, USA

Introduction

Cytokines constitute a large group of macromolecules which mediate
the development, activation and/or function of the various components
of the immune system. This system may have beneficial effects as an
essential instrument of defense against foreign microorganisms or transformed neoplastic cells, but it may also have deleterious effect in the
pathogenesis of several types of autoimmune diseases. As the finely
tuned mechanisms of immune activation and action are typified by their
multiplicity and the diversity of their characteristics, they offer unique
opportunities for selective pharmacological and immunopharmacologi-
cal intervention. In this short article, four examples are briefly discussed
which illustrate aspects of regulation of cytokine function in target cells
(Mantovani), the cooperative role of cytokines in cell differentiation
(Mertelsmann), the induction of certain cytokines by biological response modifiers (Schreiber), and the clinical application of anti-TNF
(Tumor Necrosis Factor) "humanized" antibodies in Rheumatoid
arthritis (Feldman).

The Interleukin 1(IL1) Decoy Receptor: A Novel and Unique Anti-ILI Pathway

The interleukin 1 (IL1) system consists of two polypeptide mediators,
termed IL1α and IL1β, high affinity binding sites, referred to as type I
and type II receptor (RI and RII), and an IL1 receptor antagonist

Correspondence to: Enrico Mihich, Roswell Park Cancer Institute, Grace Cancer Drug
Center, Elm and Carlton Sts., Buffalo NY 14263-0001, U.S.A.

(IL1ra) [1, 2]. IL1 is perhaps the most potent and multifunctional cell activator described in immunology and cell biology. *In vitro*, its activities encompass cells of hematopoietic origin, from undifferentiated precursors to mature leukocytes, components of the vessel wall, and cells of mesenchymal and epithelial origin. *In vivo*, IL1 mediates local and systemic responses from fever, anorexia and slow-wave sleep, immunoregulation and generation of the acute phase response and local inflammatory reactions.

The IL1 system is a prime target for developing novel therapeutic strategies to local and systemic inflammatory and immune reactions, impaired hematopoiesis, and IL1 driven neoplasias such as acute myelogeneous leukemia (AML). Understanding the role and signaling functions of the IL1 receptors is important to efforts aimed at manipulating the IL1 system [1–5].

In early studies, it was found that IL1RI expressed in CHO cells mediated functional responses to IL1. Human EC respond to minute amounts of IL1 (usually in the picomolar range) with an extremely complex and varied range of responses, yet they only express IL1RI, providing an early indication that IL1RII is dispensable for responding to IL1. Blocking and anti-RI monoclonal antibodies (MAbs) inhibited IL1 activities in cells expressing predominantly the RI (T lymphocytes and osteoblasts), equal amounts of RI and RII (hepatoma cells), or predominantly RII (monocytes, polymorphomononuclear cells, PMN and pre-B cells); in the latter cells, the biological activities of IL1 are thus mediated by small amounts of IL1RI.

There is no evidence demonstrating a signaling function for IL1RII. Blocking anti-RII mAb did not inhibit the activity of IL1 in different cells, including lymphocytes, monocytes, monocytic cell lines, polymorphomononuclear cells (PMN), hepatoma cells and pre-B cells.

The IL1-RII is released as a soluble ≈ 45 kDa IL1-binding molecule from B lymphoblastoid cells, mitogen-activated leukocytes, PMN and monocytes (S. Saccani and A. Mantovani, unpublished). The finding that IL1RII is dispensable for signaling and that this "receptor" is shed in a soluble form suggested that it may act as a decoy for IL1. This hypothesis was examined by investigating the effects of blocking mAbs on IL1-induced activities in human myelomonocytes, which express predominantly the IL1-RII.

Previous studies demonstrated that IL1 is a potent inducer of PMN survival *in vitro*. IL4 inhibited the effects of IL1 on PMN survival. The inhibitory activity of IL4 was specific for IL1, suggesting that IL4 could act on IL1R. It was found that IL4 is a potent inducer of the IL1-RII in PMN, augmenting both the surface expression and release of soluble form. The inhibitory activity of IL4 was totally reverted by blocking anti-RII mAb, thus demonstrating that the IL1-RII, upregulated by IL4, inhibits IL1. To extend these findings to another cell type expressing predominantly the RII, and to IL1 activities other than survival, human circulating monocytes were examined. Also in these cells, a blocking anti-1RII mAb augmented the activity of suboptimal concentration of IL1β on cytokine production and membrane adhesion molecule ICAM-1 expression.

In preliminary experiments, forced expression of RII in RI-expressing cells by gene transfer blunted the response to suboptimal IL1 concentrations, but did not affect the response to TNF. This effect was only observed at suboptimal agonist concentrations and did not occur at high cytokine levels (F. Re and A. Mantovani, unpublished), as also observed with myelomonocytes and blocking mAb. These results are consistent with a "decoy" rather than a "negative signaling" model of function for IL1RII.

As a "natural" anti-IL1 pathway, sRII is a candidate for anti-IL1 therapy. The main theoretical advantage of sRII over the sRI, currently under clinical evaluation, is that it is expected to block IL1β, the main released form of IL1, with little effect of IL1ra. The lack of blocking of IL1α, mainly cell-associated, would leave cell-to-cell dialogue unaffected.

In addition to itself representing a potentially useful tool for pharmacological intervention, the decoy Rr may also be a target for anti-inflammatory/immunosuppressive agents. Glucocorticoids (GC) are protypic inhibitors of inflammation and immunity, and have long been known to augment ILIR expression in a seemingly paradoxical way. Recently, GC have been shown to induce augmented expression and release of RII inmyelomonocytes, indicating a novel, and possibly important, mechanism by which GC can exert antiflammatory/immuno-suppressive activities. Immunosuppressive prostaglandins were shown to augment the expression of IL1R in fibroblast and, most notably, to increase gene and protein expression of IL1RII in myelomonocytes.

Purification and *Ex Vivo* Expansion of Human Peripheral Blood Progenitor Cells (PBPC) Recruited by Chemotherapy and Hematopoietic Growth Factors

In order to reduce the number of potentially contaminating tumor cells within PBPC preparations CD34$^+$ PBPCs in an ongoing phase I/II study in patients with advanced malignancies were positively selected.

Up to now, 20 patients have been enrolled. PBPCs were mobilized upon VP16, ifosfamide and cisplatin (VIP) chemotherapy and G CSF administration, harvested by leukapheresis, and CD34$^+$ cells were positively selected with the Ceprate^M Stem Cell Concentrator, an avidin-biotin immunoabsorption column. After column separation, the yield of CD34$^+$ cells was $67 \pm 26\%$ with a purity of $55 \pm 18\%$. The median number of CD34$^+$ cells recovered was 2.3×10^6/kg (range 0.27–9.51). *In vitro* clonogenic assays of the enriched CD34$^+$ cell population revealed myeloid (CFU-GM), erythroid (BFU-E) and multilineage (CFU-GEMM) colony formation and multicolor flow cytometry demonstrated a coexpression of the lineage-associated molecules CD33, CD38, and HLA-DR, indicating more than 94% of cells being committed progenitors. In addition, CD34$^+$ cells comprised primitive progenitor cells as demonstrated by the presence of CD34$^+$ lineage negative cells as well as long term culture initiating cells. Thus far, positively selected CD34$^+$ cells have been retransfused into 12 patients after high-dose VP16, ifosfamide, carboplatin and epirubicin, and subsequent G-CSF administration. As compared to historical control patients given comparable numbers of unseparated PBPCs and G-CSF, median time to neutrophil recovery and platelet recovery was identical in both groups. These results demonstrate the feasibility of the positive selection of peripheral blood CD34$^+$ cells and their efficacy to rapidly reconstitute hematopoiesis after high-dose chemotherapy without support of bone marrow cells. One advantage of PBPCs over the use of autologous bone marrow might be a reduced risk of tumor cell contamination. The actual level of tumor cell contamination, however, is still poorly investigated; 358 peripheral blood and 54 apheresis samples from 16 patients with Stage IV or high-risk Stage II/III breast cancer, small cell or non-small cell (NSCLC) lung cancer as well as other advanced malignancies were evaluated for the detection of circulating epithelial tumor cells. Monoclonal antibodies against cytokeratin components (AE1-3/KL1) and epithelial antigens (HEA) were used in an alkaline phosphatase-anti-alkaline phosphatase assays with a sensitivity of 1 tumor cell within 4×10^5 total cells. Before initiation of PBPC mobilization, circulating tumor cells were detected in 29% of patients with stage IV breast cancer and 20% of patients with extensive-disease SCLC, respectively. In these patients, an even higher number of tumor cells was detected after VP16, ifosfamide and cis-Platin (VIP chemotherapy) + G-CSF induced mobilization of PBPCs, an approach which had previously been shown to be highly effective in mobilizing PBPCs. In the 42 patients without circulating tumor cells during steady-state, tumor cells were mobilized in 21% of patients after VIP chemotherapy + G-CSF induced recruitment of PBPCs with an overall number of tumor cells varying between 4 and 5600 per 1.6×10^6 mononuclear cells analyzed. From these and other studies, it was concluded that there is a high proportion of patients with circulating tumor cells under steady-state conditions, and in addition, a substantial risk of concomitant tumor cell mobilization upon recruitment of PBPCs, particularly in patients with tumor cells infiltrating the bone marrow. The biological and clinical significance of this finding is unknown at present.

Cytokine Induction by Biological Response Modifiers

There is little question that much of the anti-cancer effects of chemo- or radiation therapy depends upon the direct cytotoxic effects of these treatments on the cancer cells. However, the nomalignant stroma of cancers is an additional and important target for immunologic and pharmacologic intervention [6]. The stroma not only determines the vascular supply, but the cytokine and growth-factor environment which may stimulate or inhibit tumor cell growth. In addition, cytokines produced by the tumor cells themselves may determine the type of stromal response, i.e., which cells are attached to the stroma and what factors these cytokines produce. For example, tumor cells transfected to produce TNF are not directly affected by this cytokine *in vitro*, but *in vivo* the growth of the transfectants is profoundly inhibited and the effect of the secreted TNF is clearly indirect [7]. Since the effect is on normal stromal cells, drug-resistant variants do not occur. The specificity of the strong growth inhibitory effect can be shown by reversing the inhibition *in vivo*, by giving neutralizing antibody to TNF or by giving a TNF receptor – Fc fusion protein [7, 8]. Treatment of some tumor cells with certain agents can induce TNF release even from these untransfected cells; such release could possibly cause growth inhibition *in vivo*. In addition, it has been found that tumor cell lines often secrete certain cytokines spontaneously in amounts equal or in excess of those found in cytokine gene-transfected cancer cells (e.g. [9]). Some of the factors released by cancer cells attract inflammatory cells into the tumor stroma where the attracted cells seem to release growth factor that stimulate the malignant growth. It was shown that interruption of this paracrine stimulatory loop by giving an antibody specific for granulocytes lead to significant inhibition of tumor growth, even in the absence of host T-cells, and depletion of this subpopulation of leukocytes resulted in rejection of the tumor by T-cell competent hosts [10]. It is possible that some of the chemotherpay of certain tumors is caused by interrupting paracrine stimulation by tumor stroma. Once the critical stimulatory molecules in this stimulatory circuit are identified, it is quite possible that newly developed specific inhibitors of the relevant cytokines or growth factors can be used as a more selective down-regulator of malignant growth. In addition, these therapies might synergize significantly with the T-cell-mediated immune responses of the tumor-bearing host.

Effects of Anti-TNF Humanized Antibody (AG) in Rheumatold Arthritis

Rheumatoid arthritis (RA) is an autoimmune disease with inflammatory manifestations in the peripheral synovial joints, which are infiltrated by activated T-cells, macrophages and plasma cells. The role of cytokeins in RA has been investigated and a pivotal role for the tumor necrosis factor (TNF) in the pathogenesis of this disease has been proposed. Extensive studies have led to the first clinical trial in RA patients using a chimeric anti-TNFa antibody. In addition to pro-inflammatory cytokine production, at sites of inflammation such as the RA synovial joint, there is also evidence for homeostatic immunoregulatory mechanisms which include the production of cytokine inhibitor such as soluble TNF-R and the IL1 receptor antagonist and cytokines with immunoregulatory properties like IL10.

To understand more about the importance of cytokines in RA, their expression in the RA mononuclear cell (MNC) suspension culture system were studied. In the rheumatoid synovial mononuclear cell suspension cultures it was observed that the spontaneous production of IL1 (IL1α and IL1β) mRNA and protein continued at a high level for many days in culture, without stimulation [11]. These results suggested that the factor(s) or signals responsible for this chronic production of IL1 were contained within the RA suspension culture, and thus a model was available to study this aspect of the disease process. TNF strongly induces IL1 activity in monocytes, at a level comparable to lipopolysaccharide (LPS) [12]. Thus the effect of neutralizing TNFα on IL1 production in the RA MNC cultures was determined. The surprising results indicated that in all RA cultures tested, IL1 bioactivity was markedly inhibited in the presence of anti-TNF antibodies [13]. This implied that an important signal for IL1 production in the RA synovial tissue was TNFα. It was also found that anti-TNFα antibodies inhibited the production of another pro-inflammatory cytokine, granulocyte monocyte colony stimulating factor (GM-CSF) [14, 15]. GM-CSF has been implicated in contributing to the pathogenesis of RA based on the observation that it induces and maintains HLA class II expression on RA synovial cells. It may also affect other cell types in RA tissue, as it also can augment neutrophil mediated cartilage degradation and adherence [16].

Supportive evidence concerning the major role of TNFα in RA has come from animal studies. Transgenic mice expressing human TNFα modified, in the 3' untranslated AU rich region (which yields unstable mRNA) by replacement of it with the β-globin 3' UTR, develop an RA-like arthritis from 4 weeks of age, preventable by treatment with anti-human TNFα antibodies [17]. As the disregulated TNF expression is widespread, it is not clear why the mice develop arthritis. This result suggests that joint tissue is particularly sensitive to the pro-inflammatory effects of TNFα.

There are two categories of cytokine inhibitors. The first is the IL1 receptor antagonist (IL1Ra) [18]. The second group is comprised of soluble versions of the cell surface cytokine receptors [19]. The presence of IL1ra in synovial fluid mononuclear cells and synovial fluid has been demonstrated but it is insufficient in quantity as bioactive IL1 is detectable [20]. The soluble TNF-R (p55 and p75) function as TNF blockers, as the TNF inhibitory activity detected in synovial fluid is reversed with antibodies directed against these receptors. However, despite the elevated levels of these molecules, this regulatory mechanism is insufficient as bioactive TNF is still detectable in the synovial joint [21]. It was also found that IL10 further inhibits TNF effects by augmenting the production of the soluble TNF receptors from monocyte cultures, while down regulating surface expression. Thus an immunoregulatory cytokine such as IL10 can have effects on the regulation of pro-inflammatory cytokines such as TNF and IL1.

Verification of the importance of a molecular target for therapy in a human disease requires a clinical trial. The rationale was established for a trial of blocking TNF in RA. A chimeric (mouse Fv/human IgC) high affinity antibody, has been tested in an "open" (non-placebo controlled) clinical trial initiated in May 1992. The results have verified the concept that TNF is a good therapeutic target in RA. In fact, 20 longstanding, active, severe RA patients who had failed multiple disease modifying drugs (median 4) responded markedly in the open study [22]. The response was monitored at several levels. Subjectively, there was a rapid reduction in morning swelling and pain. Semi-objective clinical parameters (assessed by a metrologist) such as swollen joint count, were reduced as was joint tenderness. There were also significant changes in composite rheumatological scores such as the Paulus score, index of disease activity, and patients' assessment of disease activity. The median duration of clinical benefit was 12 weeks. Most importantly, objective markers of disease activity also improved significantly. These include C reactive protein. A double-binding, randomized, placebo-controlled study has just been completed and has verified the results of the open study, establishing TNF as a therapeutic target [23].

References

1. Colotta F, Sironi M, Borre A, Pollicino T, Bernasconi S, Boraschi D, et al. Type II interleukin-1 receptor is not expressed in cultured endothelial cells and is not involved in endothelial cell activation. Blood 1993; 81(5): 1347–1351.
2. Munoz E, Zubiaga AM, Sims JE, Huber BT. J. IL-1 signal transduction pathways. I. two functional IL-1 receptors are expressed in T cells. J. Immunol. 1991; 146: 136–143.
3. Eastgate JA, Symons JA, Duff GW. Identification of an interleukin-1 beta binding protein in human plasma. FEBS Lett. 1990; 260(2): 213–216.
4. Mancilla J, Ikejima T, Dinarello CA. Glycosylation of the interleukin-1 receptor type I is required for optimal binding of interleukin-1. Lymphokine & Cytokine Res. 1992; 114: 197–205.
5. Colotta F, Dower SK, Sims JE, Mantovani A. The type II 'decoy' receptor: a novel regulatory pathway for interleukin 1. Immunol. Today 1994; 15(12): 562–56.
6. Singh S, Ross SR, Acena M, Rowley DA, Schreiber H. Stroma is critical for preventing or permitting immunological destruction of antigen cancer cells. J. Exp. Med. 1992; 175: 139–146.
7. Teng MN, Park BH, Koeppen HKW, Tracey KJ, Fendly BM, Schreiber H. Long-term inhibition of tumor growth by tumor necrosis factor in the absence of cachexia or T-cell immunity. Proc. Natl. Acad. Sci. USA 1991; 88: 3535–3539.

8. Teng MN, Turksen K, Jacobs CA, Fuchs E, Schreiber H. Prevention of runting and cachexia by a chimeric TNF receptor-Fc protein. Clinical Immunol. And Immunopath 1993; 69: 215–222.

9. Pekarek LA, Weichselbaum RR, Beckett MA, Nachman J, Schreiber H. Footprinting of individual tumors and their variants by constitutive cytokine expression patterns. Cancer Res. 1993; 53: 1978–1981.

10. Pekarek LA, Starr BA, Weichselbaum RR, Toledano AY, Schreiber H. (submitted) Inhibition of tumor growth by elimination of granulocytes.

11. Buchan G, Barrett K, Turner M, Chantry D, Maini RN, Feldmann M. Interleukin-1 and tumor necrosis factor mRNA expression in rheumatoid arthritis: prolonged production of IL1α. Clin Exp. Immunol 1988; 73: 449–455.

12. Portillo G, Turner M, Chantry D, Feldmann M. Effects of cytokines on HLA-DR and IL-1 production by a monocyte tumour, THP-1. Immunol. 1989; 66: 170–175.

13. Brennan FM, Chantry D, Jackson A, Maini R, Feldmann M. Inhibitory effect of TNFα antibodies on synovial cell interleukin-1 production in rheumatoid arthritis. Lancet 1989; 2: 244–247.

14. Haworth C, Brennan FM, Chantry D, Turner M, Maini RN, Feldmann M. Expression of granulocyte-macrophage colony-stimulating factor in rheumatoid arthritis: regulation by tumor necrosis factor-α. Eur. J. Immunol. 1991; 21: 2575–2579.

15. Alvaro-Garcia JM, Zvaifler NJ, Brown CB, Kaushansky L, Firestein GS. Cytokines in chronic inflammatory arthritis. VI. Analysis of the synovial cells involved in granulocyte-macrophage colony stimulating factor production and gene expression in rheumatoid arthritis and its regulation by IL-1 and TNF-alpha. J. Immunol. 1991; 146: 3365–3371.

16. Kowanko IC, A F. Granulocyte macrophage colony-stimulating factor augments neutrophil-mediated cartilage degradation and neutrophil adherence. Arthritis Rheum. 1991; 11: 1452–1460.

17. Keffer J, Probert LHC, Georgopoulos S, Kaslaris E, Kioussis D, Kollias G. Transgenic mice expressing human tumour necrosis factor: a predictive genetic model of arthritis. EMBO J. 1991; 10: 5025–4031.

18. Arend W. Interleukin 1 receptor antagonist. A new member of the interleukin 1 family. J. Clin. Invest 1991; 5: 1445–1451.

19. Fernandez-Botran R. Soluble cytokine receptors: their role in immunoregulation. FASEB J. 1991; 5: 2567–2574.

20. Roux-Lombard P, Modoux C, Vischer T, Grassi J, Dayer JM. Inhibitors of interleukin 1 activity in synovial fluids and in cultured synovial fluid mononuclear cells. J. Rheumatol 1992; 19: 517–523.

21. Cope AP, Aderka D, Doherty M, Englemann H, Gibbons D, Jones AC et al. Increased levels of soluble tumor necrosis factor receptros in the sera and synovial fluid patients with rheumatic diseases. Arthritis Rheum. 1992; 35: 1160–1169.

22. Elliot MJ, Maini RN, Feldmann M, Long-Fox A, Charles P, Katsikis P et al. Treatment of rheumatoid arthritis with chimeric monoclonal antibodies to TNFα. Arthritis Rheum. 1993; 36: 1681–1690.

23. Elliot MJ, Maini RN, Feldmann M, Kalden JR, Antoni C, Smollen JS et al. Treatment with a chimaeric monoclonal antibody to tumour necrosis factor α suppresses disease activity in rheumatoid arthritis. Lancet 1994; 344: 1105–1110.

Pharmacological Sciences: Perspectives for
Research and Therapy in the Late 1990s
ed. by A.C. Cuello and B. Collier

Targeted Therapy of Cancer and Autoimmune Diseases

Oliver W. Press[1], John Wijdenes[2], Martin J. Glennie[3]
and Kenneth D. Bagshawe[4]

[1]*The University of Washington and the Fred Hutchinson Cancer Research Center, Seattle, USA;*
[2]*Immunotherapie Laboratories, Besançon, France;* [3]*The Lymphoma Unit, Tenovus Laboratory,*
General Hospital, Southampton, U.K; [4]*Charing Cross Hospital, London, U.K.*

Summary. Antibody-conjugates represent a promising new modality for refractory malignancies and autoimmune diseases that may afford greater selectivity and lesser toxicity than conventional cytotoxic treatments. Many types of immunoconjugates have been described including antibodies conjugated to conventional chemotherapeutic agents, plant toxins, bacterial toxins, radionuclides, and enzymes. We describe four promising new approaches. First, clinical trials employing unconjugated monoclonal antibodies targeting CD4 or the IL2 receptor are described for patients with autoimmune disorders such as rheumatoid arthritis, multiple sclerosis, and acute graft versus host disease. Second, preclinical and clinical trials employing bispecific monoclonal antibodies recruiting the toxin, saporin, to idiotypic immunoglobulins or the CD22 antigen on B-cell lymphomas are discussed. Third, the use of antibody-enzyme conjugates in conjunction with prodrug substrates is presented. Finally, the use of radiolabeled anti-CD20 antibodies for treating patients with refractory B-cell lymphomas is summarized. Each of these methods has demonstrated impressive therapeutic results in preclinical models and clinical trials, warranting further investigation.

Introduction

Traditional cytotoxic therapies for malignancies and autoimmune diseases have been associated with considerable toxicity and limited efficacy. In theory, it should be possible to exploit the specificity of the immune system to devise more selective treatments capable of effectively eradicating aberrant cells while sparing normal tissues from deleterious side-effects. In this chapter we will review four new antibody-based approaches that show promise for achieving these objectives, namely unmodified monoclonal antibodies (MAbs), bispecific antibodies recruiting saporin, antibody-enzyme-prodrug systems, and radioimmunoconjugates.

Monoclonal Antibodies for Treating Auto-immune Diseases

In recent years, the utility of monoclonal antibodies (MAbs) for diagnosing and treating auto-immune diseases such as rheumatoid arthritis

Correspondence to: Dr. Oliver Press; University of Washington Cancer Center, EE122; 1959 NE Pacific St., RC-08; Seattle, WA 98195, USA.

and multiple sclerosis for preventing graft-vs-host disease and for inducing immunosuppression after solid organ transplantation has become evident. Rheumatoid arthritis is a clinical setting in which MAb therapy may be particularly valuable, though clinical results have been variable. The B-F5 MAb targeting the CD4 antigen (made in Dr. Wijdenes' laboratory) appears to be particularly promising, since it produces clinical "ameliorations" of arthritis lasting for up to 1 year in patients, consisting of improvements in joint pain and morning stiffness as quantified by the Ritchie Index [1]. Other anti-CD4 MAbs have been less effective. For example, the B-A1 anti-CD4 MAb recognizes a different CD4 epitope and has higher inhibitory activities than B-F5 in several *in vitro* bioassays, but has had no beneficial effects in patients with rheumatoid arthritis. Furthermore re-treatment of one of the B-A1 treatment failures with B-F5 led to significant disease amelioration, suggesting superiority of the latter MAb.

Side-effects of B-F5 are absent or mild in patients with rheumatoid arthritis. No depletion of CD4 cells is observed at the end of 10 days of B-F5 treatment, though transient declines lasting <24 h were seen in the numbers of CD4-, CD8-, CD16-, CD19- and CD56-positive cells. These transient decreases in circulating lymphocyte subsets have also been observed with other CD4 MAbs, and are apparently induced by release of cytokines from targeted T-cells. The B-F5 MAb provokes a human anti-mouse antibody (HAMA) response in only 20% of the treated patients, in contrast to other CD4 MAbs that evoke HAMA in almost 100% of patients.

Rheumatoid arthritis patients experiencing transient improvements with anti-CD4 therapy and exhibiting high circulating IL-6 serum levels have been treated with a neutralizing anti-IL-6 MAb (B-58) for 10 days. Improvements of $\geq 50\%$ in the Ritchie Index lasting $2-4$ months were observed in six patients and a seventh patient had a response lasting >36 months. The short-term benefits of anti-IL6 therapy have been attributed to the anti-inflammatory action of this MAb, because co-incident normalization of the sedimentation rates and C-reactive protein levels have been observed during and shortly after the treatment. The explanation for the protracted response in the seventh patient is unclear. No side-effects were observed in these seven patients. CD16- and CD56-positive cells declined during the first day of anti-IL6 therapy, and remained depressed until 2 weeks after the last injection.

Multiple sclerosis patients have also been treated with the BF-5 (anti-CD4) MAb [2]. Saturation of the CD4 positive cells was observed at doses of $20-30$ mg/day for 10 days, whereas 50 mg/day was required to achieve saturation of CD4 positive cells in rheumatoid arthritis patients. In contrast to observations in rheumatoid arthritis, the CD4 cell population was depleted for 2 months after therapy in multiple sclerosis patients. Transient decreases in CD8, CD16, CD19, and CD56 positive cells were also observed. Patients with the "remitting relapse" form of multiple sclerosis (≥ 3 relapses per year) responded much better to therapy than patients with the "chronic" form of this disease, with no relapses occurring for ≥ 30 months in "remitting relapse" patients. Patients with the chronic form of multiple sclerosis experienced responses 1 to 8 months after the end of treatment, but these improvements were short-lived. Transient side-effects of anti-CD4 therapy were more common in multiple sclerosis patients than in the rheumatoid arthritis patients, occurring in 50% of patients on the first day of therapy.

A third immune disease for which MAb therapy has proven useful is acute graft versus host disease. Whereas most trials of anti-IL2 receptor MAb therapy have failed to demonstrate clinical benefit, the B-B10 MAb demonstrated was clearly effective [3]. It is presumed that the superiority of this particular MAb derives from its weak recognition of the soluble form of the IL-2 receptor, allowing it to bind to the membrane-bound receptor despite high concentrations of circulating soluble antigen.

In conclusion, it is evident that (1) not all MAbs recognizing the same antigen are equally effective, (2) side-effects may be dependent on the underlying disease, and (3) the activities of MAbs in bioassays and animal models often do not predict clinical effectiveness in humans. It is likely that a better understanding of the mode of action of MAbs will elucidate the observed variabilities in MAb efficacy, toxicity, and immunological effects. It is also anticipated that combination therapy using several different MAbs or employing MAbs in conjunction with immunosuppressive drugs may augment the magnitude and duration of observed responses. Furthermore, the development of humanized MAbs (or human MAbs obtained by repertoire cloning) may revolutionize the field of MAb serotherapy since these reagents should substantially mitigate the risk of human anti-mouse antibody formation (HAMA).

Bispecific Antibodies for Targeting Ribosome-inactivating Proteins

Bispecific antibodies (BsAbs) represent highly versatile biological cross-linking agents that can be used to bridge two different epitopes. They were first produced by allowing reduced Fab' fragments from two different rabbit IgG antibodies to re-oxidase via their hinge region SH groups to form F (ab)$_2$ heterodimers. More recently, bispecific antibodies have been produced from hybrid-hybridomas using somatic cell fusion [4] and by chemical cross-linkers that generate pure, stable, bispecific F(ab)$_2$ derivatives from monoclonal antibodies [5].

Practical applications of BsAbs over the last 30 years have included coupling of detecting agents, such as ferritin and horseradish peroxidase, to selected antigens to allow single-step immunocytochemistry and immunoassays, and delivery of drugs, ribosome-inactivating proteins (RIPs) and radionuclides in a therapeutic setting [5]. BsAbs offer a number of attractive features for targeting cytotoxic, therapeutic compounds. In particular, they avoid the need for chemical modifications to the toxic moiety or the antibody binding site and thereby maintain full activity of the delivered compound, they yield a precisely constructed product that does not deteriorate with storage and shows very little batch-to-batch variation, they do not require cleavage of covalent bonds for release of the cytotoxic moieties at target sites, and they facilitate two-step delivery systems in which the BsAb is administered first and allowed to reach the maximum tumor-to-normal tissue localization ratio before a short-lived pharmacological moiety is given for capture by the pre-localized antibody. This latter strategy appears especially useful for immunoscintigraphy with radiometal/chelate complexes [6].

Dr Glennie's laboratory has developed a series of $F(ab')_2$ derivatives that use their dual specificity to deliver the RIPs saporin or gelonin to neoplastic B cells [7]. Initial animal studies were performed by targeting a lymphoma line via its surface IgM idiotype, and recently a small clinical trial has been initiated targeting end-stage lymphomas via the surface differentiation antigen CD22. Using the appropriate CD22-specific derivatives, the cytotoxicity of saporin and gelonin can be increased 4–5 logs for the Daudi and Raji Burkitt's cell lines, rendering $IC_{50}s$ (concentration of RIP giving 50% inhibition or protein synthesis) between 10^{-11} and 10^{-12} M. However successful targeting is achieved only if (1) a pair of BsAbs is employed to bind the RIP cooperatively to the target cell and assure high avidity binding, (2) the target antigen is constitutively internalized by the target cell to the proper intracellular compartment (CD22 is optimal), and (3) the ratio of BsAb: saporin is optimized so that saporin is administered with a molar excess of BsAb to prolong the toxin's survival time in the circulation by immune complex formation.

Recently, BsAb/saporin complexes have been tested in a preliminary clinical trial with considerable success [8]. Two anti-CD22-specific BsAbs were pre-mixed with saporin at a molar ratio of 3:1 (2–4 mg of saporin + 20–40 mg of two CD22 BsAbs) before treatment of five patients with low-grade B-cell lymphomas by 1 h intravenous infusion weekly for up to 6 weeks. Toxicity consisted of minimal weakness and myalgia for 1–2 days. One patient produced an anti-mouse Fab and anti-saporin response. All patients showed a rapid and beneficial response to treatment including eradication of circulating tumor cells (4/4 patients), disappearance of ascitic and pleural effusions (2/2 patients), striking reduction in splenomegaly (one patient), partial regressions of lymphadenopathy (6/6 patients), and clearance of tumor from marrow with impressive resolution of pancytopenia in some patients. Although responses were striking, they were short-lived. Attempts to increase the treatment regime from weekly to daily doses in two patients has met with significant increases in myalgias and troublesome vascular leak syndrome. Future clinical investigations will test pairs of BsAbs capable of delivering saporin via two different antigens; e.g., CD22 and CD38.

Antibody-enzyme Conjugates and Prodrugs in Cancer Therapy

If the action of cytotoxic drugs was restricted to tumor sites, the tumor concentration of the cytotoxic could be substantially increased and the risk of therapy-related second tumors reduced. Dr. Bagshawe has suggested that this requires (1) generation of the drug at tumor sites and (2) avoidance of drug generation at non-tumor sites (or inactivation of cytotoxic drug in the vascular compartment). In the case of cytotoxic agents for which there are effective antagonists or rescue agents, there is the option of destroying the rescue agent selectively at tumor sites. Dr. Bagshawe has developed both two-stage and three-stage systems that achieve these objectives, and has named this approach "Antibody Directed Enzyme Prodrug Therapy (ADEPT)."

The two-stage system comprises an antibody or MAb fragment conjugated to an enzyme that is not normally present in extra-cellular fluids and a non-toxic prodrug that is substrate for the enzyme [9]. After the antibody-enzyme conjugate is administered there is a time interval during which the enzyme localizes at tumor sites, but is cleared from plasma before the prodrug is administered. The time interval required before giving the prodrug varies substantially and depends on the presence of tumor antigen in plasma and on the dwell time of enzyme in the tumor. Initial studies at Charing Cross utilized the bacterial enzyme carboxypeptidase G2 (CPG2) which deglutamates folates. A series of benzoic acid mustards was synthesized by Caroline Springer, who demonstrated that they were rendered non-toxic by cleavage of a glutamic acid moiety by CPG2. The toxicity differential *in vitro* between the best of this series of prodrugs and the active drug was 50–100 fold. CPG2 conjugated to anti-hCG antibody and the prodrug CMDA eliminated a high proportion of choriocarcinoma xenografts in nude mice that were resistant to conventional cytotoxic therapy.

The same enzyme and prodrug have been used by Eccles et al. [10] with an antibody directed at c-erbB 2 to eliminate breast carcinoma xenografts that over-express the marker. Both these tumors may be atypical, however. In the choriocarcinoma model it was possible to give the prodrug 56 h after the antibody enzyme conjugate, because high levels of the target antigen, hCG, are present in the plasma and this accelerates clearance of the conjugate through immune complex formation. In the breast carcinoma model there is no circulating antigen and the target antigen binds the conjugate firmly and allows administration of the prodrug to be delayed for 11–12 days during which time natural clearance of the enzyme from plasma occurs. Attempts to eradicate colorectal carcinoma with anti-CEA antibodies and a similar regimen were ineffective because when the prodrug was given before 6 days post-conjugate, toxicity was incurred and after 6 days there was too little enzyme retained at tumor sites to be effective. This indicated the need to develop a three-stage system in which enzyme in plasma was eliminated before giving the prodrug. Surinder Sharma demonstrated this could be accomplished with a second antibody directed at the enzyme allowing instantaneous inactivation in plasma [11]. A galatosy-lated form of the anti-enzyme that is rapidly removed from the circulation by hepatocyte galactose receptors is used to avoid inactivation of enzyme at tumor sites. Using this three-stage system a growth delay of 20–25 days was achieved with LS174T colorectal xenografts.

Subsequently, a pilot trial in patients with advanced colorectal carcinomas resistant to 5-fluorouracil and folinic acid was performed. Variables encountered in this trial included dosages, timing, and rates of administration of each of the three components. Administration of the prodrug by itself confirmed its low toxicity, and HPLC studies demonstrated absence of activation of the prodrug by intestinal flora (contrary to findings in mice). Therapy with the full ADEPT protocol resulted in four partial remissions (PRs), one mixed response, and three symptomatic improvements in eight evaluable patients. Myelosuppression occurred in early patients despite initial reduction in plasma enzyme to low levels with the anti-enzyme antibody. Giving the anti-enzyme antibody in small amounts throughout the period of prodrug administration greatly ameliorated myelosuppression in later patients. Cyclosporine delayed the development of host antibody responses to both the MAb and enzyme moieties allowing up to three weekly cycles of treatment. Further studies with this CPG2 are in progress, but much more effective prodrugs with very short half-lives have been developed.

Intravascular inactivation of active drug (IVIAD) is an alternative to inactivation of residual enzyme at non-tumor sites. This approach requires an agent that discriminates between active drug and prodrug as well as attachment of the inactivating agent to a macromolecule that restricts it to the vascular compartment. An alternative method that has been studied by Dr. Bagshawe is called "antimetabolite with inactivation of rescue agent at cancer sites" (AMIRACS). Trimetrexate is an antifolate that is not inactivated by CPG2, whereas folinic acid is inactivated by the enzyme. Using an anti-tumor antibody conjugated to CPG2 and folinic acid about five times the maximum tolerated dose of trimetrexate could be given without any subjective or hematological toxicity, and with clinical responsiveness in one patient with advanced colorectal carcinoma tested.

Despite their limitations, antibody-enzyme conjugates allow a new dimension of selectivity to be introduced to cytotoxicity therapy.

Radioimmunotherapy of Human B-cell Lymphomas

An alternative approach to adoptive immunotherapy involves administration of radionuclide-conjugated MAbs to cancer patients. Radiolabeled antibody methods are advantageous because the β particles emitted by most clinically relevant radioisotopes are capable of killing tumor cells over a radius of many cell diameters. This feature permits eradication of antigen-negative tumor cells by cross-fire from neighboring antigen-positive cells, and also allows delivery of cytotoxic doses to tumor cells embedded in the center of tumor nodules. This feature is important because the limited diffusing capacity of large immunoglobulin molecules and the erratic vascularity of many tumors results in highly heterogeneous accretion of antibody, with limited MAb deposition at the center of large tumor nodules. Centrally placed tumor cells are therefore often insusceptible to antibody-conjugates requiring direct binding.

Although many candidate radioisotopes have been suggested for radioimmunotherapy (including yttrium-90, rhenium-186, copper 67, etc.), most studies have been conducted with Iodine-131 because of the simplicity of its conjugation to proteins, widespread availability, low cost, and versatility for both imaging and therapeutic applications. Several tumor types have been proposed for radioimmunotherapy, including hepatomas, renal cell carcinomas, breast cancers, and lung cancers, but the most promising tumor types are leukemias and lymphomas, because of their marked radiosensitivity and well-delineated antigenic constitutions. In contrast to immunotoxins that require internalization of toxins to kill cells, I-131-MAb conjugates are most effective when targeted to non-internalizing cell surface antigens; e.g., CD20 on B-cell lymphomas. Endocytosis of the I-131-labeled reagents results in rapid intracellular catabolism and expulsion of low molecular weight I-131 metabolites that are rapidly excreted in the urine, rendering them therapeutically ineffective.

Over the past 8 years, Dr. Press's group in Seattle has conducted a series of investigations to determine the utility of treating patients with relapsed B-cell lymphomas with radiolabeled antibodies. Iodine-131 was conjugated to a series of anti B-cell monoclonal antibodies targeting CD20 (B1, 1F5), CD37 (MB-1), DR (anti-LYM1), or surface idiotype. In a phase I dose escalation trial, patients were assessed on successive weeks with 0.5, 2.5, or 10 mg/kg of antibody trace-labeled with 5–10 mCi of I-131 and assessed by serial gamma camera imaging, tumor biopsies, and computed tomography [12]. Absorbed radiation doses to tumor sites and normal organs were estimated using the MIRD method. In 24 of the 43 patients on the study, every assessable tumor was estimated to receive more radiation than any of the normal organs (except the thyroid gland) and this circumstance was defined as a "favorable biodistribution." Nineteen patients with "favorable biodistributions" were given high dose radioimmunotherapy with 280–780 mCi of I-131-labeled antibodies according to a dose escalation scheme delivering between 1000 and 3000 cGy to normal organs in cohorts of three patients at each dose level. Moderate nausea was the only significant acute toxicity. Myelosuppression requiring reinfusion of autologous, purged marrow occurred 2–4 weeks following treatment in 15 patients. Two patients developed severe, but reversible cardiopulmonary toxicity (after absorbed lung doses of 2700 and 3100 cGy), resulting in termination of the Phase I study with a presumed maximal tolerated dose (MTD) of approximately 2700 cGy to the lungs and heart. Sixteen of the 19 treated patients achieved complete remissions, two had partial responses, and one patient had a minor response [13]. The median response duration is >21 months. Nine patients remain in continuous complete remission after 20+ to 71+ months.

A subsequent phase II study is being conducted at the maximally tolerated dose of I-131-B1 determined in the Phase I study. To date, 24 patients have been entered onto this study and 21 have achieved "favorable" antibody biodistributions and received therapeutic infusions of I-131-B1. Nearly all of them have achieved objective clinical responses. Further observation will be necessary to determine the durability of the remission induced with I-131 therapy, but preliminary results suggest that long-term remission can be routinely obtained even in patients resistant to multiple chemotherapy regimens.

Acknowledgements

This work was supported by NIH Grant CA 44991 (OWP), the Boeing Corporation (OWP), Tenovus, Cardiff (MJG) and the Cancer Research Campaign (MJG). Drs D. Wendling, L. Rumbach, and E. Racadot collaborated with Dr. J. Wijdenes on the studies reported for therapy of autoimmune diseases. Dr. Glennie's collaborators includes Drs. A. Bell, T. Hamblin, A. Tutt, and R.R. French. Dr. Bagshawe's collaborators include Carole Springer and Surinder Sharma. Dr. Press's colleagues include Drs. I. Bernstein, J. Eary, D. Matthews, F. Appelbaum, and P. Martin.

References

1. Wendling D, Racadot E, Morel-Founter, Wijdenes J. Treatment of rheumatoid arthritis with anti-CD4 monoclonal antibody. Open study of 25 patients with B-F5 clone. Clin. Rheumatology 1992; 11: 542–547.
2. Rumbach L, Racadot E, Bataillard M, Galmiche J, Henlin JL, Truttmann M et al. Essai therapeutique ouvert d'un anticorps monoclonal anti-T CD4 dans la sclerose en plaques. Rev. Neurol. (Paris) 1994; 150: 418–424.

3. Herve P, Racadot E, Wendling D, Rumbach L, Tiberghien P, Cahn J-Y et al. Use of monoclonal antibody in vivo as a therapeutic strategy for alloimmune or autoimmune reactivity: the Besancon experience. Immunological Reviews 1992; 129: 31–55.
4. Milstein C, Cuello AC. Hybrid hybridomas and their use in immunohistochemistry. Nature 1983; 305: 537–540.
5. Glennie MJ, McBride HM, Worth AT, Stevenson GT. Preparation and performance of bispecific F (ab'g)2 antibody containing thioether-linked Fab'g fragments. J. Immunol. 1987; 139: 2367–75.
6. Peltier P, Curtet C, Chatal JF, LeDoussal JM, Daniel G, Aillet G et al. Radioimmunodetection of medullary thyroid cancer using a bispecific anti-CEA/anti-indium-DTPA antibody and an indium-111-labeled DTPA dimer. J. Nucl. Med. 1993; 34: 1267–73.
7. Glennie MJ, Brennend DM, Bryden F, McBride HM, Stirpe F, Worth AT. Bispecific F(ab')$_2$ antibody for the delivery of saporin to lymphoma. J. Immunol. 1988; 141: 3662–70.
8. Bonardi MA, Bell A, French RR, Gromo G, Hamblin T, Modena D, et al. Initial experience in treating human lymphoma with a combination of bispecific antibody and saporin. Int. Cancer 1992; 7(suppl): 73–77.
9. Bagshawe KD. Antibody-directed enzyme predrug therapy (ADEPT). J. Controlled Release 1994; 28: 187–193.
10. Eccles SA, Springer CJ, Box GA, Dean CJ, Melton R. Antibody directed enzyme prodrug therapy causes regression of established breast carcinoma xenografts. Br. J. Cancer 1994; 69(suppl. xxi): 15.
11. Sharina SK, Bagshawe KD, Burke PJ, Boden JA, Rogers ST, Springer CJ et al. Galactosylated antibodies and antibody-enzyme conjugates in antibody-directed enzyme predrug therapy. Cancer 1994; 73(3): 1114–1120.
12. Press OW, Eary J, Appelbaum FR, Martin PJ, Badger CC, Nelp WB et al. Radiolabeled antibody therapy of B-cell lymphomas with autologous. Bone marrow support. New England J. of Medicine 1993; 324: 1219–1224.
13. Press OW, Eary JF, Appelbaum FR, Bernstein ID. Radiolabeled antibody therapy of lymphomas. In: De Vita V, Hellman S, Rosenberg SA (editors). Biologic therapy of cancer updates. Philadelphia: J.B. Lippincott Co., 1994; 4(4): 1–13.

The Pharmacology of Gene Expression

Pharmacological Sciences: Perspectives for
Research and Therapy in the Late 1990s
ed. by A.C. Cuello and B. Collier
© 1995 Birkhäuser Verlag Basel/Switzerland

Antisense Therapeutics

Stanley T. Crooke

Isis Pharmaceuticals, Inc., Carlsbad, CA 92008, USA

Background

Although the basic antisense concept was first clearly enunciated in 1978 [1], interest in the area developed slowly and significant progress in the discovery and development of antisense drugs was not registered until 1989. Three factors contributed to the emergence of interest in the area. First, the suggestion that modifications of oligonucleotides are feasible and could improve the properties of antisense drugs deriving from the work on phosphate modifications (for review see [2]) suggested that modifications might be introduced that improve pharmacokinetic and other properties without adversely affecting hybridization. Second, advances in the sequencing of the human genome provided an expanding choice of targets. Third, advances in automated synthesis, analysis and purification of oligonucleotides supported broad scale testing.

Scientific Principles

Oligonucleotides are designed to modulate the information transfer from the gene to protein, in essence, to alter the intermediary metabolism of RNA. Conceptually, interruption of the process at any point from transcription, i.e., transcriptional arrest, to utilization, translational arrest is feasible. However, for a wide variety of reasons [3] transcriptional arrest via DNA strand invasion or triple stranding strategies is chemically, pharmacologically and toxicologically substantially more difficult than intervening at post-transcriptional steps in the intermediary metabolism of RNA. Consequently, virtually all of the pharmacologically significant progress has been achieved with drugs that bind to pre- or mRNAs, antisense drugs.

The receptor for antisense drugs is a site in a specific mRNA species. Affinity and specificity of binding derive from hybridization interactions and are theoretically much larger than can be achieved with small molecules. In addition to the dramatic theoretical advantages in specificity, the rational design of antisense drugs is more feasible than the design of small molecules interacting with proteins. Finally, it is possible to consider the design of antisense drugs to treat a very broad range of disorders.

In short, the antisense concept is seductive. The questions about this technology have, therefore, never been whether it has potential value, but whether it would work. This question was asked in many vernaculars, but fundamentally reduces to the following key questions:
– Can oligonucleotides that have satisfactory pharmacokinetic properties be created?
– Given the complexity of the cellular processes, the kinetics of the oligonucleotide-RNA receptor interactions and other factors, can satisfactory proof be generated that the basic pharmacological principle is valid?
– Given the complexity of antisense oligonucleotides and their potential biological interactions, can the hoped-for specificity of antisense drugs be shown?
– Will antisense oligonucleotides have technology limiting toxicities?
– Is there sufficient scope for medicinal chemistry to create multiple generations of improved antisense drugs?

Additionally, a number of development related questions required resolution, e.g., can oligonucleotides be scaled up and will the cost of therapy be acceptable?

Although much remains to be done, I think we have reasonably definitive answers to these questions today that justify continuing and growing optimism about the technology.

Current Status

We now have nearly definitive answers to all of the key questions about the technology. We have made substantial advances in understanding the mechanisms of antisense drugs and in establishing rigorous experimental approaches. We have made rapid progress in developing antisense drugs.

Phosphorothioates

The first generation analogs that have proven to have the most attractive properties are the phosphorioates. We and others have demonstrated that these compounds are taken up by many cells in tissue culture, but that uptake is quite variable and influenced by cell type, tissue culture conditions and oligonucleotide sequence. Most importantly, we have shown that *in vitro* experiments do not predict for *in vivo* pharmacokinetics [4–7].

Table 1

Target	Species	Model	ID$_{50}$ Dose (mg/kg)	Route of Admin.	Collaborator
ICAM	Mouse	Cardiac Allograft	1–5	I.V. S.C.	Stepkowski et al (Univ. of TX, Houston)
	Mouse	Carrageenan soaked sponge (S.C.)	1–10	I.P.	—
	Mouse	Toxin induced lung inflammation	10	I.V.	Doerschuk et al (Indiana Univ.)
	Mouse	Ulcerative Colitis	1–5	I.V.	
	Mouse	Collagen induced arthritis	(10)	I.V.	Ciba-Geigy
p120	Human	LOX human tumor xenograph (I.P.)	0.65	I.P.	Busch et al (Baylor)
Protein Kinase C	Human	A549 human tumor xenograft (S.C.) (2 oligos)	(20 3xwk)	I.P.	—
	Human	MDA-MB 231 human tumor xenograft (S.C.) (2 oligos)	< 1	I.V.	Ciba-Geigy
	Human	Colo 205 Tumor xenograft (S.C.)	< 6 < 20 3xwk	I.V.Ciba-Geigy	
	Mouse	Northern analysis of PKC-α RNA in liver	15	I.P.	—
C RAF Kinase	Human	MDA-MB 231 human tumor xenograft (S.C.)	0.65	I.P.	Ciba-Geigy
	Human	Colo 205 human tumor xenograft	> 6	I.V.	Ciba-Geigy
RAS	Human	A549 human tumor xenograft (S.C.)	(20 3xwk)	I.P.	—
	Human	A549 human tumor xenograft (S.C.)	(20 3xwk)	S.C.	—

Table 2

Target	Animal	Reference
HSV-1	Mouse	Kulka et al, 1989 [12]
p120 Oncogene	Mouse	Perlaky et al, 1993 [13]
c-myb	Rat	Simons et al, 1992 [14]
Interleukin 1	Mouse	Burch & Mahan, 1991 [15]
NF-$^\kappa$B	Mouse	Kitajima et al, 1992 [16]
CDC-2 and CDK-2	Rat	Abe et al, 1994 [17]
N-myc	Mouse	Whitesell et al, 1991 [18]
Y-Y1 Receptors	Rat	Wahlestedt et al, 1993 [19]
NMDA-R1 Receptor channel	Rat	Wahlestedt et al, 1993 [20]
Synptosomal-Associated Protein 25	Rat	Osen-Sand et al, 1993 [21]
NF-$^\kappa$B	Mouse	Higgins et al, 1993 [22]
Intercellular Adhesion Molecule 1	Mouse	Stepkowski et al, 1994 [23]
Protein Kinase C-α	Mouse	Deak & McKay, 1994 [11]
Philadelphia[1] Leukemia	Mouse	Skorski et al, 1994 [24]
MYB	Mouse	Hijiya et al, 1994 [25]

We have performed definitive pharmacokinetic studies on several phosphorothioates in mice, rats, monkeys and humans administered by IV bolus, IV infusion, intramuscular, intraperitoneal, subcutaneous, intradermal, oral and intravitreal routes [8–10]. These studies have shown that phosphorothioates have very attractive properties for parenteral administration but are not orally bioavailable and do not cross an intact blood brain barrier.

Studies in our laboratories and many others have shown potent antisense activities and proven mechanisms in scores of cells and against scores of targets *in vitro* (for review, see [3]). More important, we and our collaborators have shown potent activities of systemically administered antisense drugs *in vivo* and directly proven mechanisms in several-studies (Table 1). Investigators in other laboratories have reported similar results (Table 2). The studies of Dean and McKay are of particular importance as they demonstrate isotypically selective inhibi-

Table 3. Limits of phosphorothioates

- Pharmacodynamic
  - Low affinity per nucleotide unit
  - Inhibition of RNase H at high concentrations
- Pharmacokinetic
  - Limited oral bioavailability
  - Limited blood brain barrier penetration
  - Dose-dependent pharmacokinetics
- Toxicologic
  - Release of cytokines
  - Complement associated effects on blood pressure?
  - Clotting effects

tion of a constituitively produced RNA with effects of single doses lasting longer than 24 h and no evidence of tachyphylaxis [11].

We have recently reported evidence of local activity of two antiviral antisense drugs in clinical trials, ISIS 2922 and ISIS 2105 [26, 27]. The effects of ISIS 2922 were clearly therapeutically important and impressive.

We now understand the potential toxicities of phosphorothioates reasonably well. The therapeutic index appears to be satisfactory and the dose limiting toxicities are likely to be secondary to thrombin binding or complement activation.

In conclusion then, we think phosphorothioates have outperformed many expectations. They will probably be useful for parenteral administration for a variety of viral infections, inflammatory disorders, cancer and selected cardiovascular indications. Nevertheless, they have significant limits as shown in Table 3.

Medicinal Chemistry

Progress in the medicinal chemistry of oligonucleotides has been particularly gratifying. Figure 1 summarizes the novel modifications that have

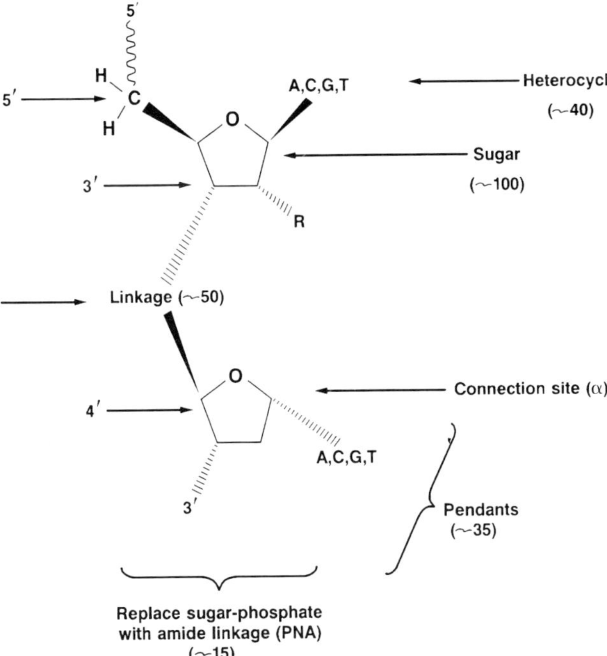

Fig. 1. Types of oligonucleotide modifications.

been made, incorporated into oligonucleotides and tested at Isis. Based on these studies, we have identified approximately 30 analog classes that appear to have more attractive properties than phosphorothioates and we have recently completed a head-to-head comparison of the first seven novel classes in animal pharmacokinetic studies. *In vivo* pharmacologic and toxicologic comparisons are in progress.

Conclusions

Antisense appears to be a broadly enabling technology that, to date, has lived up to rational expectations. However, the technology is still immature. There are many questions to be answered and much work before a definitive assessment of the therapeutic value of antisense technology can be made.

References

1. Zamencik PC, Stephenson ML. Inhibition of Rous sarcoma virus replication and cell transformation by a specific oligodeoxynucleotide. Proc. Natl. Acad. Sci. USA 1978; 75: 280.
2. Crooke ST. Therapeutic applications of oligonucleotides. Ann. Rev. Pharmacol. Toxicol. 1992; 32: 329.
3. Crooke ST, Lebleu B, editors. Antisense research and applications. Boca Raton, FL: CRC Press, 1993.
4. Crooke RM. *In vitro* toxicology and pharmacokinetics of antisense oligonucleotides. Anti-Cancer Drug Design 1991; 6: 585.
5. Crooke RM, Graham MJ, Cooke ME, Crooke ST. *In vitro* pharmacokinetic analysis of tritiated ISIS 2105 and other phosphorothioate antisense oligonucleotides. Submitted, 1994.
6. Graham MJ, Cummins L, Owens SR, Cooke ME, Crooke ST. *In vitro* metabolism of phosphorothioate antisense oligonucleotides. In press.
7. Crooke RM, Hoke GD, Shoemaker JEE. *In vitro* toxicological evaluation of ISIS 1082, a phosphorothioate oligonucleotide inhibitor of herpes simplex virus. Antimicrobe Agents Chemother. 1992; 36: 527.
8. Cossum PA, Sasmor H, Dellinger D, Truong L, Cummins L, Owens SR, et al. Disposition of the ^{14}C-labeled phosphorothioate oligonucleotide ISIS 2105 after intravenous administration to rats. J. Pharm. Exp. Ther. 1993; 267: 1181.
9. Cossum PA, Truong L, Owens SR, Markham PM, Shea JP, Crooke ST. Pharmacokinetics of a ^{14}C-labeled phosphorothioate oligonucleotide, ISIS 2105, after intradermal administration to rats. J. Pharm. Exp. Ther. 1994; 267: 89.
10. Crooke ST, Grillone LR, Tendolkar A, Garrett A, Fratkin M, Leeds J, et al. A pharmacokinetic evaluation of ^{14}C-labeled afovirsen sodium in genital wart patients. Clin. Pharm. Therap. 1994; 56: 641.
11. Dean NM, McKay R. Inhibition of PKC-α expression in mice after systemic administration of phosphorothioate antisense oligonucleotides. Proc. Natl. Acad. Sci. 1994; 91: 11762.
12. Kulka M, Smith C, Aurelian L, Fishelevich R, Meade K, Miller P, et al. Site specificity of the inhibitory effects of oligo(nucleoside methylphosphonates) complementary to the acceptor splice junction of herpes simplex virus type 1 immediately early MRNA. Proc. Natl. Acad. Sci USA 1989; 86: 6868.

13. Perlaky L, Saijo Y, Busch RK, Bennett CF, Mirabelli CK, Crooke ST, et al. Growth inhibition of human tumor cell lines by antisense oligonucleotides designed to inhibit p120 expression. Anti-Cancer Drug Design 1993; 8: 3.

14. Simons M, Edelman ER, DeKeyser J-L, Langer R, Rosenberg RD. Antisense c-*myb* oligonucleotides inhibit intimal arterial smooth muscle cell accumulation *in vivo*. Nature 1992; 359: 67.

15. Burch RM, Mahan LC. Oligonucleotides antisense to the interleukin 1 receptor MRNA block the effects of interleukin 1 in cultured murine and human fibroblasts and in mice. J. Clin. Invest. 1991; 88: 1190.

16. Kitajima I, Shinohara T, Bilakovics J, Brown DA, Xu X, Nerenberg M. Ablation of transplanted HTLV-I tax-transformed tumors in mice by antisense inhibition of NF-kB. Science 1992; 258: 1792.

17. Abe J, Zhou W, Taguchi J, Takuwa N, Miki K, Okazaki H, et al. Suppression of neointimal smooth muscle cell accumulation *in vivo* by antisense CDC2 and CDK2 oligonucleotides in rat carotid artery. Biochemical and Biophysical Research Commun. 1994; 198: 16.

18. Whitesell L, Rosolen A, Neckers LM. *In vivo* modulation of N-*myc* expression by continuous perfusion with an antisense oligonucleotide. Antisense Research and Development 1991; 1: 343.

19. Wahlestedt C, Pich EM, Koob GF, Yee F, Heilig M. Modulation of anxiety and neuropeptide Y-Y1 receptors by antisense oligodeoxynucleotides. Science 1993; 259: 528.

20. Wahlestedt C, Golanov E, Yamamoto S, Yee F, Ericson H, Yoo H, et al. Antisense oligodeoxynucleotides to NMDA-R1 receptor channel protect cortical neurons from excitotoxicity and reduce focal ischaemic infarctions. Nature 1993; 363: 260.

21. Osen-Sand A, Catsicas M, Staple JK, Jones KA, Ayala G, Knowles J, et al. Inhibition of axonal growth by SNAP-25 antisense oligonucleotides *in vitro* and *in vivo*. Nature 1993; 364: 445.

22. Higgins KA, Perez JR, Coleman TA, Dorshkind K, McComas WA, Sarmiento UM, et al. Antisense inhibition of the p65 subunit of NF-kB blocks tumorigenicity and causes tumor regression. Proc. Natl. Acad. Sci. 1993; 90: 9901.

23. Stepkowski SM, Tu Y, Condon TP, Bennett CF. Treatment with ICAM-1 antisense oligonucleotides alone or in combination with anti-LFA-1 monoclonal antibodies, anti-lymphocyte serum, rapamycin, but not cyclosporine prolong the survival of heart allografts in mice. Submitted, 1994.

24. Skorski T, Nieborowska-Skorska M, Nicolaides NC, et al. Suppression of Ph[1] leukemia cell growth in mice by BCR-ABL antisense oligodeoxynucleotide. Proc. Natl. Acad. Sci. USA 1994; 91: 4504.

25. Hijiya N, Zhang J, Ratajczak MZ, et al. Biological and therapeutic significance of NM expression in human melanoma. Proc. Natl. Acad. Sci. 1994; 91: 4499.

26. Palestine AG, Cantrill HL, Luckie AP, et al. Intravitreal treatment of CMV retinitis with an antisense oligonucleotide, ISIS 2922. Tenth International Conference on AIDS, Vol. 2, Yokohama, Japan; 1994, August 7–12 (Abstract).

27. Crooke ST. Oligonucleotide therapeutics. XII International Congress of Pharmacology, Montreal, Quebec, Canada; 1994, July 28 (Abstract) (4).

Pulmonary Pharmacology

Pharmacological Sciences: Perspectives for
Research and Therapy in the Late 1990s
ed. by A.C. Cuello and B. Collier
© 1995 Birkhäuser Verlag Basel/Switzerland

Molecular Mechanisms of Anti-Asthma Therapy

Peter J. Barnes

*Department of Thoracic Medicine, National Heart and Lung Institute, London SW3 6LY,
UK*

Introduction

Asthma is one of the commonest chronic disease world-wide and
anti-asthma medications are widely prescribed. Asthma is a complex
chronic inflammatory disease, involving many cells, mediators and
inflammatory responses. Current anti-asthma therapy is categorized as
bronchodilator (β_2-agonists, theophylline, anticholinergics) or anti-inflammatory (glucocorticosteroids, cromones, immunomodulators), although the distinction is not always clear-cut. Bronchodilators are
believed to act predominately by relaxing airway smooth muscle,
whereas anti-inflammatory drugs suppress the inflammatory response in
asthmatic airways.

There have recently been several advances in our understanding of
the mode of action of anti-asthma therapies. Increased understanding of
the molecular pathways involved in anti-asthma therapy may lead to
their more logical use alone and in combination, may increase our
understanding of the underlying pathobiology of asthma and may lead
to the development of more specific therapies in the future.

I wll review the mode of action of β_2-agonists, glucocorticosteroids
and theophylline, as these are the most widely used and most effective
anti-asthma therapies in current clinical use.

β_2-Agonists

Inhaled β_2-agonists are by far the most effective bronchodilators available, and give rapid relief of asthma symptoms. β-Agonists produce
bronchodilation by activating β_2-receptors on airway smooth muscle
cells [1]. Autoradiographic mapping studies have demonstrated that
β_2-receptors are expressed in smooth muscle of large and small airways
and β-agonists act as functional antagonists of bronchoconstriction in
airways of all sizes [2, 3]. There is a high level of expression of
β_2-receptor mRNA in airway smooth muscle, which may be indicative
of a high rate of receptor synthesis; this may contribute to the relative

resistance to β-agonist tolerance in airway smooth muscle [3]. β-Receptors are also localized to many other cell types in the airways and β-agonists may therefore have additional effects that contribute to their bronchodilatore efficacy. For example, β_2-receptors on mast cells may inhibit the release of bronchoconstrictor mediators, such as histamine and leukotriene D_4, whereas β_2-receptors on cholinergic nerves may inhibit the release of acetylcholine.

β_2-Receptor activation results in activation of adenylyl cyclase, via stimulatory G-proteins (G_s), resulting in an increase in cyclic AMP (cAMP). cAMP, through the activation of protein kinase A phosphorylates several proteins, resulting in relaxation of the airway smooth muscle cell. One of the substrates for PKA is a large conductance calcium-activated potassium channel (maxi-K channel) resulting in hyperpolarization of the cell membrane and relaxation. Airway smooth muscle has a large number of maxi-K channels. Inhibitors of maxi-K channels, such as charybdotoxin and iberiotoxin, thus inhibit the relaxant effects of β-agonists in animal and human airways [4, 5]. Recently, it has become apparent that β_2-receptors in airway smooth muscle may be directly coupled to maxi-K channels via G_s, suggesting that elevation of cAMP is not necessary for the bronchodilator response to β-agonists [6]. This would explain the discrepancy that relaxation of airway smooth muscle in response to β-agonists occurs long before a rise in cAMP is detectable, and the finding that direct activation of adenylyl cyclase with forskolin, gives a large cAMP response in airway smooth muscle, but a poor bronchodilator response. It is likely that bronchodilator responses to low doses of β-agonists may be mediated via direct coupling, whereas the response to higher doses involves cAMP-dependent mechanisms.

Many responses to β-agonists show tolerance after chronic administration. The bronchodilator response to β-agonists is curiously resistant to the development of tolerance, possibly because of a high rate of receptor transcription and a large number of spare receptors. However it has recently been observed that the protective effects of β-agonists against bronchoconstrictor challenges may demonstrate tolerance. β-Agonists have a greater protective effect against adenosine challenge than against cholinergic challenge, which may be due to the inhibitory effect of β-agonists on mast cells, since adenosine causes bronchoconstriction via mast cell mediator release rather than directly constricting airway smooth muscle. After regular treatment with an inhaled β_2-agonist over 1 week, the additional protective effect of β-agonists against adenosine is lost, indicating that tolerance of mast cell β-receptors has developed [7]. The molecular mechanisms involved in tolerance to β-agonists are now being investigated experimentally. Long-term infusion of β-agonists in animals results in down-regulation of β_2-receptors in lung, and this is greater in lung parenchyma than airway smooth muscle [8, 9]. The down-regulation of β_2-receptors is accompanied by a reduction in β_2-receptor mRNA and a reduced rate of β_2-receptor transcription. Transcription of β_2-receptors may be regulated by the transcription factor cAMP response element binding protein (CREB), and CREB activity is also reduced after prolonged β-agonist infusion [8].

By contrast, glucocorticosteroids up-regulate β_2-receptors in lung, by increasing the rate of gene transcription [10]. Indeed, the concomitant infusion of a steroid with β-agonists completely prevents the down-regulation. This may be important clinically, since inhaled steroids may prevent down-regulation of β-receptors in lung, and thus the development of tolerance to the protective effects of β-agonists, which may be important in the control of asthma symptoms.

Glucocorticosteroids

Steroids are by far the most effective therapy currently available for asthma and are the only treatment that can suppress asthmatic inflammation. Inhaled steroids are highly effective in controlling asthma and have few, if any, side-effects at the doses most patients need [11, 12]. It is important to understand the molecular mechanisms involved in the anti-inflammatory actions of steroids in asthma, as this may help our understanding of the inflammatory process. Steroids bind to an intracellular glucocorticoid receptor (GR), which then translocates to the nucleus, where it binds to specific glucocortoid recognition elements (GRE) on the upstream promoter regions of steroid-responsive target genes [13]. This may increase transcription (as in the case of β_2-receptors), or decrease transcription (as in the case of cytokine genes). Asthmatic inflammation is orchestrated by a network of cytokines that act on inflammatory and structural cells of the airway [14]. An important action of steroids in asthma is to inhibit the synthesis of genes for certain cytokines, including interleukin (IL)-5 and chemokines, such as IL-8 and RANTES. It is likely that steroids inhibit the synthesis of cytokines in many cells, but for inhaled steroids airway epithelial cells may be a particularly relevant target. Steroids inhibit the synthesis of IL-8 and RANTES in epithelial cells, via inhibition of chemokine gene transcription. Cytokines often produce their cellular effects by activating specific transcription factors, such as the *Fos-Jun* heterodimer AP-1 and NF-kB. Steroids may inhibit the effects of cytokines on transcription by blocking these transcription factors, via a direct protein-protein interaction between AP-1 and NF-kB and GR. This interaction has been demonstrated in human lung and peripheral blood monocytes [15, 16] and may be an important anti-inflammatory mechanism in asthma [17]. Indeed this mechanism may account for many of the inhibitory effects of steroids on the transcription of a variety of genes.

A small proportion of asthmatic patients are resistant to the effects of steroids [18]. The molecular mechanism of steroid-resistant asthma has recently been elucidated. In these patients, steroids appear to interact normally with GR, but there appears to be a defect in the interaction between GR and DNA [19]. Similarly, there is an abnormality in the interaction between GR and AP-1, but not other transcription factors, in these patients. This suggests an abnormality in the function of AP-1 and this may prevent the anti-inflammatory effect of steroids in these patients.

There are several steroid-responsive genes that may be important targets in asthma. An inducible form of nitric oxide synthase (iNOS) is expressed in the epithelial cells of asthmatic patients [20]. Proinflammatory cytokines increase the expression of iNOS in animal and human airway epithelial cells [21, 22], and this is inhibited by steroids. Asthmatic patients have an increased concentration of nitric oxide (NO) in exhaled air and this is reduced by steroids [23]. Steroids also have an inhibitory effect on the induction of cyclo-oxygenase-2 in airway epithelial cells [24]. Thus, inhaled steroids may inhibit the transcription of a number of genes in airway epithelial cells, thus reducing asthmatic inflammation [11].

Interaction between β-Agonists and Steroids

As discussed above, glucocorticosteroids increase the expression of β_2-receptors in lung, and may prevent down-regulation of β_2-receptors. We have recently investigated whether β-agonists have any effect on the anti-inflammatory effect of steroids. Although β-agonists are highly effective in the short-term relief of symptoms, the use of high doses of β-agonists has been linked to increases in the morbidity and mortality of asthma [25]. A possible mechanism that might underlie this adverse effect of β-agonists may be a negative interaction between high doses β-agonists and steroids. High concentrations of β-agonists activate CREB, and we have demonstrated that CREB interacts with GR, thus blocking the interaction between GR and DNA. We have demonstrated that high doses of salbutamol interfere with GR binding to DNA in mononuclear cells, human lung, T-lymphocytes and epithelial cells [26]. This would lead to a secondary resistance to the anti-inflammatory effects of steroid and this would make asthma more difficult to control. This provides a strong rationale for limiting the use of short-acting inhaled β_2-agonists to symptomatic treatment as required, since this reduces the amounts of β-agonists used. Long-acting inhaled β_2-agonists may reduce the need for short-acting β-agonists and thus may be safer as they reduce the overall daily consumption of β-agonists in patients with unstable asthma.

Theophylline

Theophylline is still the most widely used anti-asthma drug worldwide, but its mode of action in asthma remains uncertain. Theophylline is traditionally regarded as a bronchodilator, but it is relatively weak in this respect. This has led to the use of high doses, with an increased risk of unwanted effects. There is compelling evidence that theophylline may have some other actions in asthma than bronchodilation, including immunomodulation and that these effects may be seen at lower doses of theophylline than previously recommended [27].

We have investigated the effect of withdrawing theophylline in patients treated with chronic asthma, all of whom were also treated with high doses of inhaled steroids. In a placebo-controlled study theophylline withdrawal was associated with an increase in asthma symptoms and a fall in activated CD4 + and CD8 + T-lymphocytes in the circulation. Biopsy studies showed an increase in CD4 + cells in the bronchial mucosa, suggesting that theophylline may influence the traffic of lymphocytes from the circulation into the asthmatic airway, and in this way function as an immunomodulator [28]. These effects are seen at concentration of theophylline from 5–10 mg/L.

The molecular basis for the anti-inflammatory actions of theophylline are not yet known, although it has been recognized for some time that theophylline is an inhibitor of phosphodiesterase, thus increasing intracellular concentrations of cyclic nucleotides. Several isoenzyme families of PDE are now recognized, and selective inhibitors have now been developed [29]. PDE IV isoenzymes are expressed in inflammatory cells and PDE IV inhibitors, such as rolipram and denbufylline, have been found to inhibit several of these cells involved in asthmatic inflammation, including mast cells, macrophages, eosinophils and CD4 + T-cells. PDE IV inhibitors may therefore form a novel class of anti-inflammatory drugs, with the additional advantage that they appear to have bronchodilator effects [30].

Conclusions

There have been several recent advances in our understanding of anti-asthma therapies at a molecular level. This will lead to more rational use of these treatments (avoiding high concentrations of β-agonists, earlier use of inhaled steroids, use of low dose theophylline), and may lead to the use of new therapies, such as selective PDE IV inhibitors. In the future it may be possible to develop new classes of drugs that interact with transcription factors.

References

1. Nijkamp FP, Engels F, Henricks PAJ, van Oosterhout AJM. Mechanisms of β-adrenergic receptor regulation in lung and its implication for physiological responses. Phys. Rev. 1992; 72: 323–367.
2. Carstairs JR, Nimmo AJ, Barnes PJ. Autoradiographic visualization of beta-adrenoceptor subtype in human lung. Am. Rev. Respir. Dis. 1985; 132: 541–547.
3. Hamid QA, Mak JC, Sheppard MN, Corrin B, Venter JC, Barnes PJ. Localization of $beta_2$-adrenoceptor messenger RNA in human and rat lung using in situ hybridization: correlation with receptor autoradiography. Eur. J. Pharmacol. 1991; 206: 133–138.
4. Jones TR, Charette L, Garcia ML, Kaczorowski GJ. Interaction of iberiotoxin with β-adrenoceptor agonists and sodium nitroprusside on guinea pig trachea. J. Appl. Physiol. 1993; 74: 1879–1884.
5. Miura M, Belvisi MG, Stretton CD, Yacoub MH, Barnes PJ. Role of potassium channels in bronchodilator responses in human airways. Am. Rev. Respir. Dis. 1992; 146: 132–136.
6. Kume H, Hall IP, Washabau RJ, Takagi K, Kotlikoff MI. β-Adrenergic agonists regulate K_{Ca} channels in airway smooth muscle by cAMP-dependent and -independent mechanisms. J. Clin. Invest. 1994; 93: 371–379.
7. O'Connor BJ, Aikman SL, Barnes PJ. Tolerance to the non-bronchodilator effects of inhaled β_2-agonists. New Engl. J. Med. 1992; 327: 1204–1208.
8. Nishikawa M, Mak JCW, Shirasaki H, Barnes PJ. Differential down-regulation of pulmonary β_1- and β_2-adrenoceptor messenger RNA with prolonged in vivo infusion of isoprenaline. Eur. J. Pharmacol. (Molecular Section) 1993; 247: 131–138.
9. Nishikawa M, Mak JCW, Shirasaki H, Harding SE, Barnes PJ. Long-term exposure to norepinephrine results in down-regulation and reduced mRNA expression of pulmonary β-adrenergic receptors in guinea pigs. Am. J. Respir. Cell Mol. Biol. 1994; 10: 91–99.
10. Mak JCW, Nishikawa M, Barnes PJ. Glucocorticoids increase β_2-adrenoceptor transcription in human lung. Am. J. Physiol. 1995; 12: L41–L46.
11. Barnes PJ, Pedersen S. Efficacy and safety of inhaled steroids in asthma. Am. Rev. Respir. Dis. 1993; 148: S1–S26.
12. Barnes PJ. Inhaled steroids for asthma. N. Engl. J. Med. 1995; 332: 868–875.
13. Beato M. Gene regulation by steroid hormones. Cell 1989; 56: 335–344.
14. Barnes PJ. Cytokines as mediators of chronic asthma. Am. J. Resp. Crit. Care Med. 1994; 150: S42–S49.
15. Adcock IM, Brown CR, Gelder CM, Shirasaki H, Peters MJ, Barnes PJ. The effects of glucocorticoids on transcription factor activation in human peripheral blood mononuclear cells. Am. J. Physiol. 1995; 37: C331–C338.
16. Adcock IM, Shirasaki H, Gelder CM, Peters MJ, Brown CR, Barnes PJ. The effects of glucocorticoids on phorbol ester and cytokine stimulated transcription factor activation in human lung. Life Sci. 1994; 55: 1147–1153.
17. Barnes PJ, Adcock IM. Anti-inflammatory actions of steroids: molecular mechanisms. Trends Pharmacol. Sci. 1993; 14: 436–441.
18. Cypcar D, Busse WW. Steroid-resistant asthma. J. Allergy Clin. Immunol. 1993; 92: 362–372.
19. Adcock IM, Lane SJ, Brown CR, Peters MJ, Lee TH, Barnes PJ. Differences in binding of glucocorticoid receptor to DNA in steroid-resistant asthma. J. Immunol. 1995; 154: 3500–3505.
20. Hamid Q, Springall DR, Riveros-Moreno V, Chanez P, Howarth P, Redington A et al. Induction of nitric oxide synthase in asthma. Lancet 1993; 342: 1510–1513.
21. Robbins RA, Springall DR, Warren JB, Kwon OJ, Buttery LDK, Wilson AJ et al. Inducible nitric oxide synthase is increased in murine lung epithelial cells by cytokine stimulation. Biochem. Biophys. Res. Commun. 1994; 198: 1027–1033.
22. Robbins RA, Barnes PJ, Springall DR, Warren JB, Kwon OJ, Buttery LDK et al. Expression of inducible nitric oxide synthase in human bronchial epithelial cells. Biochem. Biophys. Res. Commun. 1994; 203: 209–218.
23. Kharitonov SA, Yates D, Robbins RA, Logan-Sinclair R, Shinebourne E, Barnes PJ. Increased nitric oxide in exhaled air of asthmatic patients. Lancet 1994; 343; 133–135.

24. Mitchell JA, Belvisi MG, Akarasereemont P, Robbins RA, Kwon OJ, Croxtell J et al. Induction of cyclo-oxygenase-2 by cytokines in human pulmonary epithelial cells: regulation by dexamethasone. Br. J. Pharmacol. 1994; 113: 1008–1014.
25. Barnes PJ, Chung KF. Questions about inhaled β_2-agonists in asthma. Trends Pharmacol. Sci. 1992; 13: 20–23.
26. Peters MJ, Adcock IM, Brown CR, Barnes PJ. β-Agonist inhibition of steroid-receptor DNA binding activity in human lung. Am Rev. Respir. Dis. 1993; 147: A772.
27. Barnes PJ, Pauwels RA. Theophylline in asthma: time for reappraisal? Eur. Resp. J. 1994; 7: 579–591.
28. Kidney J, Dominguez M, Taylor P, Rose M, Chung KF, Barnes PJ. Immunomodulatory by theophylline: demonstration by withdrawal of chronic therapy. Am. J. Resp. Crit. Care Med. 1995. In press.
29. Beavo JA, Reifsnyder DH. Primary sequence of cyclic nucleotide phosphodiesterase isoenzymes and the design of selective inhibitors. Trends Pharmacol. Sci. 1990; 11: 150–155.
30. Giembycz MA. Could selective cyclic nucleotide phosphodiesterase inhibitors render bronchodilator therapy redundant in the treatment of bronchial asthma? Biochem. Pharmacol. 1992; 43: 2041–2051.

Chemotherapy

Pharmacological Sciences: Perspectives for
Research and Therapy in the Late 1990s
ed. by A.C. Cuello and B. Collier
© 1995 Birkhäuser Verlag Basel/Switzerland

Membrane Transport of Anticancer Drugs and Drug Resistance

Carol E. Cass

Department of Biochemistry, University of Alberta, Edmonton, Alberta, Canada T6G 2H7

Summary. Resistance to anticancer drugs is a major problem in the clinical management of disseminated cancers. Because the pharmacological targets of most anticancer drugs are intracellular, drug resistance may arise through changes in membrane transport processes that result in reductions in the cellular accumulation of drug. This article highlights recent advances in the understanding of mechanisms of membrane transport of two major groups of anticancer drugs (plant natural products, nucleoside antimetabolites), focusing on results of studies with native and recombinant P-glycoprotein that demonstrate its ability to function as an ATP-driven pump and insights derived from the molecular cloning of cDNAs that encode nucleoside transport proteins. The role of P-glycoprotein in resistance to Taxol, an anticancer drug derived from the bark of *Taxus brevifolia*, is summarized.

Introduction

Of the total number of new cases of cancer in North America each year, approximately half are curable with surgery and radiation therapy. The remaining cases require systemic treatment because of either local spread of disease that cannot be eradicated by surgery or radiation or the occurrence of distant, and frequently multiple, metastatic lesions. Systemic cancer therapy most commonly involves the administration of various, sometimes highly complex, combinations of cytotoxic drugs to patients with disseminated disease. While chemotherapy reliably produces cures of some hematologic and pediatric cancers, the majority of metastatic human cancers are either intrinsically resistant to chemotherapy or become resistant after an initial response to drug treatment. During the past decade, a large body of work on the mechanisms of cellular uptake and release of various anticancer drugs from drug-resistant and drug-sensitive cancer cell lines has established that low levels of intracellular accumulation of drug are frequently associated with resistance to cytotoxic drugs.

There are two quite different resistance mechanisms involving membrane transport processes that can give rise to reductions in cellular accumulation of drug. The importance of these mechanisms differs among drugs, depending on the combined effects of their ability to

Correspondence to: Dr. Carol E. Cass, 356 Medical Sciences Building, Department of Biochemistry, University of Alberta, Edmonton, Alberta, Canada T6G 2H7.

permeate lipid bilayers by passive diffusion and their transportability by carrier-mediated systems. Many natural products with anticancer activity are highly lipophilic and penetrate cells primarily by passive diffusion. Resistance to natural product drugs occurs when the activity of outwardly directed transport processes exceeds the rate of inward permeation by diffusion. This type of resistance typically involves cross-resistance to other, structurally unrelated drugs and has thus been termed "multidrug resistance". In contrast, drugs that penetrate cells poorly by diffusion must be transported into cells by carrier-mediated processes and resistance is observed when the activity of inwardly directed transport is either low or absent. This type of resistance is usually limited to drugs of the same structural family. For example, highly specific transport processes are required for manifestation of chemosensitivity to most antimetabolites, and genetic loss or pharmacologic blockade of a particular transport process gives rise to resistance to only those antimetabolites that are permeants for the deficient process. This article considers the role of membrane transport proteins in drug resistance, highlighting recent advances in our understanding of membrane permeation of two groups of anticancer drugs: natural product drugs and nucleoside analogs.

Multidrug resistance was initially identified in studies with transplantable tumors that demonstrated cross-resistance to structurally unrelated drugs after the development of resistance to a particular anticancer drug. Analysis of the biochemical basis of multidrug resistance in drug-resistant cultured cells led to the discovery of a heavily glycosylated membrane protein, termed P-glycoprotein, that is now known to be responsible for the drug resistance phenotype because of its ability to selectively remove drug from resistant cells. For historical perspective and details on the biology, genetics and biochemistry of P-glycoprotein and its role in multidrug resistance, the reader is referred to published comprehensive reviews [1–4]. The role of P-glycoprotein-dependent multidrug resistance in clinical chemotherapy has not been fully determined and studies in a variety of different cancer types are currently in progress. Since the discovery of P-glycoprotein, a central question has been the mechanism by which a single membrane protein is capable of preventing cellular accumulation of a chemically heterogeneous group of substrates that freely diffuse through the plasma membrane. This article summarizes results of studies that demonstrate the ability of P-glycoprotein to function as an ATP-driven pump to transport drug molecules from the cytosolic to the extracellular side of the plasma membrane. The implications of multidrug resistance for use of Taxol, a new anticancer drug derived from the bark of the Pacific yew, *Taxus brevifolia*, are also discussed.

Nucleosides comprise an important class of antimetabolites, used in anticancer and antiviral therapies, whose mechanisms of transport have been extensively studied. A large number of functionally distinct nucleoside transport processes have been observed in mammalian cells and a major issue in the consideration of the role of membrane transport processes in cellular sensitivity to nucleosides is the biologic basis of this remarkable functional heterogeneity. A major advance in evaluating the role of membrane transport processes in the cellular uptake of nucleoside drugs is the recent identification of nucleoside transporter proteins through isolation and expression of transporter cDNAs. This article presents an overview of nucleoside transporter heterogeneity and summarizes the current status of the molecular analysis of transporter proteins. For more detailed discussions, the reader is referred to comprehensive review articles that describe characterization of nucleoside transport processes in mammalian cells [5–8].

P-Glycoprotein Dependent Multidrug Resistance

Resistance to multiple drugs can arise through several distinct mechanisms, ranging from changes in processes that control cellular accumulation of drug (the topic of this article) to processes responsible for degradation and/or inactivation of drug [9]. "Classical" multidrug resistance (MDR) involves a unique family of membrane glycoproteins in the manifestation of resistance to a chemically hetergeneous group of drugs (Table 1). In the early descriptions of multidrug resistance, cells selected for resistance to a single hydrophobic drug, most commonly of plant or fungal origin, showed cross-resistance to other structurally unrelated drugs [10]. The resistance phenotype was associated with decreased intracellular accumulation of drug that appeared to be the result of an energy-dependent process [11, 12], giving rise to the suggestion that resistance was due to an active efflux process that removed drug from cells. Subsequently, a 170-kD glycoprotein, termed P-glycoprotein [13], was identified in plasma membranes of highly drug-resistant cells and, subsequent to the isolation of cDNAs encoding

Table 1. Cytotoxic agents involved in the multidrug resistance phenotype*

Anticancer drugs	Other agents
Actinomycin D	Colchicine
Anthracyclines	Emetine
Epipodophyllotoxins	Ethidium bromide
Mithramycin	Gramicidin D
Mitomycin C	Podophyllotoxin
Taxol	Puromycin
Topotecan	Valinomycin
Vinca alkaloids	

*Adapted from ref [3].

P-glycoprotein of humans and rodents in several different laboratories (for summary, see 3), the central role of P-glycoprotein in the MDR phenotype was firmly established by gene transfer studies (for example, see [14, 15]).

Two genes (MDR1, MDR2) that encode P-glycoprotein isoforms have been identified in humans and three (*mdr1, mdr2, mdr3*) in rodents [93]. The level of multidrug resistance is correlated with the abundance of P-glycoprotein in the plasma membrane of resistant cells, and overexpression of the human MDR1 and rodent *mdr1* and *mdr3* genes consistently gives rise to P-glycoprotein dependent drug resistance [1–4]. In contrast, the proteins encoded by human MDR2 and rodent *mdr2* do not confer drug resistance [16, 17]. These proteins, which are found primarily in liver [18, 19], are thought to be involved in phospholipid metabolism; mice with a mutated *mdr2* gene produce bile that lacks phosphatidylcholine [20] and the *mdr2*-encoded protein, when expressed in yeast, exhibits phospholipid translocase activity [21]. The discussion that follows is limited to consideration of the P-glycoprotein isoforms that are implicated in multidrug resistance.

The amino acid sequences of the human and rodent P-glycoproteins have been deduced from the cloned cDNAs (MDR1, *mdr1, mdr3*) and used to produce a topological model for insertion of P-glycoprotein into the plasma membrane [1–4]. Human and mouse P-glycoprotein consist, respectively, of 1280 and 1276 amino acid residues comprising two halves that share a high degree of sequence similarity and are thought to have arisen either through tandem duplication of an ancestral gene or fusion by two closely related genes [3, 4]. Each half of P-glycoprotein includes a hydrophobic region of six predicted transmembrane segments followed by a large C-terminal hydrophilic region that includes consensus sequences that define a nucleotide-binding domain found in a variety of ATP-binding proteins [22, 23]. For example, in rodents the two halves of P-glycoprotein share up to 38% sequence indentity, with the greatest identity in the nucleotide binding domains and somewhat less in the transmembrane domains, and there is no homology between the extreme 5′ ends of the two halves [4]. The predicted topology of P-glycoprotein assumes that each of the 12 hydrophobic domains traverses the membrane such that the cluster of three potential N-linked glycosylation sites between the first and second transmembrane domain is extracellular and the two hydrophilic domains that contain the predicted nucleotide binding sites are intracellular. While several aspects of this topological model have been confirmed (for example, see [24, 25]), other models have been suggested [26] and detailed studies are required to precisely define the number and orientation of the predicted transmembrane domains.

P-glycoprotein is a member of a large superfamily of ubiquitous membrane proteins that have been termed the ATP-binding-cassette

(ABC)-transport proteins [27], or traffic ATPases [28]. Members of the ABC superfamily may be ubiquitous and include the well-characterized periplasmic permeases of bacteria and ATP-dependent export proteins of bacteria [29, 30], a protein implicated in resistance of the malarial parasite to chloroquine [31], a protein of yeast (Ste6) that mediates export of a peptide hormone involved in mating [32], the chloride transporter encoded by the cystic fibrosis gene of humans [33] and peptide transporters associated with the major histocompatibility complex of humans [34] and rodents [35]. While the overall homology between the P-glycoprotein isoforms and the various members of the ABC superfamily is modest (about 20–40%), the high homology in the nucleotide-binding domains, together with similar predicted size and secondary structures, have resulted in the placement of these proteins in a single superfamily (for discussion, see [4]). Some members of the ABC superfamily are only half the size of P-glycoprotein and are thought to exist as functional dimers. The predicted sequences of the ABC transporters, either as monomers or in some instances as dimers, give rise to hydropathy profiles that are characteristic of a large group of membrane transport proteins (> 100) from several different evolutionarily related families, whose hydropathy profiles predict 12 membrane-spanning α-helices [36]. Of these, only a few of the bacterial proteins have been shown experimentally to possess the predicted 12-helix structure.

The molecular mechanism by which P-glycoprotein functions is not known, and a variety of models have been suggested [3]. The homology of the resistance-associated human and rodent P-glycoproteins with bacterial transport proteins that mediate ATP-dependent export of specific substrates is consistent with earlier suggestions from studies of drug uptake by resistant cells that P-glycoprotein functions as an energy dependent "drug-efflux pump". A number of experimental observations provide strong support for the efflux pump model. Expression of recombinant P-glycoprotein in transfected mammalian cells results in ATP-dependent reductions in the accumulation of drug that have been directly associated with enhanced drug efflux [37]. P-glycoprotein reversibly binds drugs associated with the multidrug resistance phenotype and can be covalently labeled with photoreactive drugs [38] and chemosensitizers [39]. P-glycoprotein is specifically labeled with photoreactive analogs of ATP [40], native and recombinant P-glycoprotein exhibit ATPase activity [41, 42], and mutations in either of the two nucleotide binding domains of P-glycoprotein eliminate its ability to confer drug resistance when expressed in mammalian cells [43]. Recombinant P-glycoprotein has also been shown to mediate energy-dependent drug efflux in bacteria [44] and yeast [45]. Many of these observations can also be explained by other models in which P-glycoprotein binds and removes drugs from within the lipid bilayer [3] or somehow alters cellular retention of drug, perhaps by changing intracellular pH [46]. It has also been suggested that P-glycoprotein is able to function as a Cl⁻ channel [47] and as an ATP transporter [48].

Several recent studies have provided strong evidence favoring the model in which P-glycoprotein functions as an ATP-driven transporter. Sharom *et al.* [49] have developed procedures for purification of functionally active P-glycoprotein from multidrug-resistant Chinese hamster ovary cells and have reconstituted concentrative, energy-dependent transport activity in proteoliposomes [50]. Previous attempts to demonstrate the transport and ATPase properties of isolated preparations of P-glycoprotein had failed, apparently because of irreversible structural changes during detergent solubilization of P-glycoprotein from plasma membranes of resistant cells [51]. The key to isolation of functionally active P-glycoprotein was the identification of a detergent (CHAPS) and lipids (saturated phosphatidylethanolamines) that preserve ATPase activity during solubilization [51], thereby allowing functional reconstitution into proteoliposomes [50]. P-glycoprotein was isolated from plasma membranes of a prototypic multidrug-resistant cell line (CH^{R}C5) and inserted into lipsomes that were then shown to exhibit ATP-dependent uptake of radiolabeled colchicine. The uptake was osmotically sensitive and could be inhibited by other natural product drugs that are associated with the multidrug resistance phenotype, either as putative substrates or inhibitors. Colchicine accumulation at steady-state was saturable with increasing concentrations (half-maximal value, 50 μM) and concentrative (about 6-fold) in the presence, but not in the absence, of ATP. The reconstituted liposomes exhibited high levels of constitutive ATPase activity that were stimulated by verapamil, a compound known from earlier work to sensitize multidrug-resistant cells. Thus, isolated P-glycoprotein has the capacity to mediate ATP-driven transport with permeant selectivity and inhibitor sensitivity characteristic of the efflux process observed in intact multidrug resistant cells.

Similar conclusions have been reached from studies of expression of recombinant P-glycoprotein in the yeast *Saccharomyces cerevisiae*. This work was made possible by the discovery that P-glycoprotein encoded by the mouse *mdr3* gene can functionally complement a null mutation at the STE6 locus of *S. cerevisiae* by restoration of mating in yeast cells [52]. STE6 is a member of the ABC transporter superfamily [32] that mediates export of the pheromone a-factor, a farnesylated dodecapeptide required for mating of *S. cerevisiae*. P-glycoprotein and yeast Ste6 exhibit 50% homology, with similar predicted membrane topologies, and P-glycoprotein not only complements the Ste6 mating deficiency [52] but also confers the drug resistance phenotype on yeast [53]. These observations were followed by a series of studies in which recombinant mouse P-glycoprotein was produced in yeast membrane vesicles under conditions that allowed rigorous analysis of the levels of recombinant

P-glycoprotein and the effects of ATP and electrochemical gradients on P-glycoprotein induced fluxes of radiolabeled drugs. In the first of these studies [45], "inside-out" plasma membrane vesicles were prepared from yeast that had been transformed with plasmids that lacked an insert or that carried a full-length cDNA that encoded one of two forms of the mouse *mdr3* gene. The presence of fully active *mdr3*-encoded P-glycoprotein was associated with osmotically sensitive, ATP-dependent inward fluxes of two drugs that are associated with the multidrug resistance phenotype (colchicine, vinblastine); the drug fluxes were inhibited by verapamil. The second study [54] exploited a well characterized mutant strain of yeast (*sec 6–4*) that has a temperature-sensitive defect in the final step (plasma membrane fusion) of the vesicular secretory pathway such that large numbers of secretory vesicles accumulate at the nonpermissive temperature. Since secretory vesicles contain newly snythesized plasma membrane proteins, this system can be used to obtain vesicles with an inside-out orientation (relative to the plasma membrane) that contain exogeneously introduced recombinant membrane proteins. Expression of mouse *mdr1* and *mdr3* (but not *mdr2*) yielded vesicles capable of concentrating several drugs (e.g., vinblastine, colchicine) and this activity was shown to be independent of either membrane potential or proton gradients.

While the molecular mechanism of action of P-glycoprotein is unknown, the results of the foregoing studies, combined with extensive analyses of other membranes of the ABC superfamily that are known transporters, strongly suggest that P-glycoprotein-mediated drug resistance is due to outwardly directed, energy-dependent transport of drug. The ABC transporters were identified as a superfamily because of homologies in the nucleotide-binding domains, which in the case of P-glycoprotein, appear to be responsible for the energy transduction required for active transport. It is likely that the membrane-associated domains of P-glycoprotein are responsible for its substrate selectivity since these regions contain little, if any, homology to other ABC transporters. With the recent advances in production of recombinant P-glycoprotein amenable to functional analysis in membrane vesicles, it should now be possible to begin to identify structural determinants of substrate and inhibitor selectivity. Such information will be invaluable in the design of clinical protocols that minimize the likelihood of cancer treatment failures that are due to P-glycoprotein-dependent multidrug resistance.

Taxol: Mechanisms of Action and Resistance

Taxol and its semi-synthetic derivative, Taxotere, are members of a promising new class of clinically effective anticancer agents whose pharmacologic activity is greatly reduced in cells that exhibit the multi-drug resistance phenotype (for overview, see [55, 56]). Taxol, a complex diterpene with anti-cancer activity [57], is a natural product that was originally isolated from bark of the Pacific yew tree (*Taxus brevifolia*) and, after clinical activity was shown in two notoriously refractory solid tumors, has received much public attention because of the challenge in developing a commercially viable supply of drug [55, 58]. Like the *Vinca* alkaloids, Taxol is an "antimitotic" agent and blocks cells in the G_2/M phase of the cell cycle [59], but, in contrast to other antimitotic agents, Taxol exhibits a unique mechanism of action in that it promotes the assembly of tubulin into stable microtubules [60, 61]. A characteristic feature of Taxol-treated cells is the formation of abnormal microtubular arrays [62], resulting from reorganization of the microtubule cytoskeleton into stable bundles of microtubules [59]. Detailed analyses of the effects of Taxol on the *in vitro* polymerization of isolated tubulin has established that drug molecules bind specifically and reversibly to polymerized tubulin, thereby shifting the equilibrium that exists between tubulin dimers and microtubules in favor of microtubules [63]. The biochemical target of Taxol is the β-subunit of tubulin [64].

Because Taxol is a hydrophobic natural product, it was not surprising that a cultured mouse tumor cell line that was selected for resistance to Taxol by a series of step-wise increases in drug concentrations displayed the multidrug resistance phenotype [65]. Cross-resistance to other hydrophobic anticancer drugs reduced cellular accumulation of Taxol and overproduction of P-glycoprotein was observed in the resistant cell lines. One of the Taxol-resistant cell lines (J7.T1) exhibited an unusual characteristic among multidrug-resistant model cell lines in that both genes encoding functional P-glycoprotein isoforms (*mdr1*, *mdr3*) were amplified, resulting in overproduction of the two isoforms in roughly equal quantities [66]. Analysis of the promoter regions of the *mdr1* and *mdr3* genes suggests that the Taxol-resistant cell line experienced a recombination event involving the two genes either before or during the selection of drug-resistant cells [67]. The J7.T1 cell line is partially dependent on Taxol for growth [56], a characteristic that has been noted in Taxol-resistant cells with altered tubulin subunits [68], suggesting acquisition of multiple resistance factors during the step-wise exposures to Taxol.

The relative importance of multidrug resistance in clinical use of Taxol is unknown (for discussion, see [55]). However, the possibility has been anticipated in the design of clinical trials to test the efficacy of Taxol against refractory tumors, by including the use of combination therapy with well-established chemosensitizers, such as verapamil or cyclosporin. The use of chemosensitizers to reverse multidrug resistance is based on the observation that verapamil circumvented resistance to vincristine in a murine leukemia model [69]. A large number of antagonists of multidrug resistance have since been identified (for commentary, see [70]).

Nucleoside Transport and Drug Resistance

Nucleoside drugs have important clinical applications in therapy of hematologic cancers and viral diseases (for reviews, see [71, 72]). The most important of these (Table 2) are used for treating leukemias, lymphomas and cancers of the gastrointestinal tract or a variety of viral diseases, including human immunodeficiency virus (HIV) and herpesvirus infections. As well, a large number of nucleoside analogs with anticancer or antiviral activity in experimental systems are currently in various stages of clinical testing. Most nucleoside drugs act intracellularly, after anabolic phosphorylation, by interfering, either directly or indirectly, with DNA synthesis. For those nucleosides that are hydrophilic, mediated transport systems are required for passage across the plasma membrane. In experimental systems, there is evidence that transport activity can be an important determinant of pharmacologic action of cytotoxic nucleoside drugs. For example, cultured cells made incapable of transporting nucleosides by genetic mutations [73, 74] or treatment with transport inhibitors [75, 76] exhibit low levels of uptake of adenosine and other endogenous nucleosides and are resistant to a variety of nucleoside analogs with anticancer activity.

Cellular uptake of nucleosides is a multi-factorial process (for discussion, see [7]). Depending on the nucleoside, it involves (i) permeation across the plasma membrane, by passive diffusion and one or more mediated mechanisms, and (ii) metabolism, by either anabolic (e.g., nucleoside kinases) or catabolic enzymes (e.g., nucleoside deaminases,

Table 2. Nucleoside drugs used in anticancer and antiviral therapy*

Drug	Major use
Cladribine (chlorodeoxyadenosine)	Leukemias, lymphomas
Cytarabine (arabinosylcytosine)	Leukemias
2-Fludarabine (fluoroarabinosyladenine)	Leukemias, lymphomas
Pentostatin (2′-deoxycoformycin)	Leukemias, lymphomas
Floxidine (fluorodeoxyuridine)	Colorectal cancer
Didanosine (dideoxyinosine)	HIV
Zalcitabine (dideoxycytidine)	HIV
Zidovudine (azidothymidine)	HIV
Zovirax (acyclovir)	Herpes virus
Cytovene (ganciclovir)	Herpes virus
Vidarabine (arabinosyladenine)	Herpes virus
Idoxuridine (iododeoxyuridine)	Herpes virus
Viroptic (trifluoromethylthymidine)	Herpes virus
Virazole (ribavirin)	RNA and DNA viruses

*Adapted from ref [8].

nucleoside phosphorylases). For most nucleosides, transport and metabolism are independent events, and intracellular metabolism is usually rate-limiting in the uptake process. In general, the enzymes of nucleoside metabolism exhibit much narrower substrate selectivities than transport processes. Nucleoside kinases are almost always required for the intracellular activation of purine and pyrimidine nucleosides with antiviral or anticancer activity, although there are exceptions, and resistance to nucleoside analogs may arise through loss of kinase activity in target cells.

There are two distinct classes of nucleoside transport processes. The equilibrative processes are driven by the concentration gradient of the nucleoside(s) being transported and function in both uptake and release of nucleosides from cells. The concentrative processes are secondary-active systems that are driven by transmembrane Na^+ gradients and are inwardly directed Na^+/nucleoside contransporters or symporters. The equilibrative processes are widely distributed among mammalian cells and tissues and may be ubiquitous, whereas the concentrative processes are limited to specialized cell types, including intestine, kidney, spleen lymphocytes, macrophages and choroid plexus. Both equilibrative and concentrative transport processes have been observed in neoplastic cell types.

The total number of nucleoside transport processes in mammalian cells is uncertain, although seven functionally distinct subclasses (Table 3) have been recognized from studies of permeant fluxes in intact cells or plasma membrane vesicles (for detailed summary and original references, see [8]). The characteristics that have been used to distinguish among the different transport processes are (i) dependence on Na^+ gradients and/or ability to translocate a non-metabolized nucleoside against its concentration gradient; (ii) sensitivity to inhibition by nitrobenzylthioinosine (NBMPR), a potent and specific inhibitor of equilibrative transport, and (iii) preference for purine and/or pyrimidine nucleosides as permeants. One classification scheme involves the use of trivial names that are related to the functional characteristics of the various nucleoside transport processes [77, 78], and a second scheme, applicable only to the concentrative processes, involves numerical designations that signify the order of discovery (for examples, see [79–81]). A revised nomenclature, based on evolutionary relationships, may be more appropriate when the molecular structures of the transporter proteins are known. The classification schemes outlined in Table 3 have evolved through comparisons of transport activity in only a few mammalian species, and, while it is unlikely that all seven occur within a single species, five distinct transport processes have been observed in rats and humans. As many as three different nucleoside transport subclasses have been observed in a single cell type [82].

Table 3. Mammalian nucleoside transporter subclasses*

	Equilibrative		Concentrative				
Trivial	*es*	*ei*	*cif*	*cit*		*cib*	*cs*
Numerical			N1	N2	N4	N3	N5
Na^+-dependent	−	−	+	+	+	+	+
Inhibited by:							
NBMPR	+	−	−	−	−	−	+
dipyridamole	+	+	−	−	−	−	+
Permeants:							
adenosine	+	+	+	+	+	+	+
uridine	+	+	+	+	+	+	
guanosine	+	+	+	−	+	+	
inosine	+	+	+	−	−	+	
formycin B	+	+	+	−	−	+	+
tubercidin	+	+	−	−		+	
thymidine	+	+	−	+	+	+	

*Adapted from refs. [8, 77]. The letters are used to designate (i) the transport mechanism (e = equilibrative, c = concentrative); (ii) the sensitivity to inhibition by NBMPR (s = sensitive, i = insensitive); (iii) the nature of the nucleosides accepted as permeants (f = formycin B or purine-selective, t = thymidine or pyrimidine selective, b = broad or purine/pyrimidine-selective) of concentrative transporters. Thus, the six transport subclasses are: equilibrative NBMPR-sensitive (*es*); equilibrative NBMPR-insensitive (*ei*); concentrative NBMPR-insensitive purine-selective (*cif*); concentrative NBMPR-insensitive pyrimidine-selective (*cit*); concentrative NBMPR-insensitive broadly selective (*cib*); concentrative NBMPR-sensitive (*cs*).

The two equilibrative processes of mammalian cells are remarkably broad in their permeant selectivities, accepting all of the endogenous nucleosides as well as a diverse group of structural analogs [5–8]. Because of large differences (> 4-log) in sensitivity of equilibrative transport of nucleosides to inhibition by NBMPR, the equilibrative processes have been subdivided into two functionally distinct subclasses termed *es* and *ei* [83]. The *es* processes are inhibited by low concentrations (≤ 1 nM) of NBMPR as a direct result of a noncovalent interaction of NBMPR with high affinity ($K_d \sim 0.1$ nM) binding sites [84] that are located on the extracellular face of the plasma membrane [85] whereas the *ei* processes, which also exhibit broad permeant selectivity, are unaffected by NBMPR at all, or are inhibited only by high (> 10 μM) concentrations [83]. Both *es* and *ei* processes are inhibited by low concentrations (0.1–100 nM) of dipyridamole and dilazep [86], although there are differences among cell types and species. The *es* and *ei* NT processes exhibit different kinetic properties and different substrate specificities when present in the same cell type [87].

The tight-binding of NBMPR, coupled with its high specificity for *es* transporters and intrinsic photoreactivity, has permitted identification and isolation of transporter polypeptides (for review, see [88]). The number of high-affinity NMBPR-binding sites is assumed to be a measure of the relative abundance of *es* transporter [84], which varies considerably among various cell types, with low numbers (10^4/cell) reported for human erythrocytes [89] and extraordinarily high numbers ($> 10^7$/cell) for some cultured cell types [90]. Functional reconstitution of *es* transporter polypeptides from human erythrocytes has been achieved [91].

Electrophoretic analysis of NMBPR-photolabeled polypeptides isolated from membrane preparations from various cell types indicates considerable species and/or tissue-related variations in size of *es* transporters [8, 88], some of which are clearly due to differences in glycosylation states. However, *es* transporters from different mammalian species evidently share sequence homology since polyclonal antibodies specific for protein epitopes of the human *es* transporter also recognize *es* transporter polypeptides of pig and rabbit erythrocytes and rat liver [92]. An additional indication of structural homology among *es* transporters is the recent finding of near identity in the N-terminal sequences of the transporters of human and pig erythrocytes [93], although structural differences have also been demonstrated for these transporters. The glycosylated and deglycosylated *es* transporter polypeptides of pig erythrocytes exhibit M_r values of 66 000 and 57 000, respectively, whereas those of human erythrocytes exhibit values of 55 000 and 45 000, respectively [92, 94], and monoclonal antibodies raised against the pig transporter do not cross-react with the human or mouse transporters [95]. There is evidence that multiple *es* transporter isoforms exist within a single species and tissue. For example, brush-border and basal membranes of human placental syncytiotrophoblast contain equal quantities of NBMPR-binding sites, yet erythrocyte *es*-specific antibodies recognize polypeptides in immunoblots prepared from brush-border membranes but not from basal membranes [96, 97].

Concentrative transport of nucleosides comprises a heterogeneous group of processes, with a complex pattern of overlapping permeant selectivities, which have been defined by assessing the ability of nonlabeled test permeants to block inward transport of radioactive tracer permeant (for a comprehensive summary, see [8]). The most extensively studied processes are N1/*cif* and N2/*cit*, which share the ability to transport adenosine and uridine but otherwise exhibit selectivity for purine nucleosides (N1/*cif*) or pyrimidine nucleosides (N2/*cit*). N1/*cif* activity has been found in freshly isolated rodent splenocytes, macrophages and hepatocytes and in several human and rodent cultured cell types. N2/*cit* activity has been observed in freshly isolated mouse enterocytes and in brush border vesicles from epithelial cells of bovine, rat and rabbit kidney. Much less is known about the recently discovered N3/*cib*, N4/*cit* and N5/*cs* processes. Three distinct Na^+/nucleoside cotransport processes have been observed in human material:

(i) N3/*cib* in cultured neoplastic cells, (ii) N4/*cit* (permeant selectivity similar to that of rodent N2/*cit* except for the additional ability to transport guanosine) in brush border vesicles from human kidney, and (iii) N5/*cs* in freshly isolated human leukemic cells. The N5/*cs* process is highly sensitive to inhibition by low (< 10 nM) concentrations of NBMPR and dipyridamole whereas the N1/*cif*, N2/*cit*, N3/*cib* and N4/*cit* processes are unaffected by high concentrations (> 10 μM) of either NBMPR or dipyridamole.

Although proteins that mediate Na^+-dependent nucleoside transport processes have not yet been physically purified, cDNAs encoding two different proteins with Na^+-dependent transport activities have now been cloned and expressed [98, 99]. While the predicted proteins of the two cloned transporters (termed SNST1 and cNT1) are similar in size (73 and 71 kDa, respectively), their amino acid sequences are completely unrelated. SNST1 (from rabbit kidney) belongs to a family of Na^+/organic solute cotransporters whose members are found in bacteria and mammals, whereas cNT1 (from rat intestine), which is not related to any known mammalian transporters, has some homology with a bacterial H^+/nucleoside cotransporter and thus belongs to a previously unrecognized family of transporter proteins.

The SNST1 cDNA was isolated by low stringency hybridization with a probe derived from the Na^+/glucose cotransporter (termed SGLT1) of rabbit intestine and is 61% identical and 80% similar in sequence to rabbit SGLT1 [98]. When recombinant SNST1 was expressed in *Xenopus* oocytes, it exhibited low levels of Na^+/nucleoside cotransport activity with the characteristics of a *cib*/N3 NT-mediated process. SNST1 mRNA is expressed in rabbit kidney and heart, but not in liver or intestine, and may function in reabsorption of nucleosides from the glomerular filtrate by proximal tubules. The SNST1 cDNA sequence predicts a protein of 672 amino acids (molecular mass, 73161) with three potential N-linked glycosylation sites, twelve hydrophobic, potential transmembrane domains and two amino acid residues that may be involved in Na^+ binding.

cNT1 is encoded by a cDNA that was isolated from a rat intestine cDNA library by expression selection in *Xenopus* oocytes [99]. Intestinal epithelial cells exhibit high levels of Na^+-dependent nucleoside transport activity [78] and cNT1 was isolated by screening RNA transcripts produced *in vitro* for their ability to stimulate Na^+-dependent uptake of uridine in injected oocytes. When expressed in *Xenopus* oocytes, the recombinant cNT1 transporter exhibited high levels (> 20 000-fold stimulation) of NT activity with typical N2/*cit* characteristics, including the ability to transport the anti-HIV drugs, azidothymidine and dideoxycytidine. The cDNA sequence of cNT1 predicts a protein of 648 amino acids (molecular mass, 71 000) with 14 potential transmembrane domains, high cysteine content (3.1%), three potential N-linked and four potential O-linked glycosylation sites, and four protein kinase C-dependent phosphorylation sites. cNT1 mRNA was detected in rat intestine and kidney but not in heart, brain, spleen, lung, liver or skeletal muscle. cNT1 exhibits 27% identity in amino acid sequence with a bacterial transporter, the H^+/nucleoside symporter (NUPC) of *Escherichia coli*. The absence of sequence homologies between the amino acid sequences of cNT1 and SNST1, or of the latter with other transporter proteins of mammalian origin, suggests that there is considerable heterogeneity among the Na^+-dependent nucleoside transport proteins.

In conclusion, transport of nucleoside drugs is an important determinant of pharmacologic action. Although there is considerable knowledge about the transportability of nucleoside drugs by *es*-mediated processes, only a few nucleoside drugs have been tested directly for transportability in cell systems that possess other types of nucleoside transport processes. The presence of multiple transport processes in cells complicates the analysis of drug fluxes, if the experimental design does not allow a clear separation of the various transport activities, by use of cells that express a single process naturally or by genetic and/or pharmacologic elimination of all but one process. With the identification and characterization of two distinct nucleoside transport proteins, and the anticipated identification of others in the near future, it will be possible to determine the extent of functional and structural heterogeneity of nucleoside transport proteins. Because nucleosides penetrate cells poorly by diffusion, resistance will arise whenever inwardly directed nucleoside transport processes are absent. While *es* transporters are thought to be ubiquitous, very little is known about the *in vivo* distribution of the other types of nucleoside transporters. The importance of membrane-transport deficiencies in clinical resistance to nucleoside drugs will depend on the relative abundance and activity of particular transporter types in target cells.

Acknowledgements

The author's work was supported by the National Cancer Institute of Canada and the Medical Research Council of Canada. CEC is a Terry Fox Cancer Research Scientist of the National Cancer Institute of Canada. This article highlights the topics considered at a symposium entitled "Membrane Transport of Anticancer Drugs and Drug Resistance" held on July 27, 1994, in Montreal, Canada. The speakers were F.J. Sharom, S.B. Horwitz, H. Chan, S.M. Jarvis and C.E. Cass.

References

1. Endicott JA, Ling V. The biochemistry of P-glycoprotein mediated multidrug resistance. Annu. Rev. Biochem. 1989; 58: 137–171.

2. Roninson IB. The role of the MDR1 (P-glycoprotein) gene in multidrug resistance *in vitro* and *in vivo*. Biochem. Pharmacol. 1992; 43: 95–102.
3. Gottesman MM, Pastan I. Biochemistry of multidrug resistance mediated by the multidrug transporter. Annu. Rev. Biochem. 1993; 62: 385–427.
4. Gros P, Buschman E. The mouse multidrug resistance gene family: structural and functional analysis. Int. Rev. Cytology 1993; 137: 169–197.
5. Jarvis SM. Adenosine transporters. In: Cooper DMF, Londos C, editors. Adenosine receptors. New York: Alan R Liss, 1988: 113–123.
6. Plagemann PGW, Wohlhueter RM, Woffendin C. Nucleoside and nucleobase transport in animal cells. Biochim. Biophys. Acta 1988; 947: 405–443.
7. Paterson ARP, Cass CE. Transport of nucleoside drugs in animal cells. In: Goldman ID, editor. Membrane transport of antineoplastic agents. Oxford: Pergamon Press, 1986: 309–329.
8. Cass CE. Nucleoside transport. In: Georgopapadakou NH, editor. Drug transport in antimicrobial and anticancer chemotherapy. New York: Marcel Dekker. In press.
9. Morrow CS, Cowan KH. Drug resistance and its clinical circumvention. In: Holland JF, Frei E, Bast RC et al., editors. Cancer medicine. Philadelphia: Lea & Febiger, 1993: 618–631.
10. Biedler JL, Riehm H. Cellular resistance to actinomycin D in Chinese hamster cells *in vitro*: cross-resistance, radioautographic, and cytogenetic studies. Cancer Res. 1970; 30: 1174–1184.
11. Danö K. Active outward transport of daunomycin in resistant Ehrlich ascites tumor cells. Biochim. Biophys. Acta 1973; 323: 466–483.
12. Skovsgaard T. Mechanism of cross-resistance between vincristine and daunorubicin in Ehrlich ascites tumor cells. Cancer Res. 1978; 39: 4722–4727.
13. Juliano RL, Ling V. A surface glycoprotein modulating drug permeability in Chinese hamster ovary cell mutants. Biochim. Biophys. Acta 1976; 455: 152–162.
14. Ueda K, Cardarelli C, Gottesman MM et al. Expression of a full-length cDNA for the human *MDR1* gene confers resistance to colchicine, doxorubicin, and vinblastine. Proc. Natl. Acad. Sci. USA 1987; 84: 3004–3008.
15. Guild BC, Mulligan RC, Gros P et al. Retroviral transfer of a murine cDNA for multidrug resistance confers pleiotropic drug resistance to cells without prior drug selection. Proc. Natl. Acad. Sci. USA 1988; 85: 1595–1599.
16. Schinkel AH, Roelofs MEM, Borst P. Characterization of the human *MDR3* P-glycoprotein and its recognition by P-glycoprotein specific monoclonal antibodies. Cancer Res. 1991; 51: 2628–2635.
17. Gros P, Raymond M, Bell J et al. Cloning and characterization of a second member of the mouse *mdr* gene family. Mol. Cell Biol. 1988; 8: 2770–2778.
18. van der Valk P, van Kalken CK, Ketelaars H et al. Distribution of multi-drug resistance-associated P-glycoprotein in normal and neoplastic human tissues. Ann. Oncol. 1990; 1: 56–65.
19. Buschman E, Arceci RJ, Croop JM et al. Mouse *mdr2* encodes P-glycoprotein expressed in the bile canalicular membrane as determined by isoform specific antibodies. J. Biol. Chem. 1992; 267: 18093–18099.
20. Smit JJM, Schinkel AH, Oude-Elferink RPJ et al. Homozygous disruption of the mutant *mdr2* P-glycoprotein gene leads to a complete absence of phospholipid from bile and to liver disease. Cell 1993; 75: 451–462.
21. Ruetz S, Gros P. Phosphatidylcholine translocase: a physiological role for the *mdr2* gene. Cell 1994; 77: 1071–1081.
22. Walker JE, Sraste M, Runswick MJ et al. Distantly related sequences in the alpha- and beta-subunits of ATP synthase, myosin, kinases and other ATP-requiring enzymes and a common nucleotide-binding fold. EMBO J. 1982; 1: 945–951.
23. Higgins CF, Hyles ID, Salmond GPC et al. A family of related ATP-binding subunits coupled to many distinct biological processes in bacteria. Nature 1986; 323: 448–450.
24. Yoshimura A, Kuwazuru Y, Sumizawa T et al. Cytoplasmic orientation and two-domain structure of the multidrug transporter, P-glycoprotein, demonstrated with sequence-specific antibodies. J. Biol. Chem. 1989; 264: 16282–16291.
25. Bruggemann EP, Currier SJ, Gottesman MM et al. Characterization of the azidopine and vinblastine binding site of P-glycoprotein. J. Biol. Chem. 1992; 267: 21020–21026.

26. Zhang JT, Ling V. Membrane topology of the N-terminal half of the hamster P-glycoprotein molecule. J. Biol. Chem. 1993; 268: 15101–15110.
27. Higgins CF, Hyde SC, Mimmack MM et al. Binding protein-dependent transport systems J. Bioenerg. Biomembr. 1990; 22: 571–592.
28. Ames GF-L, Mimura C, Shyamala V. Bacterial periplasmic permeases belong to a family of transport proteins operating from Escherichia coli to human: Traffic ATPases. FEMS Microbiol. Rev. 1990; 75: 429–446.
29. Higgins CF. ABC transporters: from microorganisms to man. Annu. Rev. Cell Biol. 1992; 8: 67–113.
30. Doige CA, Ames GF-L. ATP-dependent transport systems in bacteria and humans: relevance to cystic fibrosis and multidrug resistance. Annu. Rev. Microbiol. 1993; 47: 291–319.
31. Foote SJ, Thompson JK, Cowman AF et al. Amplification of the multidrug resistance gene in some chloroquine-resistant isolates of *P. falciparum*. Cell 1989; 57: 921–930.
32. McGrath JP, Varshavsky A. The yeast STE6 gene encodes a homologue of the mammalian multidrug resistance P-glycoprotein. Nature 1989; 340: 400–404.
33. Riordan JR, Rommens JM, Kerem B-S et al. Identification of the cystic fibrosis gene: cloning and characterization of complementary DNA. Science 1989; 245: 1006–1073.
34. Trowsdale J, Hanson I, Mockbridge I et al. Sequences encoded in the class II region of the major histocompatibility complex related to the 'ABC' superfamily of transporters. Nature 1990; 348: 741–743.
35. Deverson EV, Gow IR, Coadwell J et al. Major histocompatability complex class II region encoding proteins related to the multidrug resistance family of transmembrane transporters. Nature 1990; 348: 738–741.
36. Henderson PJF. The 12-transmembrane helix transporters. Current Biol. 1993; 5: 708–721.
37. Hammond JR, Johnstone RM, Gros P. Enhanced efflux of [3H]vinblastine from Chinese hamster ovary cells transfected with a full-length complementary DNA clone for the *mdr1* gene. Cancer Res. 1989; 49: 3867–3871.
38. Safa AR, Glover CJ, Meyers MB et al. Vinblastine photoaffinity labeling of high molecular weight surface membrane glycoprotein specific for multidrug resistance cells. J. Biol. Chem. 1986; 261: 6137–6140.
39. Safa AR. Photoaffinity labeling of the multidrug resistance related P-glycoprotein with photoactived analogs of verapamil. Proc. Natl. Acad. Sci. USA 1988; 85: 7187–7191.
40. Cornwell MM, Tsuruo T, Gottesman MM et al. ATP-binding properties of P-glycoprotein from multidrug resistant KB cells. FASEB J. 1987; 1: 51–54.
41. Doige CA, Yu X, Sharom FJ. ATPase activity of partially purified P-glycoprotein from multidrug-resistant Chinese hamster ovary cells. Biochim. Biophys. Acta 1992; 1109: 149–160.
42. Sarkadi B, Price EM, Boucher RC et al. Expression of the human multidrug resistance cDNA in insect cells generates a high activity drug-stimulated membrane ATPase. J. Biol. Chem. 1992; 267: 4854–4858.
43. Azzaria M, Schurr E, Gros P. Discrete mutations introduced in the predicted nucleotide-binding sites of the *mdr1* gene abolish its ability to confer multidrug resistance. Mol. Cell Biol. 1989; 9: 5289–5297.
44. Bibi E, Gros P, Kaback HR. Functional expression of mouse *mdr1* in *Escherichia coli*. Proc. Natl. Aad. Sci. USA 1993; 90: 9209–9213.
45. Ruetz S, Raymond M, Gros P. Functional expression of P-glycoprotein encoded by the mouse *mdr3* gene in yeast cells. Proc. Natl. Acad. Sci. USA 1993; 90: 11588–11592.
46. Wei LY, Roepe PD. Low external pH and osmotic shock increase the expression of human MDR protein. Biochem. 1994; 33: 7229–7238.
47. Valverde MA, Diaz M, Sepulveda FV et al. Volume-regulated chloride channels associated with the human multidrug-resistance P-glycoprotein. Nature 1992; 355: 830–833.
48. Abraham EH, Prat AG, Gerweck L et al. The multidrug resistance (mdr1) gene functions as an ATP channel. Proc. Natl. Acad. Sci. USA 1993; 90: 312–316.
49. Doige CA, Sharom FJ. Strategies for the purification of P-glycoprotein from multidrug-resistant Chinese hamster ovary cells. Protein Expression Purification 1991; 2: 256–265.
50. Sharom FJ, Yu X, Doige CA. Functional reconstitution of drug transport and ATPase activity in proteoliposomes containing partially purified P-glycoprotein. J. Biol. Chem. 1993; 268: 24197–24202.

51. Doige CA, Yu X, Sharom FJ. The effect of lipids and detergents on ATPase-active P-glycoprotein. Biochim. Biophys. Acta 1993; 1146: 65–72.
52. Raymond M, Gros P, Whiteway M, Thomas DY. Functional complementation of yeast *ste6* by a mammalian multidrug resistance gene. Science 1992; 256: 232–234.
53. Raymond M, Ruetz S, Thomas DY et al. Functional expression of P-glycoprotein in *Saccharomyces cerevisiae* confers cellular resistance to the immunosuppressive and antifungal agent FK520. Mol. Cell Biol. 1994; 14: 277–286.
54. Ruetz S, Gros P. Functional expression of P-glycoprotein in secretory vesicles. J. Biol. Chem. 1994; 269: 12277–12284.
55. Arbuck SG, Christian MC, Fisherman JS et al. Clinical development of taxol. Monographs/J. Natl. Cancer Inst. 1993; 15: 11–24.
56. Horwitz SB, Cohen D, Rao S et al. Taxol: mechanisms of action and resistance. Monographs/J. Natl. Cancer Inst. 1993; 15: 55–61.
57. Wani MC, Taylor HL, Wall ME et al. Plant antitumor agents. VI. The isolation and structure of taxol, a novel antileukemic and antitumor agent from *Taxus brevifolia*. J. Am. Chem. Soc. 1971; 93: 2325–2327.
58. Cragg GM. Alternative sources of Taxol. Monographs/J. Natl. Cancer Inst. 1993; 15: 7–8.
59. Schiff PB, Horwitz SB. Taxol stabilizes microtubules in mouse fibroblast cells. Proc. Natl. Acad. Sci. USA 1980; 77: 1561–1565.
60. Manfredi JJ, Parness J, Horwitz SB. Taxol binds to cellular microtubules. J. Cell Biol. 1982; 94: 688–696.
61. Parness J, Horwitz SB. Taxol binds to polymerized tubulin *in vitro*. J. Cell Biol. 1981; 91: 479–487.
62. Roberts JR, Rowinsky EK, Donehower RC et al. Demonstration of the cell cycle positions for taxol-induced "asters" and "bundles" by sequential measurements of tubulin immunofluorescence, DNA content, and autoradiographic labeling of taxol-sensitive and resistant cells. J. Histochm. Cytochem. 1989; 37: 1659–1665.
63. Horwitz SB. Mechanism of action of taxol. Trends Pharmacol. Sci. 1992; 13: 134–136.
64. Rao S, Horwitz SB, Ringle I. Direct photoaffinity labeling of tubulin with Taxol. J. Natl. Cancer Inst. 1992; 84: 785–788.
65. Roy SN, Horwitz SB. A phosphoglycoprotein associated with taxol-resistance in J774.2 cells. Cancer Res 1985; 45: 3856–3863.
66. Hsu SI-H, Lothstein L, Horwitz SB. Differential over-expression of three *mdr* gene family members in multidrug resistant J774.2 mouse cells. J. Biol. Chem. 1989; 264: 12053–12062.
67. Cohen D, Higman SM, Hsu SI-H et al. The involvement of a LINE-1 element in a DNA rearrangement upstream of the *MDR1a* gene in a taxol multidrug resistant murine cell line. J. Biol. Chem. 1992; 267: 20248–20254.
68. Schibler MJ, Cabral F. Taxol-dependent mutants of Chinese hamster ovary cells with alterations in α- and β-tubulin. J. Cell. Biol. 1986; 102: 1552–1531.
69. Tsuruo T, Iida H, Tsukagoshi S et al. Overcoming of vincristine resistance in P388 leukemia *in vivo* and *in vitro* through enhanced cytotoxicity of vincristine and vinblastine by verapamil. Cancer Res. 1981; 41: 1967–1972.
70. Dalton WS. Is P-glycoprotein a potential target for reversing clinical drug resistance? Curr. Opinion Oncol. 1994; 6: 595–600.
71. Pérgaud CG, Gossselin G, Imbach J-L. Nucleoside analogues as chemotherapeutic agents: a review. Nucleosides and Nucleotides 1992; 11: 903–945.
72. De Clercq E. Antiviral agents: characteristic activity spectrum depending on the molecular target with which they interact. Adv. Virus Res. 1993; 42: 1–55.
73. Cohen A, Ullman B, Martin DW. Characterization of a mutant mouse lymphoma cell with deficient transport of purine and pyrimidine nucleosides. J. Biol. Chem. 1979; 254: 112–116.
74. Cass CE, Kolassa N, Uehara Y et al. Absence of binding sites for the transport inhibitor nitrobenzylthioinosine on nucleoside transport deficient mouse lymphoma cells. Biochim. Biophys. Acta 1981; 649: 769–777.
75. Paterson ARP, Yang S, Lau EY et al. Low specificity of the nucleoside transport mechanism of RPMI 6410 cells. Mol. Pharmacol. 1979; 16: 900–908.
76. Cass CE, King KM, Montano JT et al. A comparison of the abilities of nitrobenzylthioinosine, dilazep, and dipyridamole to protect human hematopoietic cells from 7-deazaadenosine (tubercidin). Cancer Res. 1992; 52: 5879–5886.

77. Belt JA, Marina NM, Phelps DA et al. Nucleoside transport in normal and neoplastic cells. Adv. Enzyme Regul. 1993; 33: 235–252.
78. Vijayalakshmi D, Belt JA. Sodium-dependent nucleoside transport inmouse intestinal epithelial cells. Two transport systems with differing substrate specificities. J. Biol. Chem. 1988; 263: 19419–19423.
79. Williams TC, Jarvis SM. Multiple sodium-dependent nucleoside transport systems in bovine renal brush-border membrane vesicles. Biochem. J. 1991; 274: 27–33.
80. Huang Q-Q, Harvey CM, Paterson ARP et al. Functional expression of Na$^+$-dependent nucleoside transport systems of rat intestine in isolated oocytes of *Xenopus laevis*. Demonstration that rat jejunum expresses the purine-selective system N1 (*cif*) and a second, novel system N3 having broad specificity for purine and pyrimidine nucleosides. J. Biol. Chem. 1993; 268: 20613–20619.
81. Wu X, Yuan G, Brett CM et al. Sodium-dependent nucleoside transport in choroid plexus from rabbit. Evidence for a single transporter for purine and pyrimidine nucleosides. J. Biol. Chem. 1992; 267: 8813–8818.
82. Crawford CR, Ng CYC, Noel D et al. Nucleoside transport in L1210 murine leukemia cells. J. Biol. Chem. 1990; 265: 9732–9736.
83. Belt JA. Heterogeneity of nucleoside transport in mammalian cells. Two types of transport activity in L1210 and other cultured neoplastic cells. Mol. Pharmacol. 1983; 24: 479–484.
84. Cass CE, Gaudette LA, Paterson ARP. Mediated transport of nucleosides in human erythrocytes. Specific binding of the inhibitor nitrobenzylthioinosine to nucleoside transport sites in the eyrthrocyte membrane. Biochim. Biophys. Acta 1974; 345: 1–10.
85. Agbanyo FR, Paterson ARP, Cass CE. External location of sites on pig erythrocytes that bind nitrobenzylthioinosine. Mol. Pharmacol. 1988; 33: 332–337.
86. Plagemann PG, Woffendin C. Species differences in sensitivity of nucleoside transport in erythrocytes and cultured cells to inhibition by nitrobenzylthioinosine, dipyridamole, dilazep and lidoflazine. Biochim. Biophys. Acta 1988; 969: 1–8.
87. Jarvis SM, Young JD. Nucleoside transport in rat erythrocytes: two components with differences in sensitivity to inhibition by nitrobenzylthioinosine and p-chloromercuriphenyl sulfonate. J. Membr. Biol. 1986; 93: 1–10.
88. Jarvis SM, Young JD. Photoaffinity labelling of nucleoside transport peptides. Pharmacol. Ther. 1987; 32: 339–359.
89. Jarvis SM, Hammond JR, Paterson ARP et al. Species differences in nucleoside transport. A study of uridine transport and nitrobenzylthioinosine binding by mammalian erythrocytes. Biochem. J. 1982; 208: 83–88.
90. Boumah CE, Hogue DL, Cass CE. Expression of high levels of nitrobenzyl-thioinosine-sensitive nucleoside transport in cultured human choriocarcinoma (BeWo) cells. Biochem. J. 1992; 288: 987–996.
91. Kwong FY, Davies A, Tse C-M et al. Purification of the human erythrocyte nucleoside transporter by immunoaffinity chromatography. Biochem. J. 1988; 255: 243–249.
92. Kwong FY, Fincham HE, Davies A et al. Mammalian nitrobenzylthioinosine-sensitive nucleoside transport proteins. Immunological evidence that transporters differing in size and inhibitor specificity share sequence homology. J. Biol. Chem. 1992; 267: 21954–21960.
93. Baldwin SA, Beaumont N, Barros LF et al. Antibodies as probes of nitrobenzylthioinosine-sensitive nucleoside transporters. Drug Develop. Res. 1994; 31: 245.
94. Kwong FYP, Wu J-SR, Shi WM et al. Enzymic cleavage as a probe of the molecular structures of mammalian equilibrative nucleoside transporters. J. Biol. Chem. 1993; 268: 22127–22134.
95. Good AH, Craik JD, Jarvis SM et al. Characterization of monoclonal antibodies that recognize band 4.5 polypeptides associated with nucleoside transport in pig erythrocytes. Biochem. J. 1987; 244: 749–755.
96. Barros LF, Beaumont N, Jarvis SM et al. Immunological detection of nucleoside transporters in human placental trophoblast brush-border plasma membranes and placental capillary endothelial cells. J. Physiol. (London) 1992; 452: 348P.
97. Barros LP, Bustamante JC, Yudilevich DL et al. Adenosine transport and nitrobenzylthioinosine binding in human placental membrane vesicles from brush-border and basal sides of the trophoblast. J. Membr. Biol. 1991; 119: 151–161.

98. Pajor AM, Wright EM. Cloning and functional expression of a mammalian Na$^+$/nucleoside cotransporter. A member of the SGLT family. J. Biol. Chem. 1992; 267: 3557–3560.
99. Huang QQ, Yao SYM, Ritzel MWL et al. Cloning and functional expression of a complementary DNA encoding a mammalian nucleoside transport protein. J. Biol. Chem. 1994; 269: 17757–17760.

Toxicology

Importance of Individual Enzymes in the Control of Ultimate Carcinogens

Franz Oesch[1], Barbara Oesch-Bartlomowicz[2] and Hansruedi Glatt[3]

[1,2,3]*Institute of Toxicology, University of Mainz, Mainz, Germany;* [3]*Present Address: Deutsches Institut für Ernährungsforschung, Abteilung Ernährungstoxikologie, Bergholz-Rehbrücke, Germany*

Summary. The metabolic activation of most chemical mutagens and carcinogens is a prerequisite for their mutagenic and carcinogenic activity. Reactive metabolites are under the control of activating, inactivating and precursor sequestering enzymes. These enzymes are under the long-term control of induction and repression and under the short-term control of posttranslational modification. As far as carcinogen-metabolizing enzymes are concerned, posttranslational modification has received little attention. This short-term regulation may be especially important since it works fast and may affect the enzymatic activity as well as the degradation of the enzyme. The enzymatic activity is modified by activators and inhibitors. Further crucial determinants of the control of ultimate carcinogens are the compartmentalization of these enzymes and of the target molecules. Carcinogen metabolizing enzymes differ widely between toxicological test systems, animal species and man. Careful consideration of the basic mechanisms responsible for the control of the ultimately active species derived from mutagenic tumor initiators is important, since due to the long latency time of genotoxic effects correct predictions of genotoxic risks are urgently needed.

Introduction

Enzymatic activation to the ultimately carcinogenic species is required for most chemical carcinogens. Moreover, most chemical carcinogens can be enzymatically inactivated and the precursors of the ultimate carcinogen can be enzymatically sequestered to non-carcinogenic pathways. Therefore, in almost all cases enzymes are important for the control of the ultimately carcinogenic metabolites. These enzymes differ drastically in their levels and in their specificity – some even in their presence or absence – between toxicological test systems and animal species compared with man so that for any toxicological risk evaluation these control mechanisms require careful attention.

Role of Cytochrome P-450-Dependent Monooxygenases in the Control of Mutagenic Tumor Initiators

Many individual cytochromes P-450 are inducible by appropriate foreign compounds [1]. Besides the obvious change in the rate of biotransformation this may lead to crucial shifts in metabolic routes from the generation of ultimate carcinogens to the prevention of their generation or vice-versa [2]. However, enzyme induction is generally a slow process. In order to elucidate the possible role of protein kinase mediated short-term regulation of cytochromes P-450, we investigated the phosphorylation of 14 purified cytochromes P-450 by purified protein kinases [3]. In the compartmentalized situation of the unbroken cell, however, the substrate cytochrome P-450 may or may not be available to the corresponding protein kinase. Hepatocytes were, therefore, isolated from the liver of rats which had been pretreated with phenobarbital in order to induce the two major phenobarbital-inducible cytochromes P-450 2B1 and 2B2. Among the 14 investigated isoenzymes, these two had been the best substrates of the cAMP-dependent protein kinase (PKA) in the cell-free system [3]. In order to label the intracellular ATP pools the hepatocytes were exposed to ^{32}P-orthophosphate. The incorporation of radioactive phosphate into cytochrome P-450 isoenzymes (isolated and purified after the incubation with ^{32}P-phosphate) in absence of stimulation was low. After stimulation by extracellular glucagon or by the membrane permeating cAMP-derivatives N^6, $O^{2'}$-dibutyryl-cAMP (dbcAMP) and 8-thiomethyl-cAMP [4], a high degree of incorporation of ^{32}P-phosphate into cytochrome P-450 took place (Table 1). Visualization on Western blots by specific antibodies combined with autoradiography of gel electrophoretically separated proteins from solubilized microsomes and of purified cytochromes P-450, showed that 4 cytochromes P-450 were selectively phosphorylated, namely CYP2B1 and 2B2 as well as two CYP2B1-related proteins [4]. PKA often act via an Arg-Arg-X-Ser-recognition sequence in the substrate protein. Within this recognition sequence the phosphorylation of rabbit CYP2B4 (closely related to rat CYP2B2) occurs on serine 128 [5]. In many other members of the CYP gene family 2 of rat liver this recognition sequence is also present. The presence of this recognition sequence is not always sufficient for phosphorylation to occur, e.g., CYP2C11 is not significantly phosphorylated [6].

Table 1. Stimulation of the phosphorylation of CYP2B1 related proteins by glucagon and cAMP derivatives[a]

Amount of protein	Treatment of hepatocytes		
	None	Glucagon	cAMP derivatives
25.0 mg	10 646	18 433[b]	75 219[b]
		20 729[b]	73 416[b]
12.5 mg	6 344	9 978	39 842
6.25 mg	—	—	15 753

[a]Cytochromes P-450 (partially purified, octylamino-Sepharose eluate) obtained from heptocytes incubated in the presence of (^{32}P) orthophosphate with glucagon (10^{-7} M) or N^6, $O^{2'}$-dibutyryl-cAMP (1.4 mM) and 8-thiomethyl-cAMP (0.4 mM), subjected to immunoblotting followed by autoradiography. Values are arbitrary intensity units.
[b]Sample applied in two lanes and each lane scanned.

The 0-dealkylation of 7-pentylresorufin (a selective substrate of CYP2B1 and 2B2) markedly decreased after the phosphorylation of CYP2B1 and 2B2 in intact hepatocytes [7]. The metabolism of testosterone was region and stereoselectively decreased in those positions which are attacked by CYP2B1 and 2B2 [7]. After the treatment of the hepatocytes with the agents leading to phosphorylation of CYP2B1 and 2B2 (CAMP derivatives), the hydroxylation of testosterone in position 16β which is catalyzed exclusively by CYP2B1 and 2B2 [8] was decreased by about 50%. The influence of the phosphorylation on the hydroxylation at the 16α-position was "diluted" (decrease of about 30%) [7] by those isoenzymes which, in addition to the phosphorylated CYP2B1 and 2B2, also catalyze this reaction, namely CYP2C7 and 2C13 [8]. CYP2B1 and 2B2 also catalyze the oxidation of the 17β-OH group of testosterone to a keto group. This reaction is still less specific. The decrease in activity was even more "diluted" at this position and did not reach statistical significance [7].

After pretreatment of the hepatocytes with dbcAMP the activation of cyclophosphamide and ifosfamide to mutagens (known to be mediated by CYP2B1) was markedly reduced. Cyclophosphamide and ifosfamide activation to mutagens for *Salmonella typhimurium* TA 1535 was reduced to 51% and 38% of unstimulated controls when hepatocytes were incubated for 1 h with dbcAMP in the presence of the phosphodiesterase inhibitor theophylline. Using *Salmonella typhimurium* TA 100 as target strain and after pretreatment of the hepatocytes with dbcAMP for 1.5 h without theophylline a marked reduction was observed in mutagenicity of cyclophosphamide (35% compared with unstimulated controls). Continued presence of the CYP2B1 and 2B2 inducer phenobarbital in the medium increased the mutagenicity of cyclophosphamide. This led to an even more marked reduction of mutagenicity by pretreatment with dbcAMP and theophylline [9]. The mutagenicity of the ifosfamide metabolite ifosfamide mustard (which does not require metabolic activation by cytochrome P-450) was not changed by an incubation with dbcAMP. Likewise, the metabolic formation of cytotoxic metabolites from cyclophosphamide and ifosfamide but not that of ifosfamide mustard was decreased by pretreatment with dbcAMP and theophylline [9]. Thus, the stimulation of PKA in intact cells has important consequences for the control of genotoxic and cytotoxic metabolites derived from cyclophosphamide and ifosfamide.

Control of Monofunctional Epoxides and of Vicinal Diol Epoxides by Epoxide Hydrolases and Dihydrodiol Dehydrogenase

The non-bay-region vicinal diol epoxide, benz(*a*)anthracene-8,9-dihydrodiol-10,11-epoxide(*anti*) (BA-8,9-diol-10,11-oxide) is especially useful for metabolic studies since it has a half-life of many hours. Moreover, it is often the major DNA-binding and mutagenic species formed from benz(a)anthracene *in vivo* and *in vitro* [10, 11]. This diol epoxide was metabolically inactivated by dihydrodiol dehydrogenase, but not by microsomal or cytosolic epoxide hydrolase [12]. The relative role of the three enzymes, microsomal and cytosolic epoxide hydrolase and cytosolic dihydrodiol dehydrogenase in metabolic inactivation was studied by the use of purified enzymes and bacterial mutagenicity. Test compounds (BA-8,9-diol-10,11-oxide and benz(a)anthracene 5,6-oxide (BA 5,6-oxide), the K-region epoxide) were used at concentrations of the increasing portion of the concentration-mutagenicity curve. As expected from its substrate specificity [13], microsomal epoxide hydrolase readily inactivated BA 5,6-oxide. Even with a 100-fold excess over that sufficient for complete inactivation of the K-region oxide no significant effect on the mutagenicity of BA-8,9-diol-10,11-oxide was seen. Neither microsomal nor cytosolic epoxide hydrolase inactivated the diol epoxide. In the presence of NADP$^+$ the diol epoxide was inactivated by dihydrodiol dehydrogenase [12]. High concentrations of dihydrodiol dehydrogenase were necessary for inactivation of the diol epoxide. However, the inability of NADP$^+$ or dihydrodiol dehydrogenase alone to inactivate the vicinal diol epoxide and the lack of inactivation of BA 5,6-oxide by dihydrodiol dehydrogenase with or without NADP$^+$, indicate that the inactivation was a consequence of enzymic activity and not the result of nonspecific binding of the diol epoxide to protein or to NADP$^+$. A low concentration of microsomal epoxide hydrolase was sufficient for the inactivation of BA 5,6-oxide. A 50% inactivation was achieved by 0.4 unit of purified rat microsomal epoxide hydrolase, equivalent to 1.3 mg of liver, or by 7 units of purified rabbit cytosolic epoxide hydrolase, equivalent to 28 mg of liver. However, microsomal epoxide hydrolase equivalent to 200 mg of rabbit liver did not inactivate the vicinal diol epoxide BA-8,9-diol-10,11-oxide, whereas with dihydrodiol dehydrogenase equivalent to 200 mg of liver a 50% inactivation was obtained. The K-region epoxide is inactivated noticeably more efficiently than the vicinal diol epoxide which is in line with the much stronger biological activities in mammalian systems of vicinal diol epoxides.

Perspective

The species and tissue-dependent control of mutagenic and carcinogenic metabolites is largely due to the individual pattern of different enzymes and the metabolic cooperation of activating and inactivating enzymes. In addition, *in vivo* their compartmentalization will play an important role and factors such as DNA repair and cell proliferation will profoundly influence the resulting carcinogenicity. The more precisely these individual factors will become definable the better the predictions of the true carcinogenic risk of a given compound for man will be.

References

1. Conney AH. Induction of microsomal enzymes by foreign chemicals and carcinogenesis by polycyclic aromatic hydrocarbons: G.H.A. Clowes Memorial Lecture. Cancer Res. 1982; 42: 4875–4917.
2. Bücker M, Golan M, Schassmann HU, Glatt HR, Stasiecki P, Oesch F. The epoxide hydratase inducer *trans*-stilbene oxide shifts the metabolic epoxidation of benzo(*a*)pyrene from the bay- to the K-region and reduces its mutagenicity. Mol. Pharmacol. 1979; 16: 656–666.
3. Pyerin W, Taniguchi H, Horn F, Oesch F, Amelizad Z, Friedberg T, et al. Isoenzyme-specific phosphorylation of cytochromes P-450 and other drug metabolizing enzymes. Biochem. Biophys. Res. Commun. 1987; 142: 885–892.
4. Bartlomowicz B, Waxman DJ, Utesch D, Oesch F, Friedberg T. Phosphorylation of carcinogen metabolizing enzymes: regulation of the phosphorylation status of the major phenobarbital inducible cytochromes P-450 in hepatocytes. Carcinogenesis 1989; 10: 225–228.
5. Müller R, Schmidt WE, Stier A. The site of cyclic AMP-dependent protein kinase catalyzed phosphorylation of P-450. FEBS Lett. 1985; 187: 21–24.
6. Koch JA, Waxman DJ. Posttranslational modification of hepatic cytochrome P-450. Phosphorylation of phenobarbital-inducible P-450 forms PB-4 (IIB1) and PB-5 (IIB2) in isolated rat hepatocytes and *in vivo*. Biochemistry 1989; 28: 3145–3152.
7. Bartlomowicz B, Friedberg T, Utesch D, Molitor E, Platt K, Oesch F. Regio- and stereoselective regulation of monooxygenase activities by isoenzyme-selective phosphorylation of cytochrome P-450. Biochem. Biophys. Res. Commun. 1989; 160: 46–52.
8. Waxman DJ. Interactions of hepatic cytochromes P-450 with steroid hormones. Regioselectivity and stereospecificity of steroid hydroxlation and hormonal regulation of rat P-450 enzyme expression. Biochem. Pharmacol. 1988; 37: 71–84.
9. Oesch-Bartlomowicz B, Vogel S, Arens H-J, Oesch F. Modulation of the control of mutagenic metabolites derived from cyclophosphamide and ifosfamide by stimulation of protein kinase A. Mutat. Res. 1990; 232: 305–312.
10. Wood AW, Chang RL, Levin W, Lehr RE, Schaefer-Ridder M, Karle, et al. Mutagenicity and cytotoxicity of the bay region 1,2-epoxides. Proc. Natl. Acad. Sci. USA 1977; 74: 2746–2750.
11. Vigny P, Kindts M, Duguesne M, Cooper CS, Grover PL, Sims P. Metabolic activation of benz(*a*)anthracene: fluorescence spectral evidence indicated the involvement of a non-'bay-region' diol epoxide. Carcinogenesis 1980; 1: 33–41.
12. Glatt HR, Cooper CS, Grover PL, Sims P, Bentley P, Merdes M et al. Inactivation of a diol-epoxide by dihydrodiol dehydrogenase, but not by two epoxide hydrolases. Science 1982; 215: 1507–1509.
13. Bentley P, Schassmann H, Sims P, Oesch F. Epoxides derived from various polycyclic hydrocarbons as substrates of homogeneous and microsome-bound epoxide hydratase. Eur. J. Biochem. 1976; 69: 97–103.

Molecular and Cellular Aspects of Chemical Carcinogenesis

Franz Oesch[1], John M. Essigmann[2], Chris J. Kemp[3], Toshio Kuroki[4] and Jay I. Goodman[5]

[1]*Institute of Toxicology, University of Mainz, Mainz, Germany;* [2]*Whitaker College of Health Sciences and Technology, Massachusetts Institute of Technology, Cambridge, Massachusetts, USA;* [3]*CRC Beatson Laboratories, Beatson Institute for Cancer Research, Bearsden, Glasgow, UK;* [4]*Department of Cancer Cell Research, Institute of Medical Science, University of Tokyo, Tokyo, Japan;* [5]*Department of Pharmacology and Toxicology, Michigan State University, East Lansing, Michigan, USA*

Summary. Different DNA-adducts produced by one given genotoxic agent can lead to different mutational preferences, as exemplified in this mini-review, by the *cis*-diamminedichloroplatinum(II)-induced d(GpG)-$N7(1)$-$N7(2)$ adduct, giving rise predominantly to G→T transversions targeted to the 5′-modified G, the d(ApG)-$N7(1)$-$N7(2)$ adduct to A→T transversions, while for the d(GpTpA)-$N7(1)$-$N7(3)$ adduct no crosslink-specific mutations were observed. Expression of mutated H*ras* put under the control of cell-compartment specific promoters in different layers of the mouse skin showed that the consequent expression of the malignant phenotype depended on the degree of differentiation of the cell, progression to carcinomas occurring in the basal cells of the hair follicles, but not in the suprabasal layer of the epidermis. The novel Ca^{2+}-independent protein kinase C η is expressed in differentiating and differentiated epithelial cells of the skin and was activated by cholesterol sulfate whereby tumor promotion was inhibited. Finally, the hypothesis was tested whether hypomethylation of DNA is a nongenotoxic mechanism underlying the aberrant expression of protooncogenes involved in carcinogenesis. The 5-methylcytosine content of the mouse liver tumorigenesis-relevant H*ras* and *raf* oncogenes was lower in the liver of the liver tumor-prone B6C3F1 and C3H/He mice than in the relatively resistant C57BL/6 mice, indicating that differences in DNA methylation, at least in part, account for the different susceptibilities to nongenotoxic liver carcinogens.

Mechanisms of Mutagenesis and Toxicity of DNA Adducts of the Anticancer Drug *cis*-Diamminedichloroplatinum(II)

The anticancer drug *cis*-diamminedichloroplatinum (II) (cisplatin) is a paradigm for the study of the toxic, mutagenic and carcinogenic chemicals to which humans are exposed. The therapeutic as well as the mutagenic effects of cisplatin result from its covalent binding to DNA to form a spectrum of adducts. The work reported below focused on the biological properties of individual DNA adducts produced by cisplatin.

The first part of the study examined the mutagenic and genotoxic properties of cisplatin adducts [1, 2]. These adducts included *cis*-[Pt(NH$_3$)$_2$ {d(GpG)-$N7(1)$,-$N7(2)$}] (G*G*, −65% of all binding), *cis*-[Pt(NH$_3$)$_2$ {d(ApG-$N7(1)$,-$N7(2)$}] A*G*, −25%) and *cis*-[Pt(NH$_3$)$_2$-{d(GpXpG)-$N7(1)$,-$N7(3)$}] (G*XG* (where X = A, C or T), −5%; in this work only the G*TG* adduct was studied). Synthetic oligonucle-otides containing each of the aforementioned adducts were positioned at a known site within M13 mp7L2 bacteriophage DNA. Following transfection into *Escherichia coli* cells, the genomes containing the G*G*, A*G*, and G*TG* adducts had survival levels of 5.2, 22, and 14%, respectively, compared to unmodified genomes. Upon SOS induction, the survival of genomes containing the G*G* and A*G* adducts increased to 31% and 32%, respectively. The survival of genomes containing the G*TG* adduct did not increase upon SOS induction. In SOS-induced cells, the G*G* and A*G* adducts gave rise predominantly to $G \rightarrow T$ (mutation frequency = 1.4%) and $A \rightarrow T$ (mutation frequency = 6%) transversions, respectively, targeted to the 5′ modified base. No cis-$\{Pt(NH_3)_2\}^{2+}$ intrastrand crosslink-specific mutations were observed for the G*TG* adduct. Further work focused upon a class of eukaryotic proteins that binds tightly to DNA adducts formed by cisplatin. It was found that the HMG-box protein, human upstream binding factor (hUBF), specifically recognized therapeutically effective platinum DNA adducts [3]. Moreover, it was discovered that the binding was tight (k_d(appar.) -60 pM). hUBF is a trnscription factor involved in regulation of rRNA synthesis. It was shown that hUBF binds to the rDNA promoter only three times more tightly than to a platinum adduct. On the basis of these results, two non-mutally exclusive models are proposed to explain a possible role for hUBF and related HMG box proteins in the therapeutic (cytotoxic) response of cells to cisplatin. First, it is proposed that the proteins may shield adducts from DNA repair, hence promoting the longevity of the adduct in the genome. Second, it is possible that adducts may act as decoys for critical regulatory proteins such as hUBF, disrupting the activity of these regulatory proteins *in vivo* [4].

Transgenic Approaches to the Analysis of Genetic Determinants of Tumor Progression

The multistage nature of carcinogenesis is well established. In this work the process was dissected by genetic analysis of alterations in both oncogenes and tumor suppressor genes during chemically-induced skin carcinogenesis in mice. Skin treatment with the carcinogen 7,12-dimethylbenz[a]anthracene (DMBA) followed by repeated treatments with the tumor promoter TPA gives rise to benign papillomas, a small percentage of which can progress to malignant, invasive squamous cell carcinomas. Earlier work showed that a very high percentage of these tumours contain activating mutations in the H*ras* proto-oncogene [5]. Both karyotype and allelotype analysis revealed a high frequency of trisomy of chromosomes 6 and 7 in both papillomas and carcinomas;

trisomy of 7 serving to increase the dosage of the mutant H*ras* oncogene [6]. To demonstrate that mutation of H*ras* was causally related to initiation, transgenice mice were created with the mutant H*ras* oncogene driven by a keratin 10 promoter which directs expression to the suprabasal layer of the epidermis. These mice developed papillomas but no progression to carcinomas was observed [7]. When expression of the same oncogene was directed to a different cell compartment of the skin, the basal cell compartment in the hair follicles using a modfied keratin 5 promoter, papillomas again resulted but with a much higher progression frequency to carcinomas (unpublished observations). These results demonstrate that mutant H*ras* can function as an initiating event and, moreover, that the phenotype of the resulting tumor depends on the degree of differentiation of the cell in which the oncogene is expressed.

Genetic event associated with progression to carcinomas include mutation and loss of heterozygosity and of the p53 tumor suppressor gene. To determine if loss of wild type p53 function was specifically and causally related to this transition, the normal tumor induction protocol was performed on mice lacking one or both alleles of p53 (i.e., the p53 knockout mice obtained from L. Donehower and A. Bradley) [8]. Whereas the rate of appearance and growth of papillomas was not increased in p53 deficient mice, the progression frequency to carcinomas was dramatically enhanced [9]. Moreover, the resulting carcinomas from the p53 deficient mice were markedly more malignant in terms of the level of differentiation and in the degree of metastasis. Thus, mutational inactivation of p53 is a major rate limiting step for tumor progression and may contribute to genetic instability associated with the malignant behavior of the tumors.

Role of Protein Kinase C Isoforms in Tumor Promotion and Epithelial Differentiation

Protein kinase C (PKC) exists as a family of at least 11 isoforms classified into three major groups according to their structures and activation mechanisms. These are Ca^{2+}, phosphatidylserine- and diacylglycerol or phorbol ester dependent conventional PKC (α, βI, βII, and γ), Ca^{2+}-independent novel PKC (δ, ε, η, θ, and μ) and Ca^{2+}- and diacyglycerol-independent atypical PKC (λ/ι, and ζ). The η and θ isoforms were isolated from a cDNA library of mouse skin. The η isoform has a unique tissue distribution. Unlike other members of the PKC family, it is expressed predominantly in epithelial cells of skin, and gastrointestinal and respiratory tracts. *In situ* hybridization and immunohistochemical staining indicated that this particular isoform is localized in differentiated or differentiating epithelial cells. Rough endoplasmic reticulum is the site of its intracellular localization. While the activity of other PKC isoforms is regulated by interactions with polar head-groups of membrane phospholipids such as phosphatidylserine and diacylglycerol, the η isoform was found to be preferentially activated by cholesterol sulfate, a cholesterol metabolite with a sulfonic head-group. It was further demonstrated that cholesterol sulfate, when applied at the stage of tumor promotion, inhibited skin carcinogenesis of mouse skin. Since cholesterol sulfate is formed durign squamous differentiation, the activation of the η isoform by cholesterol sulfate may mediate squamous differentiation, thereby modifying skin carcinogenesis. This is the first demonstration that cholesterol turnover, like that of phospholipids, is directly involved in a signal transduction pathway.

Hypomethylation of Protooncogenes/Oncogenes: A Possible Mechanism Involved in the Promotion Stage of Carcinogenesis

DNA methylation, i.e., the 5-methylcytosine (5MeC) content of DNA, plays a role in the regulation of gene activity. There is a persuasive body of evidence indicating that differential methylation of DNA is a determinant of higher order chromatin structure and that the methyl group provides a chemical signal which is recognized by trans-acting factors. Binding or lack of binding of these factors regulate transcription. Thus, DNA methylation appears to be a mechanism whereby cells can control the expression of genes with similar promoter regions in the presence of ubiquitous transcription factors. Decreases in DNA methylation are frequently observed in tumor tissue and aberrations in methylation might be a key factor in carcinogenesis [10]. There is a direct relationship between DNA methylation and gene silencing. Hypomethylation of a gene, i.e., low levels of 5MeC, is necessary but not sufficient for its expression. Therefore, a hypomethylated gene can be considered to possess an increased potential for expression as compared to a hypermethylated gene [11].

Hypothesis: Changes in the methylation status of a gene provide a mechanism by which its potential for expression can be altered in an epigenetic heritable manner, and it is expected that modifications in DNA methylation would result from threshold-exhibiting events. In this work the hypothesis was tested whether hypomethylation of DNA is a nongenotoxic mechanism underlying the aberrant expression of protooncogenes involved in carcinogenesis. This serves as a focal point for a mechanism of action oriented approach for considering key aspects of carcinogenesis: aberrant gene expression, heritable epigenetic events, tumor promotion, thresholds and species to species extrapolation issues.

The ability to maintain the nascent pattern of methylation is dependent on a complex relationship between the capacity and fidelity of DNA maintenance methylase (including the accessibility of CpG regions to the enzyme), the amount of S-adenosylmethionine and the level of cell proliferation. Therefore, a simple one-to-one relationship between the level of cell proliferation and hypomethylation of DNA is not anticipated. Additionally, this hypothesis is not mutally exclusive with the role for mutation in carcinogenesis. On the contrary, the involvement of these mechanisms is not only compatible, but complementary.

This work focused upon oncogenes (e.g., H*ras* and *raf*) relevant to mouse liver tumorigenesis. The liver tumor-prone B6C3F1 mouse (C57BL/6♀ × C3H/He♂), in conjunction with the more susceptible C3H/He paternal strain and relatively resistant C57BL/6 maternal strain, were used for these investigations because this provides an excellent model for the study of mechanisms involved in carcinogenesis. The B6C3F1 and C3H/He, but not the C57BL/6, are susceptible to liver tumorigenesis following treatment with nongenotoxic compounds, e.g., phenobarbital and liver tumors develop with a much shorter latency in the sensitive as compared to the resistant mouse strains following treatment with a genotoxic compound, e.g. N-nitrosodiethylamine. The results [12–14] support the hypothesis and indicate that differences in DNA methylation between the C57BL/6 and B6C3F1 mice could, in part, account for the unusually high propensity of the later strain to develop liver tumors.

Perspective

Much progress has been made in more finely defining individual steps in chemical carcinogenesis and in the molecular and cellular elements required for their occurrence. Yet it still remains a stimulating challenge to precisely recognize the chain of events from structurally defined damage and molecularly characterized mutation to consequent functional aberrations in their defined cellular context causally linked to individual steps in the process of carcinogenesis, as well as to better understand the precise nature of the contribution of nongenotoxic, nonmutational events. Such knowledge may allow the development of tools to interfere at some crucial steps of this process.

References

1. Bradley LJN, Yarema KJ, Lippard SJ, Essigmann JM. Mutagenicity and genotoxicity of the major DNA adduct of the antitumor drug *cis*-diamminedichloroplatinum (II). Biochemistry 1993; 32: 982–988.
2. Yarema KJ, Wilson JM, Lippard SJ, Essigmann JM. Effects of DNA adduct structure and distribution on the mutagenicity and genotoxicity of two platinum anticancer drugs. J. Mol. Biol. 1994; 236: 1034–1048.
3. Treiber DK, Zhai X, Jantzen H-M, Essigmann JM. Cisplatin-DNA adducts are molecular decoys for the ribosomal RNA transcription factor hUBF. Proc. Natl. Acad. Sci. USA 1994; 91: 5672–5676.

4. Donahue BA, Augot M, Bellon SF, Trieber DK, Toney JH, Lippard SJ et al. Character-ization of a DNA damage-recognition protein (DRP) from mammalian cells that binds specifically to the d(GpG) and d(ApG) DNA adducts of the anticancer drug cisplatin. Biochemistry 1990; 29: 5872–5880.
5. Quintanilla M, Brown K, Ramsden M, Balmain A. Carcinogen-specific mutation and amplification of Ha-ras during mouse skin carcinogenesis. Nature 1986; 322: 78–80.
6. Kemp CJ, Fee F, Balmain A. Allelotype analysis of mouse skin tumours using polymor-phic microsatellites: sequencial genetic alterations on chromosomes 6, 7 and 11. Cancer Res. 1993; 53: 6022–6027.
7. Bailleul B, Surani MA, White S, Barton SC, Brown K, Blessing M et al. Skin hyperker-atosis and papilloma formation in transgenic mice expressing a ras oncogene from a suprabasal keratin promoter. Cell 1990; 62: 697–708.
8. Donehower LA, Harvey M, Slagle BL, McArthur MJ, Montgomery CA, JR, Butel JS et al. Mice deficient for p53 are developmentally normal but susceptible to spontaneous tumours. Nature 1992; 356: 215–221.
9. Kemp CJ, Donehower LA, Bradley A, Balmain A. Reduction of p53 gene dosage does not increase initiation or promotion but enhances malignant progression of chemically induced skin tumors. Cell 1993; 74: 813–822.
10. Jones PA. DNA methylation and cancer. Cancer Res. 1986; 46: 461–466.
11. Vorce RL, Goodman JI. Altered methylation of *ras* oncogenes in benzidine-induced B6C3F1 mouse liver tumors. Toxical. Appl. Pharmacol. 1989; 100: 398–410.
12. Goodman JI, Ward JM, Popp JA, Klaunig JE, Fox TR. Mouse liver carcinogenesis: Mechanisms and relevance. Fund. Appl. Toxicol. 1991; 17: 651–665.
13. Ray JS, Harbison ML, McClain RM, Goodman JI. Alterations in the methylation status and expression of the raf oncogene in phenobarbital-induced and spontaneous B6C3F1 mouse liver tumors. Molec. Car. 1994; 9: 155–166.
14. Counts JL, Goodman JI. Comparative analysis of the 5′ flanking region of Ha-ras in B6C3F1, C3H/He and C57BL/6 mouse liver. Cancer Letters 1994; 75: 129–136.

Pharmacological Sciences: Perspectives for
Research and Therapy in the Late 1990s
ed. by A.C. Cuello and B. Collier

Alterations in Cell Signaling and Cytotoxicity

Pierluigi Nicotera[1] and Erik Dybing[2]

[1]*Institute of Enviromental Medicine, Karolinska Institute, S-171 77 Stockholm, Sweden;*
[2]*Department of Environmental Medicine, National Institute of Public Health, N-0462 Oslo,
Norway*

Introduction

Cells communicate by means of chemical signaling such as that mediated
by hormones, neurotransmitters, cytokines and by intercellular contacts
achieved through gap junctions and other transmembrane proteins.
During the last few years, it has become increasingly clear that environmental toxicants can interact with cellular signal transduction pathways,
affecting the ability of the cells to adequately respond to variety of
signals. The alteration of cellular responses to hormone and growth
factors, as well as the inability to deliver appropriate signals to other
cells, will result in the lack of trophic stimulation. In addition, it may also
determine a progressive alteration in cell plasticity and modify the
program for differentiation. Ultimately, toxicant-induced disturbances in
cell signaling can lead to the inappropriate activation or suppression of
the cell death program, apoptosis. Thus, exposure to chemicals rather
than causing rapid cell killing with loss of function, may eventually
induce subtle alterations in normal signaling, resulting in long-lasting
pathological effects. The potential targets on the cell signaling pathways
are numerous and disparate, as is their sensitivity to toxicant-induced
modifications. The purpose of this chapter is confined to reviewing the
issues discussed at the symposium "Alterations in Cell Signaling and
Cytotoxicity" given at the XIIth International Congress of Pharmacology, 1994. These include alterations in hormone-stimulated inositol lipid
breakdown, intracellular Ca^{2+} pools and Ca^{2+} channels, gene expression
and maintenance of intercellular communications.

Modulation of Gap Junctional Intercellular Communication During
Tissue Injury, Regeneration and Carcinogenesis

It has long been postulated that gap junctional intercellular communication (GJIC) plays a pivotal role in the maintenance of homeostasis [1].
During tissue injury and regeneration, and carcinogenesis, homeostasis

is disturbed and GGJIC is altered: tumor-promoting stimuli such as skin wounding and liver regeneration are accompanied by blocking of GJIC. After partial hepatectomy, GJIC block precedes DNA replication, suggesting a possible role of GJIC inhibition in starting cell proliferation. Many other tumor-promoting stimuli also inhibit GJIC, suggesting that one pathway of tumor promotion involves blocking of GJIC. Most rodent and human tumors show decreased levels of GJIC among tumor cells (homologous GJIC) and/or with surrounding normal cells (heterologous GJIC) [2]. Molecular mechanisms of GJIC inhibition include decreased connexin mRNA level and aberrant post-translational control of connexin proteins such as phosphorylation of connexin proteins, aberrant localization of connexin proteins (in cytoplasm rather than cytoplasmic membranes) and lack of a cell adhesion molecule which controls GJIC [3]. Mutations of cx 32 genes were not found in human gastric and liver cancers, but a rat liver tumor contained a point mutation in this gene. Recent studies have shown that transfection of connexin genes into GJIC-deficient tumor cells restores GJIC and growth control of recipient cells, suggesting an import role of connexin genes in the control of cell growth.

Alterations of Cell Signaling Alter Differentiation and Promote Apoptosis in Neural Cells

While alterations in cell growth control and suppression of apoptosis are very important in carcinogenesis, alterations in cell differentiation patterns and eventually inappropriate activation of apoptosis decrease cellular life span.

Recently, much attention has focused on the role of Ca^{2+} signaling alterations and mild oxidative stress in the development of neurotoxic disorders. Mild oxidation can cause receptor alterations, affect Ca^{2+} channel conductance, and ultimately interfere with growth factor stimulation [4, 5]. In contrast, marginally higher pro-oxidant levels can cause death [4]. Amplification of agonist-stimulated Ca^{2+} pulses in cells previously exposed to mild oxidation can potentiate physiological processes such as cell secretory activity or differentiation [5, 6]. Again, when the duration and extent of the Ca^{2+}-increase exceed the cell regulatory capacity, death may result [7].

Cellular responses to endogenous or environmental toxicants is also quite dependent on the degree of differentiation. Thus, cell death may occur in dividing cells as a result of unbalanced mitogenic stimulation, whereas the same condition may not cause cytotoxicity in differentiated cells. Recently, studies have shown that metals such as mercury and tin can affect neurotransmitter production, alter neural differentiation and plasticity, and ultimately activate apoptosis in neural cell lines [6]. In particular, we have shown that exposure of PC12 cells to inorganic mercury causes alterations of L-type Ca^{2+} channel properties, which result in potentiation of Ca^{2+} signals elicited by depolarization or agonist stimulation. Mercury potentiated intracellular Ca^{2+} responses to hormone stimulation or to KCl-induced depolarization and enhanced cell differentiation induced by NGF [5]. Conversely, Hg^{2+} concentrations which potentiate PC12 cell differentiation, resulted in the activation of apoptosis in CGC cells.

Notably, Hg^{2+} addition-elicited apoptosis also potentiated the neurotoxic effects of the glutamate analogue N-methyl-D-aspartate (NMDA) suggesting that very low Hg^{2+} concentrations in combination with endogenously produced glutamate would elicit Ca^{2+} increases high and sustained enough to promote apoptosis. The observation that Hg^{2+} caused apoptosis in CGC cells is more relevant if one considers that levels of Hg^{2+} that were non-toxic to CGC become lethal when associated with relatively low glutamate concentrations.

Thus, one obvious difference between PC12 cells and CGC cells is the stage of development, and it is well known that the implications of enhanced Ca^{2+} signals in mature and immature neurons are quite different. While in immature neural cells, increased $[Ca^{2+}]_i$ following exposure to high K^+ protects cells lacking growth stimulation, Ca^{2+} overload in differentiated neurons rapidly causes altered neurotransmitter release and can cause cell death. Ca^{2+} responses to high K^+ are amplified with cell differentiation and the types of ion channels expressed in the mature cell are higher than those found in neuroblasts or in undifferentiated cells. While high Ca^{2+} levels may be required to maintain differentiated responses in adult neural cells, further Ca^{2+} increases would result in cell killing. The presented data support this hypothesis.

Cell Signaling and Developmental Ethanol Neurotoxicity

Neurotransmitters, growth factors and hormones play multiple roles in various stages of brain development by activating specific receptors and second messenger systems that lead to intracellular modifications. Nerve cell proliferation and differentiation, in particular, appear to be influenced by cell signaling. Among second messenger systems, receptor activated hydrolysis of membrane phospholipids (phosphoinositides and phosphatidylcholine) has received particular attention, as it leads to mobilization of intracellular calcium and activation of protein kinase C. The developmental neurotoxicity of ethanol, as evident in the long-term CNS effects in children affected by the fetal alcohol syndrome (FAS), is well established. A series of *in vivo* and *in vitro* experiments has focused on the interaction of ethanol with the metabolism of membrane phospholipids, particularly that activated by cholinergic muscarinic receptors, as a possible mechanism for the developmental neurotoxicity of this compound. Administration of ethanol during the brain growth spurt (postnatal days 4 to 10 in the rat) leads to micorencephaly (a common sign in FAS) and inhibition of muscarinic receptor-stimulated phosphoinositide metabolism in cerebral cortex and hippocampus [8]. *In vitro* studies have confirmed the brain region-, neurotransmitter receptor- and age-specific sensitivity of the effects of ethanol [9, 10]. In primary cortical cultures ehtanol also inhibits calcium mobilization induced by muscarinic agonists [11]. Muscarinic agonists induce the formation of phosphatidylcholine-derived phosphatidic acid in cortical slices from immature rats and in astrocytes in culture. Formation of phosphatidic acid is antagonized by ethanol with the concomitant formation of phosphatidylethanol. Muscarinic agonists also induce proliferation of rat cortical astrocytes, by a mechanism that appear to involve activation of protein kinase C. Ethanol inhibits this mitogenic effect at concentrations of 10–50 mM. These suggest that the muscarinic receptor-stimulated phospholipid metabolism may be involved in the developmental neurotoxicity of ethanol; disruption of neurotransmitter – or growth factor – coupled with second messengers may lead to alterations in brain development and represents a novel mechanism of developmental neurotoxicity.

Mediator Networks Involved in Chemically-Induced Lung Disease

Many recent studies have emphasized that importance of cytokine production as signals mediating immune responses. Production of interleukins, tumor necrosis factor α and other cytokines can produce lethal effects in a variety of tissues. In addition, mediators usually involved in the inflammatory response, such as nitric oxide (NO) may also have cytotoxic effects. Inhalation of chemical irritants initiates a complex array of inflammatory responses resulting ultimately in both obstructive and restrictive pulmonary dysfunction. Production of cytokines and reactive intermediates by infiltrating macrophags is one of the major determinants in the tissue injury caused by a variety of chemicals. Exposure to irritants such as ozone cause an inflammatory response involving the local accumulation of cytokines such as TNF-α and IL-1 and reactive oxygen species including the superoxide anion, nitric oxide and hydrogen peroxide. Both alveolar and interstitial macrophages together with type II epithelial cells and fibroblasts are involved in this response. The inflammatory cytokines, TNF-β and IL-1β seem to elicit in the cell types listed above the generation of reactive species after ozone inhalation. However, the *in vivo* effects of ozone inhalation are not limited to the lung. Systemic toxicity, involving cytokine production by the lung and subsequent generation of cytotoxic mediators (nitric oxide and TNF-α) by the liver, is also observed following *in vivo* exposure to ozone. Thus, production of local and distantly acting effector molecules contributes to the development of tissue injury elicited by inhalation exposure to irritants such as ozone.

Signaling Pathways Involved in Tissue Responsiveness to Dioxins

Cytotoxic mediators including those mentioned above and many hormones and growth factors can ultimately modify gene expression. A number of toxic agents can also affect gene expression because of their ability to interact with DNA directly or interfere with the signaling normally associated with gene expression. For example, oxidant stress may directly activate nuclear genes or modify cytoplasmic protein such as NF-κb to gain transcriptional capability and initiate gene expression. A specific signaling system involving cytoplasmic binding of ligands seems to be instead involved in the cytotoxic effects of dioxins (2,3,7, 8-tetrachlorodibenzo-p-dioxin, TCDD). Toxic responses elicited by dioxins include the onset of chloracne in humans and the promotion of liver tumors in rats. Such manifestations appear to require the activation of a signaling system linked to the dioxin (or Ah, aryl hydrocarbon) receptor. The latter may affect growth and differentiation by controlling a network of regulatory genes. At least five dioxin responsive cDNA clones have been isolated from a human keratinocyte cell line used as *in vitro* model for chloracne. Of these clones, one was the cDNA for CYP1A1 (cytochrome P4501A1), which had previously been shown to be under the control of the dioxin receptor. Of the remaining four clones, two were identified as growth regulation genes encoding for proteins involved in the acute inflammatory response and growth regulation: plasminogen activator inhibitor-2 (PAI-2), a regulator of extracellular matrix proteolysis, and interleukin-1β. Several other effectors, in addition to dioxin, obviously regulate PAI-2 expression. These include tumor promoting phorbol esters and calcium. Thus, the similarity between dioxin and other tumor promoters (i.e., the induction of inflammatory responses and cellular differentiation) suggests that these effects of TCDD occur as the result of activation of similar signal-transduction pathways that in turn regulate overlapping gene networks expressing multifunctional proteins.

Conclusions

It is apparent that several environmental pollutants and other chemical toxicants can exert their effects on cell signaling at very low concentrations. The effects observed both *in vivo* and *in vitro* include alterations in gap junction proteins, intracellular messenger production, channel conductance, and gene expression. Local effects of such alterations include modification of cell differentiation and ultimately cell deletion. In addition, increased production of mediators during inflammation produced by chemical or environmental irritants may mediate local or distal effect via cytokine production, thereby using cell signaling molecules to mediate cytotoxic effects. This symposium restricted itself to highlighting a few of the possible implications that alterations in cell signaling caused by toxicants can have for human health. However, it became apparent that the sensitivity of signal-transduction elements to chemicals makes them promising targets for therapeutical approaches and potential indicators for early adverse effects of chemical injury.

References

1. Mesnil M, Yamasaki Y. Cell-cell communication and growth control of normal and cancer cells: evidence and hypothesis. Mol. Carcinogenesis 1993; 7: 14–17.
2. Yamasaki H, Krutovskikh V, Mesnil M, Columbano A, Tsuda H, Ito N. Gap junctional intercellular communication and cell proliferation during rat liver carcinogenesis. Env. Hlth. Perspect 1993; 101(Suppl 5): 191–198.
3. Krutovskikh V, Mazzoleni G, Mironov N, Omori Y, Aguelon AM, Mesnil M et al. Altered homologous and heterologous gap junctional intercellular communication in primary human liver tumors associated with aberrant protein localization but not gene mutation of connexin 32. Int. J. Cancer 1994; 56: 87–94.
4. Dypbukt JM, Ankarcrona MM, Burkitt M, Sjöholm Å, Ström K, Orrenius S et al. Different prooxidant levels stimulate cell growth, activate apoptosis, or produce necrosis in insulin-secreting RINm5F cells. J. Biol. Chem. 1994; 269: 30553–30560.
5. Rossi AD, Larsson O, Manzo L, Orrenius S, Vahter M, Berggren P-O et al. Modification of Ca^{2+} signaling by inorganic mercury in PC 12 cells. FASEB J. 1993; 7: 1507–1514.
6. Viviani B, Rossi AD, Chow SC, Nicotera P. Organotin compounds induce calcium overload and apoptosis in PC 12 cells. Neurotoxicol. 1995; 16: 20–27.
7. Jiang SA, Chow SC, Nicotera P, Orrenius S. Intracellular Ca^{2+} signals activate apoptosis in thymocytes: studies using the Ca^{2+}-ATPase inhibitor thapsigargin. Exper. Cell Res. 1994; 212: 84–92.
8. Balduini W, Costa LG. Effects of ethanol on muscarine receptor-stimulated phosphoinositide metabolism during brain development. J. Pharmacol. Exp. Ther. 1989; 250: 541–547.
9. Balduini W, Costa LG. Developmental neurotoxicity of ethanol: *in vitro* inhibition of muscarinic receptor-stimulated phosphoinositide metabolism in brain from neonatal but not adult rats. Brain Res. 1990; 512: 148–151.
10. Balduini W, Candura SM, Manzo L, Cattabeni F, Costa LG. Time-, concentration-, and age-dependent inhibition of muscarine receptor-stimulated phosphoinositide metabolism by ethanol in the developing brain. Neurochem. Res. 1991; 16: 1235–1240.
11. Kovacs KA, Costa LG. Inhibition of muscarinic receptor-stimulated phosphoinositide metabolism and calcium mobilization in rat primary cortical cultures by ethanol. Toxicologist 1993; 13: 172.

Pharmacological Sciences: Perspectives for
Research and Therapy in the Late 1990s
ed. by A.C. Cuello and B. Collier
© 1995 Birkhäuser Verlag Basel/Switzerland

Calcium-Dependent Mechanisms in Drug Toxicity and Cell Killing

Giorgio Bellomo[1], Ernesto Carafoli[2], Claus W. Heizmann[3], Alan Horton[4], and Sten Orrenius[5]

[1]*Clinica Medica I, Policlinico S. Matteo, Università di Pavia, I-27100 Pavia, Italy;* [2]*Laboratorium für Biochemie, ETH-Zentrum, CH-8092, Zürich, Switzerland;* [3]*Department of Pediatrics, University of Zürich, CH-8032 Zürich, Switzerland;* [4]*Department of Biochemistry, University of Birmingham, P.O. Box 363, Birmingham B15 2TT, U.K.;* [5]*Institute of Environmental Medicine, Karolinska Institutet, Box 210, S-171 77 Stockholm, Sweden*

Summary. There is now convincing evidence that the calcium ion plays a critical role in cytotoxicity and cell death. In a variety of experimental models cell killing has been found to be preceded by intracellular Ca^{2+} accumulation, and pretreatment with Ca^{2+} entry blockers or intracellular Ca^{2+} chelators has been shown to provide protection. Further, resistance to cell killing has been reported to correlate with cellular levels of the Ca^{2+}-binding protein calbindin-D_{28k} in neurons and thymoma cells. It is also clear that sustained increases in intracellular Ca^{2+} level can activate cytotoxic mechanisms which result in perturbations of cell structure and function. For example, the stimulation of Ca^{2+}-dependent proteases can cause a disruption of cytoskeletal organization, and Ca^{2+}-mediated phospholipase activation can result in an impairment of mitochondrial function and cessation of ATP synthesis. The activation of Ca^{2+}, Mg^{2+}-dependent endonuclease(s) is associated with characteristic chromatin cleavage which is a hallmark of apoptotic cell killing in the immune system and other tissues. The aim of this overview is to discuss some recent advances in our knowledge of the role of the calcium ion in cytotoxicity.

Regulation of Intracellular Ca^{2+} Homeostasis

Calcium is a fundamental intracellular messenger, transmitting information to numerous enzymes: for this reason, its intracellular ionic concentration is very precisely regulated [1, 2]. A number of injuring conditions and toxic agents cause disruption of its intracellular homeostasis and a sustained increase of cell Ca^{2+}, leading to cytotoxicity and eventually to cell death. The mechanism adopted by evolution for the control of intraceullular Ca^{2+} is its reversible complexation by classes of proteins, which are intrinsic to membranes, soluble in the cytoplasm, or organized in non-membranous structures, e.g., the myofibrils of muscle cells. The best studied non-membranous proteins (e.g., calmodulin) consist of two perpendicular alpha-helices flanking a non-helical loop where Ca^{2+} coordinates to $6-8$ oxygen atoms this structural arrangement is commonly known as the EF-hand model. The

Correspondence to: Dr. Sten Orrenius, at the above address.

buffering of cell Ca^{2+} by these proteins is quantitatively limited by their total cellular amount; for example, the concentration of calmodulin in different cells may be as high as $20\,\mu M$. Their main function, rather than the buffering of Ca^{2+}, is the processing of the Ca^{2+} signal; they express hydrophobic surfaces upon complexing Ca^{2+} and undergo a second conformational transition, which makes them more compact when complexing a target enzyme. This set of conformational changes is the essence of the process of decoding of the Ca^{2+} information.

The buffering of cell Ca^{2+} is performed essentially by membranous proteins, which transport Ca^{2+} reversibly in the plasma membrane, in the membrane of (endo) sarcoplasmic reticulum, in the inner membrane of mitochondria and in the nuclear membrane. Four basic transport modes are known: ATPases, exchangers, channels and electrophoretic uniporters: only the first mode has high Ca^{2+} affinity.

The primary structures of the Ca^{2+}-ATPases of skeletal and heart sarcoplasmic reticulum and of plasma membranes are now known [3]. The sequences of Na^{+}/Ca^{2+} exchangers of the heart, brain and retinal cells plasma membrane have also recently become known; intra-cellular organelles contribute differently to the homeostasis of Ca^{2+}. While the concerted operation of the importing and exporting systems of the plasma membrane maintains the 10^{4}-fold gradient of Ca^{2+} concentration between cells and ambient, the total amount of Ca^{2+} exchanged with the intracellular medium is a minor fraction of the total Ca^{2+} used by cells for their functional cycle. In the plasma membrane, the Na^{+}/Ca^{2+} exchanger has a low Ca^{2+} affinity and high total trans-port capacity. It plays the dominant exporting role in cells like those of the heart, where the functional cycle demands the periodic and rapid ejection of bulk amounts of Ca^{2+}. The Ca^{2+} pump has much higher Ca^{2+} affinity and much lower maximal transport capacity, and it thus satisfies the Ca^{2+} ejection requirements of most other cells. Even if the total amount of Ca^{2+} exchanged by most cells with the environment is comparatively minor, Ca^{2+} penetrating from the external spaces triggers essential events, e.g., the liberation of massive amounts of Ca^{2+} from intracellular stores. These stores are in the endo-(sarco-) plasmic ret-iculum, which performs the rapid and precise regulation of Ca^{2+} at the sub-μM level, using a Ca^{2+} pump for the uptake reaction and a channel which is operated by inositol-trisphosphate in some cells, or sensitive to Ca^{2+} itself in others. The mitochondrion is a low affinity system which transports Ca^{2+} to regulate some Ca^{2+}-sensitive matrix dehydrogenases. For the uptake leg of the "mitochondrial Ca^{2+} cycle" it uses a carrier driven by the membrane potential maintained by the respiratory chain. For the release leg it uses a Na^{+}/Ca^{2+} antiporter in a number of mitochondria, particularly those of excitable tissue, and a Ca^{2+} proton antiporter in other mitochondrial types. Although the mitochondrial Ca^{2+} transport system operates at a very low rate in normal cells, it becomes stimulated under cell injury conditions, responding to abnormal increases in cytosolic Ca^{2+} with the deposition of Ca^{2+}-phosphate granules in the matrix; in this way the matrix dehydrogenases are not deranged. When cytosolic Ca^{2+} has returned to normal, mitochondria release it at a slow rate, compatible with the capacity of the plasma membrane exporting systems. Thus, mitochondria play a key defense role in Ca^{2+}-promoted cell injury.

Cytoskeleton as a Target in Ca^{2+} Toxicity

The structural organization of the three main classes of cytoskeletal fibers – microfilaments, microtubules and intermediate filaments – can be dramatically affected by unphysiological changes in intracellular Ca^{2+} concentration via activation of different biochemical mechanisms.

Microfilaments are primarily composed of actin and several actin-binding proteins which regulate the state of actin polymerization to form fibers, the self-association of fibers to form bundles, and the association of these bundles to the plasma membrane. Several actin-binding proteins are Ca^{2+}-regulated. Among them, α-actinin is involved in the normal organization of actin filaments into regular parallel arrays. In the presence of micromolar Ca^{2+} concentration, α-actinin dissociates from the actin filaments and from other actin-binding proteins regulating the cytoskeleton-plasma membrane interaction [4].

Calcium-dependent neutral proteases catalyze the proteolysis of several cytoskeletal proteins, including spectrin, fodrin, caldesmon, adducin, tubulin, microtubule-associated protein 2(MAP-2), tau factor, vimetin, cytokeratins. Two cytoskeletal proteins that are directly involved in anchoring microfilaments to the inner surface of plasma membrane, i.e., vinculin and actin-binding protein, are preferential substrates for Ca^{2+}-dependent proteases. An increase in cytosolic free Ca^{2+} concentration to micromolar level (high enough to activate the protease) results in the proteolysis of these two polypeptides. This occurs physiologically during platelet activation and toxicologically during oxidative stress [5].

Since several cytoskeletal elements are phosphorylated by Ca^{2+}/ calmodulin-dependent protein kinases, or by other protein kinases, it is conceivable to assume that Ca^{2+}-stimulated phosphorylation may lead to cytoskeleton rearrangements [6]. On the other hand, the continuous dephosphorylation of the phosphorylated proteins will assure the physiological control. Inhibitors of type 1 and type 2A protein phosphatases, by causing an imbalance of the protein kinase/phosphatase activity, could induce abnormal phosphorylation of various cellular proteins (including cytoskeletal proteins) ultimately leading to cell injury.

Ca^{2+}-dependent cytoskeletal alterations are considered among the factors responsible for the observed changes in cell morphology preceding cell injury and death. They primarily consist of cell swelling associated with the appearance of protrusions from the plasma membrane (blebs). It has been postulated that dissociation of the actin network from the inner surface of the plasma membrane would generate sites of weakness where blebs could be formed. However, in order for blebs to be formed, an increase of the cellular water content, and thus of the cell volume should take place. A marked increase in Na^+ influx has been detected in many conditions associated with irreversible cell injury [7]. In some cases, it is linked to the activation of the Na^+/Ca^{2+} antiporter following the nonphysiological increase in cytosolic free Ca^{2+}. In these conditions, Ca^{2+} overload could contribute to the generation of surface blebs both by modifying the association of the cytoskeleton to the plasma membrane and by promoting the protrusion and rupture of blebs, thus precipitating cell death.

Ca^{2+}-Induced Arachidonic Acid Cascade in Toxicity

Early evidence for a role for eicosanoids in toxicity came from *in vivo* experiments with rats or mice with paracetamol-induced toxicity. Prior admininstration of acetyl salicylic acid provided marked protection as indicated by serum glutamate-pyruvate transaminase (SGPT) activity. An increase in concentration of cytosolic free Ca^{2+} associated with paracetamol-induced toxicity *in vivo* was reported later from several different laboratories. In 1988–89, paracetamol-induced toxicity *in vivo* was shown to cause eicosanoid synthesis in mice and rats [9, 10]. Prostaglandin E_2 (PGE_2), 6-keto $PGF_{1\alpha}$ and particularly thromboxane B_2 (TXB_2) were synthesised in substantially increased amounts relative to controls. A wide range of inhibitors of phospholipase A_2 and cyclooxygenase provided protection from toxicity when assessed by SGPT activity implying that products of cyclooxygenase are responsible for cell damage as no cytoprotective PGI_2 would be synthesized under these conditions. Inhibitors of TX synthetase were also protective although in this case, synthesis of PGI_2 and other PG's would proceed.

CCl_4-induced toxicity, which is also associated with an increase in concentration of free cytosolic Ca^{2+}, stimulated generation of TXB_2 in livers of mice relative to controls. A TX synthetase inhibitor and a TX receptor antagonist both protected, suggesting an involvement of TX receptors *in vivo*, but this effect may be attributable to inhibition of hemodynamic effects.

Perfused organs appear to respond to increased concentrations of free cytosolic Ca^{2+} as do whole animals. Lungs of animals perfused with Ca^{2+} ionophore A23187 (10^{-5} M) showed substantial increases in prostanoid production although there were marked interspecies differences. For example, while rats produced mainly 6-keto $PGF_{1\alpha}$, the guinea pig produced TXB_2 predominantly.

In whole animals and, probably, perfused organs, TX's caused vasoconstriction leading to hemodynamic effects which are opposed by PGI_2. The Ca^{2+}-induced toxic effect *in vivo* could be a consequence of hemodynamic effects alone (liver congestion is a feature of paracetamol toxicity), or hemodynamic effects plus a contribution from prostanoids which may have a direct effect on cell toxicity. For this reason, the effect of increased concentration of cytosolic free Ca^{2+} on isolated hepatocytes was investigated. Isolated rat hepatocytes incubated with A23187 were assessed for cell damage by plasma membrane bleb formation and cell viability by trypan blue exclusion. Preincubation with inhibitors of enzymes of the arachidonic acid cascade afforded marked protection from bleb formation [11], implying that the role of prostanoids in Ca^{2+}-induced toxicity *in vivo* may include a contribution other than that from hemodynamic effects. In view of this evidence, and the possibility that the prostanoids may act via receptors, the effect of TX receptor antagonists on A23187-induced bleb formation was investigated. Three TX receptor antagonists, Daltroban (BM 13505), Sulotroban (BM 13177) and SK and F 88046, all protected, thereby providing a hint that the mechanism of action of products of cyclooxygenase in Ca^{2+}-induced toxicity may not be a simple chemical effect.

Some prostanoid levels (e.g., 6-keto $PGF_{1\alpha}$) measured in isolated hamster hepatocytes after incubation with paracetamol showed a significant increase, thus demonstrating activation of the arachidonate cascade within 30 min. Incubation of isolated rat hepatocytes with CCl_4 induced synthesis of prostanoids (e.g., TXB_2, 6-keto $PGF_{1\alpha}$). Assessment of cell damage by lactate dehydrogenase (LDH) release showed that cyclooxygenase and lipoxygenase inhibitor (BW 755C) protected against CCl_4-induced damage. Anti-PG1$_1$ antibody (effective against PGI_2) partially reversed the inhibitory effect of benzyl imidazole (BI, TX synthetase inhibitor) on LDH release in peritoneal leukocytes treated with CCl_4 although TXB_2 formation was not affected. This indicated that part of the effect of BI could be due to synthesis of PGI_2, but part could be attributable to decreasing the amount of TXs.

In conclusion, inhibitors of enzymes of the arachidonic acid cascade protect cells from Ca^{2+}-induced toxicity both *in vivo* and *in vitro* but the mode of action of prostanoids in toxicity remains unresolved.

Regulation of Apoptosis by Ca^{2+}

The elimination of cells through apoptosis is important in development, normal cell turnover, hormone-induced tissue atrophy, and various pathological processes. Cells undergoing apoptosis show characteristic morphological changes, including plasma and nuclear membrane blebbing, cell shrinkage, and chromatin condensation and fragmentation [12]. These changes distinguish apoptosis from cell death by necrosis.

In most cells, the biochemical characteristics of the apoptosis response include endogenous endonuclease and protease activation as well as transglutaminase activation. All mammalian cells appear to constitutively express the basic enzymatic machinery that mediates apoptotic cell death, and Ca^{2+} is implicated in the activation of these processes in many experimental models, suggesting that it may play a major regulatory role in the response.

Several lines of evidence indicate that the cytosolic Ca^{2+} concentration can regulate apoptosis. In many cells, apoptosis is induced by treatment with thapsigargin or Ca^{2+} ionophores [13]. The process can also be triggered in thymocytes (immature T-cells) or primed mature T-cells by stimulation of the T-cell receptor and in neurons by stimulation of glutamate (NMDA) receptors, and both these responses involve sustained Ca^{2+} increases. Certain chemical toxins may also promote apoptosis by disrupting intracellular Ca^{2+} homeostasis, leading to non-physiological Ca^{2+} increases that promote endonuclease activation and apoptotic cell death [14]. Intracellular or extracellular Ca^{2+} chelators, Ca^{2+} channel blockers and calmodulin antagonists can all delay or abolish apoptosis in several model systems. Consistent with these effects, overexpression of the Ca^{2+}-binding protein calbindin D-28K can block apoptotic cell death. Finally, recent work suggests that the protective effects of the anti-apoptosis oncoprotein Bcl-2 involve alterations in Ca^{2+} compartmentalization [15]. Together, these observations indicate that Ca^{2+} is a frequent trigger of apoptosis in diverse experimental systems.

Cytoprotection by Intracellular Ca^{2+}-Binding Proteins

The calcium message is converted into an intraceullular response, in many cases by Ca^{2+}-binding proteins that are involved in a number of activities [16]. One class of these proteins shares a common Ca^{2+}-binding motif, the EF-hand. A consensus amino acid sequence for this motif has aided the identification of new members of this protein family, which has now about 200 members. A few of these proteins are present in all cells, whereas the vast majority are expressed in a tissue-specific fashion. The cytosolic Ca^{2+}-binding proteins, parvalbumin and calbindin D-28K have proved to be useful neuronal markers for a variety of functional brain systems and their circuitries. It is assumed that neurons containing cytosolic Ca^{2+}-binding proteins, such as parvalbumin and calbindin D-28K, and therefore having a greater capacity to buffer Ca^{2+}, would be more resistant to degeneration [17]. To test this hypothesis human neuroblastoma cells were used to ectopically express parvalbumin and to monitor intracellular Ca^{2+} fluxes under stimulating conditions. The results demonstrate an impaired Ca^{2+} response in parvalbumin-transfected cells. The outcome of these studies may also be relevant to medication developed to correct hyperactivity of intracellular Ca^{2+}.

The S100 proteins belong to another subfamily of EF-hand Ca^{2+}-binding proteins. Some S100 proteins are associated with apoptosis, others with tumor development and the metastatic behavior of tumors. S100 proteins are presently also used as selective markers for estimating the extent of brain damage in various neurological disorders and for classifying various tumors in children and adults. Recently, six S100 genes were found to be clustered on human chromosome 1q21, a region showing rearrangements in neoplasms at high frequency. It is suggested that the S100 protein, CAPL, may play a role in the acquisition of the metastatic potential of tumor cells and CAPL is presently evaluated as a marker for the invasivity and the prognosis of tumor cells [18].

Concluding Remarks

It is now clear that the calcium ion plays a critical role in cytotoxicity, acting on multiple cellular targets. The relative contribution of the various Ca^{2+}-dependent degradative enzymes to the development of toxicity differs between cell types and is dependent on experimental conditions. Modulation of intracellular Ca^{2+} concentration can affect toxicity in several experimental models and is now being developed as a pharmacotherapeutic principle.

References

1. Carofoli E, Penniston JT. The calcium signal. Sci. Am. 1985; 253: 70.
2. Carofoli E. Intracellular calcium homeostasis. Ann. Rev. Biochem. 1987; 56: 395–433.
3. Carofoli E. The Ca^{2+} pump of the plasma membrane. J. Biol. Chem. 1992; 267: 2115–2118.
4. Bellomo G, Mirabelli F, Richelmi P, Malorni W, Iosi F, Orrenius S. The cytoskeleton as a target in quinone toxicity. Free Rad. Res. Commun. 1990; 8: 391–399.
5. Mirabelli F, Salis A, Vairetti M, Bellomo G, Thor H, Orrenius S. Cytoskeletal alterations in human platelets exposed to oxidative stress are mediated by oxidative and Ca^{2+}-dependent mechanisms. Arch. Biochem. Biophys. 1989; 270: 478–488.
6. Eriksson JE, Paatero GIL, Meriluoto JAO, Codd GA, Kass GEN, Nicotera P et al. Rapid microfilament reorganization induced in isolated rat hepatocytes by microcystin-LR, a cyclic peptide toxin. Exp. Cell Res. 1989; 185: 86–100.
7. Carini R, Bellomo G, Dianzani MU, Albano E. Evidence for a sodium-dependent calcium influx in isolated hepatocytes undergoing ATP depletion. Biochem. Biophys. Res. Commun. 1994; 202: 360–366.
8. Whitehouse LW, Paul CJ, Thomas BH. Effect of acetylsalicylic acid on a toxic dose of acetaminophen in the mouse. Toxicol. Appl. Pharmacol. 1976; 38: 571–582.
9. Guarner F, Boughton-Smith NK, Blackwell GJ, Moncada S. Reduction by prostacyclin of acetaminophen-induced liver toxicity in the mouse. Hepatology 1988; 8: 248–253.
10. Horton AA, Wood JM. Effects of inhibitors of phospholipase A$_2$, cyclooxygenase and thromboxane synthetase on paracetamol hepatotoxicity in the rat. Eicosanoids 1989; 2: 123–129.

11. Horton AA, Wood JM. Prevention of Ca^{2+} induced hepatocyte plasma membrane bleb formation by inhibitors of eicosanoid synthesis. J. Lipid Mediator 1989; 1: 213–242.
12. Wyllie AH, Kerr JFR, Currie AR. Cell death: The significance of apoptosis. Int. Rev. Cytol. 1980; 68: 251–305.
13. Jiang S, Chow SC, Nicotera P, Orrenius S. Intracellular Ca^{2+} signals activate apoptosis in thymocytes: Studies using the Ca^{2+}-ATPase inhibitor thapsigargin. Exp. Cell Res. 1994; 212: 84–92.
14. McConkey DJ, Hartzell P, Håkansson H, Orrenius S. 2,3,7,8-Tetrachlorodibenzo-p-dioxin kills immature thymocytes by Ca^{2+}-mediated endonuclease activation. Science 1988; 242: 256–259.
15. Baffy G, Miyashita T, Williamson JR, Reed JC. Apoptosis induced by withdrawl of interleukin-3 (IL-3) from an independent hemapoietic cell line is associated with repartitioning of intracellular calcium and is blocked by enforced Bcl-2 oncoprotein production. J. Biol. Chem. 1993; 268: 6511–6519.
16. Heizmann CW, Hunziker W. Intracellular calcium-binding proteins: more sites than insights. Trends Biochem. Sci. 1991; 16: 98–103.
17. Heizmann CW, Braun K. Changes in Ca^{2+}-binding proteins in human neurodegenerative disorders. Trends Neurosci. 1992; 15: 259–264.
18. Pedrocchi M, Schafer BW, Mueller H, Eppenberger U, Heizmann CW. Expression of Ca^{2+}-binding proteins of the S100 family in malignant human breast-cancer cell lines and biopsy samples. Int. J. Cancer 1994; 57: 684–90.

Pharmacological Sciences: Perspectives for
Research and Therapy in the Late 1990s
ed. by A.C. Cuello and B. Collier

Liver Toxicity Mediated by Leukocytes and Kupffer Cells

Ruth E. Billings[1], Bernhard H. Lauterburg[2], Ronald G. Thurman[3],
Jack Uetrecht[4], and Albrecht Wendel[5]

[1]*Department of Environmental Health, Colorado State University, Fort Collins, Colorado 80523,
USA;* [2]*Department of Clinical Pharmacology, University of Berne, 3010 Berne, Switzerland;*
[3]*Department of Pharmacology, University of North Carolina, Chapel Hill, North Carolina
27599, USA;* [4]*Faculties of Pharmacy and Medicine, University of Toronto and Sunnybrook
Health Science Centre, Toronto, Canada M 5S 1A 1;* [5]*Department of Biochemical Pharmacology,
University of Konstanz, D-7750 Konstanz, Germany*

Summary. Five presentations comprised this symposium. Leukocyte-mediated drug
metabolism was discussed in one talk (J.P.U.). Myeloperoxidase is the primary system
responsible for formation of reactive metabolites in neutrophils and monocytes. Metabolites
formed directly within leukocytes can inhibit the function of these cells and mediate some of
the activities of drugs such as dapsone, propylthiouracil, and 5-aminosalicylic acid. A second
presentation (B.H.L.) focused on explaining the low plasma levels of reduced glutathione
(GSH) in HIV-infected individuals. Consequently, lymphocyte function is impaired and viral
replication is facilitated. The liver is the major source of circulating GSH, and, in AIDS
patients, the liver's capacity to synthesize GSH may be deficient. Data on several *in vivo and
in vitro* models of immunologically mediated liver injury was also discussed (A.W.). The data
suggest that, regardless of the initial stimulus and the primary effector cell, a cytokine
syndrome with tumor necrosis factor-alpha (TNF) as a major distal mediator is initiated.
Cytokine-mediated liver cell injury involves apoptosis. The role of liver macrophages (Kupffer
cells) in liver transplantation and alcoholic liver injury was also discussed (R.G.T.). The data
implicate cytokines such as TNF and interleukin-6 and deleterious free radicals in the
pathophysiology observed in both of these liver disease models. The symposium was concluded with a presentation of data in which the effects of recombinant-DNA derived
cytokines, such as TNF, were studied in cultured mouse and rat hepatocytes (R.E.B.). In these
cells, both the cytotoxic effects of TNF and the induction of nitric oxide synthase are mediated
by oxygen radicals. Glutathione also plays a major role in TNF's effects. Nitric oxide may
function in hepatocytes as a regulator of cytochrome(s) P450.

Introduction

Since interactions between immune and non-immune cells are critical to
tissue injury, this symposium focused on mechanisms of interactions
between leukocytes and hepatocytes. Although drug metabolism is
commonly attributed to hepatocytes, leukocytes also have the capacity
to metabolize drugs and other exogenous chemicals to reactive intermediates. One major function of the liver is to manufacture and secrete
substances which are transported via the systemic circulation to other
organs. Therefore, hepatic function may influence leukocyte-mediated

Correspondence to: Dr. Ruth E. Billings, Dept. of Environmental Health, B120 Microbiology
Bldg., Colorado State University, Fort Collins, CO 80523, USA.

activities throughout the body. Within the liver, Kupffer cells represent 80–90% of the resident macrophages present in the body and constitute approximately 15–20% of the total liver cell population and 3–5% of liver weight. When activated, peripheral leukocytes may infiltrate the liver and release a variety of potentially toxic mediators such as cytokines, eicosanoids and reactive oxygen intermediates. Abolition of Kupffer cell function can attenuate and sometimes eliminate irreversible liver injury in a variety of *in vivo* models. This suggests that liver injury is a two-stage process in which reversible events mainly initiated in hepatocytes lead to infiltration and activation of leukocytes which release substances, particularly pro-inflammatory cytokines such as tumor necrosis factor-α and reactive oxygen. These substances cause oxidative injury to hepatocytes and play a causative role in hepatoycte death. Each of the following sections summarize one of the five talks on how leukocytes participate in liver injury.

Metabolism of Drugs by Leukocytes and its Clinical Significance

Reactive metabolites formed in the liver, the major site of drug metabolism, are responsible for the toxicity of many compounds. However, many types of toxicity do not involve the liver and the biological half-life of most reactive metabolites precludes their escape from the liver.

Therefore, the metabolism of drugs by neutrophils and monocytes has been investigated. Compounds with easily oxidized functional groups are oxidized to reactive intermediates by these cells [1]. These include aryl amine, hydrazine and sulfhydryl/thiono sulfur groups. The major system responsible for this metabolism appears to be the myeloperoxidase system. It is proposed that such metabolites are responsible for toxicity involving leukocytes such as agranulocytosis and drug-induced lupus. Reactive metabolites generated by activated leukocytes may also play a role in the therapeutic effects of some drugs. Several of the drugs metabolized to reactive metabolites by activated neutrophils have anti-inflammatory effects, especially in those conditions in which neutrophils or macrophages play a major role in the inflammation. Examples of diseases in which neutrophils or macrophages appear to play a significant role are dermatitis herpetiformis, acute adult respiratory distress syndrome, ulcerative colitis, rheumatoid arthritis, reperfusion injury and possibly alcoholic hepatitis and Alzheimer's disease. Drugs that appear to be effective in some of these diseases are dapsone, propylthiouracil (PTU) and 5-aminosalicyclic acid. The effect of dapsone on dermatitis herpetiformis is dramatic and usually results in complete clearing of lesions within 24 h. It also has activity in rheumatoid arthritis and appears to decrease the risk of

Alzheimer's disease, possibly by inhibition of microglial cells which are the macrophages of the brain. Dapsone is oxidized to a reactive hydroxylamine by the myeloperoxidase system of neutrophils and monocytes, and it inhibits myeloperoxidase, at least in part, by converting compound I to compound II [2]. The reactive hydroxylamine is also likely to have other effects on neutrophil and/or macrophage function; one possible mechanism is inhibition of the production of cytokines such as TNFα. PTU is also oxidized to reactive metabolites by the myeloperoxidase system and inhibits neutrophil function. It increases survival in patients with alcoholic hepatitis [3], in which the histology is characterized by infiltration of neutrophils. In addition, PTU markedly decreases the hepatotoxicity of acetaminophen [4]. Thyroidectomy also decreases acetaminophen toxicity, but the effect is less than that of PTU which suggests that additional mechanisms are involved. It is proposed that reactive metabolites of drugs formed by activated neutrophils and macrophages can inhibit the function of these cells and mediate some of the anti-inflammatory effects of the drugs involved.

Liver-Lymphocyte Interactions in HIV Infection

The concentrations of GSH and cysteine are lower than normal in plasma and in peripheral blood mononuclear cells (PBMC) of HIV-infected patients [5, 6]. Since sulfhydryls play an important role in lymphocyte function, and the decreased concentrations of sulfhydryls may contribute to disease progression. Indeed, incubation of HIV-infected cells with sulfhydryls in millimolar concentrations can inhibit viral replication [7]. Oxidant stress has been proposed to account for the decreased intracellular concentrations of GSH [8]. However, the expected increase in the concentrations of extracellular glutathione disulphide (GSSG) and cysteine was not found in patients with AIDS.

Since the liver accounts for most of the circulating GSH in rats [9] and circulating GSH is an important source for cysteine to be utilized by lymphocytes for GSH synthesis, a decreased efflux of GSH from the liver could explain the findings in HIV-infected subjects. We, therefore, measured the input of GSH into the circulation, an estimate of systemic GSH production, in eight HIV-infected patients using a pharmacokinetic approach. Plasma cysteine (7.7 ± 2.6 vs 13.4 ± 4.9 nmol/l) and GSH (7.6 ± 3.1 vs 11.6 ± 3.3 nmol/l) were lower in patients than in eight healthy control subjects. Upon infusion of GSH at a constant rate the plasma concentration of GSH reached a new plateau. The input of GSH into the circulation was calculated from the basal concentration of total GSH, the new steady-state concentration and the rate of infusion. The total input of GSH (12.9 ± 5.7 vs 30.1 ± 11.7 nmol/min) and the clearance of GSH (25 ± 7 and 35 ± 7 ml/min/kg) were significantly lower in HIV-patients, suggesting that low circulating concentrations of cysteine and GSH in HIV-infection are due to decreased synthesis of the tripeptide and not to an increased consumption. In agreement with this interpretation the supplementation of N-acetylcysteine at a dose of 3 times 600 mg per day for 2 weeks did not significantly increase the concentration of GSH in PBMC or plasma of patients with AIDS.

These data indicate that the systemic production of GSH is impaired in HIV-infected patients. The liver as the major source of circulating GSH may thus in part determine the function of lymphocytes in HIV infection via its export of GSH. Furthermore, GSH precursors that do not require *de novo* synthesis of GSH, such as glutathione esters, might be better suited to explore the therapeutic potential of GSH in HIV-infection than precursors of cysteine, such as N-acetylcysteine or procysteine.

Immunologically Mediated Experimental Liver Injury

Liver disease caused by excessive host defence reactions against infection is a problem that requires understanding of the underlying mechanisms in order to develop drugs. Therefore, several *in vivo* models of immunologically mediated liver failure in mice were established, and details of these models were investigated in suitable hepatic *in vitro* systems.

Endotoxin-induced liver failure in galactosamine (GalN)-sensitized mice [10] was characterized by an early peak of TNFα (1.5 h after endotoxin) which preceded liver injury. A similar extent and time-course of liver injury was observed when animals were injected with recombinant TNFα instead of endotoxin. Pretreatment of mice with anti-TNF antibodies fully protected against hepatotoxicity induced by endotoxin or by TNFα. Only animals pretreated with GalN, which inhibits transcription in the liver, were sensitive. These findings demonstrate that *in vivo*, TNF is a terminal mediator of endotoxic liver failure when transcription is blocked.

In GalN/TNFα-treated mice, programmed cell death (apoptosis) was demonstrated in liver tissue at time points before any other signs of toxicity became overt [11]. Apoptosis was also induced in primary mouse hepatocyte cultures by TNFα when transcription was inhibited. This process started before release of lactate dehydrogenase was detected. These *in vitro* and *in vivo* data support the conclusion that apoptotic changes in hepatocytes are a cause rather than a consequence of cell death.

When GalN-sensitized mice were injected with an antibody against the T-cell receptor molecular complex (αCD3) they developed fulminant liver failure [12]. In this model, anti-TNF-antibodies were also protective as were the immunosuppressive drugs cyclosporin or dexamethasone. Formation of apoptotic bodies and DNA fragmentation were detectable early in these livers. These data provide evidence that also in this T-cell-mediated model, TNFα represents a distal pathogenic mediator responsible for induction of apoptosis.

When *non-sensitized* mice were intravenously injected with the plant lectin Concanavalin A (ConA), a selective liver injury was induced [13]. Mice devoid of functional CD4 positive T-cells were resistant to Con A. Systemic TNFα was detected in this model also, with a peak at 1.5 h. These experiments suggest that immunocompetent T-cells play a pivotal role in lectin-induced TNFα release *in vivo*. Con A-induced hepatocellular leakage was preceded by apoptosis as assessed by DNA-fragmentation and histopathological evaluation.

Taken together, these observations suggest that regardless of the initial stimulus and the primary effector cell, a cytokine syndrome with TNFα as the major distal mediator is initiated. Once cytokines circulate, they activate or maintain processes in the liver that convey apoptotic signals to liver parenchymal cells, especially to those that are metabolically compromised.

Role of Macrophages in Graft Failure Following Liver Transplantation and in Alcoholic Liver Disease

Kupffer cells, the resident hepatic macrophages, are involved in mechanisms of pathophysiology following liver transplantation [14] and alcoholic liver injury [15], possibly by release of toxic mediators such as cytokines (e.g., TNF) and eicosanoids [16]. Following transplantation of fatty livers from alcohol-treated rats, platelets and microthrombi were observed at least twice as frequently as in controls; AST levels are increased; survival is decreased. These are all parameters which are influenced by mediators produced by Kupffer cells.

When livers stored under non-survival conditions were transplanted, serum TNF levels were elevated to 15 U/ml 150 min after graft reperfusion. The calcium channel blocker nisoldipine was found to decrease phagocytotic activity of Kupffer cells, improve survival following liver transplantation and prevent the increased serum TNF levels. Serum interleukin-6 (IL-6) levels also increased after transplantation and values were diminished significantly by nisoldipine. These data indicated that release of TNF and IL-6 was increased transiently after transplantation only under non-survival conditions and that nisoldipine prevented cytokine release by blocking activation of Kupffer cells.

Elimination of Kupffer cells by gadolinium chloride ($GdCl_3$) treatment of rats prevented alcohol-induced increases in AST levels. In addition, fatty changes, inflammation and necrosis, as well as elevated rates of ethanol elimination were all minimized by GdCl$_3$ treatment. Therefore, Kupffer cells may participate in alcohol-induced liver injury via mechanisms involving parenchymal cell hypoxia secondary to activation of Kupffer cells by endotoxin. To test this hypothesis directly, bacterial endotoxin was minimized by intestinal sterilization. Under these conditions, plasma endotoxin levels were reduced, ethanol-induced increases in AST levels were diminished significantly, and the hepatic pathological score was decreased. Furthermore, hypoxia due to a hypermetabolic state was also prevented. A carbon-centered free radical adduct was detected with electron paramagnetic spin resonance spectroscopy (EPR) in bile of rats exposed to ethanol. Importantly, this adduct was diminished significantly by GdCl$_3$ treatment, supporting the hypothesis that elevated levels of endotoxin, which activate Kupffer cells to release eicosanoids and cytokines, also increase deleterious free radicals. At present, it is not clear whether these free radicals arise directly from a respiratory burst by the Kupffer cells, or indirectly as a consequence of a reperfusion injury following hypoxia in the parenchymal cells.

Since fatty livers from alcoholics transplant poorly, the effect of ethanol on free radical formation in a rearterialized rat transplant model was evaluated by EPR. Liver from rats treated with an ethanol-containing diet for 3–5 weeks exhibited characteristic pericentral lipid accumulation. Following transplantation of livers from ethanol-treated rats, postoperative AST levels were elevated and survival was decreased. Further, a robust six-line complex EPR spectrum was observed containing a mixture of three radical species, two with coupling constants similar to lipid-derived free radicals and the third most likely oxygen-derived. Taken together, these data implicate cytokines released from Kupffer cells and deleterious free radicals in the pathophysiology observed in transplanted livers from fatty grafts due to alcohol as well as in alcoholic liver injury.

Effects of Recombinant Cytokines in Primary Cultures of Hepatocytes

Activated Kupffer cells release cytokines such as TNFα, IL-1, and IL-6 which form a cytokine network in which mutual induction occurs and activities are linked. TNF is released early in this process and it induces both IL-1 and IL-6. The cytotoxic effects of TNF are associated with oxidative damage and it has been observed that TNF causes an oxidative stress in cultured mouse hepatocytes [17]. Reactive oxygen intermediates (ROI), such as superoxide anion, are also released from activated leukocytes. Thus, Kupffer cell-mediated liver toxicity may be primarily oxidative in nature.

TNF stimulates ROI formation through multiple pathways including stimulation of phospholipase A$_2$ activity [18] and xanthine oxidase activity [19]. Recently, it has been found that mitochondrial generation of oxgyen radicals in cultured rat hepatocytes treated with TNF leads to induction of nitric oxide synthase (NOS) activity [20]. Although NOS activity contributes little to the cytotoxic effects of TNF [21], inhibition abolishes its effects on cytochrome P450. In cultured rat hepatocytes, TNF in combination with IL-1 and interferon-γ, reduces P450 levels to approximately 30% of control values. The addition of the NOS competitive inhibitor, N^G-monomethyl-L-arginine, prevents this decline.

The regulation of cytokine-inducible NOS has been extensively investigated using polymerase chain reaction technology to quantify expression of the NOS gene [20]. NOS activity was determined by measuring stable NO products in the cell culture medium and it was found to correlate in all cases with NOS gene expression, suggesting that NOS is regulated primarily at the level of transcription. Anti-oxidants such a trolox, an α-tocopherol analog, decreased NOS induction by TNF. NOS was also induced by addition of an extracellular ROI generating system, which consisted of xanthine/xanthine oxidase. These results suggest that cytokine-inducible NOS is regulated by ROI, possibly as a feed-back mechanism to inactivate superoxide anion. It is well known that NO avidly binds to superoxide. Paradoxically, NOS induction is also dependent upon glutathione. Agents which deplete the intracellular concentration of glutathione, such as diethylmaleate and buthionine sulfoximine decreased NOS induction, whereas N-acetylcysteine increased induction. These results suggest that NOS induction is regulated by the reductive state of the hepatocytes, as well as by ROI.

Acknowledgements

This work was supported by the following grants. National Institutes of Health (U.S.A.): DK 44755 (R.E.B.); AA09156, AA03624, DK37034 (R.G.T.); Nationales AIDS-Forschungsprogramm (B.H.L.); Medical Research Council of Canada, MT9336 and MT10036 (J.P.U.); Deutsche Forschungsgemeinschaft We 13-1 (A.W.)

References

1. Uetrecht JP. The role of leukocyte-generated metabolites in the pathogenesis of idiosyncratic drug reactions. Drug Metab. Rev. 1992; 24: 299–366.
2. Kettle AJ, Winterbourn CC. Mechanism of inhibition of myeloperoxidase by anti-inflammatory drugs. Biochem. Pharmacol. 1991; 41: 1485–1492.
3. Orrego H, Blake JE, Blendis LM, Compton KV, Israel Y. Long-term treatment of alcoholic liver disease with propylthiouracil. N. Engl. J. Med. 1987; 317: 1421–1427.
4. Linscheer WG, Raheja KL, Cho C, Smith NJ. Mechanism of the protective effect of propylthiouracil against acetaminophen toxicity in the rat. Gastroenterol. 1980; 78: 100–107.
5. Eck HP, Gmuender H, Hartman H, Petzoldt D, Daniel V, Droege W. Low concentrations of acid-soluble thiol (cysteine) in blood plasma of HIV-1 infected patients. Biol. Chem. Hoppe Seyler 1989; 370: 101–108.

6. DeQuay B, Malinverni R, Lauterburg BH. Glutathione depletion in HIV-infected patients: Role of cysteine deficiency and effect of N-acetylcysteine. AIDS 1992; 6: 815–819.
7. Kalebic T, Kinter A, Poli A, Anderson ME, Meister A, Fauci AS. Suppression of human immunodeficiency virus expression in chronically infected monocytic cells by glutathione, glutathione ester, and N-acetylcysteine. Proc. Natl. Acad. Sci. USA 1991; 1991: 986–990.
8. Staal FJT, Ela SW, Roederer M, Anderson MT, Herzenberg LA. Glutathione deficiency and human immunodeficiency virus infection. Lancet 1992; 4: 909–912.
9. Lauterburg BH, Adams JD, Mitchell JR. Hepatic glutathione homeostasis in the rat: Efflux accounts for glutathione turnover. Hepatology 1984; 4: 586–590.
10. Tiegs G, Wolter M, Wendel A. Tumor necrosis factor is a terminal mediator in galactosamine/endotoxin-induced hepatitis in mice. Biochem. Pharmacol. 1989; 38: 627–631.
11. Leist M, Gantner F, Bohlinger I, Germann PG, Tiegs G, Wendel A. Murine hepatocyte apoptosis induced *in vitro* and *in vivo* by TNFα requires transcriptional arrest. J. Immunol. 1994; 153: 1778–1787.
12. Gantner F, Jilg S, Tiegs G. Anti-CD3 antibody-induced liver injury in D-galactosamine/endotoxin-induced hepatitis in mice. Naunyn-Schmiedeberg's Arch. Pharmacol. 1994; Suppl. 349; R 66.
13. Tiegs G, Hentschel J, Wendel A. A T-cell dependent experimental liver injury in mice inducible by concanavalin A. J. Clin. Invest. 1992; 90: 196–203.
14. Nolan JP. Endotoxin, reticuloendothelial function, and liver injury. Hepatology 1981; 1: 458–465.
15. Adachi Y, Bradford BU, Gao W, Bojes HK, Thurman RG. Inactivation of Kupffer cells prevents early alcohol-induced liver injury. Hepatology 1994; 20: 453–460.
16. Decker K. Biologically active products of stimulated liver macrophages (Kupffer cells). Eur. J. Biochem. 1990; 192: 245–261.
17. Adamson GM, Billings RE. Tumor necrosis factor-α induced oxidative stress in isolated mouse hepatocytes. Arch. Biochem. Biophys. 1992; 294: 223–229.
18. Adamson GM, Carlson TJ, Billings RE. Phospholipase A_2 activation in cultured mouse hepatocytes. J. Biochem. Toxicol. 1994; 9: 181–190.
19. Adamson GM, Billings RE. The role of xantine oxidase in oxidative damage caused by cytokines in cultured mouse hepatocytes. Life Sci. 1994; 55: 1701–1709.
20. Duval DL, Sieg DJ, Billings RE. Regulation of hepatic nitric oxide synthase by reactive oxygen intermediates and glutathione. Arch. Biochem. Biophys. 1995; 312: 699–706.
21. Adamson GM, Billings RE. Cytokine toxicity and induction of NO Synthase activity in cultured mouse hepatocytes. Toxicol. Appl. Pharmacol. 1993; 119: 100–107.

Regulatory Requirements for Drug Registration

Pharmacological Sciences: Perspectives for
Research and Therapy in the Late 1990s
ed. by A.C. Cuello and B. Collier
© 1995 Birkhäuser Verlag Basel/Switzerland

Harmonization of Drug Regulation and Trial Requirements: Clinical Pharmacological Aspects and Responsibilities

Eigill F. Hvidberg

Clinical Pharmacology, University Hospital, Copenhagen, Denmark

Summary. Regulatory requirements for drug registration and for clinical trials must be seen as an entity, as they represent two sides of the same problem. Furthermore, international harmonization of this area is highly needed. The most ambitious harmonization programmes are presently being performed within the context of International Conference on Harmonization (ICH), but also good initiatives have been taken, e.g. by WHO. A good example is Good Clinical Practice (GCP). In order to achieve the right balance in this development, contributions from clinical pharmacology are indispensable because of its central role between administrative regulators and the clinical/scientific elements. Harmonization of other related areas, e.g. measures against fraud in clinical trials, pharmaco-economics, and ethics committee systems should also be considered, as clinical pharmacological expertise can also make valuable contributions. The ultimate goals for harmonization – improvement of health care systems to the benefit for both patients and society – should not be forgotten.

Introduction

During recent years the scientific quality, the ethical performance and the data credibility of clinical drug trials have improved considerably, and coherent with that, the documentation filed with new drug applications has become much better. This is at least the experience of many regulatory agencies. However, differences between countries and regions are certainly present and they may even be great. This is mainly due to different speed of development in drug regulation and to the variability in legal systems, but also variations of cultural, economical and administrative conditions play a role. The fast growing internationalization of drug development and drug trade makes it increasingly evident that harmonization of requirements for registration of medicinal products and for drug trials are necessary and unavoidable. In order to solve the technical, legal and practical problems professional and scientific skill from various biological disciplines is needed to match the administrative and legal expertise. This is where the clinical pharmacologists come in to share the responsibility for an expedient development of the future integration and harmonization of regulatory matters, also because this

Correspondence to: Professor Eigill F. Hvidberg, Danish National Board of Health, Medicines Department, 378 Frederikkundsvej, DK-2700 Brønshøj, Denmark development will impact clinical pharmacology itself. Most importantly, however, is that clinical pharmacology is well suited, within the regulatory environment, to maintain the demand for improved drug therapy for the patient. This is fully in line with the concept of the WHO Working Group on Clinical Pharmacology in Europe [1, 2], and particularly advanced by Professor Folke Sjöqvist of Karolinska, Stockholm, as he points out that clinical pharmacology has a significant and growing role in health care delivery. Drug regulation is an important part of this function.

Principal Considerations on Drug Regulation

In order to fully appreciate the problems of harmonization, it is necessary to look at the coherence of the systems in question. Drug regulatory procedures for marketing authorization on one hand, and regulatory requirements for clinical trials that generate data to document such an authorization or registration on the other hand, form a logical continuum [3]. Consequently, the same quality standards must be applied. In the US, for example, FDA measures to control Investigational New Drugs (IND) and regulatory requirements for New Drug Applications (NDA) have many links [4]. Similarly, for drug regulation in the European Union the "Clinical Expert Report" required in common registration procedures must state whether studies included in the application comply with Good Clinical Practice (GCP) or, if not, why [5]. These examples from two of the major regions in the World clearly point to the strong link between new drug registration and clinical trial regulations. Clinical pharmacological expertise is a *sine qua non* for designing and conducting the individual trial so that it is scientifically sound and ethically immaculate, and that it meets all GCP requirements. Furthermore, it is highly necessary that clinical pharmacological expertise is integrated in the assessment of the entire dossier, both on the industry side and in the regulatory agency.

Harmonization in the European Region

The most interesting exercise in harmonization of drug regulatory requirements has been done in the European Community (now European Union). A condition for the establishment of the internal market for goods and services, including pharmaceuticals, was unified rules concerning drug registration. These were called the Future System, and went into operation January 1, 1995 [6]. Also, a common European GCP was required. Two-and-a-half years of drafting by a Committee on Proprietary Medicinal Products (CPMP) working group, resulted in the documented [7], extending previous work on the conduct of clinical trials. Several recommendations developed by national and industrial groups were taken into account. The basic professional, scientific, and ethical issues were agreed upon without major problems, although differences in medical culture and practice may affect both clinical trials and marketing authorization procedures even in the relatively homogeneous collection of EC member states.

Another example of harmonization initiatives was carried out between the five Nordic Countries (Denmark, Finland, Iceland, Norway, Sweden) under the auspices of the Nordic Council of Medicines (NLN). During the 1970s guidelines concerning drug registration were issued. Regarding clinical trials, a guideline was worked out in the mid 1980s and, furthermore, a GCTP (Good Clinical Trial Practice) recommendation [8] was published almost at the same time as the EC-guideline. The two documents (the Nordic and the European GCPs) are strikingly similar, not in appearance and structure, but in scope, principles and requirements, although the Nordic one may be more oriented towards the investigator. It has been used as a basis for others, e.g., the recent Australian guidelines.

Also the EFTA (European Free Trade Association) countries have been working for harmonizing GCP and other regulatory matters. Generally, they are accepting most of the EU requirements, and eventually this may also be the case for several of the central European countries, previously parts of the eastern block. Furthermore, several of the EFTA countries have joined the EC in 1995. The European scene, therefore, seems to face a more uniform future in relation to drug regulation, although a unified implementation of the rules may well cause problems. This perspective is a great challenge for European clinical pharmacology, which has different conditions in the many countries throughout Europe. Coordination of these conditions would highly facilitate the practical implementation of regulatory requirements in Europe. This seems, however, to be on its way. A recent report on clinical pharmacology in Europe states the demand for clinical pharmacological expertise in drug evaluation and therapeutic trials [1]. Appropriate training programmes are therefore necessary and already functioning.

International Initiatives on Harmonization

One of the most impressive international harmonization initiatives in recent years within the area of health care, is ICH. The full name behind this acronym is the International Conference on Harmonization of Technical Requirements of Registration of Pharmaceuticals for Human Use. The participants are the European Union, the United States of

America and Japan, the three major drug-developing and drug-consuming regions in the world. The initiative was started in 1989, and from the very outset the main objective of ICH was to expedite the global development and availability of new medicines without sacrificing safeguards on quality, safety or efficacy [9]. Each of the three regions is represented by their drug authority and by their pharmaceutical industry organisation, respectively. The International Federation of Pharmaceutical Manufacturers Association (IFPMA) provides the secretariat for ICH.

The organisation of ICH includes the following elements: First, the Conferences as such take place every second year, for a total of three. The first was held in Brussels, Belgium, 1991 [9], the second in Orlando, Florida, USA, 1993, and the final will take place in Yokahama, Japan, in 1995. Second, the Steering Committee, which meets two to four times a year, has 14 members, two from each region and two from IFPMA plus three observers from EFTA, WHO, and Canada, respectively. The Steering Committee prepares for the Conferences, oversees the harmonization initiatives and actually coordinates and leads the entire operation. Third, the Expert Working Groups, where the practical work is done. The EWGs are divided into Quality, Safety and Efficacy. Each of these fields has five to eight Expert Working Groups covering several specialised areas.

As the ICH process has advanced, it has become more evident that world-wide harmonization of drug regulation cannot be limited to technical and professional objectives. If viewed in the broader context, ICH should provide more than that. The concrete results of the ICH process will have the form of guidelines or similar documents harmonizing specialised and defined regulatory areas covering both registration and clinical trials. Finally, these documents will give rise to identical—or at least very similar – requirements and procedures throughout the three regions, a more economical use of resources, and have more effective and safe medicines made available faster for patients globally. However, in addition, ICH provides a forum for debate and exchange of ideas. Discussions are taking place at all levels between the involved parties, facilitating mutual understanding of professional and technical drug regulatory problems. This is probably the most important achievement of the entire ICH procedure, even if there are areas where substantial harmonization will not be encountered. It is obvious that pharmacologists and clinical pharmacologists have an important role to play in this context by securing that the outcomes of harmonization are scientifically based and that rational drug use is not lost as an aim. Furthermore, multilateral understanding of the background for various regulatory systems is necessary. Differences must be exposed, discussed and learned to be respected through ICH, a unique opportunity for transcultural interaction.

The World Health Organisation has a long tradition for issuing recommendations for the clinical evaluation of drugs in different areas. This activity does not interfere directly with national or international drug regulation, but indirectly the WHO documents may have great significance for regulatory matters. A recent example of WHO initiatives is that the headquarters in Geneva began to develop a GCP in 1991. The basis for the resulting document was the existing regulations and guidelines. The resulting WHO GCP document [10] has a more educational aspect for investigators and is also meant for countries that are not fully covered by national or supranational regulations of the kind known in industrially developed countries familiar with new drug development.

Seen in a historical perspective, both the WHO and the ICH-GCP initiatives are, each in their special way, natural continuations of the previous documents, and they will supplement each other on a global level, rather than compete with each other. They represent not only a harmonization of requirements, but also, indeed, an expansion and further development of the GCP concept together with more detailed explanations of problems that have become visible during the years.

Examples of Clinical Pharmacological Input

Two examples of how clinical pharmacology will impact the efforts to harmonize drug regulation shall be mentioned: Dose-Response and Ethics Committees.

Valid and comprehensive instructions for how to develop clinically useful dose-response curves have until recently been scarce. One of the working groups under ICH has issued a draft guideline on this topic [11]. The German clinical pharmacologist, Professor Ursula Gundert-Remy, previously with the BGA and now at the University of Göttingen in Germany, has been the driving force behind this work. The document is a clear manifestation of how important scientific insight and clinical pharmacological experience are necessary conditions for regulatory matters. The guideline points to some very important general aspects on dose-response problems such as the approach to demonstrating the safety and efficacy. Furthermore, the document gives instructions on study designs and provides several pieces of specific guidance.

The other example concerns the problems about the ethics committee systems evoked by the increasing harmonization of clinical trial requirements. All GCP documents as well as the FDA rules demand at least an independent ethical review of trial protocols and informed consent from the trial subjects [12, 13]. Consideration should be given as to whether or not harmonized requirements for the procedures, composition, competence, etc. of such committees are needed. The place of the local ethics committee in multicentre trials should also be defined more clearly [14]. The clinical pharmacologists, who have practical and scientific experience in this field, should be engaged in this process.

Other Areas to be Harmonized

During the process of harmonization some further important items have emerged, in which clinical pharmacology should also play a role.

The problem of how to prevent and handle cases of fraud and scientific misconduct. Fraudulent behaviour and other irregularities are not unknown in clinical drug trials, where falsified data are of special danger to public health and may also have great economical consequences. The problems are not limited to investigators, as also sponsors can be, and indeed have been, involved. However, one measure to solve at least some of the problems is to introduce official inspection of clinical trials, new in Europe but being applied by the FDA for decades. The experience of clinical pharmacologists would be indispensable in this function, not because they have been trained in inspections, but because they know the particulars of trials in a clinical setting. In the United Kingdom [15], Denmark [16], and a few other countries in Europe, special committees have been set up to handle such cases. It seems logical that these initiatives should be integrated with the audit and inspection functions at an international level.

Another possible target for harmonization where clinical pharmacology could be of help is pharmaco-economics in a broader sense. As the economic demands by health care systems continue to grow, interest in the consequences of new drug testing also increases. This seems to be a very complicated issue, requiring attention from skilled pharmaco-economists, clinical experts and probably clinical pharmacologists. Further, the higher cost-benefit implications of introducing new treatments, especially in the light of the regulatory requirements, are now being studied more closely. Maybe international harmonization of basic requirements would be appropriate.

Finally, it should be emphasized that drug regulation and its global harmonization is not an objective by itself. It is meant as guidance to improve health standards individually and collectively [3]. In the process of harmonizing regulatory areas clinical pharmacologists should be the guarantee that the ultimate goal is not forgotten. This goal implies that in the long term, both the patient and society (i.e. predominantly the future patients) must benefit from such actions. This aim may sometimes be lost from view during the process of inventing new and elaborate means to achieve it, possibly because so much bureaucracy is – mostly necessarily – involved.

References

1. Clinical pharmacology in Europe: an indispensable part of health service. WHO Working Group on Clinical Pharmacology in Europe. Eur. J. Clin. Pharmacol. 1988; 33: 535–9.
2. Collier J, Herxheimer A. The roles and responsibilities of Clinical Pharmacology. Br. J. Clin. Pharmacol. 1991; 31: 497–9.
3. Hvidberg EF. Regulatory implications of good clinical practice. Towards harmonization. Drugs 1993; 45: 171–6.
4. Kessler DA. The regulation of investigational drugs. New Engl. J. Med. 1989; 320: 281–8.
5. Allen ME. Good clinical practice in international pharmaceutical product registration: aspects of safety, quality and efficacy. In: Cartwright AC, Matthews BR, editors. London: Ellis Horwood, 1994.
6. Commission communication on the implementation of the new marketing authorization procedures for medicinal products for human and vetinary use in accordance with Council Regulation (EEC) No. 2309/93 of 22 July 1993 and Council Directives 93/39 EEC, 93/40 EEC and 93/41 EEC, adapted on 14 June 1993. Official Journal of the European Communities No. 82/4, 1994 March 19, vol. 37, Luxembourg.
7. Good clinical practice for trials on medicinal products in the European Community. CPMP Working Party on Efficacy of Medicinal Products. Pharmacol. Toxicol. 1990; 67: 361–72.
8. Good clinical trial practice. Nordic Guidelines. NLN Publication No. 28 Nordiska Läkemedelsnämnden (Nordic Council on Medicines), Uppsala, Sweden, 1989.
9. D'Arcy PF, Harron DWG, editors. Proceedings of The First International Conference on Harmonization, Brussels 1991, pp. xix–xxv. The Queen's University of Belfast, 1992.
10. Good clinical practice. Annex to the 6th Report of the WHO Expert Committee. WHO Technical Report Series, Geneva, 1994.
11. Dose-response information to support drug registration. Step 4 document (10 March 1994) in the ICH process.
12. Guidelines on the practice of ethics committees in medical research involving human subjects. A report of the Royal College of Physicians, second edition. The Royal College of Physicians of London, January, 1990.
13. International ethical guidelines for biomedical research involving human subjects. CIOMS. Geneva, 1993.
14. Hvidberg EF. Experiences with the ethical review process and the Danish approach. In: Bennett PN, editor. Good clinical practice and ethics in European drug research. Bath: Bath University Press, 1994.
15. Lock S, Wells F, editors. Fraud and misconduct in medical research. London: BMJ Publishing Group, 1993.
16. Andersen D, Attrup L, Axelsen N, Riis P. Scientific dishonesty and good scientific practice. The Danish Medical Research Council. Copenhagen, Denmark, 1992.

Pharmacological Methods

Pharmacological Sciences: Perspectives for
Research and Therapy in the Late 1990s
ed. by A.C. Cuello and B. Collier
© 1995 Birkhäuser Verlag Basel/Switzerland

In Vivo Magnetic Resonance in Pharmacological Research

Markus Rudin, Nicolau Beckmann, Anis Mir, and André Sauter

Preclinical Research, Sandoz Pharma Ltd, CH-4002 Basel, Switzerland

Summary. Similar to its role in clinical diagnostics, *in vivo* magnetic resonance imaging (MRI) is becoming an important tool in pharmacological research. MRI allows noninvasive morphometric measurements, typical examples being the assessment of infarct volumes in animal models of embolic stroke, of tumor volumes, or of the myocardial mass in models of cardiac hypertrophy. Using special hardware, measurements of microscopic resolution ($< 50\,\mu$m) have become feasible. Beyond these static measurements, MRI also provides information on functional parameters of the tissue such as hemodynamics (stroke volume and ejection fraction of the heart, blood flow velocities, tissue perfusion), tissue oxygenation, and, most recently, the visualization of activated brain regions. The tissue metabolism may be studied using magnetic resonance spectroscopy (MRS). Similar to clinical MRS, the applicability of the method for drug testing and development is limited, mainly because MRS studies are time-consuming due to the inherently low sensitivity of nuclear magnetic resonance. This addresses the important issue of cost effectiveness. *In vivo* morphometric measurements may be standardized and carried out in a competitive manner compared to alternative techniques. Together with the other obvious advantages of a noninvasive technology and the possibility to measure functional parameters, *in vivo* magnetic resonance will become increasingly important in pharmacological research.

In Vivo Magnetic Resonance Imaging and Spectroscopy

Biological applications of nuclear magnetic resonance fall into two categories: Magnetic resonance imaging (MRI) and spectroscopy (MRS). *In vivo* MRI provides images representing a weighted distribution of protons (usually from water and lipids) in tissues, the weighting depending on the microphysical environment of the nuclei. Contrast in MRI is governed by several parameters such as the total proton density, the various relaxation times (T_1, T_2, T_2^*, $T_{1\rho}$), the chemical composition of a tissue (water and lipid protons), and macroscopic (vascular flow) and microscopic motion (diffusion and perfusion). In order to optimize contrast for a specific application, individual parameters may be emphasized by the experimenter by applying an appropriate data-acquisition procedure.

In vivo MRS provides biochemical information on metabolite levels in a defined volume-of-interest within the body. Volume selection is usually performed using surface coils and/or image-guided techniques. Only metabolites containing certain nuclei are measurable by MRS, the most

Correspondence to: Dr. Markus Rudin at the above address.

important being ^1H, ^{31}P, and ^{13}C. Due to sensitivity constraints, only metabolite pools in the millimolar concentration range may be tapped by *in vivo* MRS under the experimental conditions commonly used.

MRI Morphometry: Application to Disease Models

The classical application of MRI is the visualization of the normal and pathological morphology. Being basically a three-dimensional technique, MRI allows volume assessment in a straightforward manner. This is of primary interest for pharmacological MRI applications, where the effects of a pharmacological intervention have to be expressed in a quantitative way.

There is an increasing number of MRI applications in drug development. Typical examples are animal models of focal cerebral ischemia [1, 2], excitoxicity [3, 4], hypertrophic cardiomyopathy [5, 6], hyperplasias

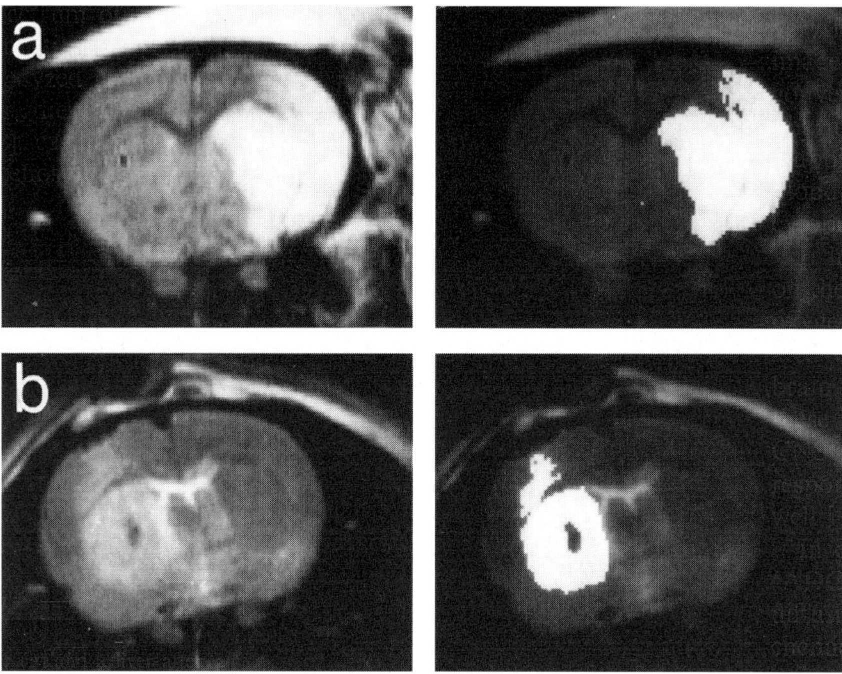

Fig. 1. MRI Morphometry: T_2 weighted transverse section through the brain of a rat 24 h after permanent unilateral occlusion of the proximal MCA (a) and 24 h after intrastriatal injection of 200 nmoles of the excitotoxin quinolinic acid (b). The representative slice is one out of eight recorded as a multislice acquisition (slice thickness 1 mm, slice distance center-center 1.5 mm). The ischemic region shows increased signal intensity (bright) due to prolonged T_2 relaxation time as a consequence of edema formation. Figures on the right side show the results of the automated segmentation procedure based on intensity thresholding overlayed to the original images.

[7] and neoplasms [8]. In these models MRI is used to identify morphological changes associated with the disease/disorder and to determine the lesion volume. For instance, in the rat middle cerebral artery occlusion (MCAO) model of focal cerebral ischemia, the accentuation of T_2 yields images with a well-defined hyperintense area due to massive edema formation as illustrated in Fig. 1a. An analogous image of an excitotoxic lesion induced by intrastriatal injection of quinolinic acid is shown in Fig. 1b. Using multislice or three-dimensional MRI procedures the lesion volume can be accurately determined.

An important issue is the validation of MRI findings using conventional techniques. It has been shown that the hyperintense areas in MR images recorded 24 h after MCAO or quinolinic acid injection match the regions of cellular necrosis, as determined by histology [2]. Analogous correlations have been reported for tumor size assessment [7, 8], and for determinations of cardiac mass [5].

Based on these validations, MRI has been applied to study the efficacy of pharmacological interventions in a variety of animal models. For example, the neuroprotective effects of a wide spectrum of drugs, such as calcium antagonists, competitive and non-competitive NMDA antagonists, and free radical scavengers, have been analyzed in the rat using quantitative MRI. Other examples of drug studies with rats involving MRI morphometry are tumor therapy [7, 8], cardiac hypertrophy [6], neurotoxic lesions [3, 4], and vascular restenosis in carotid arteries [9].

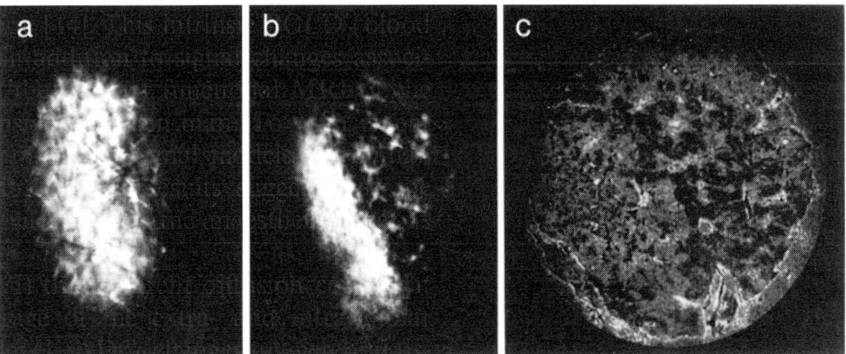

Fig. 2. MR Microscopy (MRM): *In vivo* MRM of a control rat liver (a) and a liver after administration of bromobenzene at 10% of the LD50 dose (b). The images have been recorded using a surgically implanted NMR coil, which was glued to the liver with a biocompatible adhesive. The images represent one slice out of a three-dimensional data set, the voxel resolution being $50 \times 50 \times 220 \ \mu m^3$. In (b) a dark region associated with hepatic cell damage induced by bromobenzene may be identified. A high-resolution MRM image of a liver specimen of the same animal is shown in (c), with voxel dimension of $39 \times 39 \times 39 \ \mu m^3$. The comparison with optical photomicrographs revealed that the dark regions in (b) and (c) correspond to hepatocellular necrosis and degeneration. (Adapted from [10], reproduced with permission).

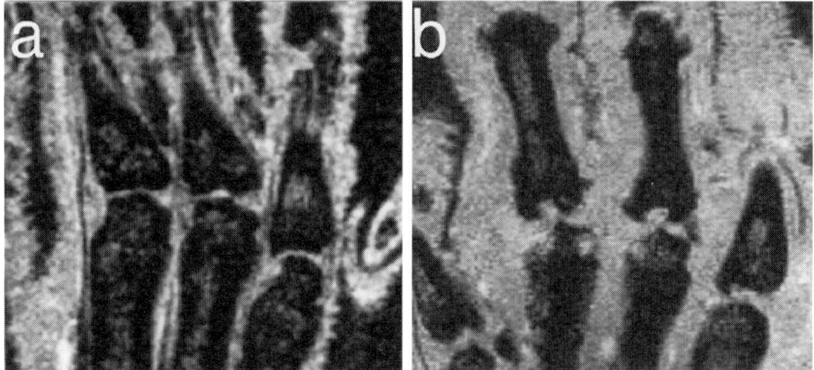

Fig. 3. MR Microscopy (MRM): High-resolution images of the rat hind paw before (a) and 60 days after immunization with bovine type II collagen (b). The images represent one slice of a three-dimensional data set, the voxel dimensions were $60 \times 80 \times 80 \, \mu m^3$, the data acquisition lasted 45 min using a specially designed resonator probe. Note the massive bone erosion in the interphalangeal joint area due to the arthritic process (b).

Potential of MR Microscopy

Developments in MRI hardware have allowed to improve spatial resolution to the "microscopic" level, typically of the order of 50 to 100 μm for *in vivo* studies and 10 to 50 μm for *in vitro* investigations. This is not very impressive when compared to other microscopic methods. However, MRI offers some distinct advantages. The method is non-destructive, inherently three-dimensional, and applicable to opaque samples. Due to the variety of MRI contrast mechanisms, different "stainings" of the sample are feasible. A limiting factor of MR microscopy is sensitivity. As a consequence, data acquisition for an image with 25 μm isotropic resolution currently requires several hours.

Figure 2 shows the use of MRI to study bromobenzene-induced hepatotoxicity [10]. The *in vivo* experiments have been carried out using a surgically implanted coil; the size of the volume elements (voxels) is $50 \times 50 \times 220 \, \mu m^3$. The corresponding *in vitro* image of a specimen of the same liver is shown in Fig. 2c. The voxel dimensions in this case are $39 \times 39 \times 39 \, \mu m^3$. A well-defined pathologic region can be identified, which corresponds to hepatocellular necrosis and degeneration on hematoxylin- and eosin-stained sections. Although *in vitro* MR microscopy has to compete with well established techniques, there seems to be a great potential for pharmacological and histochemical applications.

In vivo MR microscopy has also been applied to study bone erosion in a rat model of arthritis. Three-dimensional MRI of the interphalangeal joints was carried out with voxel dimensions of $80 \times 80 \times 60 \, \mu m^3$. Figure 3 shows images from three-dimensional data sets acquired from a rat before and 60 days after immunization with bovine type II collagen. Massive erosion of bone structures at the joint is clearly visible.

A basic problem of *in vivo* microscopy is the involuntary motion of the animal. While rhythmic movements, such as heart beat and respiration, can be controlled by gating techniques and/or by mechanical ventilation, it is virtually impossible to cope with irregular motions. Because of long measurement times and small voxel dimensions, artefacts from irregular motions become a relevant problem in MR microscopy. In this respect, we believe that a resolution of 50 to 100 μm will be close to the limit for *in vivo* microscopy studies.

Physiological Parameters Assessed with MRI: Functional MRI

Functional MRI refers to measurements of physiological parameters such as hemodynamic properties, water diffusion, or changes in blood and tissue oxygenation.

The best established examples are the measurement of hemodynamic parameters. ECG-triggered MRI allows direct assessment of cardiac chamber volume versus time profiles, the determination of end-diastolic and end-systolic ventricular volumes, and the calculation of stroke volumes and ejection fractions both in animals and humans. This can be applied in a straightforward manner to characterize functional impairments associated with a variety of heart diseases such as cardiac insufficiency, and to assess the effects of therapy.

Exploiting the inherent contrast between the MRI signals of water protons from stationary tissue and moving blood, information on blood flow velocities may be obtained. Applications are similar to those of

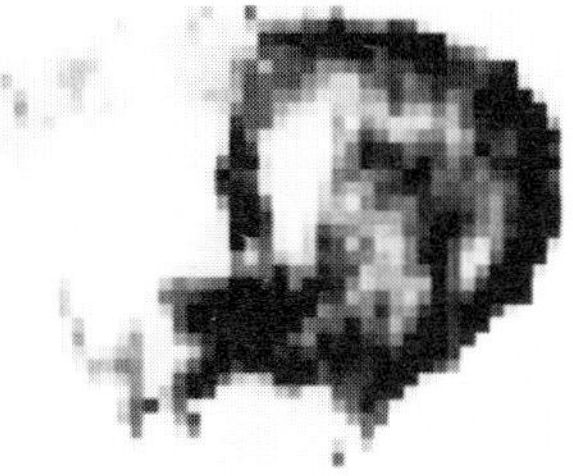

Fig. 4. Functional MRI: The image depicts the local concentration of the intravascular superparamagnetic contrast agent $(Fe_3O_4)_x$ nanoparticles coated with bovine serum albumine in the brain of a rat that underwent unilateral occlusion of the MCA. The relative concentration was estimated from the relative signal attenuation induced by the tracer. The image shows the highest tracer concentration (dark signal) in the intact cortex and little tracer present in the ischemic region (left side). Data acquisition time lasted 500 ms for an individual image, 64 images were recorded during the passage of the contrast agent.

Doppler ultrasound measurements, the advantage of MRI being that a vessel may be assessed irrespective of its location. Applications of MRI blood flow measurements are predominantly clinical; they may, however, also be utilized to estimate vascular flow velocity in small laboratory animals, such as the rat [11].

Several MRI methods to measure tissue perfusion have been described [12]. The most frequently used technique is an indicator dilution experiment using intravascular para- or superparamagnetic contrast agents. As the tracer passes through the tissue of interest, a signal attenuation is observed, the extent of which can be related to the regional tissue blood volume. Using time-resolved MRI with a temporal resolution of 1s/image or better, it is possible to measure the "mean resident time" (not the mean transit time) of the tracer in the tissue, which then allows to estimate the relative regional tissue perfusion rate. The method is applied both clinically and in animal research. Figure 4 shows a map reflecting the distribution of an intravascular tracer in the brain of a rat after unilateral MCAO. The hypoperfused area, showing very littly tracer uptake, can be clearly identified on the left side. The images have been recorded 15 min after MCAO; at this early time point, conventional T_2-weighted images do not show any abnormality. As a pharmacological application, it has been shown with MRI perfusion mapping that the calcium antagonist isradipine improves collateral blood supply to the ischemic region, a putative mechanism for its cytoprotective efficacy [13].

It has been reported recently that the tissue signal intensity depends on the oxygenation state of the blood [14]. This intrinsic BOLD (blood oxygen level dependent) contrast, in addition to signal changes associated with altered perfusion, forms the basis of functional MRI of the brain [15], which is currently extensively used in man. To what extent these results will affect the design of novel animal models for human CNS disorders remains unclear. Preliminary results suggest that the response to sensory stimuli may be visualized in the anaesthetized rat as well [16].

In diffusion-weighted MRI (DWI) the apparent diffusion coefficient (ADC), which is a weighted average of the extra- and intracellular diffusion coefficients, is measured [4]. ADC changes in cerebral ischemia occur within minutes after the insult. The most reasonable explanation for the reduced ADC is a change in extracellular versus intracellular cell space due to cellular swelling. This is also corroborated by measurements of electric conductance [4]. An obvious application of DWI is the very early (within minutes) estimation of the final volume of an infarct or excitotoxic lesion, allowing to analyze the efficacy of a pharmacological intervention in the same individual [4] (Fig. 5).

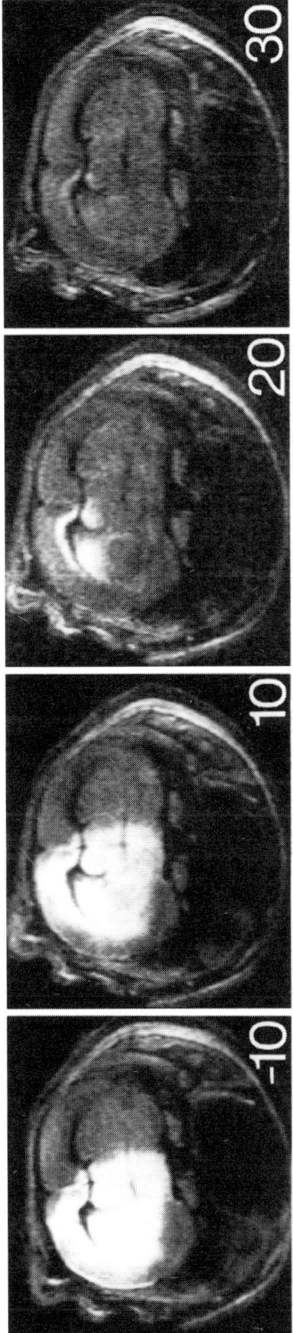

Fig. 5. Diffusion-weighted images of a 10-day-old rat pup after intrastriatal injection of 20 nmoles N-methyl-D-aspartate (NMDA). Sixty minutes after the NMDA injection the noncompetitive NMDA antagonist MK 801 (dose 1 mg/kg) was administered intraperitoneally. The labels in the images indicate the time in minutes with respect to the antagonist administration. The hyperintense area reflects the reduced apparent diffusion coefficient due to cellular swelling. The lesion was significantly reduced 30 min after the administration of MK 801.

Potential of MR Spectroscopy

MRS methods are, in general, time-consuming and therefore less suited for drug-screening programs. Nevertheless, MRS is used to study metabolic changes induced by a disease process and/or a pharmacological intervention, offering a window to metabolism.

During the first years, *in vivo* MRS mainly focused on the measurement of metabolites involved in energy metabolism such as phosphocreatine, ATP, and inorganic phosphate, as well as on the assessment of intracellular pH. Phosphorus MRS as a clinical tool, contributed significantly to the diagnosis of muscular diseases [17], while in other disorders such as stroke, cardiomyopathy, and tumors, MRS essentially did not provide new insights into the disease process.

In experimental animals changes in energy metabolism have been monitored using MRS in a number of pathological situations, and the effects of pharmacological interventions have been described. As an example, calcium antagonists were shown to reduce the ATP consumption in the rat brain [18], which may be of importance in conditions of impaired energy synthesis, such as hypoxia or ischemia.

With methodological improvements, such as efficient water suppression and volume localization procedures, the study of a number of metabolites using hydrogen MRS became feasible. Currently hydrogen MRS is explored by many research groups with the aim to identify metabolite patterns characteristic for certain diseases. Alternatively, the technique may be applied to study mechanistic aspects. It has been shown e.g. that photic stimulation causes local elevation of cerebral lactate levels in the primary visual cortex [19].

An elegant method to study specific metabolic pathways is the use of ^{13}C spectroscopy in combination with specifically ^{13}C-labeled substrates. Due to the high specificity of the technique and also to the fact that there is little background signal from endogenous compounds, the fate of the label can be studied in great detail.

MRI/MRS Routine Drug Testing

In order to be applicable to routine drug-testing programs, *in vivo* NMR has to be cost-efficient and competitive with alternative methods. Throughput is an essential cost-factor, which has to be increased by optimization of the NMR procedure, and by using standardized hardware (animal supports, radiofrequency probes), measurement protocols, and data analysis procedures. The ultimate limiting factor is the time an animal is in the magnet, which ranges from typically 10 to 15 min when assessing the lesion size in stroke and excitotoxic models, to 30 to 40 min for dynamic MRI of the heart, when recording both end-diastolic and end-systolic images. When fully automatized analysis is used (see Fig. 1), quantitative results are obtained on-line. Considering the methods for infarct volume determination, MRI seems competitive to classical histological methods, which involve many work-intensive steps such as tissue sectioning, staining, and morphometric analysis.

Besides possible cost-efficiency, *in vivo* NMR is a noninvasive technique allowing experimental animals to be monitored over prolonged time periods and/or utilized for additional tests. Thus the amount of information which may be obtained from a single animal can be significantly increased at no or very little additional stress for the animal. In addition, the use of an individual animal as its own control enhances the statistical power of an experiment.

Similar throughputs as in morphometric MRI may be achieved in functional MRI experiments. As an example, the data acquisition in MRI perfusion measurements using the bolus tracking techniques is of the order of 1 to 5 min. In this interval typically 50 to 100 images are recorded sequentially, from which functional MR images (parameter maps) are generated. In order to analyze these large data sets efficiently, automated data-processing procedures have to be applied.

The principal limitation in MRS is the low sensitivity which requires long measurement times. This disadvantage is even more serious in studies involving small laboratory animals due to small sample volumes. Moreover, since endogenous metabolism is frequently well regulated, the adaptive capacity of the metabolic processes following some physiological challenge is commonly measured, which is time consuming. Therefore, a typical spectroscopic session lasts several hours per animal. As a consequence, MRS is rarely used in routine drug testing, but rather to provide unique, supporting data, for instance, on putative mechanisms of action.

Outlook: MRI/MRS in Pharmacological Research

In vivo morphometry using MRI may be considered meanwhile an established tool in pharmaceutical research. However, pharmaceutical applications of MRI have always to compete with conventional techniques, such as histopathological analyses and other invasive methods. MRI may not always be the most cost-efficient way to solve a problem, but its noninvasiveness is becoming increasingly more important in experimental research due to the requirement to reduce the number of animals.

Methodologically, the development will focus primarily on functional MRI. The combination of the excellent anatomic definition of conventional MRI with the mapping of physiological parameters will definitely become a major application in pharmaceutical research. Similarly, new spectroscopic tools, such as spectroscopic imaging (low-resolution maps of metabolite distribution), are currently being developed. These techniques might, to some extent, close the gap between MRI and MRS and may well increase the potential of *in vivo* MRS for pharmaceutical studies.

A very important factor for a more widespread use of MRI/MRS in pharmacological research is the fact that the method is applicable both to humans and experimental animals. Thus, it is possible to monitor drug efficacy both in preclinical and clinical research using the same tools. MRI/MRS studies in animal experiments may lead to optimized clinical testing protocols. This link will definitely become much more important in the future.

References

1. Sauter A, Rudin M. Calcium antagonists reduce the extent of infarction in rat middle cerebral artery occlusion model as determined by quantitative magnetic resonance imaging. Stroke 1986; 17: 1228–1234.
2. Sauter A, Rudin M, Wiederhold KH. Reduction of neural damage in irreversible cerebral ischemia by calcium antagonists. Neurochem. Pathol. 1988; 9: 211–236.
3. Sauer D, Thedinga KH, Fagg GF, Massieu GE, Amacker H, Allegrini PR. Evaluation of quinolinic acid induced neurodegeneration in rat striatum by quantitative magnetic resonance imaging *in vivo*. J. Neurosci. Meth. 1992; 42: 69–74.
4. Verheul HB, Balazs R, Berkelbach van der Sprenkel JW, Tulleken CAF, Nicolay K, Tamminga KS et al. Comparison of diffusion-weighted MRI with changes in cell volume in a rat model of brain injury. NMR Biomed. 1994; 7: 96–100.
5. Manning WJ, Wei JY, Fossel ET, Burstein D. Measurement of left ventricular mass in rats using electrocardiogram-gated magnetic resonance imaging. Am. J. Physiol. 1990; 258: H1181–H1186.
6. Umemura K, Zierhut W, Rudin M, Novosel D, Robertson E, Pedersen B et al. Effect of spirapril on left ventricular hypertrophy due to volume overload in rats. J. Cardiovasc. Pharmacol. 1992; 19: 375–381.
7. Rudin M, Briner U, Doepfner W. Quantitative magnetic resonance imaging of estradiol-induced pituitary hyperplasia in rats. Magn. Reson. Med. 1988; 7: 285–291.
8. Siegel R, Tolcsvai L, Rudin M. Partial inhibition of growth of transplanted Dunning rat prostate tumors with the long-acting somatostatin analogue sandostatin (SMS 201–995). Cancer Res. 1988; 48: 4651–4655.
9. Cook NS, Zerwes HG, Pally C, Rudin M, Hof RP. Spirapril and cilazapril inhibit neointima lesion development but cause no detectable lumen narrowing after carotid artery balloon catheter injury in the rat. Blood Press 1993; 2: 322–331.
10. Zhou X, Maronpot RR, Cofer CP, Hedlund LW, Johnson GA. Studies of bromobenzene induced hepatotoxicity using *in vivo* MR microscopy with surgically implanted RF coils. Magn. Reson. Med. 1994; 31: 619–627.
11. Rudin M, Zierhut W, Sauter A, Cook NS. New developments in cardiovascular magnetic resonance imaging and spectroscopy. Trends Pharm. Sci. 1991; 12: 416–421.
12. Villringer A, Rosen BR, Belliveau JW, Ackerman JL, Lauffer RB, Buxton RB et al. Dynamic imaging with lanthanide chelates in normal brain: contrast due to magnetic susceptibility effects. Magn. Reson. Med. 1988; 6: 164–174.
13. Sauter A, Rudin M. Experimental studies with isradipine in stroke. Drugs 1990; 40: 44–51.
14. Ogawa S, Lee TM, Kay AR, Tank DW. Brain magnetic resonance imaging with contrast dependent on blood oxygenation. Proc. Natl. Acad. Sci. USA 1990; 87: 9868–9872.

15. Ogawa S, Tank DW, Menon R, Ellerman JM, Kim SG, Merkle M et al. Intrinsic signal changes accompanying sensory stimulation: functional brain mapping with magnetic resonance imaging. Proc. Natl. Acad. Sci. USA 1992; 89: 5951–5955.
16. Hyder F, Kaiser MG, Behar KL, Blamire AM, Chase JR, Martin MA et al. Signal time-courses of echo-planar imaging and laser doppler flowmetry during neuronal activity in rat brain. Proc. Soc. Magn. Reson. 2nd Annual Meeting (San Francisco) 1994; p. 1388.
17. Ross BD, Radda GK. Application of ^{31}P NMR to inborn errors of muscle metabolism. Biochem. Soc. Trans. 1983; 11: 627–630.
18. Rudin M, Sauter A. Dihydropyridine calcium antagonists reduce the consumption of high-energy phosphates in the rat brain. A study using combined ^{31}P/^{1}H magnetic resonance spectroscopy and ^{31}P saturation transfer. J. Pharm. Exp. Ther. 1989; 251: 700–706.
19. Prichard J, Rothman D, Novotny E, Petroff O, Kuwabara T, Avison M et al. Lactate rise detected by ^{1}H NMR in human visual cortex during physiologic stimulation. Proc. Natl. Acad. Sci. USA 1991; 88: 5829–5831.

Pharmacological Sciences: Perspectives for
Research and Therapy in the Late 1990s
ed. by A.C. Cuello and B. Collier

Application of Cell Culture Systems to the Study of Drug Transport and Metabolism

Ronald T. Borchardt[1], Harold E. Lane[2], Barry H. Hirst[3],
Philip L. Smith[4], Kenneth L. Audus[1], and Akira Tsuji[5]

[1]*Department of Pharmaceutical Chemistry, The University of Kansas, Lawrence, KS, USA;*
[2]*Corning Costar Corporation, Cambridge, MA, USA;* [3]*Department of Physiological Sciences,
University of Newcastle, Newcastle upon Tyne, England;* [4]*Drug Delivery, SmithKline Beecham
Pharmaceuticals, King of Prussia, PA, USA;* [5]*Faculty of Pharmaceutical Sciences, Kanazawa
University, Kanazawa, Japan*

Summary. In recent years, pharmaceutical scientists have begun to develop and validate cell
culture systems that mimic the biological barriers that prevent a drug from reaching its
pharmacological target. This article reviews the cell culture systems that have been developed
to mimic the intestinal, nasal and pulmonary epithelium and the brain vascular endothelium.
Examples of how these cell culture systems are used to elucidate pathways of drug transport
and metabolism and how they are used to evaluate strategies to enhance drug permeability
and minimize drug metabolism are given.

A major challenge which will confront pharmaceutical scientists in
the future is the design of drug candidates having permeability characteristics adequate for their development as clinical agents [1]. Through
rational drug design, medicinal chemists are capable of synthesizing
very potent and very specific drug candidates. These drug candidates are
developed with molecular characteristics that permit optimal interaction
with the specific macromolecules (e.g., receptor, enzyme) which mediate
their therapeutic effect. However, rational drug design does not necessarily mean rational drug delivery that strives to incorporate into drug
candidates the characteristics needed for optimal transfer between the
point of administration and the pharmacological target in the body.

For a drug candidate to reach its molecular target *in vivo*, the
molecule must circumvent various biological barriers, including epithelial (e.g., intestinal, nasal, pulmonary) and endothelial barriers (e.g.,
endothelial cells that constitute the blood-brain barrier [BBB]) [2]. In an
effort to rapidly evaluate the permeability properties of drug candidates
and to evaluate novel strategies for delivery of drug candidates, some
pharmaceutical scientists have begun to employ the technique of cell
culture to study drug transport and metabolism in specific biological
barriers [3].

In vitro cell cultures designed to mimic a biological barrier have many
advantages over conventional techniques, including (a) rapid assess-

Correspondence to: Dr. Ronald T. Borchardt, Department of Pharmaceutical Chemistry, 3006
Malott Hall, The University of Kansas, Lawrence, KS 66045, USA.

ment of the potential permeability and metabolism of a drug; (b) the opportunity to elucidate the molecular mechanism(s) of drug transport and/or the pathway(s) of drug transport and/or the pathway(s) of drug degradation (or activation); (c) rapid evaluation of strategies for achieving drug targeting, enhancing drug transport and minimizing drug metabolism; (d) the opportunity to use human rather than animal tissues; and (e) the opportunity to minimize time-consuming, expensive and sometimes controversial animal studies.

The objectives of this article are to describe some of the general factors that should be considered in developing a cell culture model and to review the recent progress that has been made in establishing, validating and using cell culture models of epithelial (e.g., intestinal, nasal and pulmonary) and endothelial barriers (e.g., BBB).

General Factors to Consider in Developing a Cell Culture Model

In order to sucessfully mimic a biological barrier with an *in vitro* cell culture system, the selection of the cell line becomes particularly important. The transport and metabolic properties of cultured cells can vary depending on (a) whether the cells are primary cultures, passaged lines or transformed lines; (b) the number of times the cells have been passaged; (c) the phenotypic stability of the cell line; (d) the heterogeneity of the cell line; and (d) the inherent ability of the cell line to undergo differentiation [4]. Once the cell line has been selected, the properties may vary depending on (a) the cell seeding density; (b) whether the cells have reached confluency; (c) the stage of cellular differentiation, and (d) the presence or absence of essential nutrients, growth factors or associated cells that produce trophic factors [4]. During transport experiments, the properties may change depending on (a) the composition of the transport media (e.g., concentration of the solute, temperature, pH, presence or absence of a metabolic source of energy or ions, presence or absence of proteins that might bind the solute, presence or absence of competing solutes); and (b) whether the solute is added to the apical or basolateral side of the monolayer [4]. All of these factors need to be carefully optimized and regulated so as to best mimic the biological barrier *in vivo*.

The development of a cell culture system that will mimic a specific biological barrier requires not only an appropriate cell line and appropriate cell culturing conditions, but also a microporous membrane, which by itself or after treatment with an appropriate matrix material (e.g., collagen) will support cell attachment and cell growth [4]. Ideally, the microporous membrane should also be (a) sufficiently translucent that the development of the cell monolayer can be verified by microscopic techniques; (b) readily permeable to hydrophilic and hydrophobic solutes, and (c) readily permeable to both low and high molecular weight solutes [4].

Finally, a particularly critical factor in the study of the transport of lipophilic molecules is the selection of the diffusion apparatus [4]. Whether the diffusion apparatus is stagnant or stirred can influence the thickness of the aqueous boundary layer on the surface of the cell monolayer and, thus, the permeability of lipophilic solutes. At least two types of diffusion apparatus are currently employed for studying transport across cultured cell monolayers. These include the unstirred cell-insert system and the side-by-side diffusion system stirred by gas lift [5]. The gas lift system was developed specifically to reduce the thickness of the aqueous boundary layer over the cell monolayer, which could be the rate-limiting barrier to the flux of lipophilic solutes.

Once the cell culture system has been developed, it must be fully characterized using cell biology techniques (e.g., morphology by EM and SEM), histochemical techniques (e.g., localization of marker enzymes by EM), biochemical techniques (e.g., demonstration of the presence and polarity of transporter systems and enzymes) and pharmaceutical techniques (e.g., demonstration of the integrity of the cell monolayer using impermeable solutes) [4]. Finally, great care must be taken in analyzing transport and metabolism data derived from these cell culture systems in order to properly correlate *in vitro* data with *in vivo* data [6].

Models of the Intestinal Epithelium

The gastrointestinal epithelium forms a barrier to the absorption of many substances, including drugs [1, 2]. However, this epithelial barrier is selective. An understanding of the nature of the barrier and the specific nutrient absorptive transport systems which may have potential for carrying drugs and secretory transport systems that may limit drug absorption is likely to lead to more rational strategies for enhancing the absorption of drugs. The availability of an *in vitro* system capable of modeling key features of the gastrintestinal epithelial barrier is a prerequisite for such mechanistic studies. Human Caco-2 intestinal epithelial cells form confluent tight epithelial layers when grown in permeable tissue culture inserts [7]. These cells demonstrate enterocytic-type differentiation, with typical brush-border hydrolase expression on the prominent apical microvilli [7]. Adjacent cells are connected by tight junctional complexes, limiting passive paracellular diffusion, with typical transepithelial electrical resistances of around $200\text{–}400\ \Omega \cdot cm^2$ [7]. On permeable tissue culture inserts, Caco-2 cells differentiate with time, expressing numerous enteroctye-specific functional absorptive transport systems, including those for glucose, fructose, amino acids, dipeptides, bile salts and cobalamin [4].

Caco-2 cells express the proton-coupled transport system for di- and tripeptides, such as Gly-Sar. Gly-Sar absorption is pH dependent, enhanced by an acidic pH at the apical surface of Caco-2 cells [8, 9], mimicking the acid microclimate at the surface of enterocytes *in vivo*. Protons are carried into the cells during Gly-Sar absorption, and this may be monitored as an intracellular acidification in Caco-2 cells preloaded with the pH-sensitive fluorochrome BCECF [8, 10]. Transport systems for dipeptides are expressed at both the apical and basolateral surfaces of Caco-2 cells, mediating both the entry and exit from the cells [8]. The apical and basolateral transport systems display distinctive characteristics, consistent with more than one exit pathway at the basolateral surface [9]. Orally available aminocephalosporins, including cephalexin and cefadroxil, but not the non-orally available cefazolin, are substrates for the peptide carrier system, as are some angiotensin-converting enzyme inhibitors, including enalapril and captopril, but not lisinopril [9, 11]. Peptides and peptide-like drugs such as thyrotropin releasing hormone and lisinopril, which are not substrates for the dipeptide carrier, may still achieve significant absorption through the passive paracellular route [12]. Determination of changes of intracellular pH in Caco-2 cells exposed to potential substrates for the dipeptide carrier and other transport systems linked to proton movements is a potentially powerful, yet simple, novel method for identifying new substrates [10]. Starting from the observation that certain amino acids cause an intracellular acidfication in Caco-2 cell, Thwaites et al. [13, 14] have identified a novel rheogenic proton-coupled amino acid cotransport system at the apical surface of Caco-2 cells. This transport system recognizes β-Ala, L-Ala, Gly and L-Pro, but not L-Thr nor L-Ser. The proton-coupled amino acid transporter may mediate the absorption of the antibiotic cyclo-D-Ser.

Caco-2 cells express the multidrug-resistance (mdr) gene product P-glycoprotein at the apical surface leading to a polarized secretion of mdr susbtrates such as vinblastine [15]. This apical expression of P-glycoprotein limits the absorption of vinblastine [15]. Inhibitors of P-glycoprotein function, such as verapamil, nifedipine and dideoxyforskolin, enhance vinblastine absorption, demonstrating a potential mechanism for enhancing absorption of some drugs [15]. Fluoroquinolone antibiotics, including ciprofloxacin and norflaxacin, are also secreted by Caco-2 cells by a P-glycoprotein-independent mechanism involving accumulative transport across the basolateral membrane followed by facilitative exit across the apical membrane [16].

Thus, human Caco-2 cells, grown as confluent polarized epithelial cell layers on permeable tissue culture inserts, provide a powerful model for investigating mechanistic features of intestinal drug absorption. Although this cell line cannot model all the features of the gastrointestinal epithelial barrier, it has already proved to be an invaluable tool in determining mechanisms of absorption and secretion of drugs and in highlighting novel transport systems.

Models of the Pulmonary Epithelium

Recently, interest has developed in systemic delivery of drugs via the airway [17]. This approach is being considered for delivery of peptides and proteins and for gene therapy. The airway epithelium provides a large surface area for absorption and also avoids first pass hepatic effects. There are a variety of methods available for evaluating pulmonary absorption of drugs, including *in vivo* techniques, isolated perfused lung, isolated tissues and cultured epithelial cells [18–21]. *In vivo* techniques commonly involve intratracheal administration of solutions or aerosols which, in combination with gamma scintigraphy, can provide information about the site of deposition, structure/absorption and bioavailability. Studies with peptides such as SK&F 110679 and SK&F 106760 indicate that absorption following pulmonary adminstration provides a route for systemic delivery of these peptides which are poorly absorbed via the oral route [20, 21]. The isolated perfused lung technique provides a method for evaluating pulmonary absorption and metabolism following either intratracheal administration of aerosols or solutions or intravascular administration of drugs [18]. Studies with this technique have provided information regarding structure/absorption and effects of formulation components [22]. *In vitro* techniques with isolated tissues and cultured cells can be employed to evaluate effects of physiological mediators on epithelial permeability and absorption of drugs from different sites of the respiratory tract (e.g., nasal, tracheal or alveolar) [23–26]. In addition, effects of formulation components on transport and morphology, structure/absorption relationships and mechanisms of transport can be determined from these studies. Conditions for growing confluent monolayers of rat alveolar cells have been developed and employed for studying mechanisms involved in transepithelial transport of drugs and effects of physiological mediators on epithelial permeability [19].

Models of the Nasal Epithelium

The air-tissue interface comprising the upper respiratory area, extending from the nasal turbinates down to the bronchiolar trees, consists of a thin lining of morphologically and functionally similar pseudostratified epithelial cells [17]. Aerosol application of drugs into this upper respiratory area has the potential for effective systems and local delivery of a number of therapeutic agents. In the absence of fundamental information on the upper airway epithelial cell uptake, metabolism and transepithelial transport processes, a number of *in vitro* models are being developed to facilitate an understanding of the pertinent mechanistic details of drug delivery strategies at the cellular level. *In vitro* models used to study the upper airway epithelial barrier include excised epithelial tissue, isolated epithelial cell membranes, and cell culture systems derived from nasal, tracheal and bronchial regions [17]. These systems have generally provided a rapid and convenient means to generate information that is complementary to *in vivo* studies, while also providing insights into cellular mechanisms that cannot be readily examined in an animal model [27].

Excised tissues, ovine nasal and bovine tracheal epithelium [27] have been developed as *in vitro* systems to characterize the efficacy and irritancy potential of absorption promoters [23, 24, 28]. Measures with the excised tissues in the presence of selected absorption promoters included permeability to inert fluorescent markers of varying molecular weights, bioelectrical properties, morphology and enzyme leakage. Agents that generally produced milder reversible effects on permeability to low molecular weight markers, bioelectrical properties, morphology and enzyme leakage included ammonium glycyrrhizinate, palmitoyl carnitine and EGTA. By contrast, agents that generally produced substantial and irreversible effects on the tissue measures included sodium lauryl sulfate and bile salts. *In vitro* results to date generally correlate well with studies reported by others in *in vivo/in situ* studies. Therefore, excised tissue systems appear to provide convenient alternative systems for evaluating intranasal drug delivery strategies.

Models of the BBB

Primary cultured brain capillary endothelial cells (BCECs) grown to confluence have recently been employed as an *in vitro* BBB model [29]. Cultured BCECs are highly viable, form polarized monolayers, retain tight junctions and express activities for carrier-mediated transport, as well as simple diffusion of drugs, and receptor- or absorptive-mediated endocytosis for peptides [29]. Using BCECs, a taurine carrier that transports one taurine molecule with cotransport of 2 Na^+ and Cl^- at both the apical and basolateral membranes has been characterized (A. Tsuji, unpublished data). From the kinetic parameters for the saturable uptake of taurine at the apical and basolateral membranes, the net transport *in vivo* is likely to be directed blood to brain. *In vivo* rat brain perfusion experiments showed a clear Na^+- and Cl^--dependent transport of taurine into the brain, which is consistent with the BCEC results (A. Tsuji, unpublished data). Carrier-mediated transport was also confirmed using BCECs for monocarboxylic acids, including the acidic forms of HMG-CoA reductase inhibitors [30]. In addition, using cultured BCECs, evidence for absorptive-mediated endocytosis of basic peptides (e.g., a dynorphin-like anesthetic peptide (DLAP) and an ACTH analog (ebriatide)) was also obtained [31].

Recently, evidence was obtained using cultured BCECs to show that P-glycoprotein acts as an active efflux pump to reduce the BBB permeability of some lipophilic drugs. P-glycoprotein was detected at the apical membrane side of BCECs by immunohistochemical techniques [32]. Tsuji et al. [36] have confirmed, based on the inhibitory effect of mdr reversing agents on the efflux from cells and from the energy dependency of cellular uptake of drugs, that P-glycoprotein functions in BCECs as an active efflux pump for cyclosporin A (CsA) [33], doxorubicin (DOX) (A. Tsuji, unpublished data) and vincristine [32]. By employing a brain ischemic and blood-recirculating model, it was shown that the *in vivo* permeabilities of CsA [34] and DOX (A. Tsjui, unpublished data) in rats depend on the brain ATP content [34]. These *in vitro* and *in vivo* data strongly indicate that P-glycoprotein is one of the functional components of the BBB which prevents the entry of lipophilic xenobiotics into the CNS. Thus, the P-glycoprotein-mediated efflux mechanism for lipophilic xenobiotics and the tight junctions, which prevent the transport of hydrophilic compounds, constitute important components of the BBB.

References

1. Lee VH, editor. Peptide and protein delivery. New York: Dekker, 1991.
2. Audus KL, Raub TJ, editors. Biological barriers to protein delivery. New York: Plenum, 1993.
3. Wilson G, Davis SS, Illum L, Zweibaum A, editors. Pharmaceutical applications of cell and tissue culture to drug transport. New York: Plenum, 1991.
4. Borchardt RT, Hidalgo IJ, Hillgren KM, Hu M. Pharmaceutical applications of cell culture – an overview. In: Wilson G, Davis SS, Illum L, Zweibaum A, editors. Pharmaceutical applications of cell and tissue culture to drug transport. New York: Plenum, 1991: 1–14.
5. Hidalgo IJ, Hillgren KM, Grass GM, Borchardt RT. A new side-by-side diffusion cell useful for studying transport across epithelial cell monolayers. *In Vitro* Cell Dev. Biol. 1992; 28A: 578–80.
6. Ho NF, Raub TR, Burton PS, Barsuhn CL, Adson A, Audus KL, Borchardt, RT. Quantitative approaches to delineate transport mechanisms in cell culture monolayers. In: Himmelstein KJ, Amidon GL, Lee PI, editors. Transport processes in pharmaceutical systems. New York: Dekker. In press.
7. Hidalgo IJ, Raub TJ, Borchardt RT. Characterization of human colon carcinoma cell line (Caco-2) as a model system for intestinal epithelial permeability. Gastroenterology 1989; 96: 736–749.
8. Thwaites DT, Brown CDA, Hirst BH, Simmons NL. Transepithelial Gly-Sar transport in intestinal Caco-2 cells mediated by expression of H^+-coupled carriers at both apical and basal membranes. J. Biol. Chem. 1992; 268: 7640–7644.
9. Thwaites DT, Brown CDA, Hirst BH, Simmons NL. H^+-coupled dipeptide (glycylsarcosine) transport across apical and basal borders of human intestinal Caco-2 cell monolayers display distinctive characteristics. Biochim. Biophys. Acta 1993; 1152: 237–245.

10. Thwaites DT, Hirst BH, Simmons NL. Direct assessment of dipeptide/H^+ symport in intact human intestinal (Caco-2) epithelium: a novel method utilizing continuous intracellular pH measurement. Biochem. Biophys. Res. Commun. 1993; 194: 432–438.
11. Thwaites DT, Hirst BH, Simmons NL. Substrate specificity of the di/tripeptide transporter in human intestinal epithelia (Caco-2): identification of substrates that undergo H^+-coupled absorption. Br. J. Pharmacol. 1994; 113: 1050–1056.
12. Thwaites DT, Hirst BH, Simmons NL. Passive transepithelial absorption of thyrotropin-releasing hormone (TRH) via a paracellular route in cultured intestinal and renal epithelial cell lines. Pharm. Res. 1993; 10: 674–680.
13. Thwaites DT, McEwan GTA, Brown CDA, Hirst BH, Simmons NL. Na^+-independent, H^+-coupled transepithelial β-alanine absorption by human-intestinal Caco-2 cell monolayers. J. Biol. Chem. 1993; 268: 18438–18441.
14. Thwaites DT, McEwan GTA, Brown CDA, Hirst BH, Simmons NL. L-Alanine absorption in human intestinal Caco-2 cells driven by proton electrochemical gradient. J. Membr. Biol. 1994; 140: 143–151.
15. Hunter J, Jepson MA, Tsuruo T, Simmons NL, Hirst BH. Functional expression of P-glycoprotein in apical membranes of human intestinal Caco-2 cells: kinetics of vinblastine secretion and interaction with modulators. J. Biol. Chem. 1993; 268: 14991–14997.
16. Griffiths NM, Hirst BH, Simmons NL. Active intestinal secretion of the fluoroquinolone antibacterials ciprofloxacin, norfloxacin and pefloxacin: a common secretory pathway? J. Pharm. Exp. Ther. 1994; 269: 496–502.
17. Patton JS, Platz RM. Pulmonary delivery of peptides and proteins for systemic action. Adv. Drug Del. Res. 1992; 8: 179–96.
18. Brazzell A, Smith RB, Kostenbauder HB. Isolated perfused rabbit lung as a model for intravascular and intrabronchial administration of bronchodialator drugs. I. Isoproterenol. J. Pharm. Sci. 1982; 71: 1268–1274.
19. Morimoto K, Yamahara H, Lee VHL, Kim K-J. Dipeptide transport across rat alveolar epithelial cell monolayers. Pharm. Res. 1993; 10: 1668–1674.
20. Smith PL, Yeulet SE, Citerone DR, Drake F, Cook M, Wall DA, Marcello J. SK&F 110679: comparison of absorption following oral or respiratory administration. J. Contr. Rel. 1994; 28: 67–77.
21. Smith PL, Marcello J, Chiossone DC, Orner D, Hidalgo IJ. Absorption of an RGD peptide (SK&F 106760) following intratracheal administration in rats. Int. J. Pharm. 1994; 106: 95–101.
22. Niven RW, Byron PR. Solute absorption from the airways of the isolated rat lung. II. Effect of surfactants on absorption of fluorescein. Pharm. Res. 1990; 7: 8–13.
23. Reardon PM, Wall DA, Hart TK, Smith PL, Gochoco CH. Lack of effect of ammonium glycyrrhizinate on the morphology of ovine nasal mucosa *in vitro*. Pharm. Res. 1993; 10: 1301–1307.
24. Reardon PM, Gochoco CH, Audus KL, Smith PL. *In vitro* nasal transport across ovine mucosa: effects of ammonium glycyrrhizinate on electrical properties and permeability of growth hormone-releasing peptide, mannitol and lucifer yellow. Pharm. Res. 1993; 10: 553–561.
25. Peirdomenico D, Madonna-Langan M, Smith PL, Wall DA. An *in vitro* model for pulmonary epithelial permeability. Proc. Intern. Symp. Contr. Rel. Bioact. Mater. 1992; 19: 228–229.
26. Welsh MJ, Smith PL, Frizzell RA. Chloride secretion by canine tracheal epithelium. III. Membrane resistances and electromotive forces. J. Membr. Biol. 1983; 71: 209–2018.
27. Audus KL, Tavokoli-Saberi MR. Aminopeptidases of newborn bovine nasal turbinate epithelial cell cultures. Int. J. Pharm. 1991; 76: 247–255.
28. Reardon PM, Audus KL. Ammonium glycyrrhizinate (AMGZ) effects on membrane integrity. Int. J. Pharm. 1993; 94: 161–170.
29. Miller D, Audus K, Borchardt RT. Application of cultured endothelial cells of the brain microvasculature in the study of the blood brain barrier. J. Tissue Cult. Meth. 1992; 14: 217–224.
30. Terasaki T, Takakuwa S, Moritani S, Tsuji A. Transport of monocarboxylic acids at the blood brain barrier: studies with monolayers of primary cultured bovine brain capillary endothelial cells. J. Pharm. Exp. Ther. 1991; 258: 932–937.

31. Terasaki T, Takakuwa S, Saheki A, Moritani S, Shimura T, Tabata S et al. Absorptive-mediated endocytosis of an adrenocorticotropic hormone (ACTH) analogue, ebiratide, into the blood brain barrier: studies with monolayers of primary cultured bovine brain capillary endothelial cells. Pharm. Res. 1992; 9: 528–534.
32. Tsuji A, Terasaki T, Takabatake Y, Tenda Y, Tamaj I, Yamashima T et al. P-glyco-protein as the drug efflux pump in primary cultured bovine brain capillary endothelial cells. Life Sci. 1992; 51: 1427–1437.
33. Tsuji A, Tamai I, Sakata A, Tenda Y. Terasaki T. Restricted transport of cyclosporin A across the blood brain barrier by a multidrug transporter, P-glycoprotein. Biochem. Pharmacol. 1993; 46: 1096–1099.
34. Sakata A, Tamai I, Kawazu K, Daguchi Y, Ohnishi T, Saheki A et al. *In vivo* evidence for ATP-dependent and P-glycoprotein-mediated transport of cyclosporin A at the blood-brain barrier. Biochem. Pharmacol. 1994; 48: 1989–1992.

Instructional Methods

Pharmacological Sciences: Perspectives for
Research and Therapy in the Late 1990s
ed. by A.C. Cuello and B. Collier

Providing Quality Education in Pharmacology: The Affordable Options

P.K. Rangachari[1], P.B. Williams[2], D.J. Crankshaw[3], J.R. Carpenter[4], D.W. Williams[5], A.B. Ebeigbe[6] and E.K.I. Omogbai[6]

[1]*Department of Medicine, McMaster University, Hamilton, Ontario, Canada;* [2]*Department of Pharmacology, Eastern Virginia Medical School, Norfolk, Virginia, USA;* [3]*Department of Obstetrics and Gynecology, McMaster University, Hamilton, Ontario, Canada;* [4]*Gardiner-Caldwell Communications Ltd., Macclesfield, England;* [5]*Department of Pharmacology, University of Melbourne, Parkville, Australia;* [6]*Departments of Physiology and Pharmacology, University of Benin, Benin City, Nigeria*

Summary. The provision of quality education in pharmacology faces two major hurdles – burgeoning information and dwindling resources. Teachers can no longer act as mere purveyors of information but need to shift the locus of control to students. Student-centered learning can be encouraged by use of problem-based or problem-solving exercises as well as interactive computer systems. These approaches require resources that may burden many countries, particularly in the Third World. IUPHAR should play a significant role in ensuring that the next generation of pharmacologists is educated appropriately.

The provision of quality education in any field is difficult at the best of times, and pressures on the modern university are making it particularly difficult in disciplines that rely heavily on experimental approaches to produce new information. Whether or not the current format of the research university can even survive the fiscal pressures is being debated [1]. Pharmacology merely suffers along with the rest.

Pharmacology is, essentially, an experimental discipline that draws from many other fields including anatomy, physiology, biochemistry, organic chemistry, etc., but retains its unique perspective centred around drugs. This clearly distinguishes it from many other disciplines which seek to use drugs as tools to understand physiological and biochemical processes. The pharmacologist, on the other hand, may use the natural processes to define the diverse ways in which drugs act on living systems. Although pharmacology appears to utilise sophisticated reductionist approaches, its goals are syncretic – to understand how the drug acts on the body and how in turn the body acts on the drug [2–4].

Since such knowledge forms the basis of therapeutics, the teaching of pharmacology has been discussed and debated in the professional schools of medicine, dentistry, nursing, veterinary medicine and pharmacy [5–10]. It is important to emphasise that pharmacologists need to

Correspondence to: Dr. P.K. Rangachari, McMaster University, HSC-3N5C, 1200 Main Street West, Hamilton, ON L8N 3Z5, Canada.

train the next generation of teachers and investigators as well, though the kind of training that the professional pharmacologist requires is different both in degree and kind. Nevertheless all teachers of pharmacology face common problems. They have to steer a careful course between the Scylla of burgeoning information and the Charybdis of ever-dwindling resources.

Remarkable advances in many fields have led to an explosion of knowledge in pharmacology. This can be gauged by comparing two editions of a classical text, Goodman and Gilman's *The Pharmacological Basis of Therapeutics*. The first edition (1940) [11] had two authors and 1305 pages (minus the index). By 1990, the book had expanded to 1736 pages (minus the index) and chapters by 61 authors [12]. Substantial portions are in fine print as well. This raises the pertinent question as to whether anyone can meaningfully absorb this much information. Although "G and G" is considered by many to be a reference text, it defines the vast scope of pharmacologic knowledge. Teachers attempting to convey this information face resentment from the students who are subjected to a similar barrage from other disciplines. Students then question whether such information is relevant. For teachers, the traditional role of disseminating information is clearly untenable given the burden of information. To add to their woes is the current ethos of the university, which operates in an increasingly fiscally-conscious mode.

The other major hurdle to the provision of quality education is the problem of limited resources. Throughout the world, fiscal problems have forced universities to institute cutbacks. This usually occurs at the expense of teaching. Added to this is the ubiquitous "publish or perish" syndrome where university faculty are rewarded not for their educational contributions but for their so-called "productivity" defined as "publication output", or research support obtained. Administrators who function largely as glorified bookkeepers, fail to realise that such cutbacks do not really save money. The average time between the discovery of a new drug to its eventual appearance on the market is greater than 12 years. The average time it takes to produce a trained pharmacologist is about 10 years. So the impact of inadequate training will not be realised for at least 20 years [13]. That period is long enough for short-sighted administrators to retire gracefully, leaving a future generation to pay the price. Clearly then, when discussing efficient utilisation of resources, it may be wise to pay heed to the needs not only of the present but of the future as well.

The solution to the problem of information overload is obvious. Teachers should not function merely as disseminators of information. They must return to their original roles as guides, permitting students to select the most appropriate path to enlightenment. The teacher of today is beginning to realise what our ancestors knew quite well: that it may be wiser to train students to fend for themselves rather than to inform them about every peril in their path. Active learning has been re-discovered, yet again. In the current context, this implies that students should become skilled in seeking, synthesising and integrating information. A critical attitude toward information may be far more relevant than an excessive familiarity with detail. The mechanics of promoting "active learning" have been much debated of late [14]. Two popular approaches involve the use of problems in different formats [15–19] and the use of computers [20–23]. Both these approaches have the potential ability to shift the locus of control from the teacher to the student and thus defuse the problem of information overload.

Problem-Based Learning has become a popular mechanism to foster student-centred learning. Although there are many different variants of PBL, they share certain common elements. Learning begins not with a formal didactic lecture but with a "problem" which is used as a starting point for exploration. The "problem" need not be "solved" in a conventional sense, but serves merely as a springboard for learning. Such problems can stem from clinical cases, laboratory data, or even newspaper clippings. In the small group variant, students meet in small groups of five or six to discuss such problems. They identify key issues, seek, find and critically assess the required information. In a subsequent tutorial, they discuss the information obtained. The faculty member functions as a guide ensuring that objectives are met, information is critically analysed and shared. In this model, much emphasis is placed on developing a critical attitude and a sense of responsibility [18, 19, 24, 25]. Other variants have been developed, but the small group format seems the most paradigmatic. Although PBL has become very fashionable as a novel approach to learning, the difficulty of marshalling adequate resources in the form of problems as well as faculty time has been a deterrent to widespread acceptance.

In this context, Patient-Oriented Problem Solving Exercises (POPS) afford a viable alternative. The original system was developed by Small and others in immunology [26] and has been subsequently expanded to deal with a number of problems in pharmacology and therapeutics. The purpose of these exercises is "to teach students problem-solving by applying the concepts and principles of pharmacology to therapeutic problems and to find ways to engage students actively in their learning of the material" [27]. Amongst the packages developed have been those dealing with issues such as cancer chemotherapy, and treatment of cardiac arrhythmias. In principle, any therapeutic problem can be designed as a POPS exercise. Students are encouraged to work in groups and interact with each other. The problems are so designed that sharing of information becomes critical. The POPS system has some of the components of the standard PBL format but differs in the greater emphasis on seeking solutions and in that students are generally applying prior knowledge obtained through didactic lectures. The merits of the POPS approach are that a single faculty member can supervise a larger number of students, prepared materials can be obtained and resource-strapped faculties can promote active learning [27].

Computer-aided instruction affords an exciting way of getting students involved in their own learning and can be used in various ways to enhance the quality of learning. At the bare minimum, interactive media can add colour to dull lectures and provide a modicum of interest to a moribund class. Such aids do not foster much student interaction, unless used carefully. Poor lecturing techniques cannot be completely glossed over with interactive media glitz. It is important to train faculty in the appropriate use of these media or much effort will come to naught. When computers are used within the context of small groups, their potential for active learning can be considerably enhanced. Computer-based tutorials have been developed to illustrate crucial concepts in pharmacology, for example pharmacokinetics. Students can learn at their own pace and can test their mastery of concepts by undertaking programmed tests. In the UK, for instance, a Pharma-CAL-ogy Project has been set up which consists of a consortium of 21 universities. This consortium aims to deliver technology-based courseware to undergraduate pharmacology students. Common courseware is being developed in topics such as the biotransformation of drugs, receptor transduction mechanisms, CNS pharmacology, etc. [28].

Since pharmacology is essentially an experimental subject, the ideal way to learn the subject is through laboratory-based exercises. However fostering such approaches is fraught with problems – resources, both technical and human, are a major problem, and in many countries, strident animal activists pose severe threats [29–31]. Furthermore, it is not entirely clear whether students interested in learning pharmacology for the practice of therapeutics benefit much from animal-based laboratory exercises. For such reasons, attempts have been made to develop interactive multi-media presentations.

Thus at the University of Melbourne, an online reference work of 36 experimental protocols has been developed to train science, medical, optometry and dental students in the principles of drug action. The interactive software, written in HyperTalk and C, provides important information on each experiment using a combination of multimedia resources including sound, video, animation, text and graphics. The programme also incorporates self-assessment tasks and an online procedure for writing and submission of practical reports [32].

Whether systems that promote active learning, be it through tutorials or computer-aided instruction, are universally applicable is a moot point. Particularly poignant is the plight of pharmacology teachers in the Third World where all problems appear grossly exaggerated due to extremely meagre resources, obsolete equipment, exceptionally high student-teacher ratios. The only driving force to counter this escalating build-up of entropy appears to be "will-power" [33]. Textbooks are scarce and often obsolete. Even when available, the context is missing. This is particularly relevant for pharmacology and therapeutics, since it is pointless to learn about drugs that are unavailable. Clearly most countries have a rich store of tradition and herbal lore which, used judiciously, may solve many therapeutic problems. Encouragement of ethnopharmacological approaches would be profitable. This information could be incorporated into texts tailored to the particular needs of a country. Journals could be available to promote and enhance local knowledge. Unfortunately, attempts to develop first rate journals to promote research done within a national context runs afoul of bureaucracy. Even getting an African journal listed in Current Contents to promote wider dissemination of information has proven difficult, since the proponents of such ventures are asked to provide evidence for the impact of their journal before it can be listed. If the developed world is serious about promoting Third World autonomy, genuine action is needed rather than mere lip service at International Meetings. This involves much attention to detail, a burden that most scientists in the developed countries are reluctant to assume, given their already heavy load of teaching and administrative responsibilities.

In this context, problem-based learning appears the ideal format since the students will be required to learn within the context in which they will eventually function [4]. But the paucity of trained faculty poses a problem. More than in the developed countries, there is a crucial need to invest in student-centred learning. The student who is able to set personal goals, and who can critically assess the information available, may be far more resilient and less dependent on the vicissitudes of fortune. Unfortunately this cannot happen without support and direction. Here again, international bodies should provide the required support to encourage and train faculty on site.

If technologies are the sum of what we do, educational enterprises can be thought of as technologies [34]. We are becoming acutely conscious of the appropriateness of specific technologies to particular climates and situations, since these do not often transfer well. Attempts made to impose a system that works well under one situation to others where the appropriate mix of motivation and resources is absent, may lead to much embarrassment. Too often proponents of a particular approach become so wedded to it that they fail to realise that the same ends could be attained by entirely different means. Programmes and processes are found to be deficient when evaluated by criteria that may be entirely inappropriate to the situation at hand. It is worth remembering that much of what passes for evaluation is ideology masquerading as technology. At all times the goals of such endeavours must be kept in view. This requires a willingness on the part of all involved in teaching pharmacology to consider carefully the options that are affordable within their local context. Sharing of such experiences with others may prove valuable provided it is accompanied by a willingness to learn from each other's mistakes and successes. The International Union of Pharmacology can provide a forum where the exchanges of such information can occur with minimal cost. Such efforts can be considerably amplified by the willingness of the current generation of teachers and transduced into effective training of the next.

Acknowledgement

Support for the symposium on which this review is based was provided by IUPHAR.

References

1. Maddox J. Can the research university survive? Nature 1994; 369: 703.
2. Page CP, Sutter MC, Walker MJA. Wither, whether and whither pharmacology. TiPS 1994; 15: 17–19.
3. Fuller RW. Industry needs for pharmacology training. The Pharmacologist 1992; 34: 248–251.
4. Rangachari PK. Quality education for undergraduates in pharmacology: a Canadian experiment. TiPS 1994; 15: 211–214.
5. Burford HJ. Innovative approaches to teaching clinical pharmacology: problems of an emerging curriculum. J. Clin. Pharmacol. 1991; 31: 423–428.
6. Ingenito AJ, Noble BG, Wooles WR. The case conference approach to teaching clinical pharmacology. J. Clin. Pharmacol. 1992; 32: 502–510.
7. Nierenberg DW. Teaching clinical pharmacology: a process of 'lifelong learning'. J. Clin. Pharmacol. 1993; 33: 311–315.
8. Orme M, Sjoqvist F, Bircher J, Bogaert M, Dukes MNG, Eichelbaum M, Gram LF, Huller H, Lunde I, Tognoni G. The teaching and organisation of clinical pharmacology in European medical schools. Eur. J. Clin. Pharmacol. 1990; 38: 101–105.
9. Vestal RE, Benowitz NL. Workshop on problem-based learning as a method for teaching clinical pharmacology and therapeutics in medical school. J. Clin. Pharmacol. 1992; 32: 779–797.
10. Williams PB. Incorporation of clinical pharmacology into the fourth year of the medical curriculum: teachng clinical pharmacology without a clinical pharmacologist. J. Clin. Pharmacol. 1990; 30: 1065–1073.
11. Goodman LS, Gilman A. The pharmacological basis of therapeutics. New York: MacMillan, 1940.
12. Gilman AG. Goodman and Gilman's the pharmacological basis of therapeutics. 8th ed. New York: Pergamon Press, 1990.
13. Carpenter JR. Do cutbacks really save money? Can. J. Physiol. Pharmacol. 1994; 72, Suppl. 1: 59 (abstract).
14. Richardson D. Active learning: a personal view. Am. J. Physiol. 1994; 10: S79.
15. Boud D, Felleti G. The challenge of problem-based learning. New York: St. Martin's Press, 1991.
16. Norman GR, Schmidt HG. The psychological basis of problem-based learning: a review of the evidence. Acad. Med. 1992; 67: 557–565.
17. Wolf FM. Problem-based learning and meta-analysis: can we see the forest through the trees? Acad. Med. 1993; 68: 542–554.
18. Ferrier BM. Problem-based learning: does it make a difference? J. Dent. Edu. 1990; 54: 550–551.
19. Branda LA. Implementing problem-based learning. J. Dent. Educ. 1990; 54: 1–2.
20. Williams DW. The use of interactive multimedia in teaching behavioral pharmacology. In: Johnson NE, editor. Proceedings Animal Welfare Conference; 1993: 74–77.

21. Eisenberg R. MacPharmacology[R]: A series of computerized study and review programs. J. Clin. Pharmacol. 1992; 32: 10–17.
22. Hollingsworth M, Foster RW. Computer-assisted learning (CAL) programs in drug disposition. Br. J. Pharmac. 1991; 102: 216P (abstract).
23. Hutcheon DE, Wadie El-Gawly H. A computer-based, problem-solving system of instruction in clinical pharmacology. J. Clin. Pharmacol. 1991; 31: 198–204.
24. Barrows HS. A taxonomy of problem-based learning methods. Med. Educ. 1986; 20: 481–486.
25. Rangachari PK. Design of a problem-based undergraduate course in pharmacology: implications for the teaching of physiology. Am. J. Physiol. 1991; 260: S14–S21.
26. Small PA, Jr. Science education: simulation methods for teaching process and content. Fed. Proc. 1974; 33: 2008–2013.
27. Burford HJ, Ingenito A, Williams PB. Development and evaluation of patient-oriented problem-solving materials in pharmacology. Acad. Med. 1990; 65: 689–693.
28. Dewhurst DG, Hughes IE. The pharma-CAL-ology project: utilising technology to improve learning in students of pharmacology. Can. J. Physiol. Pharmacol. 1994; 72, Suppl. 1: 59 (abstract).
29. Modell HI. Can technology replace live preparations in student laboratories? Am. J. Physiol. 1989; 256: S18–S20.
30. Hansen PA. Changing attitudes about using animals [editorial]. Am. J. Physiol. 1993; 265: S1.
31. Crankshaw DJ. Providing quality education in pharmacology: the dimensions of the problem. Can. J. Physiol. Pharmacol. 1994; 72, Suppl. 1: 59 (abstract).
32. Williams DW. Teaching practical pharmacology using interactive multimedia. Can. J. Physiol. Pharmacol. 1994; 72, Suppl. 1: 59 (abstract).
33. Ebeigbe AB, Omogbai EKI. Teaching pharmacology in the Third World. Can. J. Physiol. Pharmacol. 1994; 72, Suppl. 1: 59 (abstract).
34. Franklin U. The real world of technology. Concord, Ontario: House of Anansi Press Ltd., 1990.

Subject index